TECHNOLOGIES FOR SMART SENSORS AND SENSOR FUSION

Devices, Circuits, and Systems

Series Editor
Krzysztof Iniewski
CMOS Emerging Technologies Research Inc.,
Vancouver, British Columbia, Canada

FORTHCOMING TITLES:

Gallium Nitride (GaN): Physics, Devices, and Technology
Farid Medjdoub and Krzysztof Iniewski

High Frequency Communication and Sensing: Traveling-Wave Techniques
Ahmet Tekin and Ahmed Emira

High-Speed Devices and Circuits with THz Applications
Jung Han Choi and Krzysztof Iniewski

Labs-on-Chip: Physics, Design and Technology
Eugenio Iannone

Laser-Based Optical Detection of Explosives
Paul M. Pellegrino, Ellen L. Holthoff, and Mikella E. Farrell

Metallic Spintronic Devices
Xiaobin Wang

Microfluidics and Nanotechnology: Biosensing to the Single Molecule Limit
Eric Lagally and Krzysztof Iniewski

Mobile Point-of-Care Monitors and Diagnostic Device Design
Walter Karlen and Krzysztof Iniewski

Nanoelectronics: Devices, Circuits, and Systems
Nikos Konofaos

Nanomaterials: A Guide to Fabrication and Applications
Gordon Harling and Krzysztof Iniewski

Nanopatterning and Nanoscale Devices for Biological Applications
Krzysztof Iniewski and Seila Selimovic

Optical Fiber Sensors and Applications
Ginu Rajan and Krzysztof Iniewski

Organic Solar Cells: Materials, Devices, Interfaces, and Modeling
Qiquan Qiao and Krzysztof Iniewski

Power Management Integrated Circuits and Technologies
Mona M. Hella and Patrick Mercier

Radio Frequency Integrated Circuit Design
Sebastian Magierowski

Semiconductor Device Technology: Silicon and Materials
Tomasz Brozek and Krzysztof Iniewski

Smart Grids: Clouds, Communications, Open Source, and Automation
David Bakken and Krzysztof Iniewski

Soft Errors: From Particles to Circuits
Jean-Luc Autran and Daniela Munteanu

TECHNOLOGIES FOR SMART SENSORS AND SENSOR FUSION

Edited by
Kevin Yallup
Krzysztof Iniewski

CRC Press
Taylor & Francis Group
Boca Raton London New York

CRC Press is an imprint of the
Taylor & Francis Group, an **informa** business

MATLAB® is a trademark of The MathWorks, Inc. and is used with permission. The MathWorks does not warrant the accuracy of the text or exercises in this book. This book's use or discussion of MATLAB® software or related products does not constitute endorsement or sponsorship by The MathWorks of a particular pedagogical approach or particular use of the MATLAB® software.

CRC Press
Taylor & Francis Group
6000 Broken Sound Parkway NW, Suite 300
Boca Raton, FL 33487-2742

First issued in paperback 2017

© 2014 by Taylor & Francis Group, LLC
CRC Press is an imprint of Taylor & Francis Group, an Informa business

No claim to original U.S. Government works

Version Date: 20140210

ISBN 13: 978-1-138-07574-0 (pbk)
ISBN 13: 978-1-4665-9550-7 (hbk)

Library of Congress Cataloging-in-Publication Data

Technologies for smart sensors and sensor fusion / editors, Kevin Yallup, Krzysztof Iniewski.
     pages cm -- (Devices, circuits, and systems)
    Includes bibliographical references and index.
    ISBN 978-1-4665-9550-7 (hardback)
    1. Detectors. 2. Multisensor data fusion. 3. Smart materials. I. Yallup, Kevin.

TK7872.D48T43 2014
681'.2--dc23
                                                        2014001303

Visit the Taylor & Francis Web site at
http://www.taylorandfrancis.com

and the CRC Press Web site at
http://www.crcpress.com

Contents

PART I   Microfluidics and Biosensors

PART II   Chemical and Environmental Sensors

Preface

Microsystems and micro electro mechanical systems (MEMS) have revolutionized the world of sensors and detectors. A whole range of miniaturized sensors has emerged that can add a wide range of capabilities to portable devices. For example, MEMS-based accelerometers and gyroscopes can add a number of motion-sensing capabilities to portable devices, enabling motion to be used to control the device or shake to be corrected for while capturing a digital image. Optical sensors, such as CCDs and more recently CMOS image sensors, have transformed photography and enabled low-cost image capture for a variety of applications.

So far, most sensor applications have focused on single sensors and relatively simple processing to extract specific information from the sensor. However, there are two exciting new developments that are enabling sensors to go beyond the realm of simple sensing of movement or capture of images to deliver more advanced information such as location in a built environment, the sense of touch, the presence of chemicals such as toxins or pathogens in fluids, and many more applications. These new sensors offer the potential for machines to interact with the world around them in a more intelligent and sophisticated way and could lead to smarter systems such as autonomous robots able to independently carry out complex and difficult tasks. This book has been designed to give an overview of these exciting new developments.

The first development we will examine in this book is the increasingly varied number of sensors that can be integrated into arrays. The growing diversity means that more and more different stimuli can be sensed reliably and accurately, leading to ever more complex ways of sensing the environment. This book contains parts that deal with sensors for applications in fields such as biotechnology, medical science, chemical detection, environmental monitoring, automotive sensing, and control and sensors for industrial applications. Improvements in sensing capability in these areas promise many exciting new applications for sensors.

The second trend is the increasing availability and computational power of communication devices that will both support the algorithms needed to reduce the raw sensor data from multiple sensors and convert it into the information needed by the sensor array to enable rapid transmission of the results to the required point. We have assembled chapters within this book that discuss software and sensor systems and the issues around sensor fusion.

In this book, we have built a collection of chapters from authors working at the leading edge of sensor technology that shows the richness and diversity of development work in the world of smart sensors and sensor fusion. We hope the reader will be able to gain fresh insights into the work of sensors of the future and be able to go on to develop their own new smart sensors and sensor systems.

MATLAB® is a registered trademark of The MathWorks, Inc. For product information, please contact:

The MathWorks, Inc.
3 Apple Hill Drive
Natick, MA 01760-2098 USA
Tel: 508-647-7000
Fax: 508-647-7001
E-mail: info@mathworks.com
Web: www.mathworks.com

Editors

Dr. Kevin Yallup is the CTO for ACAMP, an organization based in Alberta, Canada, that focuses on bringing new technology-based products to market. ACAMP is engaged with a number of companies developing sensors for a wide range of commercial applications in fields as diverse as biomedicine, energy, and navigation and tracking.

Dr. Yallup has over 15 years experience in MNT product development in a number of industries. He has held engineering management roles in Technology for Industry Ltd. (TFI), CDT Ltd., Kymata Ltd., BCO Technologies (NI) Ltd., National Semiconductor, and Analog Devices. He has worked with various universities and companies taking a technology through to full commercialization. Dr. Yallup graduated from Cambridge University, United Kingdom, with a degree in natural sciences, specializing in solid state physics, and received his PhD from the University of Leuven in Belgium.

Prior to ACAMP, Dr. Yallup was the technical director in TFI and specialized in providing services and consultancy for the commercialization of emerging technologies such as micro- and nanotechnologies, which enabled him to broaden his experience of the transfer of technology from university to commercialization. While working for TFI, Dr. Yallup was the project leader for the study of the commercialization of MNT in the Alberta region and developed the initial business case for the ACAMP. The early part of his career was in the semiconductor industry, where he worked on the development and commercialization of leading edge analog CMOS and BiCMOS processes.

Dr. Krzysztof (Kris) Iniewski is managing R&D at Redlen Technologies Inc., a start-up company in Vancouver, Canada. Redlen's revolutionary production process for advanced semiconductor materials enables a new generation of more accurate, all-digital, radiation-based imaging solutions. Kris is also a president of CMOS Emerging Technologies Research Inc (www.cmosetr.com), an organization of high-tech events covering communications, microsystems, optoelectronics, and sensors. During the course of his career, Dr. Iniewski has held numerous faculty and management positions at the University of Toronto, the University of Alberta, SFU, and PMC-Sierra Inc. He has published over 100 research papers in international journals and conferences. He holds 18 international patents granted in the United States, Canada, France, Germany, and Japan. He is also a frequent invited speaker and has consulted for multiple organizations internationally. He has written and edited several books for CRC Press, Cambridge University Press, IEEE Press, Wiley, McGraw-Hill, Artech House, and Springer. His personal goal is to contribute to healthy living and sustainability through innovative engineering solutions. In his leisurely time, Kris can be found hiking, sailing, skiing, or biking in beautiful British Columbia. He can be reached at kris.iniewski@gmail.com.

Contributors

Orazio Aiello
Department of Electronics and
    Telecommunications
Politecnico di Torino
Torino, Italy

L. Alfonta
Department of Biotechnology Engineering
Avram and Stella Goldstein-Goren
Ben-Gurion University of the Negev
Beer-Sheva, Israel

Josep Altet
Electronic Engineering Department
Universitat Politècnica de Catalunya
Barcelona, Spain

H. García Arellano
División de Ciencias Biológicas y de la Salud
Departamento de Ciencias Ambientales
Universidad Autónoma Metropolitana-Lerma
Lerma, Mexico

John Berring
Department of Electrical and Computer
    Engineering
The University of British Columbia
Vancouver, Canada

Russell Binions
School of Engineering and Materials Science
Queen Mary, University of London
London, United Kingdom

Robert D. Black
Scion NeuroStim
Raleigh, North Carolina

John Robert Busch
Department of Electrical and Computer
    Engineering
The University of British Columbia
Vancouver, Canada

Donald P. Butler
Department of Electrical Engineering
Nanotechnology Research and Education Center
University of Texas at Arlington
Arlington, Texas

Alfonso Carlosena
Department of Electrical and Electronic
    Engineering
Public University of Navarra
Pamplona, Spain

Zeynep Çelik-Butler
Department of Electrical Engineering
Nanotechnology Research and Education Center
University of Texas at Arlington
Arlington, Texas

Alexandr I. Chernomorsky
Department of Automated Complexes of
    Navigation and Orientation Systems
Moscow Aviation Institute
Moscow, Russia

Xuan Du
Department of Electrical and Computer
    Engineering
University of Victoria
Victoria, British Columbia, Canada

Leonardo Tomazeli Duarte
School of Applied Sciences
University of Campinas
Limeira, Brazil

W.R. Fahrner
Institute of Photovoltaics
Nanchang University
Nanchang, People's Republic of China

Atena Roshan Fekr
Department of Electrical and Computer
    Engineering
McGill University
Montréal, Québec, Canada

Dietmar Fink
Division de Ciencias Naturales e Ingeneria
Departamento de Procesos y Tecnologia
Universidad Autónoma
    Metropolitana-Cuajimalpa
México

and

Nuclear Physics Institute
Řež, Czech Republic

Franco Fiori
Department of Electronics and
    Telecommunications
Politecnico di Torino
Torino, Italy

Elain Fu
School of Chemical, Biological, and
    Environmental Engineering Oregon State
    University
Corvallis, Oregon

G. Muñoz Hernandez
Division de Ciencias Naturales e Ingeneria
Universidad Autónoma
    Metropolitana-Cuajimalpa
Cuajimalpa, México

and

Departamento de Fisica
Universidad Autónoma
    Metropolitana-Iztapalapa
Iztapalapa, México

K. Hoppe
South Westfalia University of Applied Sciences
Hagen, Germany

Ashraf B. Islam
Department of Electrical Engineering and
    Computer Science
The University of Tennessee
Knoxville, Tennessee

Syed K. Islam
Department of Electrical Engineering and
    Computer Science
The University of Tennessee
Knoxville, Tennessee

Majid Janidarmian
Department of Electrical and Computer
    Engineering
McGill University
Montréal, Québec, Canada

Christian Jutten
Université Joseph Fourier
Grenoble, France

Ryan T. Kelly
Environmental Molecular Sciences
    Laboratory
Pacific Northwest National Laboratory
Richland, Washington

A. Kiv
Department of Materials Engineering
Ben-Gurion University of the Negev
Beer-Sheva, Israel

John Kosinski
Senior Consulting Scientist
MacAulay-Brown, Inc.
Advanced Technology Group (ATG)
Dayton, Ohio

Tadahiro Kuroda
Department of Electronics and Electrical
    Engineering
Keio University
Tokyo, Japan

Antonio J. López-Martín
Department of Electrical and Electronic
    Engineering
Public University of Navarra
Pamplona, Spain

Tao Lu
Department of Electrical and Computer
    Engineering
University of Victoria
Victoria, British Columbia, Canada

Barry Lutz
Department of Bioengineering
University of Washington
Seattle, Washington

Kimberly C. MacArthur
Department of Electrical Engineering and
    Computer Science
The University of Tennessee
Knoxville, Tennessee

Suresha Mahadeva
Department of Mechanical Engineering
The University of British Columbia
Vancouver, Canada

Khandaker A. Mamun
Department of Electrical Engineering and
    Computer Science
The University of Tennessee
Knoxville, Tennessee

Diego Mateo
Electronic Engineering Department
Universitat Politècnica de Catalunya
Barcelona, Spain

Nicole McFarlane
Department of Electrical Engineering and
    Computer Science
The University of Tennessee
Knoxville, Tennessee

Benjamin Nahill
Department of Electrical and Computer
    Engineering
McGill University
Montréal, Québec, Canada

Jun Nishimura
Department of Electronics and Electrical
    Engineering
Keio University
Tokyo, Japan

Kent B. Pfeifer
Microsystems-Enabled Detection Department
Sandia National Laboratories
Albuquerque, New Mexico

Nikolaos P. Preve
School of Electrical and Computer
    Engineering
National Technical University of Athens
Athens, Greece

Craig Priest
Ian Wark Research Institute
University of South Australia
Mawson Lakes, South Australia,
    Australia

Kirill V. Poletkin
Department of Microsystems
    Engineering–IMTEK
University of Freiburg
Freiburg, Germany

Katarzyna Radecka
Department of Electrical and Computer
    Engineering
McGill University
Montréal, Québec, Canada

Omid Sarbishei
Department of Electrical and Computer
    Engineering
McGill University
Montréal, Québec, Canada

Thomas Schlegl
Institute of Electrical Measurement
    and Measurement Signal
    Processing
Graz University of Technology
Graz, Austria

Christopher Shearwood
School of Mechanical and Aerospace
    Engineering
Nanyang Technological University
Singapore, Singapore

Christoph Sielmann
Department of Electrical and Computer
    Engineering
The University of British Columbia
Vancouver, Canada

Jose Silva-Martinez
Department of Electrical and Computer
    Engineering
Texas A&M University
College Station, Texas

Boris Stoeber
Department of Mechanical Engineering and
    Department of Electrical and Computer
    Engineering
The University of British Columbia
Vancouver, Canada

Brett Y. Smolenski
Institute of Technology
State University of New York
Utica, New York

Xuefei Sun
Biological Sciences Division
Pacific Northwest National Laboratory
Richland, Washington

Shiquan Tao
Department of Mathematics, Chemistry
    and Physics
West Texas A&M University
Canyon, Texas

Steven M. Thornberg
Materials Reliability Department
    (Retired)
Sandia National Laboratories
Albuquerque, New Mexico

Fahmida S. Tulip
Department of Electrical Engineering and
    Computer Science
The University of Tennessee
Knoxville, Tennessee

J. Vacik
Nuclear Physics Institute
Řež, Czech Republic

Serge Vincent
Department of Electrical and Computer
    Engineering
University of Victoria
Victoria, British Columbia, Canada

Catherine M. Vannicola
Emergent Technologies
Austin, Texas

Ulrike Wallrabe
Department of Microsystems
    Engineering–IMTEK
University of Freiburg
Freiburg, Germany

Konrad Walus
Department of Electrical and Computer
    Engineering
The University of British Columbia
Vancouver, Canada

Paul Yager
Department of Bioengineering
University of Washington
Seattle, Washington

Hubert Zangl
Institute of Smart System Technologies
Sensors and Actuators
Alpen-Adria-Universitaet Klagenfurt
Klagenfurt, Austria

Zeljko Zilic
Department of Electrical and Computer
    Engineering
McGill University
Montréal, Québec, Canada

Part I

Microfluidics and Biosensors

1 Droplet-Based Microfluidics for Biological Sample Preparation and Analysis

Xuefei Sun and Ryan T. Kelly

CONTENTS

1.1 INTRODUCTION

Modern biological research often requires massively parallel experiments to analyze a large number of samples in order to find biomarkers, screen drugs, or elucidate complex cellular pathways. These processes frequently involve time-consuming sample preparation and expensive biochemical measurements. Another constraint frequently encountered in bioanalysis is limited amounts of available sample. Microfluidics or lab-on-a-chip platforms offer promise for addressing the challenges encountered in biological research because a large number of small samples can be handled and processed with different functional elements in an automated fashion.

Droplet-based microfluidics, in which reagents of interest are compartmentalized within femtoliter-to-nanoliter-sized aqueous droplets or plugs that are encapsulated and dispersed in an immiscible oil phase, has emerged as an attractive platform for small-volume bioanalysis [1–9]. This new platform elegantly addresses challenges encountered with conventional continuous flow systems by, for example, limiting reagent dilution caused by diffusion and Taylor dispersion and minimizing cross contamination and surface-related adsorptive losses [10]. The microdroplets isolated by the immiscible liquid can serve as microreactors, allowing for high-throughput chemical reaction screening and extensive biological research [1]. Droplet-based microfluidics also offers great promise for reliable quantitative analysis because monodisperse microdroplets can be generated with controlled sizes and preserve temporal information that is easily lost to dispersion in continuous flow systems [11,12].

Biological analysis begins with sample selection and preparation. The initial sampling can comprise cell sorting, tissue dissection, or extraction of protein or other analytes of interest from cells or tissues [13]. The biological samples are then prepared by, for example, combining reagents, mixing, incubating, purifying, and/or enriching. Depending on the complexity of the sample, the subsequent analytical measurements can be very simple employing, for example, laser-induced fluorescence (LIF) to detect a single labeled analyte. With more complex samples having multiple analytes of interest, chemical separations including capillary electrophoresis (CE) and liquid chromatography (LC) and information-rich detection methods such as mass spectrometry (MS) become necessary. To date, many operational components for microdroplets have been well developed to perform most of these basic operations. For example, stable aqueous droplets dispersed in an oil phase can be generated using various droplet generator designs for sampling in a confined small volume, the most common of which are the T-junction [14,15] and flow-focusing [16,17] geometries. Addition of reagents to existing droplets can be realized by fusion with other droplets, enabling the initiation and termination of the compartmentalized reactions confined in the microdroplets [18,19]. Rapid mixing of fluids within droplets enables a homogeneous reactive environment to be achieved and can be enhanced by means of chaotic advection [20]. In addition, droplets can be incubated in delay lines [21] or stored in reservoirs [22,23] or traps [24,25] for extended periods of time to complete reactions or facilitate the biological processes.

Droplet-based microfluidic platforms have been successfully applied in a variety of chemical and biological research areas. For example, a droplet-based platform for polymerase chain reaction (PCR) amplification has proven able to significantly improve amplification efficiency over conventional microfluidic formats [26], which is mainly due to the elimination of both reagent dilution and adsorption on the channel surfaces. Droplets have also been employed to encapsulate, sort, and assay single cells [12,27] or microorganisms [24], study enzyme kinetics [11] and protein crystallization [28], and synthesize small molecules and polymeric micro- and nanoparticles.

Although droplet-based microfluidic technology has developed to a degree where droplets can be generated and manipulated with speed, precision, and control, some real challenges still exist that limit the widespread use of these systems. One challenge is how to extract and acquire the enormous chemical information that may be contained in the picoliter-sized droplets. Detection of droplet contents has historically been limited to optical methods such as LIF, while coupling with chemical separations and nonoptical detection has proven difficult. Combining the advantages of droplet-based platforms with more information-rich analytical techniques including LC, CE, and MS can greatly extend their reach. This often requires that the droplets be extracted from the oil phase for downstream analysis and detection.

This chapter focuses primarily on the integrated droplet-based microsystems having the ability to couple with chemical separations and nonoptical detection, allowing for *ex situ* analysis and identification of the biochemical components contained in the microdroplets. Some unit operations for microdroplets will be briefly introduced, including droplet generation, fusion, and incubation. All approaches and techniques developed for droplet detection, droplet extraction, coupling CE separation, and electrospray ionization (ESI)-MS detection will be reviewed. An example of integrated droplet-based microfluidics, including on-demand droplet generation and fusion, robust and efficient droplet extraction, and a monolithically integrated nanoelectrospray ionization (nanoESI) emitter, will be given to demonstrate its potential for chemical and biological research.

1.2   DROPLET-BASED OPERATIONS

1.2.1   DROPLET GENERATION

Currently, most planar microfluidic droplet generators are designed using T-junction [14,15] and flow-focusing [16,17] geometries, in which small droplets are spontaneously formed at an intersection taking advantage of the interface instability between oil and aqueous streams. Using these approaches, droplets can be generated over a broad range of frequencies ranging from ~0.1 Hz to 10 kHz and using flow rates on the order of 0.1–100 μL/min [29]. Droplet volume and generation frequency depend on several factors, including the physical properties of the immiscible phases, flow rates, and intersection geometry. For a given geometry and solvent composition, flow-focusing and T-junction interfaces exhibit interdependence between flow rate and droplet generation frequency and cannot be easily modulated over short time scales.

For lower frequencies and applications for which the ability to rapidly change droplet size and generation frequency is desirable, on-demand droplet generation strategies become more favorable, as they ensure precise control and fine manipulation of individual droplets. Various approaches have been developed to generate droplets on demand, for example, by carefully balancing the pressure and flow in the system [27], as well as electrical [30] or laser pulsing [31] and piezoelectric actuation [32]. Pneumatic valving has also been explored and has been found to provide facile, independent control over both droplet size and generation frequency [33–36]. Galas et al. utilized a single pneumatic valve that was embedded in an active connector and assembled close to a T-junction to regulate the flow of the dispersed phase [33]. Constant pressures were applied on the inlets of two immiscible liquids to drive the flow in the microchannel. Individual droplets were created by briefly opening the valve. The aqueous droplet size depended on the valve actuation time and frequency, as well as the pressure applied at the oil inlet. Therefore, the droplet volume, spacing, and speed could be controlled accurately and independently. This device not only generated periodic sequences of identical droplets but also enabled the production of nonperiodic droplet trains with different droplet sizes or spacing. Lin et al. also reported a similar platform for pneumatic valve–assisted on-demand droplet generation [34]. Negative pressure was applied at the outlet of the device to drive the flow of the two immiscible liquids through the microchannel. The dependence of droplet size on the valve actuation time and applied pressure was investigated. In addition, they utilized several aqueous flow channels, each with independently controlled microvalves, to generate arrays of droplets containing different compositions by alternately actuating the valves.

We have also investigated valve-controlled on-demand droplet generation. To minimize the dead volume and control droplet volume precisely, the pneumatic valve was placed over the side channel exactly at the T-junction (Figure 1.1a). Carrier oil flow was driven by a syringe pump, and the dispersed aqueous phase was injected by finely controlled air pressure. Figure 1.1b shows the generation process of an individual droplet. The valve is initially closed and the aqueous fluorescein solution is confined in the side channel. When opening the valve briefly, a small volume of aqueous solution is dispensed into the oil channel to form a droplet, which is then flushed downstream by the carrier oil flow. Compared with conventional microfluidic droplet generation techniques based on a T-junction or flow focusing, the valve-integrated system can generate droplets with precise control over droplet volume, generation frequency, and velocity. Droplet velocity is determined by the syringe pump driving the oil stream, while the droplet generation rate is controlled by the valve

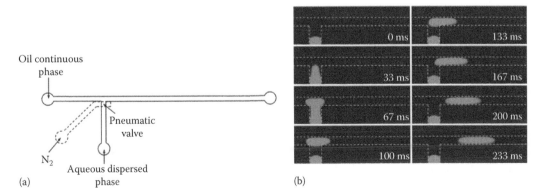

FIGURE 1.1   (a) Schematic depiction of a T-junction droplet generator controlled by a pneumatic valve. (b) Micrograph sequences depicting pneumatic valve-controlled generation of an individual fluorescein droplet. The width of the oil flow channel was 100 μm, the oil flow rate was 0.5 μL/min, and the sample injection pressure was 8 psi. Valve actuation time and pressure were 33 ms and 25 psi, respectively.

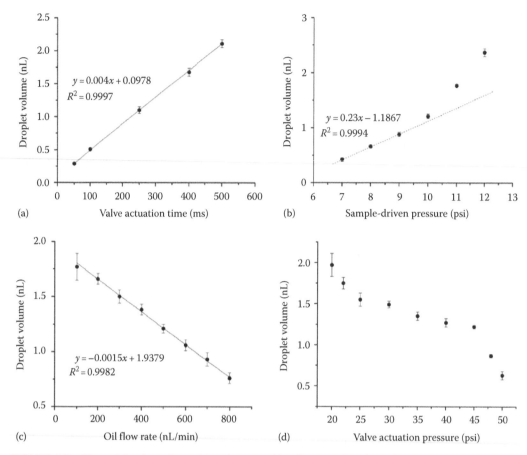

FIGURE 1.2 Plots of droplet volume dependence on (a) valve actuation time, (b) sample-driven pressure, (c) oil flow rate, and (d) valve actuation pressure. Channel dimensions were the same as for Figure 1.1.

actuation frequency as defined in the software. The droplet spacing is determined by the interval between valve openings and the oil flow velocity. Droplet volume depends on several parameters including the valve actuation time, the pressure of the aqueous solution, the oil phase flow rate, and the valve control pressure (Figure 1.2).

1.2.2 In-Droplet Reagent Combination and Mixing

Besides controlled droplet generation, droplet fusion is of crucial importance in relation to the development of microreactors because it allows precise and reproducible mixing of reagents at well-defined points to initiate, modify, and terminate reactions [6]. Ismagilov et al. carried out pioneering work to combine different reagents into individual droplets by flowing two reagent solutions in a microchannel as two laminar streams [37]. To prevent the prior contact of the reagents before droplet generation, an inert center stream was used to separate them. Thus, three streams were continuously injected into an immiscible carrier oil phase to form droplets. The gradient droplets in the reagent concentrations were achieved by varying the relative flow rates of the streams [11,38]. A subsequent winding channel was designed to accelerate mixing by chaotic advection [20,37]. This approach has been widely employed to control networks of chemical reactions [37], study reaction kinetics [11], screen protein crystallization conditions [38], and investigate single-cell-based enzyme assays [39] and protein expression [12].

Recently, Weitz et al. presented a robust picoinjector to add reagents to droplets in microfluidic systems [40]. The picoinjector was controlled by an electric field to trigger the injection of a controlled volume of reagents into each droplet. The injection volume was precisely controlled by adjusting the droplet velocity and injection pressure. Selective injection was realized by switching the electric field on and off at kilohertz frequencies.

In-channel droplet fusion is another attractive approach to combine different reagents in individual droplets to initiate or terminate the confined reactions. The process of droplet merging introduces convective flows into the system, resulting in far more rapid mixing than relying on diffusion alone [41]. In-channel droplet fusion is readily achieved by bringing two or more surfactant-free droplets into contact. Both passive and active methods have been developed to control droplet fusion. For passive fusion devices, droplet coalescence is usually initiated by utilizing specially designed fusion elements in the channel. For example, Bremond et al. incorporated an expanded coalescence chamber in the channel network in which two droplets were brought into close proximity and merged together before entering a narrow channel [42]. Fidalgo et al. reported a method for droplet fusion based on a surface energy pattern inside a microfluidic channel where the segmented flow was disrupted and the droplets were trapped and fused together [43]. In this case, full control of droplet fusion could be achieved by varying channel and pattern dimensions, as well as the fluid flow. This surface-induced droplet fusion method enabled the merging of multiple droplets containing different reagents to form a large droplet. However, this approach could potentially cause cross contamination between droplets from the patterned surface. Niu et al. developed a pillar-induced droplet merging device, in which rows of pillars were constructed in the channel network serving as passive fusion elements or chambers [44]. The pillar array trapped droplets and drained the carrier oil phase through the apertures between pillars. The first trapped droplet was suspended and merged with succeeding droplets until the surface tension was overwhelmed by the hydraulic pressure. The merging process depended on the droplet size, and the number of droplets that could be merged relied on the mass flow rate and volume ratio between the droplets and merging chamber.

Active fusion methods that can be controlled externally and selectively have also been developed using, for example, electric fields [45–48] and laser pulses [49] to trigger coalescence. To perform active droplet fusion effectively, synchronization of droplets is a key factor because fusion efficiency relies on the droplets being in very close proximity [50]. Currently, special designs are often employed to synchronize droplets in two parallel channels, which then merge into a single channel downstream to realize droplet coalescence [34,42,51]. However, this system can potentially be disturbed by a few factors such as flow rate and back pressure in the channel, which may reduce the fusion efficiency. Recently, Jambovane et al. used valve-based droplet generation for multiple reagents to perform controlled reactions and establish chemical gradients among arrays of droplets [52]. Droplets were generated at a valve-controlled side channel, and then different reagents were added to the droplets as they passed by similar side channels downstream.

An efficient method for reagent combination that we have developed recently employed two pneumatic valves integrated at a double-T intersection (Figure 1.3a). Reagents were introduced through the different side channels, each controlled by a separate valve, and simultaneous opening of the valves resulted in the creation of an aqueous plug containing both reagents. Upon actuation, the oil between the two side channels is quickly displaced, and the two aqueous streams collide and combine. No sample cross contamination was observed because of the applied pressures, the rapid valve actuation, and the offset between the two side channels. The two liquids mixed together by a combination of diffusion and convection caused by an equalization of internal pressures following the combining of the aqueous streams. The linear dependence of droplet volume on valve actuation time and the independent valving for the two aqueous streams provide a high degree of control over droplet composition. Figure 1.3b shows arrays of six droplets containing different ratios of two dyes that were created by controlling the operation of two valves. Such control should be useful for optimizing or screening reactions and for studying reaction kinetics.

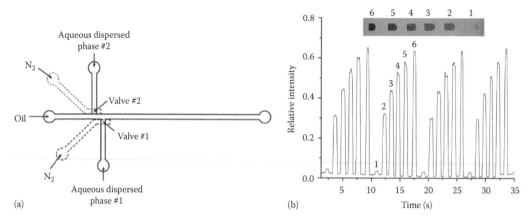

FIGURE 1.3 (a) Schematic depiction of the droplets generation, fusion, and mixing portion of the device. (b) Relative intensity of an array of six droplets containing different volume ratios of colored dyes.

1.2.3 DROPLET INCUBATION

Many biological assays involving, for example, enzymatic reactions have relatively slow kinetics, requiring microdroplets to be incubated for minutes to hours for efficient reaction. Similarly, studies involving cell incubation or protein expression require extended incubations. A straightforward method for microdroplet incubation is to simply increase the channel length following droplet generation [53,54], but increased back pressure and disruption of droplet formation can quickly become an issue. Frenz et al. incorporated deeper and wider delay lines following the droplet generation section, which enabled reactions in the droplets to increase from 1 min to >1 h [21]. Similarly, Kennedy et al. have interfaced capillaries or Teflon tubing with the droplet generation devices to collect and store sample plugs for 1–3 h [55]. For longer-term online incubation, droplets can be stored in reservoirs, traps, or dropspot arrays. For example, Courtois et al. fabricated a large reservoir for droplets storage over periods up to 20 h to study the retention of small molecules in droplets [23]. Huebner et al. designed a droplet trapping array to store and incubate picoliter-sized droplets for extended periods of time to investigate the encapsulated cells and enzymatic reactions [25]. Weitz et al. introduced a *Dropspots* device to immobilize and store thousands of individual droplets in a round chamber array over 15 h incubation period [56]. Droplets can also be incubated off-chip from minutes to several days when using appropriate surfactants to stabilize the droplets [57,58]. The incubated droplets can then be reinjected into microfluidic devices for further processing and detection.

1.2.4 DROPLET READOUT STRATEGIES

To date, in-droplet fluorescence detection remains the most widely used method for analyzing the contents of droplets due to its ability to measure in real time and with high sensitivity. Fluorescence detection has been implemented to study enzyme kinetics within droplets [11,59,60], characterize the behavior of encapsulated single cells [12,27], detect PCR products [61,62], and investigate the interactions between biological samples [63]. Fluorescence detection is ideally suited to rapid, sensitive detection of a small number of distinct species. For cases in which a large number of analytes need to be detected and identified (e.g., proteomics and metabolomics) and where fluorescent labeling is not desirable, alternative measurement strategies are needed.

In-droplet Raman spectroscopy has recently been used to detect and analyze droplet contents [64,65]. It is a nondestructive and label-free detection approach with high molecular selectivity, which can track the droplets in real time to provide fundamental droplet properties and chemical contents within the droplets, including droplet sizes, encapsulated species, structures, and concentrations.

Surface-enhanced Raman spectroscopy (SERS) can offer higher sensitivity and reproducible quantitative analysis of the droplets due to the enhancement of the Raman signal intensity [66]. Electrochemical detection is an inexpensive and label-free approach to collect information on physical and chemical properties of droplets and can monitor droplet production and measure droplet length, frequency, and velocity [67]. It can provide the chemical information when the reaction within the droplets involves an electrochemically active reactant or product [68]. Another advantage of the electrochemical measurements is its compatibility with alternative chip materials, including opaque substrates, which are difficult to implement for conventional optical detection strategies such as fluorescence. Nuclear magnetic resonance (NMR) has also been used for droplets or segmented flow analysis. Karger et al. developed a microcoil NMR probe for high-throughput analysis of sample plugs in dimethyl sulfoxide [69].

While the earlier detection strategies can be employed *in situ*, others require the contents to be removed from the droplet for subsequent analysis. Once extracted to an aqueous stream, the droplet contents can be analyzed using more information-rich techniques including LC, CE, and MS. MS is an especially attractive technique for in-depth, label-free biological analysis because of its ability to identify and provide structural information for hundreds or more unique species in a given analysis [70]. In the following, we detail methods used for droplet extraction and subsequent analysis.

Ismagilov et al. used a microfluidic system to screen and optimize organic reaction conditions in microdroplets detected using matrix-assisted laser desorption ionization mass spectrometry (MALDI-MS) [71]. The incubated reaction plugs were deposited onto a sample plate for MALDI-MS analysis. Kennedy et al. directly pumped nanoliter plugs of sample into a mass spectrometer for analysis through a metal-coated capillary nanospray emitter, separating the analyte from the carrier at the emitter itself [72,73]. Teflon tubing was placed close to the emitter tip to siphon the accumulated oil away from the tip, which could maintain stable electrospray at flow rates as high as 2000 nL/min. However, it is generally necessary to extract the aqueous droplet from the oil phase for further separation or online MS analysis to avoid contamination of the mass spectra with peaks from the oil and to maintain the electrospray Taylor cone in the most efficient cone–jet mode of operation.

Edgar et al. first reported the extraction of aqueous droplet contents into a channel for CE separation [74]. A femtoliter-volume aqueous droplet was directly delivered to fuse with the aqueous phase in the separation channel for CE separation. Niu et al. employed a similar method to inject the droplets in which the LC eluent was fractionated into a CE channel for comprehensive 2D separations in both time and space [75]. A pillar array was constructed at the interface to evacuate the carrier oil phase prior to loading samples into the separation channel. In these two cases, it was very difficult to maintain a robust extraction because the segmented flow was perpendicular to the CE separation channel.

Kennedy et al. exploited a surface modification method to form a stable interface at the junction between two immiscible phases in the microchannel [76–78]. They selectively patterned glass surfaces in the segmented flow channel to be hydrophobic in order to stabilize the oil–water interface and facilitate droplet extraction. But in some cases, only part of each droplet was extracted due to the presence of a *virtual wall*, which was not suitable for quantitative analysis because of irreproducibility and loss of information [76]. Fang and coworkers employed a similar surface modification technique to obtain a hydrophilic tongue-based droplet extraction interface, which could control the droplet extraction by regulating the waste reservoir height [79]. The extracted droplet contents were then detected by MS through an integrated ESI emitter. More recently, Filla et al. used a corona treatment to hydrophilize a portion of a polydimethylsiloxane (PDMS) chip to establish an extraction interface [80]. Aqueous droplets were transferred into the hydrophilic channel when the segmented flow encountered the interface. The droplet contents were subsequently analyzed by electrochemistry or microchip-based electrophoresis with electrochemical detection.

Huck et al. employed electrocoalescence to control droplet extraction [81,82]. The segmented flow and continuous aqueous flow met at a rectangular-shaped chamber where an interface between the immiscible phases was established. A pulsed electric field was applied over the chamber to force

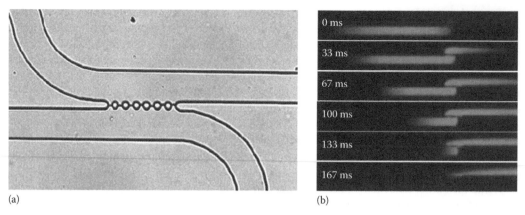

(a)   (b)

FIGURE 1.4 (a) Photograph of the droplet extraction region of the device. Water and oil fill the top and bottom channels, respectively, and the interface for the two liquids can be seen between the circular posts. (b) Micrograph sequences depicting the extraction of an individual fluorescein droplet. The flow rate in both channels was 400 nL/min.

droplets to coalesce with the continuous aqueous stream, which then delivered the droplet contents to a capillary emitter for ESI-MS detection [82]. This droplet extraction approach required careful adjustments of the flow of two immiscible phases to maintain a stable interface in the extraction chamber and avoid cross contamination of the aqueous and oil streams. In addition, the severe dilution of the droplet contents resulted in high detection limits (~500 μM bradykinin). Lin et al. used an electrical-based method to control the droplet breaking and extraction at the stable oil–water interface [83]. One reported issue in this case was the difficulty of achieving complete extraction with high efficiency, which limited its compatibility with quantitative analysis.

Kelly et al. invented a droplet extraction interface, which was constructed with an array of cylindrical posts to separate the segmented flow channel and the continuous aqueous phase channel [84]. When the aqueous stream and carrier oil phase flow rates were well controlled to balance the pressure at the junction, a stable oil–aqueous interface based on interfacial tension alone was formed to prevent bulk crossover of the two immiscible streams. The droplets could be transferred through the apertures to the continuous aqueous stream and finally detected by ESI-MS with virtually no dilution, enabling nanomolar detection limits.

Most of the reported methods and techniques for droplet extraction, as mentioned earlier, need to adjust two immiscible liquid flow rates to stabilize the interface and extract entire droplets. It is desirable to perform effective and complete droplet extraction independent of the flow rates, which would provide added flexibility for device operation. Recently, we have developed a robust interface for reliable and efficient droplet extraction, which was integrated in a droplet-based PDMS microfluidic assembly. The droplet extraction interface consisted of an array of cylindrical posts (Figure 1.4a), the same as was previously reported [84], but the aqueous stream microchannel surface was selectively treated by corona discharge to be hydrophilic. The combination of different surface energies and small flow-through apertures (~3 μm × 25 μm) enabled a very stable liquid interface between two immiscible steams to be established over a broad range of aqueous and oil flow rates. All aqueous droplets were entirely transferred to the aqueous stream (Figure 1.4b) and detected by MS following ionization at a monolithically integrated nanoESI emitter.

1.3 PERSPECTIVES FOR DROPLET-BASED MICROFLUIDICS

As mentioned earlier, droplet-based microfluidics has been employed for a wide range of analyses, and due to its unique advantages, its use will undoubtedly grow. We outline in the following a few promising applications that will leverage the strengths of the platform.

1.3.1 ENHANCED LC/MS-BASED PROTEOMIC ANALYSIS

MS-based proteomics studies are vital for biomarker discovery, identification of drug targets, and fundamental biological research. In a typical *bottom-up* proteomics workflow [85], proteins are extracted from a sample, purified, and enzymatically digested into peptides. The peptides are then separated by LC, ionized by ESI, identified by MS, and those identified peptides are then matched to their corresponding proteins based on genomic information. Alternatively, for *top-down* proteomics [86], intact proteins are separated and identified directly by MS, providing potentially more complete sequence information and the ability to characterize posttranslational modifications. However, MS identification of intact proteins is far more challenging and lower in throughput, limiting the widespread use of top-down approaches at present.

As top-down and bottom-up proteomics approaches each has unique and complimentary advantages, it would be especially attractive to obtain both intact protein and peptide-level information from a single analysis. We propose that this could be achieved by encapsulating separated proteins into droplets as they elute from an LC column, thus preserving temporal information and separation resolution while enabling further processing. For example, using our droplet-on-demand and droplet merging technologies, we could encapsulate eluting proteins into droplets and selectively add reagents for digestion to alternating droplets. The droplets could then be incubated in a delay line to allow sufficient reaction time prior to extracting and ionizing the droplets. The result would be that each droplet containing unreacted protein would be followed by a droplet containing digested peptide such that conventional bottom-up MS would be complemented with the intact molecular mass.

To this end, we have begun combining proteins with proteases in droplets to evaluate the conditions needed for digestion. The platform incorporated our integrated droplet-on-demand interface that enabled controlled in-droplet reactions, incubation in the oil stream, extraction from the aqueous stream, and ionization of the droplet contents at an integrated nanoESI emitter [87] for MS analysis (Figures 1.5 and 1.6). This integrated microfluidic platform has been successfully utilized to combine myoglobin and pepsin from separate aqueous streams into droplets to perform rapid in-droplet digestions that were detected and identified online by nanoESI-MS following droplet extraction (Figure 1.7). Given the short incubation time (18 s), the digestion did not go to completion such that peaks from the intact protein are still evident in the mass spectrum, but numerous peptides are confidently identified based on their *m/z* ratio as well (Table 1.1). We expect that simply extending the incubation time will dramatically improve digestion efficiency and enable the application of the platform to combined top-down/bottom-up proteomic analyses.

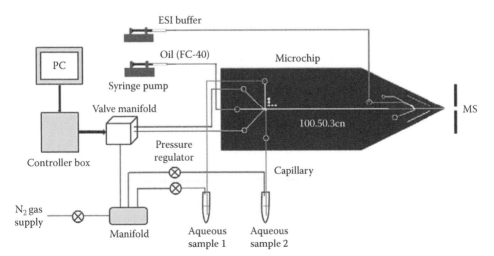

FIGURE 1.5 Schematic of the experimental setup for droplet generation, fusion, mixing, extraction, and MS detection.

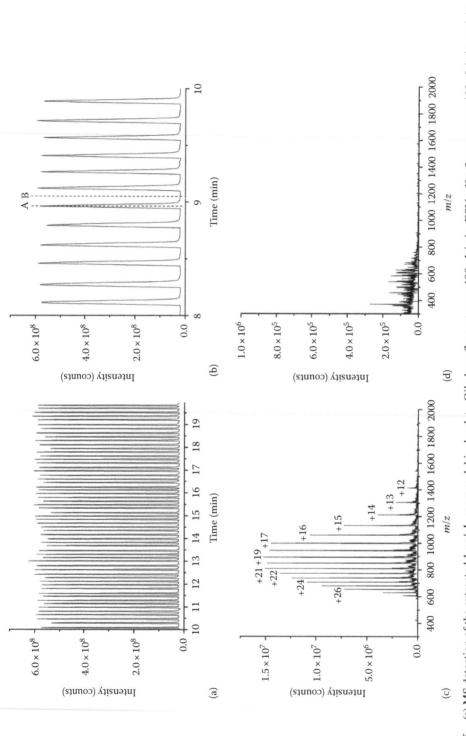

FIGURE 1.6 (a) MS detection of the extracted 1 µg/µL apomyoglobin droplets. Oil phase flow rate was 100 nL/min, ESI buffer flow rate was 400 nL/min, and droplet generation frequency was 0.1 Hz. (b) Detailed view of the MS-detected extracted apomyoglobin droplets. (c) and (d) Mass spectra obtained from the peak and baseline indicated as (A) and (B) in Figure 1.6b, respectively.

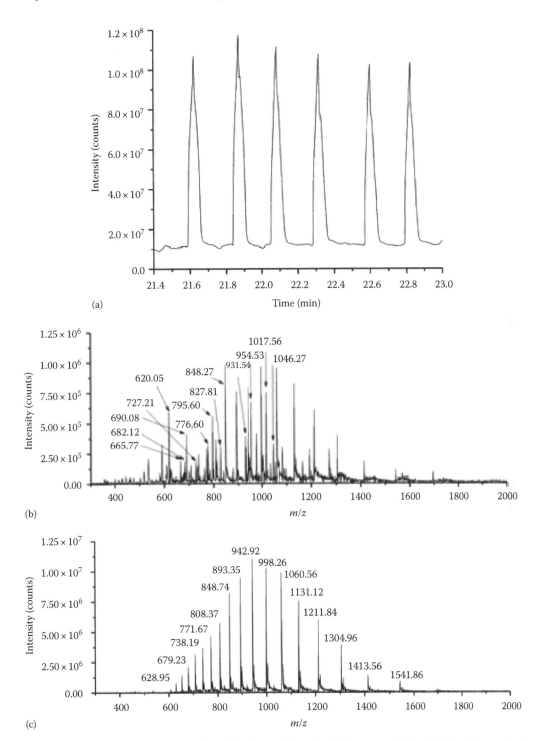

FIGURE 1.7   (a) MS detection of the fused droplets mixing 1 μg/μL apomyoglobin with 1 μg/μL pepsin in water containing 0.1% formic acid (pH ~3). The flow rates of oil and ESI buffer streams were 0.1 μL/min and 0.4 μL/min, respectively. (b) and (c) MS spectra of the fused droplet and 1 μg/μL apomyoglobin, respectively. The sequences of some digested peptide fragments labeled in (b) are listed in Table 1.1.

TABLE 1.1

Sequence of Apomyoglobin and Identification of Peptide Fragments from In-Droplet Digested Apomyoglobin Shown in Figure 1.7b

GLSDGEWQQVLNVWGKVEADIAGHGQEVLIRLFTGHPETLEKFDKFKHLKTEAEMKASEDLKKHGTGHHEAELKP
LAQSHATKHKIPIKYLEFISDAIIHVLHSKHPGDFGADA QGAMTKALELFR NDIAAKYKELGFQG
(Apomyoglobin Sequence)

m/z	Mass	z	Position	Sequence
620.05	1856.0	3+	138–153	FRNDIAAKYKELGFQG
690.08	4133.9	6+	70–106	TALGGILKKKGHHEAELKPLAQSHATKHKIPIKYLEF
827.81		5+		
665.77	4653.4	7+	30–69	IRLFTGHPETLEKFDKFKHLKTEAEMKASEDLKKHGTVVL
776.60		6+		
931.54		5+		
682.12	4767.5	7+	110–153	AIIHVLHSKHPGDFGADAQGAMTKALE
795.60		6+		LFRNDIAAKYKELGFQG
954.53		5+		
727.21	5082.8	7+	107–153	ISDAIIHVLHSKHPGDFGADAQGAMTK
848.27		6+		ALELFRNDIAAKYKELGFQG
1017.56		5+		
1046.27	3133.6	3+	1–29	GLSDGEWQQVLNVWGKVEADIAGHGQEVL

1.3.2 SINGLE-CELL CHEMICAL ANALYSIS

Sensitivity limitations on biochemical measurements typically dictate that large samples are required comprising populations of cells. These ensemble measurements average over important cell-to-cell differences. Direct chemical analysis at the single-cell level will enable the heterogeneity that is currently obscured to be better understood. The sensitivity of MS instrumentation used for proteomic and metabolomic studies has increased to the point that such single-cell measurements are now feasible. For example, while inefficient ionization and transmission of ions generated at atmospheric pressure to the high vacuum region of the mass spectrometer were previously prohibitive, recent improvements have produced combined efficiencies that can exceed 50% in some cases [88]. Indeed, around 50 proteins have been identified from samples containing just 50 pg of protein [89], which is as much protein as is contained in an average eukaryotic cell [90]. However, despite having adequate analytical sensitivity, existing methods for sample preparation, involving manual pipetting and multiple reaction vessels, are incompatible with single cells. This is another area where droplet-based microfluidics should be able to meet the need. Droplets have been previously used for single-cell encapsulation, and cells have also been lysed within droplets, with the surrounding oil preventing further dilution of the contents. Using such technologies for encapsulation and lysis in combination with our approaches for reagent mixing and droplet compatibility with ultrasensitive MS should enable us to dig deeper into the proteome and metabolome of single cells than has been accomplished previously.

1.4 CONCLUSIONS

Droplet-based microfluidics has developed substantially as a technology and will likely assume a higher profile role in biological analyses moving forward. Not only are much smaller amounts of reagents and samples consumed but also thousands of reactions and screening experiments can be performed within droplets simultaneously. Perhaps more importantly, droplet-based microfluidics is a promising tool to help us understand some fundamental biological questions such as enzymatic reactions in a confined and crowding environment, protein–protein or protein–ligand interactions, interfacial functions in biological systems, and single-cell proteomics and metabolomics. A number of operational units have been well developed for droplet-based microfluidics, including droplet generation, fusion, and incubation. Others, such as droplet extraction for subsequent analysis of the contents, have been developed recently and promise to add versatility to the platform. Robust integration of multiple functions to create a true *lab-on-a-chip* continues to be a challenge, but the unique advantages of droplets for sample-limited biological analyses will undoubtedly spawn further development, and we anticipate significant growth in the number of applications that rely on this technology in the coming years.

REFERENCES

1. Song, H., D.L. Chen, and R.F. Ismagilov, Reactions in droplets in microfluidic channels. *Angew. Chem. Int. Ed.*, 2006. **45**: 7336–7356.
2. Huebner, A. et al., Microdroplets: A sea of applications? *Lab Chip*, 2008. **8**: 1244–1254.
3. Teh, S.-Y. et al., Droplet microfluidics. *Lab Chip*, 2008. **8**: 198–220.
4. Chiu, D.T., R.M. Lorenz, and G.D.M. Jeffries, Droplets for ultrasmall-volume analysis. *Anal. Chem.*, 2009. **81**: 5111–5118.
5. Chiu, D.T. and R.M. Lorenz, Chemistry and biology in femtoliter and picoliter volume droplets. *Acc. Chem. Res.*, 2009. **42**(5): 649–658.
6. Theberge, A.B. et al., Microdroplets in microfluidics: An evolving platform for discoveries in chemistry and biology. *Angew. Chem. Int. Ed.*, 2010. **49**(34): 5846–5868.
7. Yang, C.-G., Z.-R. Xu, and J.-H. Wang, Manipulation of droplets in microfluidic systems. *Trends Anal. Chem.*, 2010. **29**: 141–157.
8. Kintses, B. et al., Microfluidic droplets: New integrated workflows for biological experiments. *Curr. Opin. Chem. Biol.*, 2010. **14**: 548–555.
9. Casadevall i Solvas, X. and A.J. deMello, Droplet microfluidics: Recent developments and future applications. *Chem. Commun.*, 2011. **47**: 1936–1942.
10. Roach, L.S., H. Song, and R.F. Ismagilov, Controlling nonspecific protein adsorption in a plug-based microfluidic system by controlling interfacial chemistry using fluorous-phase surfactants. *Anal. Chem.*, 2005. **77**: 785–796.
11. Song, H. and R.F. Ismagilov, Millisecond kinetics on a microfluidic chip using nanoliters of reagents. *J. Am. Chem. Soc.*, 2003. **125**: 14613–14619.
12. Huebner, A. et al., Quantitative detection of protein expression in single cells using droplet microfluidics. *Chem. Commun.*, 2007. **12**: 1218–1220.
13. Aebersold, R. and M. Mann, Mass spectrometry-based proteomics. *Nature*, 2003. **422**(6928): 198–207.
14. Thorsen, T. et al., Dynamic pattern formation in a vesicle-generating microfluidic device. *Phys. Rev. Lett.*, 2001. **86**: 4162–4166.
15. Garstecki, P. et al., Formation of droplets and bubbles in a microfluidic T-junction-scaling and mechanism of break-up. *Lab Chip*, 2006. **6**: 437–446.
16. Anna, S.L., N. Bontoux, and H.A. Stone, Formation of dispersions using "flow focusing" in microchannels. *Appl. Phys. Lett.*, 2003. **82**: 364–366.
17. Ward, T. et al., Microfluidic flow focusing: Drop size and scaling in pressure versus flow rate driven pumping. *Electrophoresis*, 2005. **26**: 3716–3724.
18. Baroud, C.N., F. Gallaire, and R. Dangla, Dynamics of microfluidic droplets. *Lab Chip*, 2010. **10**(16): 2032–2045.
19. Gu, H., M.H.G. Duits, and F. Mugele, Droplets formation and merging in two-phase flow microfluidics. *Int. J. Mol. Sci.*, 2011. **12**(4): 2572–2597.
20. Song, H. et al., Experimental test of scaling of mixing by chaotic advection in droplets moving through microfluidic channels. *Appl. Phys. Lett.*, 2003. **83**: 4664–4666.
21. Frenz, L. et al., Reliable microfluidic on-chip incubation of droplets in delay lines. *Lab Chip*, 2009. **9**: 1344–1348.
22. Courtois, F. et al., An integrated device for monitoring time-dependent *in vitro* expression from single genes in picolitre droplets. *ChemBioChem*, 2008. **9**: 439–446.
23. Courtois, F. et al., Controlling the retention of small molecules in emulsion microdroplets for use in cell-based assays. *Anal. Chem.*, 2009. **81**: 3008–3016.

24. Shi, W. et al., Droplet-based microfluidic system for individual *Caenorhabditis elegans* assay. *Lab Chip*, 2008. **8**: 1432–1435.

25. Huebner, A. et al., Static microdroplet arrays; a microfluidic device for droplet trapping, incubation and release for enzymatic and cell-based assays. *Lab Chip*, 2009. **9**: 692–698.

26. Schaerli, Y. et al., Continuous flow polymerase chain reaction of single copy DNA in microfluidic microdroplets. *Anal. Chem.*, 2009. **81**: 302–306.

27. He, M. et al., Selective encapsulation of single cells and subcellular organelles into picoliter- and femtoliter-volume droplets. *Anal. Chem.*, 2005. **77**: 1539–1544.

28. Lau, B.T.C. et al., A complete microfluidic screening platform for rational protein crystallization. *J. Am. Chem. Soc.*, 2007. **129**: 454–455.

29. Yobas, L. et al., High performance flow focusing geometry for spontaneous generation of monodispersed droplets. *Lab Chip*, 2006. **6**: 1073–1079.

30. He, M., J.S. Kuo, and D.T. Chiu, Electro-generation of single femtoliter- and picoliter-volume aqueous droplets in microfluidic systems. *Appl. Phys. Lett.*, 2005. **87**: 031916.

31. Park, S.-Y. et al., High-speed droplet generation on demand driven by pulse laser-induced cavitation. *Lab Chip*, 2011. **11**: 1010–1012.

32. Bransky, A. et al., A microfluidic droplet generator based on a piezoelectric actuator. *Lab Chip*, 2009. **9**: 516–520.

33. Galas, J.C., D. Bartolo, and V. Studer, Active connectors for microfluidic drops on demand. *New J. Phys.*, 2009. **11**: 075027.

34. Zeng, S. et al., Microvalve-actuated precise control of individual droplets in microfluidic devices. *Lab Chip*, 2009. **9**: 1340–1343.

35. Choi, J.-H. et al., Designed pneumatic valve actuators for controlled droplet breakup and generation. *Lab Chip*, 2010. **10**: 456–461.

36. Abate, A.R. et al., Valve-based flow focusing for drop formation. *Appl. Phys. Lett.*, 2009. **94**: 023503.

37. Song, H., J.D. Tice, and R.F. Ismagilov, A microfluidic system for controlling reaction networks in time. *Angew. Chem. Int. Ed.*, 2003. **42**: 768–772.

38. Zheng, B., L.S. Roach, and R.F. Ismagilov, Screening of protein crystallization conditions on a microfluidic chip using nanoliter size droplets. *J. Am. Chem. Soc.*, 2003. **125**: 11170–11171.

39. Huebner, A. et al., Development of quantitative cell-based enzyme assays in microdroplets. *Anal. Chem.*, 2008. **80**: 3890–3896.

40. Abate, A.R. et al., High throughput injection with microfluidics using picoinjectors. *Proc. Natl. Acad. Sci. U S A*, 2010. **107**: 19163–19166.

41. Rhee, M. and M.A. Burns, Drop mixing in a microchannel for lab on a chip platforms. *Langmuir*, 2008. **24**: 590–601.

42. Bremond, N., A.R. Thiam, and J. Bibette, Decompressing emulsion droplets favors coalescence. *Phys. Rev. Lett.*, 2008. **100**: 024501.

43. Fidalgo, L.M., C. Abell, and W.T.S. Huck, Surface-induced droplet fusion in microfluidic devices. *Lab Chip*, 2007. **7**: 984–986.

44. Niu, X. et al., Pillar-induced droplet merging in microfluidic circuits. *Lab Chip*, 2008. **8**: 1837–1841.

45. Priest, C., S. Herminghaus, and R. Seemann, Controlled electrocoalescence in microfluidics: Targeting a single lamella. *Appl. Phys. Lett.*, 2006. **89**: 134101.

46. Link, D.R. et al., Electric control of droplets in microfluidic devices. *Angew. Chem. Int. Ed.*, 2006. **45**: 2556–2560.

47. Zagnoni, M. and J.M. Cooper, On-chip electrocoalescence of microdroplets as a function of voltage, frequency and droplet size. *Lab Chip*, 2009. **9**: 2652–2658.

48. Niu, X. et al., Electro-coalescence of digitally controlled droplets. *Anal. Chem.*, 2009. **81**: 7321–7325.

49. Baroud, C.N., M.R. de Saint Vincent, and J.-P. Delville, An optical toolbox for total control of droplet microfluidics. *Lab Chip*, 2007. **7**: 1029–1033.

50. Thiam, A.R., N. Bremond, and J. Bibette, Breaking of an emulsion under an ac electric field. *Phys. Rev. Lett.*, 2009. **102**: 188304.

51. Frenz, L. et al., Microfluidic production of droplet pairs. *Langmuir*, 2008. **24**: 12073–12076.

52. Jambovane, S. et al., Creation of stepwise concentration gradient in picoliter droplets for parallel reactions of matrix metalloproteinase II and IX. *Anal. Chem.*, 2011. **83**: 3358–3364.

53. Agresti, J.J. et al., Ultrahigh throughput screening in drop based microfluidics for directed evolution. *Proc. Natl. Acad. Sci. U S A*, 2010. **107**: 4004–4009.

54. Brouzes, E. et al., Droplet microfluidic technology for single-cell high throughput screening. *Proc. Natl. Acad. Sci. U S A*, 2009. **106**: 14195–14200.

55. Slaney, T.R. et al., Push-pull perfusion sampling with segmented flow for high temporal and spatial resolution *in vivo* chemical monitoring. *Anal. Chem.*, 2011. **83**: 5207–5213.
56. Schmitz, C.H.J. et al., Dropspots: A picoliter array in a microfluidic device. *Lab Chip*, 2009. **9**: 44–49.
57. Mazutis, L. et al., Multi-step microfluidic droplet processing: Kinetic analysis of an *in vitro* translated enzyme. *Lab Chip*, 2009. **9**: 2902–2908.
58. Clausell-Tormos, J. et al., Droplet based microfluidic platforms for the encapsulation and screening of mammalian cells and multicellular organisms. *Chem. Biol.*, 2008. **15**: 427–437.
59. Damean, N. et al., Simultaneous measurements of reactions in microdroplets filled by concentration gradients. *Lab Chip*, 2009. **9**: 1707–1713.
60. Bui, M.P.N. et al., Enzyme kinetic measurements using a droplet based microfluidic system with a concentration gradient. *Anal. Chem.*, 2011. **83**: 1603–1608.
61. Beer, N.R. et al., On chip, real time, single copy polymerase chain reaction in picoliter droplets. *Anal. Chem.*, 2007. **79**: 8471–8475.
62. Beer, N.R. et al., On chip single copy real time reverse transcription PCR in isolated picoliter droplets. *Anal. Chem.*, 2008. **80**: 1854–1858.
63. Srisa-Art, M. et al., Monitoring of real time streptavidin biotin binding kinetics using droplet microfluidics. *Anal. Chem.*, 2008. **80**: 7063–7067.
64. Marz, A. et al., Droplet formation via flow through microdevices in Raman an surface enhanced Raman spectroscopy-concepts and applications. *Lab Chip*, 2011. **11**: 3584–3592.
65. Cristobal, G. et al., On line laser Raman spectroscopic probing of droplets engineered in microfluidic devices. *Lab Chip*, 2006. **6**: 1140–1146.
66. Strehle, K.R. et al., A reproducible surface enhanced Raman spectroscopy approach. Online SERS measurements in a segmented microfluidic system. *Anal. Chem.*, 2007. **79**: 1542–1547.
67. Liu, S. et al., The electrochemical detection of droplets in microfluidic devices. *Lab Chip*, 2008. **8**: 1937–1942.
68. Han, Z. et al., Measuring rapid enzymatic kinetics by electrochemical method in droplet based microfluidic devices with pneumatic valves. *Anal. Chem.*, 2009. **81**: 5840–5845.
69. Kautz, R.A., W.K. Goetzinger, and B.L. Karger, High throughput microcoil NMR of compound libraries using zero-dispersion segmented flow analysis. *J. Comb. Chem.*, 2005. **7**: 14–20.
70. Liu, T. et al., Accurate mass measurements in proteomics. *Chem. Rev.*, 2007. **107**(8): 3621–3653.
71. Hatakeyama, T., D.L. Chen, and R.F. Ismagilov, Microgram-scale testing of reaction conditions in solution using nanoliter plugs in microfluidics with detection by MALDI-MS. *J. Am. Chem. Soc.*, 2006. **128**: 2518–2519.
72. Pei, J. et al., Analysis of samples stored as individual plugs in a capillary by electrospray ionization mass spectrometry. *Anal. Chem.*, 2009. **81**: 6558–6561.
73. Li, Q. et al., Fraction collection from capillary liquid chromatography and off-line electrospray ionization mass spectrometry using oil segmented flow. *Anal. Chem.*, 2010. **82**: 5260–5267.
74. Edgar, J.S. et al., Capillary electrophoresis separation in the presence of an immiscible boundary for droplet analysis. *Anal. Chem.*, 2006. **78**(19): 6948–6954.
75. Niu, X.Z. et al., Droplet based compartmentalization of chemically separated components in two dimensional separations. *Chem. Commun.*, 2009. (41): 6159–6161.
76. Roman, G.T. et al., Sampling and electrophoretic analysis of segmented flow streams using virtual walls in a microfluidic device. *Anal. Chem.*, 2008. **80**: 8231–8238.
77. Wang, M. et al., Microfluidic chip for high efficiency electrophoretic analysis of segmented flow from a microdialysis probe and *in vivo* chemical monitoring. *Anal. Chem.*, 2009. **81**: 9072–9078.
78. Pei, J., J. Nie, and R.T. Kennedy, Parallel electrophoretic analysis of segmented samples on chip for high-throughput determination of enzyme activities. *Anal. Chem.*, 2010. **82**: 9261–9267.
79. Zhu, Y. and Q. Fang, Integrated droplet analysis system with electrospray ionization-mass spectrometry using a hydrophilic tongue-based droplet extraction interface. *Anal. Chem.*, 2010. **82**: 8361–8366.
80. Filla, L.A., D.C. Kirkpatrick, and R.S. Martin, Use of a corona discharge to selectively pattern a hydrophilic/hydrophobic interface for integrating segmented flow with microchip electrophoresis and electrochemical detection. *Anal. Chem.*, 2011. **83**: 5996–6003.
81. Fidalgo, L.M. et al., From microdroplets to microfluidics: Selective emulsion separation in microfluidic devices. *Angew. Chem. Int. Ed.*, 2008. **47**: 2042–2045.
82. Fidalgo, L.M. et al., Coupling microdroplet microreactors with mass spectrometry: Reading the contents of single droplets online. *Angew. Chem., Int. Ed.*, 2009. **48**(20): 3665–3668.
83. Zeng, S. et al., Electric control of individual droplet breaking and droplet contents extraction. *Anal. Chem.*, 2011. **83**: 2083–2089.

84. Kelly, R.T. et al., Dilution-free analysis from picoliter droplets by nano-electrospray ionization mass spectrometry. *Angew. Chem. Int. Ed.*, 2009. **48**(37): 6832–6835.
85. Swanson, S.K. and M.P. Washburn, The continuing evolution of shotgun proteomics. *Drug Discov. Today*, 2005. **10**(10): 719–725.
86. Zhou, H. et al., Advancements in top-down proteomics. *Anal. Chem.*, 2012. **84**(2): 720–734.
87. Sun, X. et al., Ultrasensitive nanoelectrospray ionization-mass spectrometry using poly(dimethylsiloxane) microchips with monolithically integrated emitters. *Analyst*, 2010. **135**: 2296–2302.
88. Marginean, I. et al., Achieving 50% ionization efficiency in subambient pressure ionization with nano-electrospray. *Anal. Chem.*, 2010. **82**(22): 9344–9349.
89. Shen, Y. et al., Ultrasensitive proteomics using high-efficiency on-line micro-SPE-NanoLC-NanoESI MS and MS/MS. *Anal. Chem.*, 2004. **76**(1): 144–154.
90. Zhang, Z.R. et al., One-dimensional protein analysis of an HT29 human colon adenocarcinoma cell. *Anal. Chem.*, 2000. **72**(2): 318–322.

2 Tailoring Wettability for Passive Fluid Control in Microfluidics

Craig Priest

CONTENTS

2.1   INTRODUCTION

Interactions of solid surfaces with droplets, streams, and films of liquid occur in a wide variety of natural processes and are exploited in countless industrial processes and commercial devices. These surfaces, however, are generally heterogeneous, rough, or structured and exhibit a diversity of wetting behaviors. At the microscale, these wetting interactions may dominate the other forces acting on the liquid phase, making them central to many microfluidic applications. The focus of this chapter is the interplay of geometry and chemistry in determining wetting behavior and the implications for passive control of fluids in microfluidic systems where immiscible fluids meet.

When a liquid comes in contact with a solid surface in the presence of a vapor or an immiscible liquid, the competition of the two fluids for the solid surface causes one to spread and the other to retreat. The spontaneity of the process is a key factor in many wetting applications, as only the initial contact between the solid and fluid phases is necessary to trigger wetting behavior. The application of spontaneous wetting behavior in microfluidic devices is often termed *passive*, due to the lack of moving parts or active switching, and is exploited in autonomous, capillary-driven microfluidic devices.[1–4]

Wettability is a nontrivial phenomenon, with detailed information about the solid surface being a prerequisite to predicting wetting behavior.[5–10] Depending on what combination of surface geometry, micro- or nanoscopic roughness, and chemistry (homogeneous or heterogeneous) is present, one can observe very different wetting phenomena ranging from superhydrophobicity and superhydrophilicity,[11] wetting hysteresis (e.g., Refs. [5,8,9,12–16]) (including the so-called asymmetric hysteresis[17–21]), and velocity dependence of the contact angle.[22–25] Working at the small length scales found in microfluidic devices invariably leads to large surface-to-volume ratios, pressures, and velocity ranges, which can result in very different wetting behaviors compared with wetting of planar (open) surfaces. Even for a microchannel with a square profile and relatively *large* width, $w = 100$ µm, the surface-to-volume ratio ($4w/w^2$) is 40,000 m^{-1}. This raises the importance of the surface tensions (and consequently wettability) above body forces, for example, gravity, that may act on the multiphase flow. Spontaneous capillary rise of a liquid against gravity in a porous solid, particle bed, or capillary is a classic example of this dominant interfacial behavior.[26] While the surface wettability of the channel does not affect all microfluidic systems beyond the initial filling of the device with liquid, the proliferation of multiphase microfluidics[27] and the potential for autonomous operation of microchips[1–4] have brought wettability to the fore of design and operation of many microfluidic devices.

In this chapter, the fundamentals of surface wettability are revisited (Section 2.2) with respect to ideal and nonideal surfaces, metastable wetting behavior, and wetting dynamics. In Section 2.3, several approaches to modifying microchannel wettability are given to provide context (not a review) for Section 2.4, which is dedicated to a discussion of several key applications of wetting in microfluidic devices and structures. These include wetting-controlled spontaneous filling, valving, flow stability, phase separation, and the role of wettability in droplet (or bubble)-based microfluidics.

2.2   THEORY OF WETTING

2.2.1   THERMODYNAMIC EQUILIBRIUM

When a droplet of liquid is placed on a solid surface in the presence of a second fluid (either liquid or vapor), the liquid will temporarily spread over the surface until the liquid front comes to rest. The final state of the liquid may be a thin film (complete wetting) or a partially wetting droplet, depending on the relative magnitude of the three interfacial tensions involved.[28] In the absence or insignificance of gravity, a partially wetting droplet will form a spherical cap on the solid surface bounded by the so-called contact line, where the all three phases meet (see Figure 2.1a). The characteristic angle measured through the droplet phase between the solid–liquid interface and the plane of the liquid–vapor

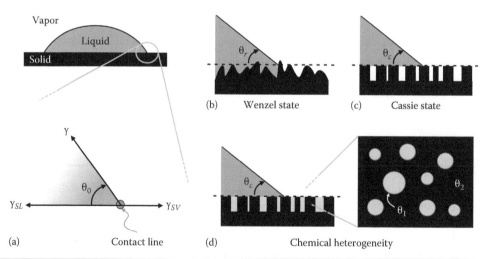

FIGURE 2.1   (a) Illustration of a droplet resting on a solid surface in vapor, showing the contact line and the interfacial tensions acting on it to yield Young's (equilibrium) contact angle. (b) and (c) show two wetting scenarios for a rough surface. (b) A droplet wetting the full surface area of the rough surface, that is, the droplet is in the *Wenzel state*. (c) A droplet with the ambient fluid remaining underneath the droplet in the rough topography of the surface, that is, the droplet is in the *Cassie state*. (d) A droplet in contact with a flat chemically heterogeneous surface.

interface at the contact line is referred to as the contact angle and is the primary measure of wettability. For a simple liquid on a flat, homogeneous, rigid, and chemically inert solid surface (i.e., an ideal surface), the contact line will come to rest only when the three interfacial tensions are perfectly balanced in the plane of the solid surface, according to the well-known Young equation[28]:

$$\gamma \cos \theta_0 + \gamma_{SL} - \gamma_{SV} = 0 \tag{2.1}$$

where

γ, γ_{SL}, and γ_{SV} are the liquid–vapor, solid–liquid, and solid–vapor interfacial tensions, respectively
θ_0 is the *equilibrium* contact angle

While the earlier discussion is based on a force balance acting on the contact line, Young's equation can also be derived via minimization of the surface free energy.

In practice, microfluidic channels, microelectromechanical devices, porous media, and a wide range of natural surfaces exhibit remarkable differences to the ideal surface, which is the basis of Equation 2.1. In particular, microfluidic devices are increasingly moving away from the traditional chemically homogeneous channel with a simple geometry to more complex surface designs that are tailored to specific microfluidic applications. The nonideal nature of these highly functional surfaces can be accounted for by derivation of modified Young's equations that take into account roughness or multiple surface components. For a rough surface, the increase in surface area relative to the projected surface area will enhance the contributions from γ_{SL} and γ_{SV} by a factor r, which is equal to the ratio of the actual to projected surface area, without impacting γ. The result is Wenzel's equation,[7] which predicts the equilibrium contact angle on a rough (but otherwise ideal) surface, θ_r:

$$\gamma \cos \theta_r + r(\gamma_{SL} - \gamma_{SV}) = 0 \tag{2.2}$$

or

$$\cos \theta_r = r \cdot \cos \theta_0 \tag{2.3}$$

At thermodynamic equilibrium, Wenzel's equation predicts that roughness will increase the contact angle of water on a hydrophobic material ($\theta_r > \theta_0 > 90°$) and decrease the contact angle on a hydrophilic material ($\theta_r < \theta_0 < 90°$). Wenzel's equation assumes that both fluid phases perfectly fill the cavities of the rough surface (see Figure 2.1b) so that the second fluid (vapor or liquid) is not trapped beneath the droplet, as shown in Figure 2.1c. This assumption does not always hold in practice and may result in very different behaviors (as discussed in the following).

For a flat and chemically heterogeneous surface, Cassie and Baxter[6] modified Young's equation to account for the different solid–liquid and liquid–vapor interfacial tensions present, weighted by their respective surface area fractions (Figure 2.1d). For a two-component solid surface, free energy minimization gives Cassie's equation:

$$\gamma \cos \theta_c = \phi_1 \left(\gamma_{S_1V} - \gamma_{S_1L} \right) + \phi_2 \left(\gamma_{S_2V} - \gamma_{S_2L} \right) \tag{2.4}$$

or

$$\cos \theta_c = \phi_1 \cos \theta_1 + \phi_2 \cos \theta_2 \tag{2.5}$$

where

θ_c is the equilibrium contact angle on the composite solid surface
ϕ_1 and ϕ_2 are the area fractions of components 1 and 2, respectively

The two solid components are differentiated in Equation 2.4 by subscripts 1 and 2. Cassie's equation has been applied to a vast number of composite surfaces with variable success, for example, Refs. [12,13,20,29–36]. Perhaps the most prominent example is the very high contact angle and low hysteresis observed on the Lotus leaf[37] and on synthetic superhydrophobic[38] surfaces. In these cases, the liquid rests on a composite surface of solid and vapor, as shown in Figure 2.1c. A surface is defined as superhydrophobic when the observed contact angle is greater than 150°, largely due to the contribution of the vapor component, and droplets move very easily over the surface, that is, a low adhesion and contact angle hysteresis. Contact angle hysteresis is an important, yet often overlooked, consideration in the design and function of microfluidic devices, and its origins are discussed in the following section.

2.2.2 Wetting Hysteresis

2.2.2.1 Advancing and Receding Contact Angles

Wenzel's and Cassie's equations are derived by minimizing the surface free energy of the three-phase system, with no consideration of local energy barriers associated with the scale or design of the surface features. In other words, the equations consider a droplet that is free to explore the whole surface energy landscape without concern for the path to that free energy minimum. Energy barriers arise when the assumption of an ideal solid surface breaks down; the surface is rough, heterogeneous, elastic, or reactive. In practice, these energy barriers can *pin* the contact line locally in a metastable state, preventing the droplet from achieving the free energy minimum (equilibrium).[5,8–10,39] The result is a significant (and sometimes large) difference between the contact angle observed after the liquid has advanced over the surface (the *static advancing contact angle*) and that observed after the liquid has receded (the *static receding contact angle*). Contact angle hysteresis is ubiquitous to wetting measurements, and no method is available to reliably access the equilibrium contact angle, despite methods being proposed (e.g., the application of mechanical energy to overcome pinning effects[40,41]). For this reason, reporting a static contact angle without specifying whether the droplet has advanced or receded over the surface is less meaningful for interpretation of wetting behavior and should be avoided.

Contact angle hysteresis by definition involves a deviation from the thermodynamic equilibrium predicted by Wenzel's and Cassie's equations (and Young's equation), despite hysteresis also originating from roughness[8,16,17,42–44] and heterogeneity.[15,18,20,45,46] Figure 2.2 shows droplet profiles for water advanced and receded on a flat (planar) surface and a structured (pillars) surface.[43] According to Wenzel's equation, the contact angles should be reduced by the presence of roughness because both contact angles on the flat surface are less than 90° (static advancing and receding contact angles are 72° and 59°, respectively). In practice, the advancing contact angle dramatically increases and the receding contact angle goes to zero due to pinning of the contact line on the array of pillars. This behavior is inconsistent with the Cassie and Wenzel equations, as well as the qualitative expectation of low hysteresis for Cassie state wetting. These results[43] and similar results from Dorrer et al.[42] were shown to be qualitatively consistent with the effect of contact line pinning on the pillar arrays for liquids resting in the Wenzel state.

This nonideal behavior, however, has not limited the application of these equations to real systems. It is common practice, although not strictly correct, to use the static advancing or receding contact angles in Equations 2.3 and 2.5 to estimate the observed wettability on a rough or heterogeneous surface. While this is a helpful approach, one should keep in mind that the degree of roughness or heterogeneity is also responsible for the magnitude of the deviation from these equations, and that the latter does not have a quantitative model. Furthermore, these deviations may be contrary to even the qualitative trends predicted by these equations. In many cases, this non-Cassie and non-Wenzel behavior has significant potential for exploitation in applications including microfluidics.

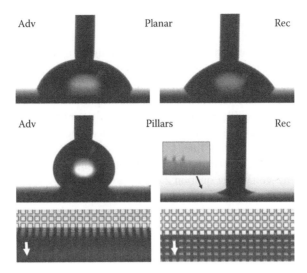

FIGURE 2.2 Static advancing and receding droplet profiles for water droplets on flat (planar) and structured (pillars) surfaces. The contact angle hysteresis is very large on the structured surface due to contact line pinning on the surface features and cannot be explained using equilibrium wetting theory. Optical microscopy of the contact line pinned on an individual row of pillars is shown below for the static advancing and receding cases. (Reprinted with permission from Forsberg, P.S.H., Priest, C., Brinkmann, M., Sedev, R., and Ralston, J., Contact line pinning on microstructured surfaces for liquids in the Wenzel state, *Langmuir*, 26, 860. Copyright 2010 American Chemical Society.)

2.2.2.2 Quantifying Hysteresis Behavior

A detailed understanding of wetting hysteresis has been pursued by researchers since Pease[39] set about explaining the departure from equilibrium on a chemically heterogeneous surface. This early work addressed the importance of more and less wettable regions on a surface in relation to the motion of the advancing and receding contact line. In essence, Pease suggested that an advancing contact line will be locally *pinned* on less wettable regions, which raises the observed static advancing contact angle. Pinning of the receding contact line on more wettable regions lowers the observed static receding contact angle. This concept has been elaborated on over the years and has been debated at length in the literature[47–49]; however, Pease's simple explanation remains conceptually relevant. In practice, this view requires replacement of the area fraction used in Cassie's equation with a *line fraction* (i.e., the local surface coverage along the contact line), which has proven an effective approximation in some instances.[20,36,43,45] For the simplest case of a droplet resting on a single circular region, this approach is intuitively simple because the line fraction is 0 or 1 depending on whether the contact line is within or outside of the region's boundary.[45] For more complex surfaces containing microscopic chemical heterogeneity (well-defined domains), the use of line fraction in Cassie's equation may resemble the experimental results.[20,36] A similar approach to rough and structured surfaces can be applied, although the complexity of these systems is even greater due to the 3D geometries involved that may or may not pin the contact line.[17,42–44,50–53] For these cases, theoretical approaches offer greater insight regarding the liquid behavior on a given surface geometry or chemical heterogeneity, revealing detailed information about the meniscus morphology and contact angle hysteresis.[42,52,54–60] The freely available software, Surface Evolver,[61] is able to incrementally modify a meniscus shape until the interfacial free energy of the three-phase system is minimized. The method has proven very effective in replicating wetting behavior and meniscus morphologies (i.e., including the contact angle) on complex solid surfaces, including between particles, on pillar arrays, and in channels.[42,52,54–56] Despite the power of these techniques, it remains prudent to use experimental approaches to verify theoretical predictions, particularly where theoretical models do not capture all the details of the experimental system, for example, a finite curvature at step edges.[42,43]

2.2.3 Dynamic Wetting

The theory discussed thus far relates exclusively to static wetting, where the contact line is at rest at thermodynamic equilibrium or in a metastable state. In many applications of wetting behavior, the contact line is moving at a finite velocity, and therefore, dynamic wetting must also be understood. Examples relevant to microfluidics include spontaneous filling of capillaries and porous materials via the Laplace pressure generated at the liquid–vapor or liquid–liquid interface (consider *paper microfluidics*[1] and other capillary-driven microfluidics[2–4]) or the displacement of one fluid by another in a channel using an external pressure. In either case, the dynamic contact angle, θ_d, can be related to the velocity of the contact line, U, according to a hydrodynamic model or molecular kinetic model. The most complete description of the hydrodynamic model was presented by Cox[62]; however, simplification by Voinov[63] (Equation 2.6) has proven useful and is correct to less than 1% error for contact angles less than $\theta \cong 135°$:

$$\theta_d^3 = \theta_0^3 + \frac{9\mu U}{\gamma}\ln\left(\frac{L}{l}\right) \tag{2.6}$$

Known as the Cox–Voinov equation, Equation 2.6 includes the viscosity of the liquid, μ; a macroscopic length scale, L; and a small length scale, l, which accounts for the violation of the no-slip boundary condition near the contact line (see Figure 2.3a). While Equation 2.6 accounts for viscous dissipation near the contact line, the molecular kinetic model considers finite molecular-scale displacements at the contact line, where K_0 and λ are the net displacement frequency (at $U = 0$) and length of these molecular displacements, respectively (see Figure 2.3b).[64] According to this model, the variation of the dynamic contact angle as a function of contact line velocity is given by

$$\cos\theta_d = \cos\theta_0 - \frac{2k_B T}{\gamma\lambda^2}\sinh^{-1}\left(\frac{U}{2K_0\lambda}\right) \tag{2.7}$$

where
k_B is Boltzmann's constant
T is the absolute temperature

Viscous and molecular effects do not act exclusively, and consequently, combined models for dynamic wetting have been proposed.[65,66] For a more thorough discussion of dynamic wetting theory, the reader is directed to Ref. [67].

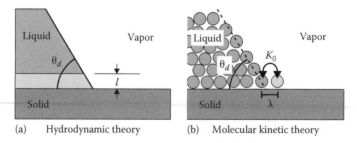

(a) Hydrodynamic theory   (b) Molecular kinetic theory

FIGURE 2.3 (a) Illustration of hydrodynamic model's small length scale, l, accounting for the violation of the no-slip boundary condition at contact line and (b) the molecular kinetic model showing molecular-scale liquid displacements with length scale, λ, and frequency, K_0.

Dynamic wetting is rarely considered in the design and operation of microfluidic chips, despite the possible consequences being quite remarkable. In the vast majority of microfluidic applications, contact lines are mobile and travel along channels (e.g., transport of droplets[68,69]), cycle their position locally (e.g., drop formation at a channel junction[70–73]), or are created and expanded rapidly (e.g., digital microfluidics[74]). The velocity in these microfluidic systems is dictated by a wettability gradient, capillary pressure, or an externally applied pressure drop. In the case where wetting is forced by an external pressure, there are limits to the maximum contact line velocities achievable, beyond which entrainment of the dewetting phase will occur.[75] While these limits are important in the success of coating technologies,[76] their role in limiting the performance of high-speed microfluidic processes has rarely been considered in the literature, as discussed in Section 2.4.

2.3   TAILORING MICROCHANNEL WETTABILITY

For a given liquid, tuning the wetting behavior is carried out via modification of the surface roughness (i.e., the roughness factor, r, in Wenzel's equation) and/or the solid–fluid interfacial tensions, γ_{SV} and γ_{SL}. There is a large body of literature that deals with the modification of planar surfaces to influence wetting behavior, which covers thin-film deposition (metal or polymer), self-assembled monolayers (SAMs), gas-phase plasma treatments, and very well-defined fabrication of structures at the micro- and nanoscale using advanced techniques. It is not the intention to present a comprehensive review of these methods here, but rather to highlight several examples of suitable fabrication methods for the modification of microchannel wettability. Additional attention is given to microplasmas as an emerging and potentially very powerful technique for channel modification.

SAMs are well-ordered molecular films that spontaneously adsorb on a surface.[77] The adsorbed molecules typically behave as a surfactant, with the SAM formation driven by a strong interaction between the head group of the molecule and the surface. Simply changing the outermost (terminal) functional group on the molecule, which contacts the fluid phase, can alter wetting behavior dramatically. Common examples are alkanethiols on gold substrates and silanes on silica.[77,78] SAMs are also inherently simple to prepare, involving immersion of the sample in a solution of the SAM molecule for a suitable time. The efficacy of SAMs in wetting studies on planar surfaces has easily transferred to the surface modification of microchannels.[79] The wettability of SAMs can span the full range of accessible contact angles on planar surfaces, from complete wetting (OH- or CO_2H-terminated SAMs, $\theta = 0°$)[80] to very hydrophobic surfaces (CH_3- or CF_3-terminated SAMs, $\theta \cong 120°$)[80] and tuneable contact angles using *mixed* multicomponent SAMs.[34]

The photosensitivity of some SAMs and other thin films make them ideal candidates for photo-initiated modification of surface wettability.[79,81] The layers themselves may be removed or replaced completely or partially to generate a pattern,[79,81] or their conformation may be manipulated to induce a switch in wettability.[82] The widespread use of optically transparent chip materials permits the local modification of *closed* microchannels (i.e., after sealing the microchip).[83] Several examples of controlled wetting described in Section 2.4 rely on photo-initiated surface modification and related patterning techniques.

While SAMs and photopatterning have proven very effective in modifying the wettability of microchannels, these methods typically rely on solution-based chemistry and several processing steps, for example, rinsing and development. In contrast, plasma processing offers a dry (gas phase) and often single-step alternative. Surface modification is achieved by exposure of the surface to an excited, ionized gas, which is generated by an electric field. Plasmas contain a variety of highly reactive species, including ions, photons, and radicals, which are able to either treat (e.g., oxidize) the surface or, where monomer is added to the plasma, deposit polymers.[84–87] It is well known that plasma treatment is widely used in the fabrication of polydimethylsiloxane

(PDMS) chips, where the hydrophobic surface is activated by an oxygen plasma prior to bonding. During this process, the surface is rendered hydrophilic, due to the formation of silanol (Si–OH) surface groups. After some time, however, the polymer becomes hydrophobic again.[88] While the majority of reported surface modifications in microchannels have been homogeneous, there has been a growing interest in plasma patterning of microfluidic channels.[89–91] Dixon and Takayama[89] guided a plasma along one side of a linear PDMS channel by offsetting the electrodes positioned at the inlet and outlet. After 5 s of treatment (at high potential; up to 50 kV), parallel regions of treated (hydrophilic, 50 μm wide) and untreated (hydrophobic) surfaces were generated. This method is quite limited in terms of channel geometry and the wettability pattern generated. In earlier work by Klages et al.,[91,92] localized plasma patterning of channels was demonstrated using electrodes that were positioned part way along the channel. In this case, plasmas at the millimeter scale could be generated containing hexamethyldisiloxane, which selectively hydrophobized the channel. More recently, Priest et al.[90] demonstrated highly localized helium plasma treatment of sub-100 μm long regions in 50 μm wide and deep microchannel between embedded gallium electrodes (*injected electrodes*) (Figure 2.4). Using this approach, a regular array of hydrophilic regions could be generated along the length of the hydrophobic PDMS channel, shown in Figure 2.4c for 300–800 μm long channel regions. The authors later showed that the technique is also effective in glass microchannels, where the surface modification is more robust than in PDMS channels.[93]

Despite many surface treatments being available, few have the ability to locally modify the chemistry and wettability of an already bonded (closed) microfluidic channel. These techniques generally rely on solution-based adsorption/reaction, oxidation (e.g., by plasma techniques), photo-induced surface modification, and electrochemical deposition (provided electrodes can be prepared during chip fabrication). For a dedicated discussion focused on surface modification techniques for already bonded microchannels, the reader is referred to Ref. [83].

The methods discussed earlier predominantly alter the chemistry, and thus the wettability of the surface, without changing the surface roughness significantly. However, as discussed in Section 2.2, the physical landscape of the solid surface is particularly important in determining the wettability observed in rough or structured microfluidic channels. Generating roughness and designed structures in microchannels can be achieved using particle deposition, wet or dry (e.g., plasma or laser) etching, polymer molding/hot embossing, and surface stress.

Particle deposition is perhaps the most straightforward due to the simplicity of the method. Where the particles are chosen to have an affinity for the microchannel walls, for example, electrostatic attraction, a particle dispersion can be flowed through the microchannel until a sufficient surface coverage is achieved, followed by rinsing and a curing (baking) step.[94] The result is a random array of adsorbed particles that introduce nano- or microscale roughness to the channel, depending on the particle size and the nature of the adsorbed layer (submonolayer, monolayer, or multilayer). The method is limited to random roughness at a small scale with respect to the channel dimensions, however is very effective in modifying microchannel wettability.[94]

Contact line pinning on microstructures plays a major role in determining the wettability of a surface, as discussed in Section 2.2.2, and, consequently, the channel geometry and structures therein must be fabricated for optimal wettability control over multiphase microfluidic flow. A variety of microscale surface features have been employed, ranging from wet-etched *guide structures* in microsolvent extraction[95] to pillar arrays to comb-like structures[96] and channel constrictions.[97] Most structures of this kind are generated during fabrication of the microchip using standard photolithography, etching, micromilling, or embossing techniques, due to the difficulty in accessing the microchannel after bonding (in contrast to particle deposition).

The following section will discuss how the theory discussed in Section 2.2 and channel structuring and surface modification can be applied to control the flow of multiple fluids in microchannels. The focus will be on passive applications of wetting in microfluidics, including capillary-driven flow, valves, flow guides, and multiphase flow stability.

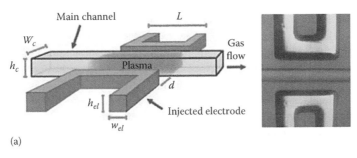

(a)

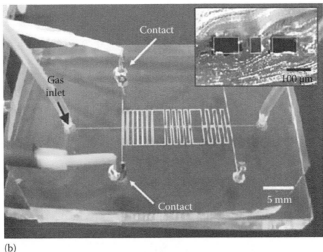

(b)

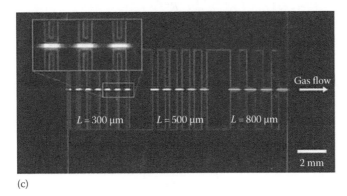

(c)

FIGURE 2.4 Plasma patterning of microchannel wettability. (a) Microchannel configuration including *injected* gallium electrodes, with an image of gallium-filled electrodes in PDMS (right). (b) Image of the microchip and, shown inset, the cross section of the gas channel (center) and electrode channels (left and right). (c) Localized helium microplasmas with the length of the plasma regions, L, noted on the figure. The gas microchannel was 50 μm wide. (Adapted from Priest, C., Gruner, P.J., Szili, E.J., Al-Bataineh, S.A., Bradley, J.W., Ralston, J., Steele, D.A., and Short, R.D., Microplasma patterning of bonded microchannels using high-precision "injected" electrodes, *Lab Chip*, 11, 541–544, 2011. Reproduced by permission of The Royal Society of Chemistry.)

2.4   FLUID CONTROL IN MICROCHANNELS

2.4.1   CAPILLARY FLOW

The introduction of fluid into microfluidic channels can be achieved through the application of a positive pressure to the fluid phase or, alternatively, through spontaneous penetration of fluid driven by the capillary pressure. The latter is particularly important for autonomous microfluidic devices, which do not require external pumping to be operated and can therefore be applied in remote areas, at the point-of-care, and using inexpensive equipment. Capillary-driven flow is simply the result of an imbalance of the Laplace pressure (the driving force) against the pressure drop along the length of the liquid filament (the sum of the hydrostatic and hydrodynamic pressures).[26] The Laplace pressure, P_L, is inversely proportional to the radius of the capillary, R, and proportional to the liquid–vapor interfacial tension, γ, and the wettability of the solid–liquid–fluid system, $\cos \theta_0$, which is given by Equation 2.8 for a cylindrical capillary[98]:

$$P_L = 2\gamma \frac{\cos \theta_0}{R} \tag{2.8}$$

The Laplace pressure may be positive or negative, depending on the wettability of the capillary wall. Where the Laplace pressure opposes filling of a microchannel, a positive pressure must be applied at the device inlet to force the liquid into the channel. Irrespective of whether the Laplace pressure induces or opposes flow, the effect is magnified by the small dimensions encountered in microfluidic channels.

Where the capillary pressure is positive (spontaneous filling), no external pressure is required to induce flow. In this case, and assuming gravity is not important (e.g., flow is perpendicular to the gravitational force), the resistance to fluid flow is the hydrodynamic pressure drop along the filament of liquid, P_μ, which, for Poiseuille flow in a cylindrical capillary, is given by the Hagen–Poiseuille equation:

$$P_\mu = \frac{8\mu l}{R^2} U \tag{2.9}$$

where
   μ is the dynamic viscosity
   U is the average velocity of the liquid through the capillary
   l is the length of the liquid filament

Equations 2.8 and 2.9 can be combined to describe the dynamics of capillary-driven filling of a capillary by a liquid[26]:

$$l^2 = \frac{\gamma R \cos \theta_0}{2\mu} t \tag{2.10}$$

Equation 2.10, known as Washburn's equation, has become a reliable basis for any study of liquid penetration in porous media and capillaries and has proven to be relevant down to the nanoscale.[99,100] It is clear from the Hagen–Poiseuille equation that the wettability of the capillary or microchannel dictates the direction of flow, either filling or emptying. Washburn's equation is fundamentally important for describing the dynamics of spontaneous capillary-driven flow in microfluidic devices, whether in open microchannels, bonded (closed) micro- and nanofluidic channels, or in the so-called paper microfluidics.[101] The following discussion will focus on several features of paper microfluidics, due to its recent emergence in the literature; however, for a comprehensive review, the reader is referred to Ref. [101].

While capillary-driven transport of liquids through porous media is not a new concept, the coupling of low cost, autonomy, and vast potential for commercialization and social benefits has popularized paper as a microfluidic tool.[101] The basic concept consists of capillary-driven transport of liquid samples to reaction sites along hydrophilic channels that are bounded by a hydrophobic material, for example, wax, photoresist, or ink, which is embedded in the paper (Figure 2.5b). Two wettability effects are at work in these devices. The first is capillary-driven flow, which is well described by Washburn's equation (Equation 2.10) and highly dependent on the physical and chemical properties of paper used. The second is the guiding of liquid streams using wettability boundaries of nonwetting material. It is the flexibility of these boundary designs that is novel, as merging and branching of streams is possible[102,103] and, consequently, the delivery of samples to multiple detection sites.[1,104] The paper approach can also embed valves[104] and separate samples[105] in the device and can include 3D channel networks using stacked[106] or folded[107] paper. Figure 2.5c shows the simultaneous detection of glucose and bovine serum albumin (BSA) using a branched channel design, as demonstrated by Martinez et al.[1] In their device, the concentration of glucose and BSA was successfully correlated with the intensity of a color change (colorimetric detection),[1] although a variety of other detection methods may be used.[101] Whichever detection method is chosen, it is widely accepted that unambiguous interpretation of the readout is vitally important for the envisaged applications of the technology. Smart phones offer one solution,[108] due to their ability to capture optical images and either send image data for analysis or interpret the data using

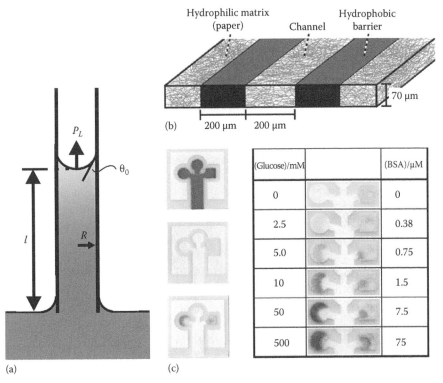

FIGURE 2.5 (a) Illustration of spontaneous capillary rise of water in a hydrophilic capillary. (b–c) An example of *paper microfluidics*, where liquid is guided via patterned wettability to reaction sites (circular and square regions) by capillary-driven flow in the porous paper. (b) The construction of the paper microfluidic device is shown. The hydrophobic barriers consist of printed, then melted wax. (c) Detection of glucose and BSA at different concentrations (indicated by the table) using two branches of the same paper microfluidic device. (Adapted with permission from Martinez, A.W., Phillips, S.T., Whitesides, G.M., and Carrilho, E., Diagnostics for the developing world: Microfluidic paper-based analytical devices, *Anal. Chem.*, 82(1), 3–10. Copyright 2009 American Chemical Society.)

on-board software. Li et al.[109] proposed a very convenient method for blood type analysis in which the result appears as text on the paper itself, enabling unambiguous interpretation of the results by any user without additional technology. Paper microfluidics remains a developing area of research and new approaches to fabrication, functionality, and detection will continue to emerge. However, optimization of capillary flow will remain important to the development of the technology. In a recent review by Li et al.,[101] several limitations of the technique were raised, including the inability of some hydrophobic surface treatments to successfully guide samples with a low surface tension. Therefore, precisely tailoring the wettability of the paper (channels and guides) will be central to the ongoing development and ultimate performance of these devices.

2.4.2   Capillary (Laplace) Valves

The ability to drive liquid penetration in microchannels using capillarity is not, however, restricted to the spontaneous filling of microfluidic devices. Local changes in wettability or geometry can be used to manipulate fluids for precise loading and sampling of nanoliter and picoliter volumes,[94,110] timing and sequencing of reactions,[97] and other triggered processes.[68,111] The so-called capillary (or Laplace) valves rely on a spatially abrupt change in channel wettability or geometry to control the capillary pressure.[112]

A simple, yet effective, approach is to couple a large and small hydrophobic microchannel in series. Using PDMS microchannels, Yamada and Seki[113] demonstrated the dispensation of 3.5 nL droplets using this technique. The smallest channel dimension (i.e., of the valve channel) was 5 μm, providing almost 20 kPa of capillary pressure for water against its vapor ($\theta_0 = 110°$ for PDMS). In a similar approach by Lai et al.,[110] many small hydrophobic *channels* in between micropillars were fabricated at the sides of a large (millimeter-scale) channel. The pressure required to drive the liquid into the smaller channels was several kilopascals and could be used as a capillary valve in bioanalytical applications.

The most commonly studied type of capillary valve is represented in Figure 2.6, where an expansion in the channel dimensions pins the contact line at a sharp edge. As the contact line cannot advance until the contact angle on the flat surface beyond the edge reaches a value of θ_0, which corresponds to an increase in interface curvature (shown in Figure 2.6a), the liquid must overcome an additional Laplace pressure to release the valve. The additional pressure can be provided by the centrifugal force on a rotating disk[114–118] or by applying a positive pressure at the liquid inlet.[94,110,119–121] The strength of the capillary valve can be tuned by modifying the expansion angle, β, and changing the wettability of the microchannel. Studies using a combination of theory and experiment have shown that a 3D model gives the most accurate predictions of the valve strength (maximum pressure).[115,117] The maximum pressure is often called a burst pressure, as it is the pressure that triggers the release of the fluid from the capillary valve. For the simplest case of a cylindrical channel opening into a conical channel (Figure 2.6a), the burst pressure, P_b, is given by

$$P_b = -2\gamma \frac{\cos(\theta_0 + \beta)}{R} \tag{2.11}$$

which can be obtained from Equation 2.8 by including an apparent (or effective) contact angle, $\theta_0 + \beta$, observed at the boundary between the cylindrical and conical segments of the channel. The negative sign accounts for the burst pressure opposing the Laplace pressure. The magnitude of the burst pressure can be several kilopascals, depending on the chosen geometry and contact angle.[116,117,122] For the most common channel geometries (e.g., a rectangular cross section), Equation 2.11 should be modified to account for the 3D geometry of the meniscus that is formed to avoid

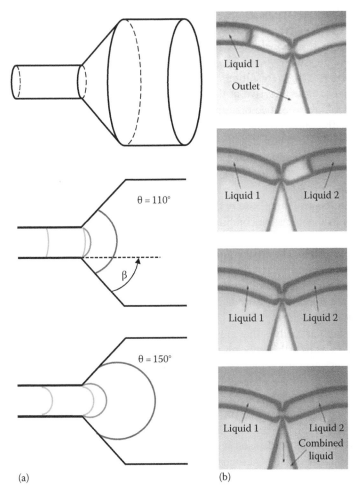

(a)

(b)

FIGURE 2.6 Geometry-based capillary valves: (a) capillary burst valve for the symmetrical case of a conical expansion at angle β for a hydrophobic, $\theta = 110°$, and superhydrophobic, $\theta = 150°$, channel. The meniscus is shown to pin at the edge separating the cylindrical and conical regions of the valve, before *bursting* into the conical region. (b) Liquid-initiated release sequence (top to bottom) from a capillary valve as demonstrated by Melin et al.[97] Liquid 1 arrives at the capillary valve and is stopped until Liquid 2 meets the liquid, which releases the valve. In the last frame, the liquids are shown to flow together in the outlet channel. (Reprinted from *Sens. Actuat. B: Chem.*, 100(3), Melin, J., Roxhed, N., Gimenez, G., Griss, P., van der Wijngaart, W., and Stemme, G., A liquid-triggered liquid microvalve for on-chip flow control, 463–468, Copyright 2004, with permission from Elsevier.)

considerable deviation from experimental results.[115,117] For example, the expansion of the meniscus may not be possible in all dimensions, due to the planar cover used to seal the microchip. As Glière and Delattre[121] have shown, this can lead to a dramatic reduction in burst pressure by more than a factor of two ($\beta = 90°$, $\theta_0 = 60°$, channel height = 15 μm, channel width = 30–115 μm) when compared with the symmetrical case. Thus, in practice, the geometry of capillary valves may be complex and care should be taken when applying predictive models.

While the pursuit of high burst pressures has been the target of many studies, the magnitude of the pressure change required to release the valve has a particularly important role and is therefore addressed here. The transition from the *closed* to *open* state for a capillary valve may be small or large, depending on the pressure required to drive the liquid to the valve (which may be negative for spontaneous filling) and the burst pressure defined in Equation 2.11. Consider the two examples depicted in Figure 2.6a, where the geometry of the valve is fixed ($\beta = 45°$, $R = 50$ µm) and water is displacing its vapor in the channel. In the first case, the main channel is hydrophobic ($\theta = 100°$), and therefore, a positive pressure of 0.5 kPa is required to fill up to the capillary valve. The magnitude of the burst pressure is 2.4 kPa, resulting in 1.9 kPa more pressure being required to burst the valve above that required to flow to arrive at the valve. The second microchannel is superhydrophobic ($\theta = 150°$), and therefore a positive pressure of 2.5 kPa is required to fill up to the capillary valve. The magnitude of the burst pressure is 2.8 kPa, resulting in only 0.3 kPa more pressure being required to burst the valve. Thus, despite the higher burst pressure, a modest (>0.3 kPa) overpressure during filling of the main channel could result in premature release of the valve, which is unlikely in the former case due to the more than six times larger pressure difference between the *closed* and *open* states (1.9 kPa).

When a pressure-controlled release of the valve is undesirable, for example, where precise pressure control is not available or the combination of two liquid streams should be precisely timed, a liquid-triggered approach reported by Melin et al.[97] offers an elegant solution. Figure 2.6b shows their multichannel design. The first liquid to arrive at the junction stops at the geometry-based capillary valve and remains pinned until the second liquid arrives. The release of the first liquid is initiated by contact with the second liquid (and so on, potentially for several different streams of liquid), breaking the Laplace pressure barrier to release both liquids simultaneously and avoiding trapped air bubbles in the channels.

Thus far, only capillary valves that are based on a local change in channel geometry have been discussed. Capillary valves that rely on inhomogeneous surface treatments, however, may have significant advantages over the geometry-based valves, especially where a large difference between the pressure required to fill the channel and the burst pressure (to release the valve) is required. Wettability patterns in microchannels can be achieved using a variety of techniques, including some that are carried out after sealing the microchip.[83] One of the most versatile methods is photolithography, which has been used extensively for making discrete regions of a channel less wettable, typically more hydrophobic. Andersson et al.[119,120] used a photoresist to expose a segment of a linear, deep reactive ion-etched silicon microchannel for subsequent coating via plasma polymerization (C_4F_8 monomer). The fluorinated region had a final contact angle for water of 105° and a burst pressure magnitude of 760 Pa. This hydrophobic region proved sufficient for valving a wide range of liquids, including solutions of surfactants and biorelevant molecules.

Where larger differences between *closed* and *open* states are required, chemical patterning can be coupled with roughness for enhanced wetting transitions at the valve location. The theoretical basis for enhancing wetting behavior using roughness was given in Section 2.2.1 (see Equation 2.3). Takei et al.[94] achieved a superhydrophobic ($\theta > 150°$) to superhydrophilic ($\theta < 9°$) wettability contrast based on nanoroughness and chemical modification. The technique relied on the electrostatic adsorption of titanium dioxide nanoparticles on the silica (Pyrex™) microchannel, surface modification using a hydrophobic trichlorosilane, and ultraviolet (UV) irradiation through a photomask. The authors report burst pressures up to ~12 kPa and the ability to dispense picoliter volumes of liquid using discrete regions of different wettabilities in a stepwise process.

The relatively high pressures achievable using simple geometry and surface-modified capillary valves illustrate the importance of tailoring the wettability of microchannels for the desired application. The precise control demonstrated earlier, including the dispensation and combination of picoliter volumes on demand, is only achievable with well-designed wetting behavior. Furthermore, the phenomenon of contact angle hysteresis, discussed in Section 2.2.2, has the potential to add uncertainty or greater control to capillary valve actuation. It is worth noting here that very few of these examples specifically characterized the static advancing and receding contact angles or considered the influence, if any, of the velocity dependence of the contact angle.

2.4.3   WETTABILITY FLOW GUIDES

Spatially controlled wettability of microchannels is not only useful in valving flow, as described in Section 2.4.2, but can also be used to guide multiphase flow along microchannels. The physical principles are unchanged from our discussion on capillary valves—the Laplace pressure acts to prevent flow in a particular direction—however, flow is now parallel to the wettability boundary (which may be a combination of surface chemistry and geometry).[79,123–128] The boundaries in this configuration have been referred to as *virtual walls* due to the absence of a solid barrier to flow in this design.[79] The strength of these guides is determined by the Laplace pressure. The general expression for the Laplace pressure is:

$$P_L = \gamma\left(\frac{1}{r_1} + \frac{1}{r_2}\right) \tag{2.12}$$

which accounts for the different radii of curvature present at a the nonspherical meniscus, r_1 and r_2. For the parallel streams shown in Figure 2.7a, r_2 approaches infinity and $P_L = \gamma/r_1$. This value is half the Laplace pressure of a spherical meniscus ($r_1 = r_2 = R$), as for the capillary valves discussed in Section 2.4.2, $P_L = 2\gamma/R$. Despite its diminished value, P_L can be sufficient to guide a liquid along the channel.[79] In practice, the Laplace pressure can be related to the contact angle of the bounding wettability, θ, and the height of the channel, h, according to

$$P_L = \frac{2\gamma\cos\theta}{h} \tag{2.13}$$

The contact angle used in Equation 2.13 is not the equilibrium contact angle but rather the static advancing contact angle on the surface beyond the wettability boundary (as this is the condition for entry to the adjacent stream).

Several examples of guiding multiphase streams using wettability have been reported by Zhao et al.[79,125] The surface modification of the microchannels was carried out using laminar streams of solvent with and without a reagent for hydrophobizing the walls of the microchannel (e.g., octadecyltrichlorosilane) or photolithography. Using laminar flow, the hydrophobic regions corresponded to the path of the stream containing the reagent (which forms a SAM) and were used as boundaries to guide the flow of water (containing dye) (see Figure 2.7b). Using photolithography, the whole channel was first modified by a photosensitive SAM, before irradiation with UV through a photomask to

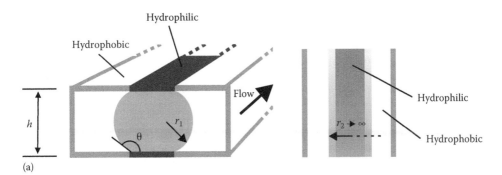

FIGURE 2.7 Guiding fluids in microchannels using *virtual* walls. (a) Side (left) and top (right) views of the *virtual wall* concept showing a liquid following a hydrophilic region along the center of the channel. The radii of curvature, contact angle, and height of the channel used in Equations 2.12 and 2.13 are noted on the figure.

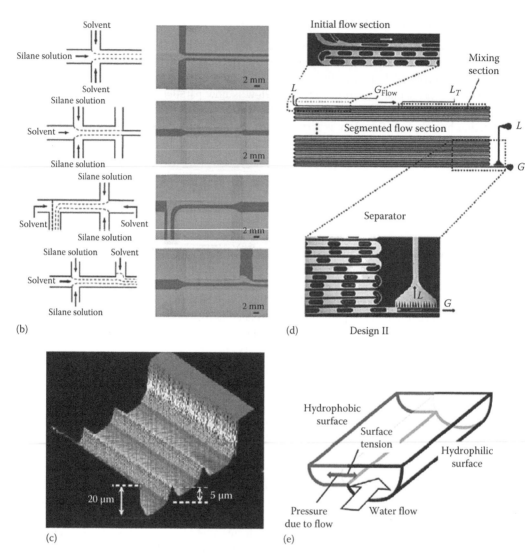

(b)

(d)   Design II

(c)

(e)

FIGURE 2.7 (continued)   Guiding fluids in microchannels using *virtual* walls. (b) Illustrations of several flow regimes for silane solution (octadecyltrichlorosilane in hexadecane; yielding θ = 112°) and solvent (hexadecane) during surface modification of the microchannel (sketches) and the corresponding flow of an aqueous solution of Rhodamine B dye, which is confined to the unmodified regions of the channel (images). (From Zhao, B., Moore, J., and Beebe, D.J., Surface directed liquid flow inside microchannels, *Science*, 291, 1023, 2001. Reprinted with permission from AAAS.) (c) *Guide structures* used to pin contact lines and, thereby, guide parallel streams of immiscible liquids (shown are two 5 μm high guide structures designed to accommodate three parallel streams). (Reprinted from *Adv. Drug Deliv. Rev.*, 55, Sato, K., Hibara, A., Tokeshi, M., Hisamoto, H., and Kitamori, T., Microchip-based chemical and biochemical analysis systems, 379, Copyright 2003, with permission from Elsevier.) (d) Phase separation of ethanol or water from gas through a comb-like structure (a series of small side channels) in a PDMS chip. (Reprinted with permission from Günther, A., Jhunjhunwala, M., Thalmann, M., Schmidt, M.A., and Jensen, K.F., Micromixing of miscible liquids in segmented gas–liquid flow, *Langmuir*, 21(4), 1547–1555. Copyright 2005 American Chemical Society.) (e) An illustration of the membrane-free phase separation microchip of Hibara et al.[127] The shallow channel in this glass chip is hydrophobized with octadecyltrichlorosilane to maximize the Laplace pressure that opposes entry by the liquid phase (from the larger channel). (Reprinted with permission from Hibara, A., Iwayama, S., Matsuoka, S., Ueno, M., Kikutani, Y., Tokeshi, M., and Kitamori, T., Surface modification method of microchannels for gas-liquid two-phase flow in microchips, *Anal. Chem.*, 77, 943. Copyright 2005 American Chemical Society.)

render the exposed regions hydrophilic. By choosing different laminar flow regimes or photomasks for the surface modification step, Zhao et al. demonstrated that the wettability boundaries, or *virtual walls*, could be used to control a stream of liquid (or multiple streams) within a single channel. The wettability boundaries could withstand critical pressures of more than 300 Pa—much less than the burst pressures of the valves described earlier, but sufficient to guide the aqueous flow. While using laminar flow to pattern wettability in a microchannel is rather restrictive in terms of the pattern design, Kenis et al.[129] demonstrated that discontinuous patterns can be achieved via partial removal of a preexisting surface coating, that is, a thin gold electrode, by laminar flow of an etch solution. However, photolithography is much more versatile where discontinuous and complex designs are desired.[79,125]

In the earlier examples, the surface wettability (resulting from chemical modification) is the guide; however, as discussed previously in this chapter, surface geometry can play an important, sometimes leading, role in guiding fluids. An example is the so-called guide structure, which has been used to stabilize parallel flow of multiple streams of liquid for a variety of microfluidic solvent extractions (see Figure 2.7c).[95,130–132] While the guide structure provides sufficient contact line pinning to stabilize flow without surface modification,[133,134] the stability window provided by the Laplace pressure is significantly larger when, in addition to channel geometry, one channel is hydrophobized.[126–128,135] The guide structure profile is readily achieved using isotropic etching (e.g., hydrofluoric acid etching of silica) and relies on the pinning of the contact line at the sharp ridge of the guide structure (similar to the geometric capillary valves described in Section 2.4.2). The pinning effect magnifies the local contact angle hysteresis at the guide (and θ in Equation 2.13), which provides greater stability to the interfaces than a planar channel wall. In terms of the contact angle hysteresis discussion in Section 2.2.2, this amounts to a geometrically induced energy barrier to contact line motion (into the adjacent fluid stream).

While the principles are similar to the capillary valve, guiding and stabilizing parallel streams of fluid rely on balancing the pressure difference between the two fluid streams over the full length of contact. This may be several hundred millimeters or more, depending on the application, and therefore the hydrodynamic pressure drops along the length of the two (or more) streams may differ greatly. Assuming that the outlets are at ambient pressure and the two streams are cylindrical, the positive pressure above ambient pressure required to flow each of the streams is given by Equation 2.9. Consequently, the pressure drop along a given length of channel for dissimilar liquids may be very different, and it is the maximum difference at any single point along the streams, ΔP, that determines whether the flow is guided successfully. The condition for guiding and stabilizing parallel flow in this configuration is therefore $\Delta P < P_L$, where both ΔP and P_L may be positive or negative depending on the geometry, surface wettability, and fluid properties.

These *virtual* walls, however, are not impenetrable to any second phase of liquid. While one liquid may not be able to pass through due to the Laplace pressure barrier involved, the other may pass through freely. In addition, a dispersed phase may pass through the wettability *wall* via coalescence (i.e., coalescence eliminates the interface responsible for the Laplace pressure), leaving immiscible phases behind. Thus, the two fluids in an emulsion phase can readily be separated under continuous flow in a well-designed microchannel (geometry and contact angle). While alternative—and very elegant—phase separation of droplets may be triggered by *active* methods, for example, electro-coalescence,[136] the spontaneous coalescence of droplets with a flowing stream segregated by wettability alone reduces device complexity. Several examples of this microfluidic phase separation mechanism have been reported, based on patterned wettability[123] or channel structure with appropriate surface wettability (including channels separated by membranes).[96,127,137] Logtenberg et al.[123] used physisorption of two polymer solutions under laminar flow to generate parallel regions of different wettabilities along a PDMS microchannel. The pattern was terminated part way along the channel via introduction of a gas bubble. Using the side-by-side wettability difference generated by the two polymers, the authors demonstrated phase separation of slug flows of 1-octanol and water.

In general, though, a combination of channel (or membrane) geometry and an appropriate contact angle provides the most robust method for microphase separations. For example, Günther et al.[96] fabricated a comb-like structure (a series of small channels) through which a gas phase could selectively escape, leaving a liquid phase reaction mixture in the main channel (Figure 2.7d). The capillary pressure opposing entry of the liquid phase into the gas outlet was more than 7 and 2 kPa for water and ethanol phases, respectively. In a similar approach using a fluoropolymer membrane with submicron pore dimensions, Kralj et al.[137] estimated a capillary pressure as high as 20 kPa opposing entry of the aqueous phase in a liquid–liquid separation. While effective, porous membranes and fine channel structures may suffer from fouling in particular applications, and therefore, the membrane-free phase separation chip presented by Hibara et al.[127] is an attractive solution (Figure 2.7e). In this chip, a hydrophobic shallow channel (8.6–39 μm deep) and a hydrophilic deep channel (100 μm deep) meet in parallel for a contact length of tens of millimeters. The Laplace pressure opposing entry of the aqueous phase into the shallow channel (against a gas phase) was up to ~8 kPa (depending on the depth of the shallow channel), in agreement with theoretical estimates and sufficient to achieve rapid phase separation of a gas phase from water.

2.4.4 Dispersed Phase Microfluidics

Dispersed phase microfluidics refers to the generation, manipulation, and combination (coalescence) of droplets or bubbles within microfluidic channels to carry out chemical, physical, and biological processes in discrete fluid volumes.[138–140] The various manipulations possible may include capture and storage,[141] mixing,[142,143] sorting,[139] spatial reorganization,[144,145] solidification (particle formation),[146,147] and encapsulation.[148] While the dispersed phase approach (sometimes termed *droplet-based*) is a powerful technique to increase sample throughput and screening, the stability and flow behavior of these multiphase systems are very susceptible to the microchannel wettability. This section first discusses the influence of wetting phenomena as a passive approach to achieving these processes.

The use of droplets in microfluidic applications has grown rapidly since the first demonstrations of extremely monodispersed bubble/droplet generation at microfluidic junctions (e.g., T-junction,[146] flow-focusing junction,[71,149] and abrupt changes in microchannel geometry[72]). While several droplet formation mechanisms exist, the wettability of the microchannel surface remains a critical factor for success. In the vast majority of cases, the dispersed phase should travel along the microchannel without strong interaction with the walls to avoid cross contamination between droplets, minimize liquid loss through entrainment (note that the flow velocities may be very high), and reduce the pressure drop through the channel. It is therefore desirable for the continuous phase to completely wet the solid wall, that is, dispersed phase contact angle equal to 180°, which is usually achieved by adding a surfactant to the continuous phase. While the addition of surfactant is simple and effective, applications where the interfacial properties of the dispersed phase are to be studied may require surfactant-free systems,[150] surfactant in both phases,[70,151] or additional stability above that provided by the surfactant. These applications may require the microchannel wettability to be modified in a well-controlled and robust manner, which, for example, is straightforward for glass (using silane chemistry) but more challenging for PDMS.[88] Perhaps one of the best illustrations of the need to control wettability in microfluidic emulsification is a study by Nisisako et al.[70,151] The authors used two junctions situated in series to prepare double emulsions, where one or more droplets are contained within a larger droplet that is dispersed in a continuous phase. By modifying the surface chemistry so that one of the junctions was hydrophilic and the other hydrophobic, water-in-oil and oil-in-water double emulsions could be generated with a high degree of control (Figure 2.8a through c). In contrast, wettability patterns can also be used to fuse droplets, where the role of the channel wall is now to trap one droplet until a subsequent droplet arrives and the two droplets fuse together.[152] Figure 2.8d shows a time sequence for fusion of droplets with and without dye at a hydrophilic region of the microchannel, as demonstrated

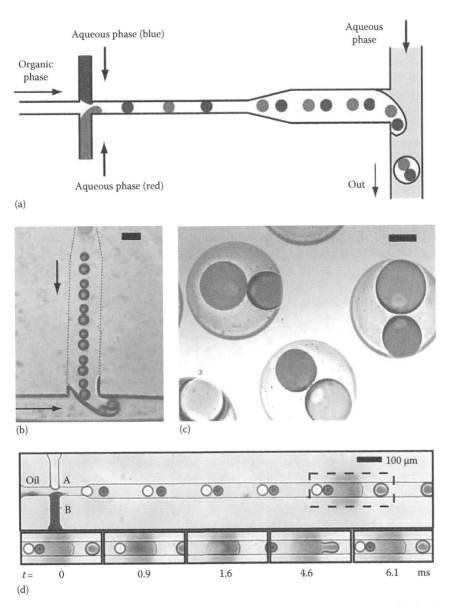

(a)

(b)   (c)

(d)

FIGURE 2.8 (a–c) Generation of double emulsions at sequential junctions in a microfluidic chip.[70] The first junction is hydrophobic to force the aqueous phases to be dispersed, while the second is hydrophilic to disperse the organic phase (containing the aqueous phase droplets). (c) The authors demonstrated precise loading of the double emulsion with pairs of droplets. (a–c: Adapted from Nisisako, T., Okushima, S., and Torii, T., Controlled formulation of monodisperse double emulsions in a multiple-phase microfluidic system, *Soft Matter*, 1, 23–27, 2005. Reproduced by permission of The Royal Society of Chemistry.) (d) Droplet fusion: time sequence for surface-induced droplet fusion in a 50 μm wide and 25 μm deep channel. An approximately 100 μm long hydrophilic region of the channel (within the dashed rectangle) was generated to trap (time = 0.9 ms) and fuse (time = 1.6 ms) droplets at the microchannel wall. The fused droplet is released from the interface between 4.6 and 6.1 ms. (From Fidalgo, L.M., Abell, C., and Huck, W.T.S., Surface-induced droplet fusion in microfluidic devices, *Lab Chip*, 7, 984, 2007. Reproduced by permission of The Royal Society of Chemistry.)

by Fidalgo et al.[152] The authors showed that after fusion, the combined droplet is released from the hydrophilic site via the viscous drag on the immobilized liquid. This method, however, may be limited in applications by the presence and concentration of surfactant in the liquid phases, which may prevent the wetting and/or fusion events.

The importance of microchannel wettability extends beyond the ability to generate droplets (or bubbles) at the microscale to the downstream flow behavior in these systems. Downstream flow behavior of droplet- and bubble-based systems is important in practical applications such as heat exchangers and oil recovery, where the type of flow directly affects performance.[153,154] Several different flow regimes are possible, depending on the wettability of the microchannel and the volume fraction of the dispersed phase.[69,153–158] These regimes are shown for water–air systems flowing in a hydrophilic and hydrophobic channel in Figure 2.9. For hydrophilic channels, shown in Figure 2.9a, Cubaud et al.[69] showed several flow regimes from *bubbly flow, wedging flow, slug flow,* and *annular flow* to *dry flow* for increasing volumetric flow fraction of the gas phase. Bubbly flow refers to bubbles that are smaller than the channel dimensions and therefore move freely to temporarily interact with one another and the walls. Wedging flow and slug flow are collectively referred to as *segmented flow* and exist when bubbles are elongated due to the confinement of the channel dimensions.[157] Fewer flow regimes and less regularity are observed in hydrophobic channels due to wetting interactions with the microchannel walls (see Figure 2.9b). The gas bubbles spread on the hydrophobic Teflon-coated channels ($\theta \approx 120°$), and the flow becomes unsteady, with gas temporarily (or even permanently) held up in corners and at surface defects, consistent with the discussion in Section 2.2.2. Similar results were observed by Fang et al.[154] where condensate flow was studied for hydrophobic ($\theta = 123°$) and hydrophilic ($\theta = 25°$) microchannels; however, for intermediate contact angle ($\theta = 91°$), dropwise condensation coexisted upstream with stratified flow on the side walls. While the differences in wettability are rather large in these cases, experiments by Salim et al.[153] revealed that changing from a rounded Pyrex to near-rectangular quartz channel, which exhibits similar wettability, can change the flow behavior from parallel streams to slug flow for a range of flow rate ratios. Thus, both the contact angle and the channel geometry should be measured precisely to ensure that the desired flow regime will be achieved.

A precise characterization of wettability in multiphase microfluidic systems must include an evaluation of contact angle hysteresis. As discussed earlier, static and dynamic advancing and receding contact angles may be vastly different, depending on the nature of the surface and the geometry of the microfluidic environment. Nonetheless, contact angle hysteresis is seldom considered in microfluidic studies with many authors quoting a single measured value for the contact angle (usually termed the *static* or, incorrectly, *equilibrium* contact angle). Contact angle hysteresis is an important factor in determining the pressure drop along channels containing droplets, due to the different contact angles (and therefore meniscus curvature) at the leading and trailing interfaces (see Figure 2.9c).[159–161] Figure 2.9c illustrates a characteristic droplet shape in segmented flow with a finite contact angle hysteresis. The capillary pressures at the lead and trailing interfaces of the droplet are therefore distinct and related to the dynamic (or static, when there is no droplet motion) advancing, θ_a, and receding, θ_r, contact angles according to Equation 2.8 (for a cylindrical geometry). Thus, the pressure drop across a single droplet due to wetting alone (i.e., excluding hydrodynamic flow resistance, i.e., $P_2 - P_3$), ΔP_θ, is

$$\Delta P_\theta = P_1 - P_4 = \frac{2\gamma}{R}\left(\cos\theta_a - \cos\theta_r\right) \tag{2.14}$$

For the condition $\theta_a = \theta_r$, ΔP_θ is zero, and excluding any influence of the wettability on hydrodynamic slip at the microchannel walls, wetting effects do not contribute to the flow resistance. This condition, however, is never practically observed, as the dynamic advancing and receding contact angles diverge with increasing velocity (see Equations 2.6 and 2.7). Measured pressure drops due

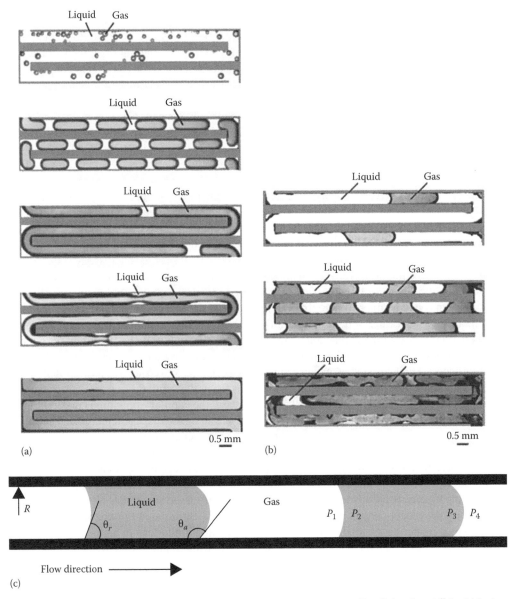

FIGURE 2.9 Flow regimes for the gas–water multiphase system, as reported by Cubaud et al.[69] for (a) hydrophilic channels and (b) hydrophobic channels. From top to bottom, with increasing flow ratio of the gas phase: (a) *bubbly flow, wedging flow, slug flow, annular flow*, and *dry flow* and (b) irregular flow patterns and entrainment of fluid at the walls and corners of the channel. (a and b: Reprinted from Cubaud, T., Ulmanella, U., and Ho, C.-M., Two-phase flow in microchannels with surface modifications, *Fluid Dyn. Res.*, 38(11), 772–786, Copyright 2006. With permission from IOP Publishing.) (c) Illustration of droplets under conditions where the liquid partially wets the microchannel walls. Contact angle hysteresis causes the profile of the lead and trailing interfaces to differ in curvature, leading to a wetting-induced pressure drop opposing flow. (Redrawn from Lee, C.Y. et al., *Exp. Therm. Fluid Sci.*, 34(1), 1–9, 2010.)

to wetting behavior for segmented flows are significant, with values ranging from several 100 Pa to 2 kPa.[159–161] The magnitude of the overall pressure drop increases, however, with the number of droplets[159] and is therefore not limited to a maximal value. As a consequence, these effects are likely to be significant where channels are partially wet and very long (e.g., in heat transfer applications), while in other applications, the effect may be negligible.

2.5 SUMMARY AND OUTLOOK

Wetting behavior is a fundamental consideration in the design of any microfluidic device where multiple fluid phases meet. The complexity of wetting behavior, including the effects of chemical heterogeneity and roughness, coupled with different channel geometries, can lead to enhanced wetting or dewetting behavior and substantial energy barriers to reversibility of contact line motion (hysteresis). While much of the theoretical considerations of wettability reflect thermodynamic equilibrium, the measured wettability of real, rough, and heterogeneous solid surfaces often reflects large energy barriers that prevent the strict application of wetting theory, for example, Wenzel's or Cassie's equation. Nonetheless, these theories may be used as approximations in microfluidics for nonideal planar surfaces, provided their limitations are fully understood, accounted for, and deficiencies supplemented by wettability measurements. This chapter has shown how the wettability of the solid can be employed as a major driver for the microhandling of fluids. Tailored wettability via controlled chemistry and well-designed geometries can be used to generate a capillary (Laplace) pressure that drives, valves, guides, and phase separates fluids in a passive manner.

The so-called paper microfluidics and other autonomous point-of-care microfluidic devices rely on the capillary pressure for their function, and in many cases, they have a patterned surface wettability to guide the liquid. In valve applications, the capillary pressure may be several kilopascals, depending on the design of the chip, and may be used to dispense nano- and picoliter volumes of liquid in a controlled manner. Where this valving action is applied perpendicular to the flow direction, the capillary pressure acts to maintain the stability of the laminar streams against mixing. This is particularly useful for reactions and separations that occur at the liquid–liquid (or fluid) interface. Fluid dispersions (droplets and bubbles) that partially wet the microchannel wall generate a wetting-induced flow resistance, which can be directly related to wetting hysteresis. Finally, droplets and bubbles can be phase separated from streams using the capillary pressure generated using surface chemistry and a particular geometry. In this case, the capillary pressure is different for the dispersed and continuous phases (due to preferential wetting by the dispersed phase) and selectively allows the dispersed phase to pass through a particular channel geometry.

The use of wettability as a passive tool in microfluidic devices is already well established; however, there remains significant scope to expand beyond the current applications. In particular, contact angle hysteresis is neglected in most microfluidics studies, yet there is a potential richness in exploiting hysteresis for advanced applications. The same assertion applies to wetting dynamics (the velocity-dependent contact angle and the onset of fluid entrainment) in microchannels, which may lead to a greater understanding of both the potential and limitations in high-velocity fluid processing. Nonetheless, even on planar surfaces, studying wetting hysteresis and dynamics has proven challenging to date, and harnessing these wetting phenomena in the confinement and complexity of microfluidic channels is unlikely to be straightforward.

ACKNOWLEDGMENTS

The author thanks Rossen Sedev and Catherine Whitby for their kind assistance in the preparation of this chapter. Financial support from the Australian Research Council (DP1094337 and LP100100272) is acknowledged.

REFERENCES

1. Martinez, A. W.; Phillips, S. T.; Whitesides, G. M.; Carrilho, E., Diagnostics for the developing world: Microfluidic paper-based analytical devices. *Analytical Chemistry* 2009, 82(1), 3–10.
2. Gervais, L.; Delamarche, E., Toward one-step point-of-care immunodiagnostics using capillary-driven microfluidics and PDMS substrates. *Lab on a Chip* 2009, 9(23), 3330–3337.
3. Juncker, D.; Schmid, H.; Drechsler, U.; Wolf, H.; Wolf, M.; Michel, B.; de Rooij, N.; Delamarche, E., Autonomous microfluidic capillary system. *Analytical Chemistry* 2002, 74(24), 6139–6144.

4. Zimmermann, M.; Hunziker, P.; Delamarche, E., Autonomous capillary system for one-step immunoas-says. *Biomedical Microdevices* 2009, 11(1), 1–8.

5. Good, R. J., Thermodynamic derivation of Wenzel's modification of Young's equation for contact angles; Together with a theory of hysteresis. *Journal of the American Chemical Society* 1952, 74, 5041–5042.

6. Cassie, A. B. D.; Baxter, S., Wettability of porous surfaces. *Transactions of the Faraday Society* 1944, 40, 547–551.

7. Wenzel, R. N., Resistance of solid surfaces to wetting by water. *Industrial and Engineering Chemistry* 1936, 28(8), 988.

8. Johnson, R. E., Jr.; Dettre, R. H., Contact angle hysteresis I. Study of an idealized rough surface. *Advances in Chemistry Series* 1964, 43, 112–135.

9. Johnson, R. E., Jr.; Dettre, R. H., Contact angle hysteresis III. Study of an idealized heterogeneous sur-face. *Journal of Physical Chemistry* 1964, 68, 1744.

10. Neumann, A. W.; Good, R. J., Thermodynamics of contact angles I. Heterogeneous solid surfaces. *Journal of Colloid and Interface Science* 1972, 38, 341.

11. Bico, J.; Thiele, U.; Quéré, D., Wetting of textured surfaces. *Colloids and Surfaces A* 2002, 206, 41–46.

12. Dettre, R. H., Johnson, R. E. J., Contact angle hysteresis—Porous surfaces. *SCI Monograph* 1967, 25, 144–163.

13. Dettre, R. H.; Johnson, R. E. J., Contact angle hysteresis. IV. Contact angle measurements on heteroge-neous surfaces. *The Journal of Physical Chemistry* 1965, 69(5), 1507–1515.

14. Shanahan, M. E. R.; Di Meglio, J. M., Wetting hysteresis: Effects due to shadowing. *Journal of Adhesion Science and Technology* 1994, 8, 1371.

15. Joanny, J. F.; de Gennes, P. G., A model for contact angle hysteresis. *Journal of Chemical Physics* 1984, 81, 552–562.

16. Huh, C., Mason, S. G., Effects of surface roughness on wetting (theoretical). *Journal of Colloid and Interface Science* 1977, 60, 11.

17. Priest, C.; Albrecht, T. W. J.; Sedev, R.; Ralston, J., Asymmetric wetting hysteresis on hydrophobic microstructured surfaces. *Langmuir* 2009, 25(10), 5655.

18. Priest, C.; Sedev, R.; Ralston, J., Asymmetric wetting hysteresis on chemical defects. *Physical Review Letters* 2007, 99, 026103.

19. De Jonghe, V.; Chatain, D., Experimental study of wetting hysteresis on surfaces with controlled geo-metrical and/or chemical defects. *Acta Metallurgica et Materialia* 1995, 43, 1505.

20. Naidich, Y. V.; Voitovich, R. P.; Zabuga, V. V., Wetting and spreading in heterogeneous solid surface-metal melt systems. *Journal of Colloid and Interface Science* 1995, 174, 104.

21. Anantharaju, N.; Panchagnula, M. V.; Vedantam, S., Asymmetric wetting of patterned surfaces composed of intrinsically hysteretic materials. *Langmuir* 2009, 25(13), 7410–7415.

22. McHale, G.; Newton, M.; Shirtcliffe, N., Dynamic wetting and spreading and the role of topography. *Journal of Physics: Condensed Matter* 2009, 21(46), 464122.

23. Semal, S.; Voué, M.; de Ruijter, M. J.; Dehuit, J.; De Coninck, J., Dynamics of spontaneous spreading on heterogeneous surfaces in a partial wetting regime. *The Journal of Physical Chemistry B* 1999, 103(23), 4854–4861.

24. Fetzer, R.; Ralston, J., Influence of nanoroughness on contact line motion. *Journal of Physical Chemistry C* 2010, 114(29), 12675–12680.

25. Fetzer, R.; Ralston, J., Dynamic dewetting regimes explored. *Journal of Physical Chemistry C* 2009, 113(20), 8888–8894.

26. Washburn, E. W., The dynamics of capillary flow. *Physical Review* 1921, 17, 273–283.

27. Gunther, A.; Jensen, K. F., Multiphase microfluidics: From flow characteristics to chemical and materials synthesis. *Lab on a Chip* 2006, 6(12), 1487–1503.

28. Young, T., An essay on the cohesion of fluids. *Philosophical Transactions of the Royal Society of London* 1805, 95(1), 65.

29. Kitaev, V.; Seo, M.; McGovern, M. E.; Huang, Y.-J.; Kumacheva, E., Mixed monolayers self-assembled on mica surface. *Langmuir* 2001, 17(14), 4274–4281.

30. Crawford, R.; Koopal, L. K.; Ralston, J., Contact angles on particles and plates. *Colloids and Surfaces* 1987, 27, 57–64.

31. Diggins, D., Fokkink, L. G. J., Ralston, J., The wetting of angular quartz particles: Capillary pressure and contact angles. *Colloids and Surfaces* 1990, 44, 299–313.

32. Woodward, J. T.; Gwin, H.; Schwartz, D. K., Contact angles on surfaces with mesoscopic chemical het-erogeneity. *Langmuir* 2000, 16(6), 2957–2961.

33. Folkers, J. P.; Laibinis, P. E.; Whitesides, G. M., Self-assembled monolayers of alkanethiols on gold: Comparisons of monolayers containing mixtures of short- and long chain constituents with CH_3 and CH_2OH terminal groups. *Langmuir* 1992, 8(5), 1330–1341.
34. Imabayashi, S.-I.; Gon, N.; Sasaki, T.; Hobara, D.; Kakiuchi, T., Effect of nanometer-scale phase separation on wetting of binary self-assembled thiol monolayers on Au(111). *Langmuir* 1998, 14(9), 2348–2351.
35. Rousset, E.; Baudin, G.; Cugnet, P.; Viallet, A., Screened offset plates: A contact angle study. *Journal of Imaging Science and Technology* 2001, 45, 517.
36. Cubaud, T.; Fermigier, M., Advancing contact lines on chemically patterned surfaces. *Journal of Colloid and Interface Science* 2004, 269(1), 171–177.
37. Barthlott, W.; Neinhuis, C., The purity of sacred lotus or escape from contamination in biological surfaces. *Planta* 1997, 202, 1–8.
38. Roach, P.; Shirtcliffe, N. J.; Newton, M. I., Progress in superhydrophobic surface development. *Soft Matter* 2008, 4(2), 224–240.
39. Pease, D. C., The significance of the contact angle in relation to the solid surface. *Journal of Physical Chemistry* 1945, 49, 107–110.
40. Sedev, R.; Fabretto, M.; Ralston, J., Wettability and surface energetics of rough fluoropolymer surfaces. *Journal of Adhesion* 2004, 80, 497–520.
41. Decker, E. L.; Garoff, S., Using vibrational noise to probe energy barriers producing contact angle hysteresis. *Langmuir* 1996, 12(8), 2100–2110.
42. Dorrer, C.; Rühe, J., Drops on microstructured surfaces coated with hydrophilic polymers: Wenzel's model and beyond. *Langmuir* 2007, 24, 1959.
43. Forsberg, P. S. H.; Priest, C.; Brinkmann, M.; Sedev, R.; Ralston, J., Contact line pinning on microstructured surfaces for liquids in the Wenzel state. *Langmuir* 2010, 26, 860.
44. Dorrer, C.; Rühe, J., Advancing and receding motion of droplets on ultrahydrophobic post surfaces. *Langmuir* 2006, 22, 7652.
45. Extrand, C. W., Contact angles and hysteresis on surfaces with chemically heterogeneous islands. *Langmuir* 2003, 19, 3793.
46. Marmur, A., Contact angle hysteresis on heterogeneous smooth surfaces. *Journal of Colloid and Interface Science* 1994, 168, 40–46.
47. Gao, L.; McCarthy, T. J., How Wenzel and Cassie were wrong. *Langmuir* 2007, 23(7), 3762–3765.
48. Marmur, A.; Bittoun, E., When Wenzel and Cassie are right: Reconciling local and global considerations. *Langmuir* 2009, 25(3), 1277–1281.
49. McHale, G., Cassie and Wenzel: Were they really so wrong? *Langmuir* 2007, 23(15), 8200–8205.
50. Spori, D. M.; Drobek, T.; Zürcher, S.; Spencer, N. D., Cassie-state wetting investigated by mean of a hole-to-pillar density gradient. *Langmuir* 2010, 26, 9465–9473.
51. Priest, C.; Forsberg, P. S. H.; Sedev, R.; Ralston, J., Structure-induced wetting of liquid in micropillar arrays. *Microsystem Technologies* 2012, 18, 167.
52. Semprebon, C.; Herminghaus, S.; Brinkmann, M., Advancing modes on regularly patterned substrates. *Soft Matter* 2012, 8(23), 6301–6309.
53. Spori, D. M.; Drobek, T.; Zürcher, S.; Ochsner, M.; Sprecher, C.; Mühlebach, A.; Spencer, N. D., Beyond the lotus effect: Roughness influences on wetting over a wide surface energy range. *Langmuir* 2008, 24, 5411–5417.
54. Choi, W.; Tuteja, A.; Mabry, J. M.; Cohen, R. E.; McKinley, G. H., A modified Cassie–Baxter relationship to explain contact angle hysteresis and anisotropy on non-wetting textured surfaces. *Journal of Colloid and Interface Science* 2009, 339(1), 208–216.
55. Gögelein, C.; Brinkmann, M.; Schröter, M.; Herminghaus, S., Controlling the formation of capillary bridges in binary liquid mixtures. *Langmuir* 2010, 26(22), 17184–17189.
56. Seemann, S.; Brinkmann, M.; Kramer, E. J.; Lange, F. F.; Lipowsky, R., Wetting morphologies at microstructured surfaces. *Proceedings of the National Academy of Sciences of the United States of America* 2005, 102, 1848.
57. Kusumaatmaja, H.; Pooley, C. M.; Girardo, S.; Pisignano, D.; Yeomans, J. M., Capillary filling in patterned channels. *Physical Review E* 2008, 77, 067301.
58. Kusumaatmaja, H.; Yeomans, J. M., Modeling contact angle hysteresis on chemically patterned and superhydrophobic surfaces. *Langmuir* 2007, 23, 6019.
59. Mognetti, B. M.; Kusumaatmaja, H.; Yeomans, J. M., Drop dynamics on hydrophobic and superhydrophobic surfaces. *Faraday Discussions* 2010, 146, 153–165.

60. Lundgren, M.; Allan, N. L.; Cosgrove, T.; George, N., Molecular dynamics study of wetting of a pillar surface. *Langmuir* 2003, 19(17), 7127–7129.
61. Brakke, K., The surface evolver. *Experimental Mathematics* 1992, 1, 141.
62. Cox, R. G., The dynamics of the spreading of liquids on a solid surface. Part 1. Viscous flow. *Journal of Fluid Mechanics* 1986, 168, 169–194.
63. Voinov, O. V., Hydrodynamics of wetting. *Mekhanika Zhidkosti i Gaza* 1976, 5, 714–721.
64. Blake, T. D.; Haynes, J. M., Kinetics of liquid/liquid displacement. *Journal of Colloid and Interface Science* 1969, 30(3), 421–423.
65. Petrov, P. G.; Petrov, J. G., A combined molecular-hydrodynamic approach to wetting kinetics. *Langmuir* 1992, 8(7), 1762–1767.
66. Brochard-Wyart, F.; de Gennes, P. G., Dynamics of partial wetting. *Advances in Colloid and Interface Science* 1992, 39(0), 1–11.
67. Blake, T. D., The physics of moving wetting lines. *Journal of Colloid and Interface Science* 2006, 299(1), 1–13.
68. Takahashi, K.; Mawatari, K.; Sugii, Y.; Hibara, A.; Kitamori, T., Development of a micro droplet collider; the liquid–liquid system utilizing the spatial–temporal localized energy. *Microfluidics and Nanofluidics* 2010, 9(4), 945–953.
69. Cubaud, T.; Ulmanella, U.; Ho, C.-M., Two-phase flow in microchannels with surface modifications. *Fluid Dynamics Research* 2006, 38(11), 772–786.
70. Nisisako, T.; Okushima, S.; Torii, T., Controlled formulation of monodisperse double emulsions in a multiple-phase microfluidic system. *Soft Matter* 2005, 1, 23–27.
71. Anna, S. L.; Bontoux, N.; Stone, H. A., Formation of dispersions using "flow focusing" in microchannels. *Applied Physics Letters* 2003, 82, 364.
72. Priest, C.; Herminghaus, S.; Seemann, R., Generation of monodisperse gel emulsions in a microfluidic device. *Applied Physics Letters* 2006, 88, 024106.
73. Sugiura, S.; Nakajima, M.; Seki, M., Prediction of droplet diameter for microchannel emulsification. *Langmuir* 2002, 18, 3854.
74. Choi, K.; Ng, A. H. C.; Fobel, R.; Wheeler, A. R., Digital microfluidics. *Annual Review of Analytical Chemistry* 2012, 5(1), 413–440.
75. Bertrand, E.; Blake, T. D.; De Coninck, J., Dynamics of dewetting. *Colloids and Surfaces A: Physicochemical and Engineering Aspects* 2010, 369(1–3), 141–147.
76. Blake, T. D.; Clarke, A.; Ruschak, K. J., Hydrodynamic assist of dynamic wetting. *AIChE Journal* 1994, 40(2), 229–242.
77. Bain, C. D.; Evans, S. D., Laying it on thin. *Chemistry in Britain* 1995, 31(1), 46–48.
78. Bain, C. D.; Troughton, E. B.; Tao, Y.-T.; Evall, J.; Whitesides, G. M.; Nuzzo, R. G., Formation of monolayer films by the spontaneous assembly of organic thiols from solution onto gold. *Journal of the American Chemical Society* 1989, 111(1), 321–335.
79. Zhao, B.; Moore, J.; Beebe, D. J., Surface directed liquid flow inside microchannels. *Science* 2001, 291, 1023.
80. Chidsey, C. E.; Loiacono, D. N., Chemical functionality in self-assembled monolayers: Structural and electrochemical properties. *Langmuir* 1990, 6(3), 682–691.
81. Tarlov, M. J.; Burgess, D. R. F., Jr.; Gillen, G., UV photopatterning of alkanethiolate monolayers self-assembled on gold and silver. *Journal of the American Chemical Society* 1993, 115(12), 5305–5306.
82. Lake, N.; Ralston, J.; Reynolds, G., Light-induced surface wettability of a tethered DNA base. *Langmuir* 2005, 21, 11922.
83. Priest, C., Surface patterning of bonded microfluidic channels. *Biomicrofluidics* 2010, 4(3), 032206–032213.
84. Kreitz, S.; Penache, C.; Thomas, M.; Klages, C.-P., Patterned DBD treatment for area-selective metallization of polymers-plasma printing. *Surface and Coatings Technology* 2005, 200, 676.
85. Siow, K. S.; Britcher, L.; Kumar, S.; Griesser, H. J., Plasma methods for the generation of chemically reactive surfaces for biomolecule immobilization and cell colonization—A review. *Plasma Processes and Polymers* 2006, 3, 392.
86. Dai, L.; Griesser, H. J.; Mau, A. W. H., Surface modification by plasma etching and plasma patterning. *Journal of Physical Chemistry B* 1997, 101, 9548.
87. Aizawa, H.; Makisako, T.; Reddy, S. M.; Terashima, K.; Kurosawa, S.; Yoshimoto, M., On-demand fabrication of microplasma-polymerized styrene films using automatic motion controller. *Journal of Photopolymer Science and Technology* 2007, 20(2), 215.

88. Zhou, J.; Ellis, A. V.; Voelcker, N. H., Recent developments in PDMS surface modification for microfluidic devices. *Electrophoresis* 2010, 31, 2.

89. Dixon, A.; Takayama, S., Guided corona generates wettability patterns that selectively direct cell attachment inside closed microchannels. *Biomedical Microdevices* 2010, 12, 769.

90. Priest, C.; Gruner, P. J.; Szili, E. J.; Al-Bataineh, S. A.; Bradley, J. W.; Ralston, J.; Steele, D. A.; Short, R. D., Microplasma patterning of bonded microchannels using high-precision "injected" electrodes. *Lab on a Chip* 2011, 11, 541–544.

91. Klages, C.-P.; Hinze, A.; Lachmann, K.; Berger, C.; Borris, J.; Eichler, M.; von Hausen, M.; Zänker, A.; Thomas, M., Surface technology with cold microplasmas. *Plasma Processes and Polymers* 2007, 4, 208.

92. Klages, C.-P.; Berger, C.; Eichler, M.; Thomas, M., Microplasma-based treatment of inner surfaces in microfluidic devices. *Contributions to Plasma Physics* 2007, 47(1–2), 49.

93. Szili, E. J.; Al-Bataineh, S. A.; Priest, C.; Gruner, P. J.; Ruschitzka, P.; Bradley, J. W.; Ralston, J.; Steele, D. A.; Short, R. D., Integration of microplasma and microfluidic technologies for localised microchannel surface modification. In *Smart Nano-Micro Materials and Devices*, Juodkazis, S.; Gu, M., eds. SPIE: 2011; Vol. 8204, p. 82042J, December 4–7, 2011.

94. Takei, G.; Nonogi, M.; Hibara, A.; Kitamori, T.; Kim, H.-B., Tuning microchannel wettability and fabrication of multiple-step Laplace valves. *Lab on a Chip* 2007, 7, 596.

95. Sato, K.; Hibara, A.; Tokeshi, M.; Hisamoto, H.; Kitamori, T., Microchip-based chemical and biochemical analysis systems. *Advanced Drug Delivery Reviews* 2003, 55(3), 379–391.

96. Günther, A.; Jhunjhunwala, M.; Thalmann, M.; Schmidt, M. A.; Jensen, K. F., Micromixing of miscible liquids in segmented gas–liquid flow. *Langmuir* 2005, 21(4), 1547–1555.

97. Melin, J.; Roxhed, N.; Gimenez, G.; Griss, P.; van der Wijngaart, W.; Stemme, G., A liquid-triggered liquid microvalve for on-chip flow control. *Sensors and Actuators B: Chemical* 2004, 100(3), 463–468.

98. Adamson, A. W.; Gast, A. P., *Physical Chemistry of Surfaces*, 6th edn.; John Wiley & Sons, New York, 1997.

99. Haneveld, J.; Tas, N. R.; Brunets, N.; Jansen, H. V.; Elwenspoek, M., Capillary filling of sub-10 nm nanochannels. *Journal of Applied Physics* 2008, 104(1), 014309–014306.

100. Han, A.; Mondin, G.; Hegelbach, N. G.; de Rooij, N. F.; Staufer, U., Filling kinetics of liquids in nanochannels as narrow as 27 nm by capillary force. *Journal of Colloid and Interface Science* 2006, 293(1), 151–157.

101. Li, X.; Ballerini, D. R.; Shen, W., A perspective on paper-based microfluidics: Current status and future trends. *Biomicrofluidics* 2012, 6(1), 011301–011313.

102. Osborn, J. L.; Lutz, B.; Fu, E.; Kauffman, P.; Stevens, D. Y.; Yager, P., Microfluidics without pumps: Reinventing the T-sensor and H-filter in paper networks. *Lab on a Chip* 2010, 10(20), 2659–2665.

103. Rezk, A. R.; Qi, A.; Friend, J. R.; Li, W. H.; Yeo, L. Y., Uniform mixing in paper-based microfluidic systems using surface acoustic waves. *Lab on a Chip* 2012, 12(4), 773–779.

104. Li, X.; Tian, J.; Shen, W., Progress in patterned paper sizing for fabrication of paper-based microfluidic sensors. *Cellulose* 2010, 17(3), 649–659.

105. Yang, X.; Forouzan, O.; Brown, T. P.; Shevkoplyas, S. S., Integrated separation of blood plasma from whole blood for microfluidic paper-based analytical devices. *Lab on a Chip* 2012, 12(2), 274–280.

106. Martinez, A. W.; Phillips, S. T.; Whitesides, G. M., Three-dimensional microfluidic devices fabricated in layered paper and tape. *Proceedings of the National Academy of Sciences of the United States of America* 2008, 105(50), 19606–19611.

107. Liu, H.; Crooks, R. M., Three-dimensional paper microfluidic devices assembled using the principles of origami. *Journal of the American Chemical Society* 2011, 133(44), 17564–17566.

108. Delaney, J. L.; Hogan, C. F.; Tian, J.; Shen, W., Electrogenerated chemiluminescence detection in paper-based microfluidic sensors. *Analytical Chemistry* 2011, 83(4), 1300–1306.

109. Li, M.; Tian, J.; Al-Tamimi, M.; Shen, W., Paper-based blood typing device that reports patient's blood type "in writing". *Angewandte Chemie International Edition* 2012, 51(22), 5497–5501.

110. Lai, H.-H.; Xu, W.; Allbritton, N. L., Use of a virtual wall valve in polydimethylsiloxane microfluidic devices for bioanalytical applications. *Biomicrofluidics* 2011, 5(2), 024105–024113.

111. Takahashi, K.; Sugii, Y.; Mawatari, K.; Kitamori, T., Experimental investigation of droplet acceleration and collision in the gas phase in a microchannel. *Lab on a Chip* 2011, 11(18), 3098–3105.

112. Oh, K. W.; Ahn, C. H., A review of microvalves. *Journal of Micromechanics and Microengineering* 2006, 16(5), R13.

113. Yamada, M.; Seki, M., Nanoliter-sized liquid dispenser array for multiple biochemical analysis in microfluidic devices. *Analytical Chemistry* 2004, 76(4), 895–899.

114. Moore, J.; McCuiston, A.; Mittendorf, I.; Ottway, R.; Johnson, R., Behavior of capillary valves in centrifugal microfluidic devices prepared by three-dimensional printing. *Microfluidics and Nanofluidics* 2011, 10(4), 877–888.

115. Chen, J.; Huang, P.-C.; Lin, M.-G., Analysis and experiment of capillary valves for microfluidics on a rotating disk. *Microfluidics and Nanofluidics* 2008, 4(5), 427–437.

116. Duffy, D. C.; Gillis, H. L.; Lin, J.; Sheppard, N. F.; Kellogg, G. J., Microfabricated centrifugal microfluidic systems: Characterization and multiple enzymatic assays. *Analytical Chemistry* 1999, 71(20), 4669–4678.

117. Leu, T.-S.; Chang, P.-Y., Pressure barrier of capillary stop valves in micro sample separators. *Sensors and Actuators A: Physical* 2004, 115(2–3), 508–515.

118. Madou, M. J.; Lee, L. J.; Daunert, S.; Lai, S.; Shih, C.-H., Design and fabrication of CD-like microfluidic platforms for diagnostics: Microfluidic functions. *Biomedical Microdevices* 2001, 3(3), 245–254.

119. Andersson, H.; van der Wijngaart, W.; Griss, P.; Niklaus, F.; Stemme, G., Hydrophobic valves of plasma deposited octafluorocyclobutane in DRIE channels. *Sensors and Actuators B: Chemical* 2001, 75(1–2), 136–141.

120. Andersson, H.; van der Wijngaart, W.; Stemme, G., Micromachined filter-chamber array with passive valves for biochemical assays on beads. *Electrophoresis* 2001, 22(2), 249–257.

121. Glière, A.; Delattre, C., Modeling and fabrication of capillary stop valves for planar microfluidic systems. *Sensors and Actuators A: Physical* 2006, 130–131, 601–608.

122. Cho, H.; Kim, H.-Y.; Kang, J. Y.; Kim, T. S., Capillary passive valve in microfluidic systems. In *NSTI-Nanotech 2004*, 2004; Vol. 1, pp. 263–266.

123. Logtenberg, H.; Lopez-Martinez, M. J.; Feringa, B. L.; Browne, W. R.; Verpoorte, E., Multiple flow profiles for two-phase flow in single microfluidic channels through site-selective channel coating. *Lab on a Chip* 2011, 11(12), 2030–2034.

124. Watanabe, M., Microchannels constructed on rough hydrophobic surfaces. *Chemical Engineering and Technology* 2008, 31(8), 1196–1200.

125. Zhao, B.; Moore, J. S.; Beebe, D. J., Pressure-sensitive microfluidic gates fabricated by patterning surface free energies inside microchannels. *Langmuir* 2003, 19, 1873.

126. Aota, A.; Nonaka, M.; Hibara, A.; Kitamori, T., Countercurrent laminar microflow for highly efficient solvent extraction. *Angewandte Chemie International Edition* 2006, 45, 1.

127. Hibara, A.; Iwayama, S.; Matsuoka, S.; Ueno, M.; Kikutani, Y.; Tokeshi, M.; Kitamori, T., Surface modification method of microchannels for gas-liquid two-phase flow in microchips. *Analytical Chemistry* 2005, 77, 943.

128. Hibara, A.; Nonaka, M.; Hisamoto, H.; Uchiyama, K.; Kikutani, Y.; Tokeshi, M.; Kitamori, T., Stabilization of liquid interface and control of two-phase confluence and separation in glass microchips by utilizing octadecylsilane modification of microchannels. *Analytical Chemistry* 2002, 74, 1724.

129. Kenis, P. J. A.; Ismagilov, R. F.; Whitesides, G. M., Microfabrication inside capillaries using multiphase laminar flow patterning. *Science* 1999, 285, 83.

130. Minagawa, T.; Tokeshi, M.; Kitamori, T., Integration of a wet analysis system on a glass chip: Determination of Co(II) as 2-nitroso-1-naphthol chelates by solvent extraction and thermal lens microscopy. *Lab on a Chip* 2001, 1, 72.

131. Tokeshi, M.; Minagawa, T.; Kitamori, T., Integration of a microextraction system on a glass chip: Ion-pair solvent extraction of Fe(II) with 4,7-diphenyl-1,10-phenanthrolinedisulfonic acid and tri-n-octyl-methylammonium chloride. *Analytical Chemistry* 2000, 72(7), 1711–1714.

132. Tokeshi, M.; Minagawa, T.; Uchiyama, K.; Hibara, A.; Sato, K.; Hisamoto, H.; Kitamori, T., Continuous-flow chemical processing on a microchip by combining microunit operations and a multiphase flow network. *Analytical Chemistry* 2002, 74, 1565–1571.

133. Priest, C.; Zhou, J.; Klink, S.; Sedev, R.; Ralston, J., Microfluidic solvent extraction of metal ions and complexes from leach solutions containing nanoparticles. *Chemical Engineering and Technology* 2012, 35(7), 1312–1319.

134. Priest, C.; Zhou, J.; Sedev, R.; Ralston, J.; Aota, A.; Mawatari, K.; Kitamori, T., Microfluidic extraction of copper from particle-laden solutions. *International Journal of Mineral Processing* 2011, 98(3–4), 168–173.

135. Aota, A.; Mawatari, K.; Kitamori, T., Parallel multiphase microflows: Fundamental physics, stabilization methods and applications. *Lab on a Chip* 2009, 9(17), 2470–2476.

136. Kralj, J. G.; Schmidt, M. A.; Jensen, K. F., Surfactant-enhanced liquid-liquid extraction in microfluidic channels with inline electric-field enhanced coalescence. *Lab on a Chip* 2005, 5, 531.

137. Kralj, J. G.; Sahoo, H. R.; Jensen, K. F., Integrated continuous microfluidic liquid-liquid extraction. *Lab on a Chip* 2007, 7, 256.

138. Song, H.; Chen, D. L.; Ismagilov, R. F., Reactions in droplets in microfluidic channels. *Angewandte Chemie International Edition* 2006, 45, 7336.
139. Tan, Y.-C.; Fisher, J. S.; Lee, A. I.; Christini, V.; Lee, A. P., Design of microfluidic channel geometries for the control of droplet volume, chemical concentration, and sorting. *Lab on a Chip* 2004, 4, 292.
140. Teh, S.-Y.; Lin, R.; Hung, L.-H.; Lee, A. P., Droplet microfluidics. *Lab on a Chip* 2008, 8, 198.
141. Boukellal, H.; Selimovic, S.; Jia, Y.; Cristobal, G.; Fraden, S., Simple, robust storage of drops and fluids in a microfluidic device. *Lab on a Chip* 2009, 9(2), 331–338.
142. Song, H.; Bringer, M. R.; Tice, J. D.; Gerdts, C. J.; Ismagilov, R. F., Experimental test of scaling of mixing by chaotic advection in droplets moving through microfluidic channels. *Applied Physics Letters* 2003, 83, 4664.
143. Tice, J. D.; Lyon, A. D.; Ismagilov, R. F., Effects of viscosity on droplet formation and mixing in microfluidic channels. *Analytica Chimica Acta* 2004, 507, 73.
144. Evans, H. M.; Surenjav, E.; Priest, C.; Herminghaus, S.; Seemann, R., In situ formation, manipulation, and imaging of droplet-encapsulated fibrin networks. *Lab on a Chip* 2009, 9, 1933.
145. Surenjav, E.; Priest, C.; Herminghaus, S.; Seemann, R., Manipulation of gel emulsions by variable microchannel geometry. *Lab on a Chip* 2009, 9, 325–330.
146. Nisisako, T.; Torii, T., Microfluidic large-scale integration on a chip for mass production of monodisperse droplets and particles. *Lab on a Chip* 2008, 8, 287.
147. Nisisako, T.; Torii, T.; Takahashi, T.; Takizawa, Y., Synthesis of monodisperse bicoloured janus particles with electrical anisotropy using a microfluidic co-flow system. *Advanced Materials* 2006, 18, 1152.
148. Priest, C.; Quinn, A.; Postma, A.; Zelikin, A. N.; Ralston, J.; Caruso, F., Microfluidic polymer multilayer adsorption on liquid crystal droplets for microcapsule synthesis. *Lab on a Chip* 2008, 8, 2182.
149. Gañán-Calvo, A. M.; Gordillo, J. M., Perfectly monodisperse microbubbling by capillary flow focusing. *Physical Review Letters* 2001, 87(27), 274501.
150. Priest, C.; Reid, M. D.; Whitby, C. P., Formation and stability of nanoparticle-stabilised oil-in-water emulsions in a microfluidic chip. *Journal of Colloid and Interface Science* 2011, 363(1), 301–306.
151. Okushima, S.; Nisisako, T.; Torii, T.; Higuchi, T., Controlled production of monodisperse double emulsions by two-step droplet breakup in microfluidic devices. *Langmuir* 2004, 20(23), 9905–9908.
152. Fidalgo, L. M.; Abell, C.; Huck, W. T. S., Surface-induced droplet fusion in microfluidic devices. *Lab on a Chip* 2007, 7, 984.
153. Salim, A.; Fourar, M.; Pironon, J.; Sausse, J., Oil–water two-phase flow in microchannels: Flow patterns and pressure drop measurements. *The Canadian Journal of Chemical Engineering* 2008, 86(6), 978–988.
154. Fang, C.; Steinbrenner, J. E.; Wang, F.-M.; Goodson, K. E., Impact of wall hydrophobicity on condensation flow and heat transfer in silicon microchannels. *Journal of Micromechanics and Microengineering* 2010, 20(4), 045018.
155. Dreyfus, R.; Tabeling, P.; Willaime, H., Ordered and disordered patterns in two-phase flows in microchannels. *Physical Review Letters* 2003, 90(14), 144505.
156. Ody, C., Capillary contributions to the dynamics of discrete slugs in microchannels. *Microfluidics and Nanofluidics* 2010, 9(2), 397–410.
157. Kim, N.; Evans, E.; Park, D.; Soper, S.; Murphy, M.; Nikitopoulos, D., Gas–liquid two-phase flows in rectangular polymer micro-channels. *Experiments in Fluids* 2011, 51(2), 373–393.
158. Huh, D.; Kuo, C. H.; Grotberg, J. B.; Takayama, S., Gas–liquid two-phase flow patterns in rectangular polymeric microchannels: Effect of surface wetting properties. *New Journal of Physics* 2009, 11(7), 075034.
159. Lee, C. Y.; Lee, S. Y., Pressure drop of two-phase dry-plug flow in round mini-channels: Effect of moving contact line. *Experimental Thermal and Fluid Science* 2010, 34(1), 1–9.
160. Yu, D.; Choi, C.; Kim, M., The pressure drop and dynamic contact angle of motion of triple-lines in hydrophobic microchannels. *ASME Conference Proceedings* 2010, 2010(54501), 1453–1458.
161. Rapolu, P.; Son, S., Characterization of wettability effects on pressure drop of two-phase flow in microchannel. *Experiments in Fluids* 2011, 51(4), 1101–1108.

3 Two-Dimensional Paper Networks for Automated Multistep Processes in Point-of-Care Diagnostics

Elain Fu, Barry Lutz, and Paul Yager

CONTENTS

3.1 NEED FOR ASSAYS WITH IMPROVED PERFORMANCE IN LOW-RESOURCE SETTINGS

Gold-standard diagnostic assays are often high-performance laboratory-based tests that require multistep protocols for complex sample processing. Trade-offs for the high performance include long sample processing times, long times for samples to be transported to the lab and for results to be transmitted back to the patient/caregiver, the need for trained personnel to run the test and interpret the results, and the need for specialized instrumentation for processing samples and detecting analytes. Also assumed is access to electricity to power the instrumentation, to maintain strict environmental conditions, and to refrigerate reagents until use in the assay. The requirements of laboratory-based tests are often incompatible with the constraints of resource-limited settings. Constraints in these settings include patients with limited access to clinics and a short amount of contact time while there, limited training of test providers, testing environments with uncontrolled temperatures and humidity levels, and limited local infrastructure, including the absence of supporting lab equipment and a lack of cold chain for refrigeration of reagents [1–3]. The World Health Organization has coined an acronym for the characteristics of point-of-care (POC) diagnostics that are appropriate for even the lowest-resource global health settings: ASSURED (affordable, sensitive, specific, user-friendly, rapid and robust, equipment-free, and deliverable to users) [4]. Thus, the overall challenge has been and continues to be to create high-performance assays that are *appropriate* for the various multiconstraint settings relevant for global health applications, including the lowest-resource settings.

High performance in high-resource settings (e.g. ELISA)	Appropriate for low-resource settings, but are lacking in performance
X Not rapid + transit times	✓ Rapid (<20 min)
X Requires trained personnel	✓ Easy to use
X Requires instrumentation	✓ No instrumentation
X Requires electricity	✓ No electricity/refrigeration
✓ Cost varies, but can be low	✓ *Very low* cost
✓ High sensitivity	X Lacks sensitivity
✓ Quantitative	X Not quantitative

FIGURE 3.1 Two classes of diagnostic devices available today. There is a need for higher performance testing that is appropriate for use in low-resource settings.

3.2 PAPER-BASED DIAGNOSTICS ARE A POTENTIAL SOLUTION

An especially compelling need in the lowest-resource settings is for equipment-free diagnostics for which ongoing maintenance and repair are not required. Simple lateral flow tests have been used in low-resource settings for decades. Though lateral flow tests fulfill many of the ASSURED criteria, they have been criticized for (1) their limited ability to multiplex (i.e., perform an assay for multiple analytes from a single biosample) and (2) their lack of sensitivity for many analytes of clinical importance [5,6]. The contrasting characteristics of the high-resource laboratory-based tests and the low-resource lateral flow tests are summarized in Figure 3.1. Recently, there has been a resurgence in work in the area of paper-based* diagnostics with the goal of bringing high-performance testing to low-resource settings. In 2008, the Whitesides group pioneered the use of microfluidic paper-based analytical devices (μPADS), 2D and 3D paper-based structures that enable colorimetric assays (e.g., for detection of glucose and protein) with multiplexing capability [7,8]. The original μPAD structures were created by photolithography [9], but since then, numerous alternative fabrication methods have been demonstrated, including wax printing [10,11], cutting [12], and inkjet printing [13]. Additional work in the area of paper-based assay development has focused on implementing multiplexed assays for the detection of additional biomarkers using one-step colorimetric reactions (e.g., nitrite, uric acid, and lactate) [14,15] or performing the simultaneous analysis of multiple controls for on-device calibration [16]. A review by Martinez et al., published in 2010 [17], described the field with respect to fabrication methods and progress toward paper-based tests for multianalyte detection. In this brief review, we focus on a discussion of recent work from the collaboration of Fu, Lutz, and Yager, using 2D paper networks (2DPNs) for automated multistep sample processing for high-sensitivity assays.

3.3 PAPER NETWORKS FOR AUTOMATED MULTISTEP SAMPLE PROCESSING

As described previously, a significant limitation of lateral flow devices is their low sensitivity with respect to the clinically relevant detection ranges for a number of analytes. This limitation effectively derives from an inability of these devices to perform multistep sample processing characteristic of high-performance gold-standard assays. For example, the case of poor sensitivity of rapid lateral flow tests for influenza has been highlighted recently. Those tests generally have an acceptable clinical specificity of >90% but have poor clinical sensitivity of 11%–70% [18–21].

---

* Note that we use the term *paper* broadly and include related porous materials.

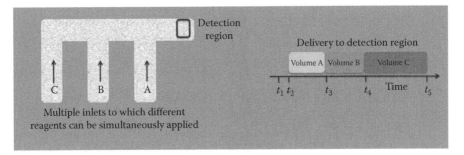

FIGURE 3.2 Two-dimensional paper networks with multiple inlets per detection region for automated multistep sample processing. The sequential delivery of multiple fluid volumes has been preprogrammed into the configuration of the network.

The Center for Disease Control even issued a statement during the influenza pandemic of 2009 that recommended discontinuing use of those tests [22].

The strength of the 2DPN is its ability to automatically perform multistep processes for increased performance while still maintaining the advantages of conventional lateral flow tests, namely, a rapid time to result, ease of use, and low cost. A key feature of the 2DPN assay is the configuration of the network, composed of multiple inlets per outlet, which can function as a program for the controlled delivery of multiple reagent volumes within the network. The example of Figure 3.2 illustrates a multi-inlet paper network that can be used for the automated sequential delivery of reagents to a downstream detection region (left schematic). Upon the simultaneous application of reagents to the three inlets, the geometry of the network performs the automated delivery of the multiple reagent volumes to the detection region in the following order: first A, then B, then C (right schematic). Critical to the automated operation of multistep paper-based assays is a set of paper fluidic tools, that is, analogs to the pump controls and valves of conventional microfluidics, to perform the desired manipulation of fluids within the network.

3.4 PAPER FLUIDIC TOOLBOX: PUMP CONTROLS AND VALVES IN THE PAPER NETWORKS

As in conventional lateral flow assays, properties of the porous materials, including pore size, pore structure, surface treatments, and properties of the fluid, can be used to tune the flow rate for the assay time and sensitivity requirements of a given application. Additionally, to enable automated multistep processing, there is a need for tools for precise control of transport of multiple fluids within the paper networks to perform processing of the sample [23]. These tools serve to replace the costly and often complicated valves and pump controls in conventional microfluidics. The next section will discuss a set of the paper fluidic tools that have been developed, including controls for flow rate, on-switches for flow, and off-switches for flow.* This last category is especially important to be able to independently meter discrete volumes of reagents within the paper networks.

One key paper fluidic tool is the use of simple geometries to control the flow rates of fluids in paper networks [24,25]. One can investigate flow in the simplest 2D structures to create some basic design rules for transport in paper networks. For example, what happens to the fluid front

---

* The authors discovered that Bunce et al. [24], in a patent issued in 1994, had disclosed many similar ideas on controlling flows in paper networks, but to our knowledge, these methods were not carried forward.

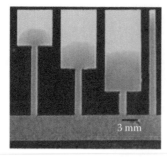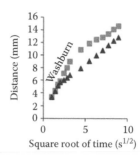

FIGURE 3.3 Transport of the fluid front in simple expansion geometries. The fluid front is slowed relative to a constant width strip and the degree of slowing depends on the location of the expansion. Squares represent data from the leftmost strip and triangles represent data from the strip second from the left. (Reproduced from Fu, E. et al., *Microfluid Nanofluidics*, 10, 29, 2011. With permission.)

FIGURE 3.4 Transport of the fluid front for different width expansions. The width of the expansion affects the degree of slowing of the fluid front.

in regions of expanding or contracting geometry? The image of Figure 3.3 shows strips that contain a simple expansion at different locations downstream and a constant width strip for comparison. Flow initially follows the Lucas–Washburn relation (i.e., the distance that the fluid front travels is proportional to the square root of time, with the proportionality factor dependent on the surface tension, the contact angle, the average pore size of the material, and the viscosity of the fluid) [26,27] in all the strips. Transition to a greater width results in a deviation from the Lucas–Washburn relation and a greater degree of slowing of the fluid front. The plot of Figure 3.3 shows the distance versus the square root of time for the two leftmost strips. Here one can see the initial Lucas–Washburn flow and then a further slowing that starts at the point of the expansion. The image in Figure 3.4 shows that for a greater width expansion, there is a greater degree of slowing of the fluid front. Thus, simple control parameters for slowing down the transport time of the fluid front are the downstream location of the expansion and the width of the expansion.

For the case of a contraction geometry (a transition to a smaller width), as shown in Figure 3.5, the flow starts out following the Lucas–Washburn prediction, increases transiently at the constriction, and then resumes Lucas–Washburn flow as the larger width section serves as a nonlimiting source for the smaller width section. The result is that the transport time of the fluid front, that is, the time that the fluid front takes to travel the length of the strip, is decreased relative to a constant width strip. The downstream location of the constriction can be used to control the transport time of the fluid front, and this time is minimized when the lengths of the two sections of different widths are equal.

In the case of fully wetted flow (i.e., when the fluid front has reached the wicking pad), one can use the electrical circuit analogy to Darcy's law [28] for fluidic circuits as shown schematically in Figure 3.6. The pressure difference across the circuit is analogous to potential difference, the volumetric flow rate is analogous to current, and the fluidic resistance depends on physical properties of the system and geometric factors of the paper circuit. Fluidic resistances in series are summed, while fluidic resistances in parallel are added in reciprocal. Using this electrical circuit analogy for

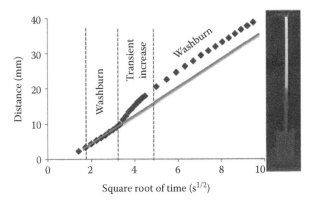

FIGURE 3.5 Transport of the fluid front in the case of a contraction geometry. The transport time of the fluid front is decreased relative to a constant width strip.

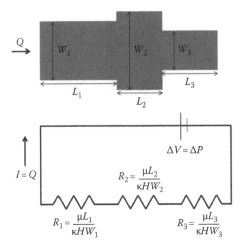

FIGURE 3.6 Electrical circuit analogy to Darcy fluid flow. Fluidic resistances in series are summed and fluidic resistances in parallel are summed in reciprocal. (Reproduced from Fu, E. et al., *Microfluid Nanofluidics*, 10, 29, 2011. With permission.)

resistances in series, one can calculate the relative resistances of simple structures, such as those shown in Figure 3.7. Since resistance is proportional to length over cross-sectional area, the resistance of A is the greatest and the resistance of B is the smallest. For a uniform pressure difference across all three structures, the volumetric flow rate in A is the smallest and in B is the greatest. The transport time for flow through a strip with a multisegment geometry can be calculated from $t = V/Q$, where V is the volume of the geometry and Q is the volumetric flow rate. Assuming that permeability and viscosity are constant, differences in the transport times of fluids in two strips will be solely due to geometric factors. The prediction is that the transport time will be the fastest in the constant width strip A, and for the strips of varying widths, the transport time should be faster in strip C than in strip B. Figure 3.7 shows a time series comparison of experimental and simulation results (COMSOL Multiphysics) for flow in the strips of different geometries. The transport times show good quantitative agreement, demonstrating the ability to predict and control flow rates for simple changes in geometry [25].

Another key paper fluidic tool is an on-switch for fluid flow. One type of on-switch for flow uses dissolvable barriers. Specifically, sugar barriers can be used to create delays in the transport of the fluid within a paper network. Both the extent of the sugar barrier and the concentration of the sugar solution used to form the barrier within the porous material can be used to control the delay time. In the top series of Figure 3.8, the presence of a barrier in the right leg creates a delay of the fluid

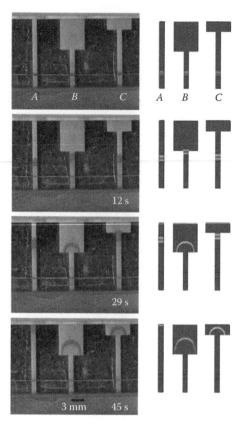

FIGURE 3.7 Comparison of experimental and model results for Darcy's law flow shows good agreement in both the shapes and the locations of the bands. (Reproduced from Chin, C.D. et al., *Lab Chip*, 7, 41, 2007. With permission.)

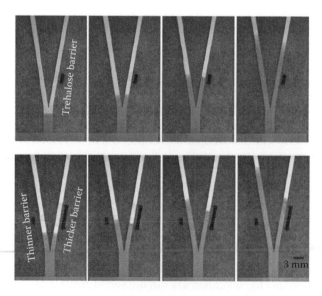

FIGURE 3.8 Sugar barriers can be used as delays for fluid transport. Both the extent of the barrier and the concentration of the sugar solution used to create the barrier can be used to tune the delay time. (Reproduced from Yager, P. et al., *Annu. Rev. Biomed. Eng.*, 10, 107, 2008. With permission.)

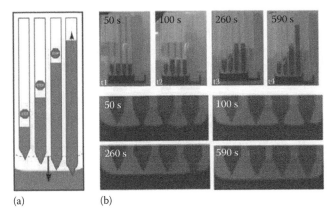

(a)   (b)

FIGURE 3.9 Inlet legs were immersed to different depths into the well (a). As fluid was wicked from the well, each leg broke contact with the fluid in a timed sequence to provide automated volume metering (b). (Data is courtesy of Philip Trinh.)

front in that leg. In the lower series of Figure 3.8, barriers were created in both legs, and the longer sugar barrier in the right leg results in a greater delay in the fluid front in the right leg compared to the left leg. These fluid delays can be critical tools in paper networks in cases in which longer fluid delays are needed, and it is no longer practical to use geometry alone.

A complementary tool for the manipulation of fluids in paper networks is an off-switch for fluid flow. One method to control the shutoff of multiple flows independently is the use of inlet legs that are submerged by varying distances into a common well [29]. The level of the fluid in the well drops as fluid wicks into the paper inlets. Fluid shuts off from the inlets at different times, in order of shortest to longest submerged lengths as shown in Figure 3.9. Thus, different volumes of multiple fluids are automatically input. This method has been characterized for reproducibility and the coefficient of variation in shutoff times is between 5% and 20% [29].

An alternative method to turn off flow from multiple inlets independently uses pads of varying fluid capacities that are prefilled to saturation [30]. Contact between the pads and inlets then activates the flows. Figure 3.10 shows the release profiles from source pads of two different fluid capacities. The properties of the pad, including the bed volume and the surface treatment,

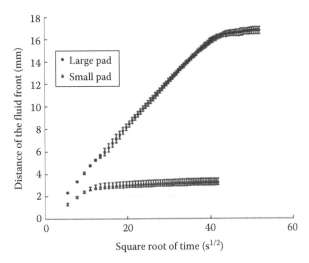

FIGURE 3.10 Examples of the release profiles from glass fiber source pads of two different fluid capacities. As expected, flow is Lucas–Washburn-like in that the distance traveled by the fluid front is proportional to the square root of time. (Reproduced from Fu, E. et al., *Anal. Chem.*, 84, 4574, 2012. With permission.)

determine the release profile of fluid from the pad to the inlet. The coefficient of variation for this method is better than 12% in the volumes delivered [30].

Other interesting tools for controlling flow have been demonstrated including modification of the wetting properties of the substrate and user-activated mechanical switches. In the context of a fluidic timer, the Phillips group has used wax to slow flow within the paper channel [31]. The Whitesides group has demonstrated the use of buttons that can be mechanically depressed by the user with a pen to activate flow between two initially disconnected fluidic paths in one of their 3D µPADs [32]. Finally, the Shen group demonstrated the analogous mechanical on-switch in a paper device composed of a single layer, using tabs that could be manipulated by the user to complete a fluidic pathway [33].

3.5   APPLICATIONS OF 2DPNs

The ability to perform automated multiple processing steps enables a host of capabilities that can be implemented in 2DPNs for higher performance testing at the point of care. In this section, three examples, sample dilution [34], small-molecule extraction [34], and signal amplification [30,35,36], are described.

3.5.1   SAMPLE DILUTION AND MIXING

Precise sample dilution is a particular type of mixing, and one often required to perform chemical reactions and binding-based assays. In conventional microfluidic systems, continuous dilution requires combining two fluid streams in a channel using expensive pumps and providing some means to mix the two fluids. A 2DPN can be used to create a paper dilution circuit that mixes two fluids and allows control over the dilution factor by simply changing the shape of the paper [34]. Figure 3.11 shows dilution of a fluid (top leg) with a buffer (right leg). The dilution factor is determined by the relative flow rates of the two fluids. In this case, the flow rates are set not by pumps, but rather by the relative fluidic *resistances* of the two inlet legs according to Darcy's law. For a given paper material, the resistance is simply proportional to the length of the leg and the viscosity of the fluid. As the length of the dilution arm increases, the volumetric flow rate of the diluent decreases, leading to a reduced dilution factor in the common channel downstream. Serial dilutions are also possible by adding multiple dilution arms, allowing a wide range of dilutions without a single pump or pipetting step. Dilution and mixing of input reagents is one useful class of applications that can be automated in a 2DPN.

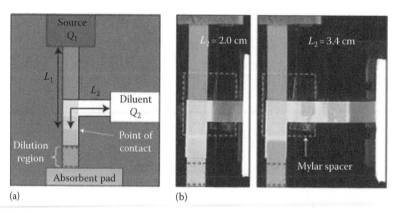

(a)   (b)

FIGURE 3.11 A sample can be diluted accurately by controlling the relative addition of sample and diluent (a). By modifying the relative resistance of two inlet channels (in this case, by the length of each channel arm), their relative contributions can be controlled, allowing the creation of a dilution circuit (b). (Reproduced from Osborn, J. et al., *Lab Chip*, 10, 2659, 2010. With permission.)

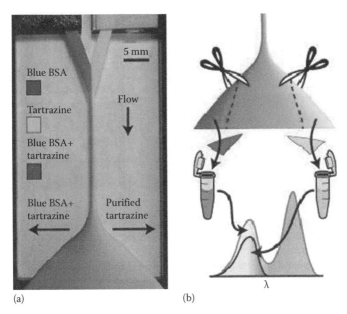

FIGURE 3.12 Small-molecule extraction in a 2DPN. In this proof-of-concept demonstration, small molecule dye was separated from a larger component, dye labeled BSA (a). The yellow extract was recovered by simply cutting out the yellow part of the 2DPN outlet. 2DPN H-filters could be used to extract analytes from complex samples for downstream analysis (b). (Reproduced from Osborn, J. et al., *Lab Chip*, 10, 2659, 2010. With permission.)

3.5.2 SMALL-MOLECULE EXTRACTION

Another sample pretreatment application that has been demonstrated in a 2DPN is extraction of small molecules from higher molecular weight components in a complex sample [34]. Previously, the Yager group developed a pump-driven microfluidic device called the H-filter that allowed extraction of small analytes from complex samples [37–39]. Separation of two species and subsequent extraction of a purified solution of the faster-diffusing species can be achieved when one inlet contains a mixed sample and the other a collection buffer. The efficiency of the extraction depends on the diffusion coefficient of each species, the contact time, and the dimensions of the common channel; no intervening membrane is required as long as the Reynolds number in the device is low. In conventional microfluidic devices, this requires a stable diffusion interface and requires multiple pumps. Figure 3.12 shows the classic H-filter recreated in a 2DPN. The significant advantages of the 2DPN H-filter are that it is pumpless, fully disposable, and much lower cost compared to any previous implementations using conventional microfluidics.

3.5.3 SIGNAL AMPLIFICATION

Chemical signal amplification has been used in many systems to improve the sensitivity of an assay. The tradeoff in the case of well-known laboratory ELISA is the requirement for many steps for labeling, washing, and amplifying the signal; these multiple critical steps are performed by either trained users or expensive laboratory robots. Using 2DPNs, we can program the structure of the paper device to automate the rinse and amplification steps in a paper-based disposable device [30,35,36]. Specifically, the simple three-inlet paper network of Figure 3.2 can be used to perform a basic three-step signal-amplified assay based on a conventional sandwich assay format.

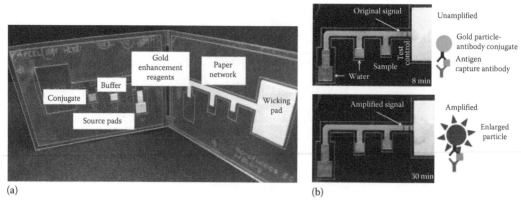

(a)   (b)

FIGURE 3.13 Easy-to-use 2DPN card format demonstrated for an amplified immunoassay. (a) Conjugate, buffer, and the gold enhancement reagent components were stored dry on the card for rehydration at the time of use. The user steps are comparable to those required to run a conventional unamplified lateral flow strip test. (b) 2DPN assay results for a high analyte concentration of 200 ng/mL. The top panel shows the original signal after 8 min due to formation of a conventional sandwich structure with a gold particle label in the detection region. The bottom panel shows the amplified, significantly darkened signal after 30 min. (Reproduced from Fu, E. et al., *Anal. Chem.*, 84, 4574, 2012. With permission.)

An example of an amplified 2DPN assay for the detection of the malaria parasite protein *Pf*HRP2 is presented in Figure 3.13. The 2DPN card (shown on the left) was designed to perform two additional processing steps—automated delivery of rinse and signal amplification reagents—for improved limit of detection (LOD) in an easy-to-use format. The three-inlet network and wicking pad were located on one side of the folding card, while the source pads were located on the opposite side of the folding card. The source pads contained dry reagents, conjugate, buffer, and gold enhancement reagents, from left to right, respectively. The user steps were to simply add sample and water to appropriate places on the card, then fold the device in a single activation step. This set of user steps is comparable in ease of use to commercially available conventional lateral flow tests and is much less complex than the many timed user steps required to operate alternate microfluidic formats proposed for performing signal amplification steps [40]. The schematic (shown on the right) outlines the capture sequence in the assay. After the activation of the multiple flows in the 2DPN card, the sample plus antibody conjugated to a gold particle label was first delivered to the detection region. The signal produced at this stage is comparable to the signal from a conventional lateral flow test. Following this was the delivery of a rinse buffer to the detection region to remove nonspecifically bound label. Finally, the signal amplification reagent was delivered to the detection region to produce an amplified signal. In this case, application of a gold enhancement solution (Nanoprobes), consisting of gold salt and a reducer, resulted in the deposition of metallic gold onto the original gold particle labels. This enlargement of the gold particles produced a significant darkening of the original signal.

The detection region of the amplified assay for a concentration series of the analyte is shown in Figure 3.14a. The signal-versus-concentration curves for the 2DPN amplified and unamplified assays are shown in Figure 3.14b. The limit of detection of the amplified 2DPN malaria card using the gold enhancement reagent was 2.9 ± 1.2 ng/mL, an almost fourfold improvement over the unamplified case (10.4 ± 4.4). For context, the LOD of the amplified assay is similar to that reported for a *Pf*HRP2 ELISA of 4 ng/mL [41].

The commercially available gold enhancement system used here was chosen for ease of use and has shown promise in other microfluidic formats [42,43]. However, the 2DPN card format demonstrated here can also be used to implement other signal amplification methods. Other amplification methods tested include silver enhancement of gold nanoparticles [44,45] and the enzymatic system of horseradish peroxidase/tetramethylbenzidine [46], both of which have been reported to be useful in lateral flow and other formats and have great potential for further improvement of the assay LOD.

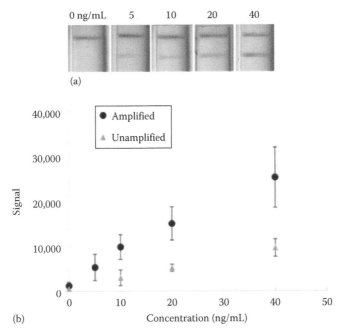

FIGURE 3.14 Sensitivity improvement in the amplified 2DPN assay. (a) Image series of the detection region for different concentrations of *Pf*HRP2. (b) Plot of the average signal for each concentration of the 2DPN card with rinse and amplification at 40 min (amplified assay, black circles). Also shown for comparison is a control case in which additional water rather than gold enhancement reagent was run in the 2DPN card format (unamplified assay, gray triangles). The error bars represent the standard deviation. The limit of detection of the amplified 2DPN malaria card was improved almost fourfold over the unamplified case. (Reproduced from Fu, E. et al., *Anal. Chem.*, 84, 4574, 2012. With permission.)

3.5.4 SELECTED ADVANCES THAT COMPLEMENT THE PAPER-BASED MICROFLUIDICS TECHNOLOGY

Recently, there have been complementary advances to address the equipment-free challenge of the lowest-resource settings. We briefly highlight two lines of work: robust methods for electricity-free temperature control and quantitative detection while minimizing dedicated instrumentation.

Recently, the Weigl group at PATH has demonstrated the use of chemical heating, for example, hydration of CaO, and phase change materials to perform loop-mediated nucleic acid amplification [47]. Their device achieved a controlled elevated temperature of 65°C ± 1.5°C for over an hour [47]. The specific combination of exothermic reactants and the composition of the phase change material can be used to tune the thermal properties of the instrument-free heater for numerous applications including other isothermal nucleic acid amplification methods, cell lysis protocols, and sample concentration methods based on temperature-responsive polymers [47]. Building on this work of controlled chemical heating using phase change materials, the Bau group has demonstrated a microfluidic self-heating cartridge for performing nucleic acid amplification [48].

A second challenge to developing high-performance tests for low-resource settings is to achieve quantitative detection with minimal dedicated instrumentation. The use of a compact reader in conjunction with fluorescent or colorimetric labels has been a common strategy for providing quantitative readout of conventional lateral flow tests (e.g., from ESE GmbH/Qiagen) [49]. The Whitesides group has also demonstrated the use of a transmission-based reader for measurements in index-matched paper devices [50]. More recently, the use of cell phones for the acquisition, analysis, and transmission of assay data has become an area of active research and development. Challenges include the acquisition of high-quality image data given the expected wide range of lighting conditions and user variability of camera positioning [51]. The Whitesides group has demonstrated the use of a cell phone camera for direct acquisition of end point intensity measurements from a colorimetric paper assay [52], while the Shen group has demonstrated quantitative detection of chemiluminescence [53]. A related approach has been to develop an adapter module to interface a standard cell phone. The Ozcan group has developed a compact adapter consisting of LEDs, lens, and filter, which couples to a cell phone camera for wide-field fluorescent and dark-field imaging capability [54].

3.6 SUMMARY

The overall challenge in point-of-care diagnostics development continues to be to create high-performance assays that are *appropriate* for the various multiconstraint settings relevant for global health applications. For the lowest-resource settings, the requirements for high performance with a rapid, instrument-free, easy-to-use, and very-low-cost device bring specific design and implementation challenges. Paper-based microfluidics is especially well-suited to addressing these challenges. Specifically, the 2DPN is an enabling technology for implementing multistep assays, characteristic of gold-standard laboratory tests, in an automated disposable package. Coupled with advances to develop electricity-free methods for temperature control and cell phones for data acquisition, analysis, interpretation, and transmission, there is great potential for creating sophisticated assays appropriate for low-resource settings. A main challenge in the field moving forward is to develop a set of robust and precise paper fluidic tools to achieve the performance specifications needed to have a positive impact in low-resource communities.

REFERENCES

1. Chin, C.D., V. Linder, and S.K. Sia, Lab-on-a-chip devices for global health: Past studies and future opportunities. *Lab Chip*, 2007. **7**: 41–57.
2. Yager, P., G.J. Domingo, and J. Gerdes, Point-of-care diagnostics for global health. *Annu Rev Biomed Eng*, 2008. **10**: 107–144.
3. Urdea, M., L.A. Penny, S.S. Olmsted, M.Y. Giovanni, P. Kaspar, A. Shepherd, P. Wilson, C.A. Dahl, S. Buchsbaum, G. Moeller, and D.C. Hay Burgess, Requirements for high impact diagnostics in the developing world. *Nature*, 2006. **444**(Suppl 1): 73–79.
4. Kettler, H., K. White, and S. Hawkes, Mapping the landscape of 524 diagnostics for sexually transmitted infections: Key findings and 525 recommendations. The World Health Organization, 2004.
5. Posthuma-Trumpie, G.A., J. Korf, and A. van Amerongen, Lateral flow (immuno) assay: Its strengths, weaknesses, opportunities and threats. A literature survey. *Anal Bioanal Chem*, 2009. **393**: 569–582.
6. O'Farrell, B., Evolution in lateral flow-based immunoassay systems, in *Lateral Flow Immunoassay*, R. Wong and H. Tse, eds. 2009, Humana Press: New York. pp. 1–33.
7. Martinez, A.W., S.T. Phillips, M.J. Butte, and G.M. Whitesides, Patterned paper as a platform for inexpensive, low-volume, portable bioassays. *Angew Chem Int Ed*, 2007. **46**: 1318–1320.
8. Martinez, A.W., S.T. Phillips, and G.M. Whitesides, Three-dimensional microfluidic devices fabricated in layered paper and tape. *Proc Natl Acad Sci U S A*, 2008. **105**: 19606–19611.
9. Martinez, A.W., S.T. Phillips, B.J. Wiley, M. Gupta, and G.M. Whitesides, FLASH: A rapid method for prototyping paper-based microfluidic devices. *Lab Chip*, 2008. **8**: 2146–2150.
10. Lu, Y., W.W. Shi, J.H. Qin, and B.C. Lin, Fabrication and characterization of paper-based microfluidics prepared in nitrocellulose membrane by wax printing. *Anal Chem*, 2010. **82**: 329–335.
11. Carrilho, E., A.W. Martinez, and G.M. Whitesides, Understanding wax printing: A simple micropatterning process for paper-based microfluidics. *Anal Chem*, 2009. **81**: 7091–7095.
12. Fenton, E.M., M.R. Mascarenas, G.P. Lopez, and S.S. Sibbett, Multiplex lateral-flow test strips fabricated by two-dimensional shaping. *Acs Appl Mater Interfaces*, 2009. **1**: 124–129.
13. Abe, K., K. Kotera, K. Suzuki, and D. Citterio, Inkjet-printed paperfluidic immuno-chemical sensing device. *Anal Bioanal Chem*, 2010. **398**: 885–893.
14. Li, X., J.F. Tian, and W. Shen, Quantitative biomarker assay with microfluidic paper-based analytical devices. *Anal Bioanal Chem*, 2010. **396**: 495–501.

15. Dungchai, W., O. Chailapakul, and C.S. Henry, Use of multiple colorimetric indicators for paper-based microfluidic devices. *Anal Chim Acta*, 2010. **674**: 227–233.
16. Wang, W., W.Y. Wu, W. Wang, and J.J. Zhu, Tree-shaped paper strip for semiquantitative colorimetric detection of protein with self-calibration. *J Chromatogr A*, 2010. **1217**: 3896–3899.
17. Martinez, A.W., S.T. Phillips, G.M. Whitesides, and E. Carrilho, Diagnostics for the developing world: Microfluidic paper-based analytical devices. *Anal Chem*, 2010. **82**: 3–10.
18. Drexler, J.F., A. Helmer, H. Kirberg, U. Reber, M. Panning, M. Muller, K. Hofling, B. Matz, C. Drosten, and A.M. Eis-Hubinger, Poor clinical sensitivity of rapid antigen test for influenza A pandemic (H1N1) 2009 virus. *Emerg Infect Dis*, 2009. **15**: 1662–1664.
19. Hurt, A.C., R. Alexander, J. Hibbert, N. Deed, and I.G. Barr, Performance of six influenza rapid tests in detecting human influenza in clinical specimens. *J Clin Virol*, 2007. **39**: 132–135.
20. Uyeki, T., Influenza diagnosis and treatment in children: A review of studies on clinically useful tests and antiviral treatment for influenza. *Pediatr Infect Dis J*, 2003. **22**: 164–177.
21. Vasoo, S., J. Stevens, and K. Singh, Rapid antigen tests for diagnosis of pandemic (swine) influenza A/H1N1. *Clin Infect Dis*, 2009. **49**: 1090–1093.
22. Center for Disease Control, Interim guidance for detection of novel influenza A virus using rapid influenza testing. 2009. http://www.cdc.gov/h1n1flu/guidance/rapid_testing.htm
23. Bunce, R., G. Thorpe, J. Gibbons, L. Keen, and M. Walker, Liquid transfer devices, in *United States Patent Office*, U.S.P. Office, Ed. 1994, University of Birmingham: Birmingham, U.K.
24. Fu, E., B. Lutz, P. Kauffman, and P. Yager, Controlled reagent transport in disposable 2D paper networks. *Lab Chip*, 2010. **10**: 918–920.
25. Fu, E., S.A. Ramsey, P. Kauffman, B. Lutz, and P. Yager, Transport in two-dimensional paper networks. *Microfluid Nanofluidics*, 2011. **10**: 29–35.
26. Washburn, E.W., The dynamics of capillary flow. *Phys Rev*, 1921. **17**: 273–283.
27. Lucas, R., Ueber das Zeitgesetz des Kapillaren Aufstiegs von Flussigkeiten. *Colloid Polym Sci*, 1918. **23**: 15–22.
28. Darcy, H., *Les Fontaines Publiques de la Ville de Dijon*, 1856, Dalmont, Paris, France.
29. Lutz, B.R., P. Trinh, C. Ball, E. Fu, and P. Yager, Two-dimensional paper networks: Programmable fluidic disconnects for multi-step processes in shaped paper. *Lab Chip*, 2011. **11**: 4274–4278.
30. Fu, E., T. Liang, P. Spicar-Mihalic, J. Houghtaling, S. Ramachandran, and P. Yager, Two-dimensional paper network format that enables simple multistep assays for use in low-resource settings in the context of malaria antigen detection, *Anal. Chem.*, 2012. **84**: 4574–4579.
31. Noh, N. and S.T. Phillips, Metering the capillary-driven flow of fluids in paper-based microfluidic devices. *Anal Chem*, 2010. **82**: 4181–4187.
32. Martinez, A.W., S.T. Phillips, Z.H. Nie, C.M. Cheng, E. Carrilho, B.J. Wiley, and G.M. Whitesides, Programmable diagnostic devices made from paper and tape. *Lab Chip*, 2010. **10**: 2499–2504.
33. Li, X., J.F. Tian, and W. Shen, Progress in patterned paper sizing for fabrication of paper-based microfluidic sensors. *Cellulose*, 2010. **17**: 649–659.
34. Osborn, J., B. Lutz, E. Fu, P. Kauffman, D. Stevens, and P. Yager, Microfluidics without pumps: Reinventing the T-sensor and H-filter in paper networks. *Lab Chip*, 2010. **10**: 2659–2665.
35. Fu, E., P. Kauffman, B. Lutz, and P. Yager, Chemical signal amplification in two-dimensional paper networks. *Sens Actuat B Chem*, 2010. **149**: 325–328.
36. Fu, E., T. Liang, J. Houghtaling, S. Ramachandran, S.A. Ramsey, B. Lutz, and P. Yager, Enhanced sensitivity of lateral flow tests using a two-dimensional paper network format. *Anal Chem.*, 2011. **83**: 7941–7946.
37. Hatch, A., E. Garcia, and P. Yager, Diffusion-based analysis of molecular interactions in microfluidic devices. *Proc IEEE*, 2004. **92**: 126–139.
38. Helton, K.L., K.E. Nelson, E. Fu, and P. Yager, Conditioning saliva for use in a microfluidic biosensor. *Lab Chip*, 2008. **8**: 1847–1851.
39. Helton, K.L. and P. Yager, Interfacial instabilities affect microfluidic extraction of small molecules from non-Newtonian fluids. *Lab Chip*, 2007. **7**: 1581–1588.
40. Cho, I.-H., S.-M. Seo, E.-H. Paek, and S.-H. Paek, Immunogold-silver staining-on-a-chip biosensor based on cross-flow chromatography. *J Chromatogr B*, 2010. **878**: 271–277.
41. Kifude, C.M., H.G. Rajasekariah, D.J. Sullivan, V.A. Stewart, E. Angov, S.K. Martin, C.L. Diggs, and J.N. Waitumbi, Enzyme-linked immunosorbent assay for detection of *Plasmodium falciparum* histidine-rich protein 2 in blood, plasma, and serum. *Clin Vaccine Immunol*, 2008. **15**: 1012–1018.
42. Lei, K.F. and Y.K.C. Butt, Colorimetric immunoassay chip based on gold nanoparticles and gold enhancement. *Microfluid Nanofluidics*, 2010. **8**: 131–137.

43. Lei, K.F. and K.S. Wong, Automated colorimetric immunoassay microsystem for clinical diagnostics. *Instrum Sci Technol*, 2010. **38**: 295–304.

44. Yan, J., D. Pan, C.F. Zhu, L.H. Wang, S.P. Song, and C.H. Fan, A gold nanoparticle-based microfluidic protein chip for tumor markers. *J Nanosci Nanotechnol*, 2009. **9**: 1194–1197.

45. Yeh, C.H., C.Y. Hung, T.C. Chang, H.P. Lin, and Y.C. Lin, An immunoassay using antibody-gold nanoparticle conjugate, silver enhancement and flatbed scanner. *Microfluid Nanofluidics*, 2009. **6**: 85–91.

46. Kolosova, A.Y., S. De Saeger, S.A. Eremin, and C. Van Peteghem, Investigation of several parameters influencing signal generation in flow-through membrane-based enzyme immunoassay. *Anal Bioanal Chem*, 2007. **387**: 1095–1104.

47. LaBarre, P., K. Hawkins, J. Gerlach, J. Wilmoth, A. Beddoe, J. Singleton, D. Boyle, and B. Weigl, A simple, inexpensive device for nucleic acid amplification without electricity—Toward instrument-free molecular diagnostics in low-resource settings. *PLOS ONE*, 2011. **6**: e19738.

48. Liu, C.C., M.G. Mauk, R. Hart, X.B. Qiu, and H.H. Bau, A self-heating cartridge for molecular diagnostics. *Lab Chip*, 2011. **11**: 2686–2692.

49. Faulstich, K., R. Gruler, M. Eberhard, D. Lentzsch, and K. Haberstroh, Handheld and portable reader devices for lateral flow immunoassays, in *Lateral Flow Immunoassay*, R. Wong and H. Tse, eds. 2009, Humana Press: New York. pp. 75–94.

50. Ellerbee, A., S. Phillips, A. Siegel, K. Mirica, A. Martinez, P. Striehl, N. Jain, M. Prentiss, and G. Whitesides, Quantifying colorimetric assays in paper-based microfluidic devices by measuring the transmission of light through paper. *Anal Chem*, 2009. **81**: 8447–8452.

51. Stevens, D., Development and optical analysis of a microfluidic point-of-care diagnostic device, *Department of Bioengineering*. 2010, University of Washington: Seattle, WA. p. 230.

52. Martinez, A.W., S.T. Phillips, E. Carrilho, S.W. Thomas, H. Sindi, and G.M. Whitesides, Simple telemedicine for developing regions: Camera phones and paper-based microfluidic devices for real-time, off-site diagnosis. *Anal Chem*, 2008. **80**: 3699–3707.

53. Delaney, J.L., C.F. Hogan, J.F. Tian, and W. Shen, Electrogenerated chemiluminescence detection in paper-based microfluidic sensors. *Anal Chem*, 2011. **83**: 1300–1306.

54. Zhu, H., O. Yaglidere, T. Su, D. Tseng, and A. Ozcan, Cost-effective and compact wide-field fluorescent imaging on a cell-phone. *Lab Chip*, 2011. **11**: 315–322.

4 Carbon Nanofibers as Amperometric Biosensors on a Silicon-Compatible Platform

*Fahmida S. Tulip, Syed K. Islam, Ashraf B. Islam,
Kimberly C. MacArthur, Khandaker A. Mamun,
and Nicole McFarlane*

CONTENTS

4.1   INTRODUCTION

Biosensors based on carbon nanomaterials have been drawing attention due to their enhanced electrical conductivity, better stability, excellent structural and catalytic properties, and high loading of biocatalysts. Carbon electrodes have been proven successful when used as enzymatic biosensors in terms of a wide range of functionality and cost-effectiveness. Among the carbon electrodes, carbon nanostructures (cylindrical or conical structures) demonstrate the best potential. Carbon nanofibers (CNFs) have excellent conductive and structural properties when compared with nanotubes, which make them excellent candidates as electrodes as well as immobilizing substrates. Enzymes are plugged with the metal electrode through the CNFs by enzyme wiring technique, which facilitates an effective way to transfer electrons from the electrode to the electrochemical reaction centers. Mediator- or membrane-free operation of this biosensor can potentially result in the application of these sensors in environmental monitoring, healthcare, as well as in varieties of scientific experiments. If realized on a silicon-compatible platform, they can be easily integrated with existing sensor technologies for the development of a fully integrated bioChem lab-on-a-chip.

4.2 BACKGROUND

A biosensor is a device that is used to detect and/or monitor a target biochemical analyte. It comprises of two closely associated elements:

1. *A biological recognition element*: It produces a detection signal in the presence of the target analyte and determines the selectivity of the biosensor. Different biological elements such as enzymes, antibodies, antigens, receptors, organelles, microorganisms, and DNA/RNA can act as the recognition element in a biosensor.
2. *A physicochemical transducer*: It converts the detection signal to an electrical, chemical, or physical signal and transmits it to the readout circuit. It influences the degree of sensitivity of the biosensor.

Depending on the parameter measured by the transducer, biosensors can be classified as optical, electrochemical, acoustic, piezoelectric, or thermal biosensors. Among these, electrochemical detection-based biosensors are the most popular because of the low cost, ease of use, portability, and simplicity of construction [1]. Electrochemical detection depends on a surface technique known as electrochemistry, which can be used for very small volumes of the samples and is not affected by the particles that often interfere with spectrophotometric detection.

Electrochemical biosensors can be divided into three subcategories based on the output of the reaction associated with biological detection. In case of amperometric biosensors, the output is a measurable current. For potentiometric and conductometric biosensors, the output either represents a measurable potential or changes the conductive properties of the medium, respectively.

Most commercial biosensors manufactured to date are amperometric type operating on the basis of monitoring the electron transfer. Amperometric method maintains a constant potential at the working electrode with respect to a reference electrode, and the changes in the current generated by the electrochemical oxidation or reduction are monitored directly with time [2]. In case of amperometric sensors, electron exchange occurs between the biological component and the electrode in the presence of the target analyte, which generates the measurable detection signal. The working principle of these sensors involves application of a potential between the two electrodes in presence of the target analyte. At a fixed potential, the analyte undergoes a redox reaction at the electrode, which results in a change in the current of the electrochemical cell. The electron-transfer signal and hence the resulting current is proportional to the amount of the redox-active species at the electrode and can be used to detect or monitor the presence and amount of the target analyte.

Amperometric biosensors offer several advantages over their counterparts, which have led to their widespread applications. This type of biosensors provides specific quantitative analytical information by monitoring the current resulting from the oxidation or the reduction of the target electroactive biological element. The potential applied for the oxidation or the reduction is specific to the biological analyte in question, which results in the additional selectivity of the sensors. The sensors also have minimum background signal as the fixed potential during amperometric detection results in a negligible charging current. In addition, hydrodynamic amperometric techniques can provide significantly enhanced mass transport to the electrode surface [3]. Amperometric biosensors have evolved over the years in terms of their monitoring of electron transport mechanism. Three distinct generations of amperometric biosensors can be identified as follows [2,4]:

4.2.1 First-Generation Amperometric Sensors

The first-generation biosensors measure the decrease in the concentration of the electroactive substrate or the increase in the concentration of the electroactive product. Typically, these biosensors measure the oxygen reduction or the generation of hydrogen peroxide. As such, oxidases and dehydrogenases are the two main classes of enzymes used in this type of biosensors. Since most analytes of clinical interest are not readily available as natural substrates of a redox enzyme, a variety of strategies have been developed to transform a nonredox reaction into a redox one, through the use of coupled enzymatic reactions. By coupling one reaction where the analyte participates to another one, a species detectable by amperometry can be produced, and the amperometric sensors can be employed. The analytes that can be detected in this method include creatine, urea, lactate, and pyruvate [5].

4.2.2 SECOND-GENERATION AMPEROMETRIC SENSORS

The second-generation biosensors employ a mediator to transport electrons between the center of the redox reaction of the enzyme and the working electrode. This method is particularly suitable for enzymes that have their redox center embedded in the core of the protein and is electrically insulated, which is true for most of the enzymes used in biosensor technology. The redox mediators facilitate the transport of the electrons between the active site of the enzyme and the electrode surface. Ferrocene and its derivatives are the most well-known and widely used mediators. The ExacTECH™ biosensor from Abbott-Medisense uses derivative as mediator to monitor blood glucose level [6,7]. Other applications of soluble mediators reported in literature include detection of infection-marker antibodies, acetylcholine, biological oxygen demand, glutathione, L-alanine, pyruvate, lactate, and cholesterol [8].

4.2.3 THIRD-GENERATION AMPEROMETRIC SENSORS

Third-generation biosensors bypass the need for external mediators by facilitating direct electron transfer between the enzyme and the electrode surface. Due to unfavorable orientation of the protein molecules, it is usually difficult for the redox proteins in solution to transfer electrons directly to bare electrodes on the electrode surface. Depending on the orientation, the electron exchange between the electrode and the electroactive center of proteins can be hindered. If the distance between the redox center and the electrodes is large, it will result in a decreased rate of direct electron transfer. To overcome these problems without the use of mediators, different methods are applied to enhance direct electron transfer. These methods include different immobilization techniques, surface modification of electrodes to make the electrode surfaces more compatible with the redox enzymes, and protein engineering in order to achieve the most favorable orientation of the enzyme.

4.3 CARBON NANOFIBERS AS ELECTRODES FOR AMPEROMETRIC BIOSENSORS

Carbon electrodes have been proven to be an excellent choice for enzymatic amperometric biosensors in terms of the range of functionality and the cost-effectiveness. Among the carbon electrodes, carbon nanostructures (cylindrical or conical) exhibit the best potential. Carbon nanomaterials demonstrate enhanced electrical conductivity, good stability, excellent structural and catalytic properties, and high loading of biocatalysts, all of which are desirable for a biosensor platform [9–14]. The dimension of these carbon nanostructures varies in diameter from a few nanometers to hundreds of nanometers and in length ranging from less than a micron to a few millimeters [15]. Among the carbon nanostructures, carbon nanotubes (CNTs) have been widely used in biosensor research, although CNFs are emerging as a potential candidate to replace CNTs. CNFs have excellent conductive and structural properties compared to CNTs, which make them better candidates as electrodes as well as immobilizing substrates [16,17]. CNTs and CNFs can be used as electron field-emission sources, electrochemical probes, functionalized sensor elements, scanning probe microscopy tips, hydrogen and charge storage devices, catalyst support, and nanoelectromechanical systems (NEMS). Due to their extremely regular shapes, only the tips of the CNTs are electrochemically active. In contrast, irregularities and defect sites with exposed carboxyl groups throughout surfaces of the CNFs serve as excellent locations for electrochemical charge transfer [18]. These result in better sensitivity and responsiveness in CNFs compared to CNTs [19]. Because of their superior structural properties, CNFs also have better mechanical stability compared to CNTs [20].

CNFs are typically grown on silicon substrates and thus provide compatibility with CMOS and thin film transistor technologies [21]. This facilitates integration of the sensor elements with associated signal processing circuits realized on a single chip [22].

4.3.1 CARBON NANOFIBER GROWTH TECHNIQUES

Different techniques have been reported in literature to grow CNFs, which include direct-current plasma-enhanced chemical vapor deposition (DC PECVD), hot filament DC PECVD, microwave PECVD (MPECVD), inductively coupled PECVD (IC PECVD), radio-frequency PECVD (RF PECVD), electron cyclotron resonance PECVD (ECR PECVD), magnetron-type radio frequency, and hollow cathode plasma. Among these techniques, MPECVD, IC PECVD, and DC PECVD have been widely used for successful growth of CNF. DC PECVD is the most popular method for nanofiber growth [15] since both forests of fibers as well as single nanofibers can be grown using this method. Ni, Fe, and Co typically act as the catalyst for the growth process. The substrate needs to serve as a cathode, and therefore a conducting surface is required as a substrate. For a substrate of insulating type, a thin metal film is generally deposited under the catalyst to maintain connectivity with the electrode. In this process, acetylene is used as the carbon source, and ammonia is used as the etchant gas. Plasma power, C_2H_2/NH_3 gas ratio, flow rate, growth time, and catalyst size influence the morphology of the fiber. Deterministic synthesis is possible in DC PECVD process. Patterned catalyst material on wafer can fix the location of the CNF growth. Diameter of the fiber depends on the catalyst size while the length depends on the growth rate and the duration of the growth process. Alignment of the fiber is controlled by the electric field present in the plasma sheath of the PECVD environment. Sidewall composition of fiber is controlled by gas composition, substrate material, and plasma power.

Figure 4.1 shows the DC PECVD growth process. Catalyst prepatterned substrate acts as the cathode in the reaction chamber. The sample is then mounted on a heater plate, and when the base pressure

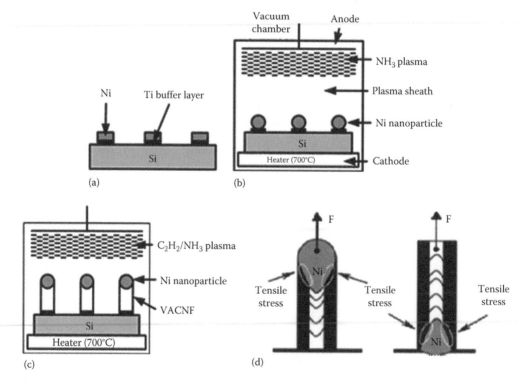

FIGURE 4.1 DC PECVD growth process (a) catalyst-patterned substrate, (b) catalyst pretreatment, (c) C_2H_2/NH_3 plasma for CNF growth, and (d) VACNF growth.

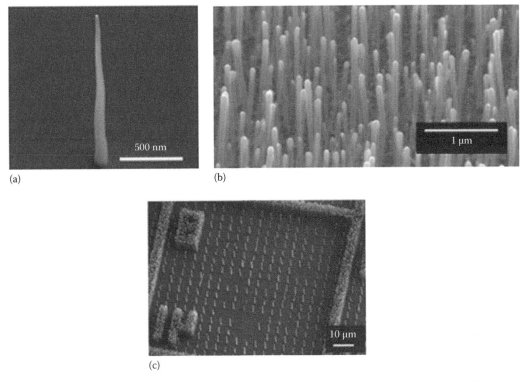

FIGURE 4.2   (a) SEM image of a single VACNF produced by DC PECVD. (b) SEM images of forests of densely spaced VACNF grown by DC PECVD. (c) SEM image of regular arrays of freestanding VACNFs and forests of VACNFs.

is reached in the chamber, ammonia is introduced from an overhead showerhead. Ammonia plasma and the chamber temperature influence the formation of catalyst nanoparticle from the deposited catalyst dots. These nanoparticles act as the seeds for growth of the nanofibers. After the pretreatment with ammonia, acetylene is introduced into the chamber. Acetylene breaks into carbon and deposits under/ over the catalyst depending on the pretreatment and the tensile strength of the catalyst. Scanning electron microscope (SEM) images of stand-alone and forests of fibers are shown in Figure 4.2.

One major advantage of DC PECVD is that both vertically aligned CNF (VACNF) forests and individual fibers can be grown on a predetermined position defined by the catalyst nanoparticle. The type of fiber depends on the size of prepatterned dots. If the dot approaches a minimum critical value, the individual single fibers begin to grow following the catalyst patterns. The drawback of the process includes plasma instability and its requirement of conductive substrate, which limits the choice of the substrate in this process.

4.3.2 FUNCTIONALIZATION OF NANOFIBER ELECTRODES

In order to develop VACNF-based biosensor systems, it is crucial to chemically modify the nanofiber electrodes for utilizing their specific properties. This is known as functionalization of nanofibers for biosensor development. CNFs can be biochemically functionalized in different ways, such as physical adsorption, covalent bonding, and electrochemical techniques [23].

In 2004, Lee et al. [24] reported a method to functionalize VACNF/single-wall nanotubes (SWNT) with nitro groups, which is reduced to amino groups through electrochemical reaction. The resulting primary amine group is used as the starting point to covalently link DNA to only nanostructures of VACNFs. The method is explained in Figure 4.3. This method can be utilized

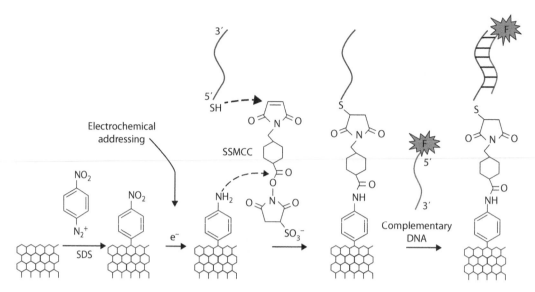

FIGURE 4.3 Schematic illustration of addressable biomolecular functionalization of CNTs. The procedure for CNFs is identical. (From Lee, C.S. et al., *Nano Lett.*, 4(9), 1713, 2004.)

to immobilize DNA, peptides, and antibodies to the surface of the nanostructure. Since the electron-transfer step in electrochemistry can typically only occur for species within one nanometer of an electrode surface [25], this process has the potential for extension down to near-atomic length scales. DNA hybridization shows that the method works well, and DNA-modified nanoscale structures provide excellent biological selectivity.

Baker et al. demonstrated two different methods for covalently modifying CNFs with biomolecules: one involving electrochemical reaction and the other involving photochemical reaction. The photochemical method is suitable for molecules that are highly insulating while the electrochemical method uses conductive molecules. As such, these two methods can be used to prepare biologically modified nanofibers with a range of electrical properties for electrical sensing of the specific biomolecules in solution [26]. The photochemical method begins with a photochemical reaction between as-grown nanofibers and molecules bearing both a terminal olefin group and a protected amine group illuminated with light at 254 nm [27–29]. After completion of the photochemical reaction, a deprotection method was followed to reveal the surface terminated with primary amine groups, as shown in Figure 4.4 [26]. The electrochemical method involves functionalizing CNFs by covalent grafting with the diazonium salt, 4-nitrobenzene diazonium tetrafluoroborate. Electrochemical reduction of the nitro groups leaves the surface terminated with primary amino groups [24,29,30]. The selective reduction of the nitro group to an amino group under electrochemical control provides a pathway toward electrically addressable functionalization. The method is shown in Figure 4.5.

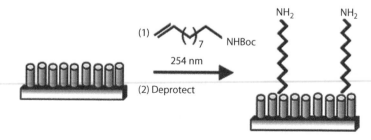

FIGURE 4.4 Schematic of photochemical functionalization method to produce amino-terminated surface. (From Baker, S.E. et al., *Chem. Mater.*, 17(20), 4971, 2005.)

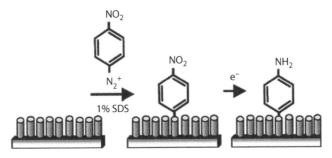

FIGURE 4.5 Schematic of electrochemical functionalization to produce amino-terminated surface. (From Baker, S.E. et al., *Chem. Mater.*, 17(20), 4971, 2005.)

McKnight introduced a flexible heterogeneous biochemical functionalization technique of VACNF arrays by a photoresist-blocking method. Instead of functionalizing the entire nanofiber, this method can be used for site-specific physical, chemical, and electrochemical functionalization of nanofiber arrays both spatially across regions of the device as well as along the length of the vertical nanofibers [31]. In this method, resist layers may be used to block functionalization sites specifically along the nanofiber height (two fibers depicted at the left of each drawing in Figure 4.6) or site specifically at different regions of an array (single fiber depicted at the right of each drawing in Figure 4.6). This site-specific binding method can be facilitated to afford additional complexity of the VACNF-based devices, as such limiting the gene delivery applications to the tip of the nanofiber and modifying nanofibers only at discrete locations for additional complexity in the microfluidic systems.

The general scheme of the photoresist-blocking method is shown in Figure 4.6. First, nanofiber arrays are deterministically grown using a DC PECVD method. Then they are spun with photoresist, which is followed by deprotection of the desired region using a photolithographic pattern.

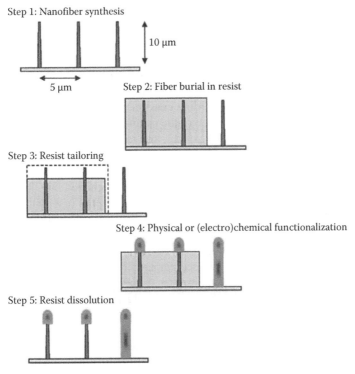

FIGURE 4.6 General scheme for photoresist-based blocking of chemical or electrochemical functionalization of arrays of VACNFs. (From Boussaad, S. et al., *Chem. Commun.*, 13, 1502, 2003.)

Using oxygen reactive ion etch (RIE), the photoresist protective layer is tailored to a desired height while exposing only the tip. These exposed regions can then be physically, chemically, or electrochemically functionalized followed by removal of the blocking resist by chemical dissolution.

Fletcher et al. developed two attachments of biomolecule schemes onto VACNF surfaces, one using a class of heterocyclic aromatic dye compounds for specific adsorption onto VACNFs, and the other using covalently coupling biomolecules through cross-linking to carboxylic acid sites on the sidewalls of the CNFs [32].

The adsorption method exploits the structure of the aromatic binding dyes. Studies suggest that the graphene outer surface of the CNTs allows for stacking interactions with aromatic molecules like pyrene [33]. The sidewalls of CNFs also possess graphene carbon, but the sheets are present in a stacked configuration rather than the smooth outer sheathing of the nanotubes. Unlike nanotubes, avidins do not strongly physisorb to VACNFs. Despite facilitating protein immobilization using fairly gentle immobilization conditions, the reversibility of the dye-binding method under stringent wash conditions may not be suitable for some applications. Nevertheless, noncovalent adsorption is straightforward and potentially useful where stringent wash conditions are not present.

Covalent coupling is a more robust technique for immobilizing biomolecules. Standard biomolecular immobilization chemistries, followed by oxygen plasma cleaning, provide effective and reproducible functionalization exploiting carbodiimide-based cross-linking for immobilization to surfaces. Both DNA and proteins can be immobilized by coupling pendant amine groups to the exposed carboxylic acid residues (resulting from the oxygen plasma treatment) on the VACNF surface. Weakly bound, physisorbed materials can be removed by stringent wash conditions without removing tethered biomolecules, thus reducing background fluorescence when performing fluorescence-based assays. These investigations have extended the application of functionalized CNFs for biosensor development.

4.4 SPECIFIC APPLICATIONS OF CARBON NANOFIBER (CNF)-BASED BIOSENSORS

Potentially, CNF-based biosensors have extensive applications such as glucose detection, alcohol detection, nicotinamide adenine dinucleotide (NADH) detection, and K562 cell detection. Among these, extensive research works have been carried out for glucose detection although detection of other analytes also shows significant promise. Vamvakaki et al. reported a glucose biosensor that exploits the larger functionalized area of the CNFs compared to the nanotubes [34]. The sensor is developed using highly activated CNFs with direct enzyme immobilization. Since the number and type of functional groups on the outer surface of the nanotube can be well controlled, this method promises the development of very sensitive, stable, and reproducible electrochemical biosensors. Vamvakaki immobilized glucose oxidase (GOx) by adsorption method to functionalize the surface of the nanofibers. Using a three-electrode system with Ag/AgCl double junction reference electrode and Pt counter electrode, electrochemical measurements were carried out. The sensitivity of the biosensor was monitored amperometrically at the working potential of +600 mV versus Ag/AgCl electrode. The larger functionalization area of the nanofibers provided a good chemical environment for enzyme immobilization whereas the high conductivity ensured excellent electrochemical signal transduction.

Wu et al. reported a CNF–GOx/Nafion-based amperometric glucose sensor [35] where GOx is used as an enzyme model. The sensor demonstrates excellent electrocatalytic activity of soluble CNF toward the electroreduction of dissolved oxygen, which is produced by GOx as a model for preparation of pertinent oxidase-based biosensors. The CNF membrane showed good stability and provided fast response to dissolved oxygen, which extends its use in different fields such as food, beverages, and fermentation liquor for glucose monitoring. The use of a low operating potential (−0.3 V) and a Nafion membrane also produced good selectivity toward glucose detection even in the presence of other interferers such as ascorbic acid and uric acid, an improvement over the sensor proposed by Vamvakaki.

The biosensor can quickly detect glucose ranging from 10 to 350 µM with a detection limit of 2.5 µM. A simple nitric acid treatment improved the solubility and biocompatibility by producing a large number of different oxygen-containing groups on the CNF surface. The soluble CNF attributes to the excellent conductivity for accelerating the electron transfer of the electroactive compounds and good preparation reproducibility of the biosensor. It also shows excellent catalytic activity toward the reduction of the dissolved oxygen, which can be used for continuous monitoring of the dissolved oxygen and represents a biocompatible platform for the development of amperometric glucose sensors.

Islam et al. [36] reported a mediator-less highly selective glucose sensor based on VACNF platform using bienzyme wiring technique. GOx and horseradish peroxidase (HRP) were used to functionalize the surface of the nanofibers. The detection of glucose is confirmed in a two-step electrochemical redox operation. In the first stage, glucose is oxidized by the ambient oxygen in the presence of GOx and produces hydrogen peroxide (H_2O_2). In the second stage, H_2O_2 is reduced to water in the presence of HRP. The two electrons required for the reduction are provided by the CNFs, and the resulting change in the current is picked up by the sensor. Depending on proper immobilization of enzymes onto the sensor surface, the sensor can have very small (as low as 0.4 µm) [36] to high detection range (1–6.6 mM, consistent with human blood glucose level) [37] and can be used for both environmental monitoring and medical applications.

Weeks et al. reported a reagent-less amperometric enzymatic biosensor constructed using VACNFs for the detection of ethanol [38]. Yeast alcohol dehydrogenase (YADH), an oxidoreductase, and its coenzyme nicotinamide adenine dinucleotide (NAD+) are immobilized by adsorption and covalent attachment to the VACNF. The VACNFs grown by plasma-enhanced chemical vapor deposition (PECVD) are chosen as the electrode material due to their excellent structural and electrical properties. The storage stability, the reusability, and the response time of the biosensor are also examined. The VACNFs are found to be an effective strategy for building a biosensor platform.

Li et al. studied three different types of CNFs, platelet type, fishbone type, and tube type, to find the effect of nanofiber structure on the electrochemical sensing of hydrogen peroxide [39]. Depending on the differences of the CNFs in morphology, texture, and crystalline structure, significant diversity of electrocatalytic activity of these CNFs toward the oxidation of hydrogen peroxide was observed.

Arvinte et al. used CNFs to develop an electrochemical biosensor platform based on the electrocatalytic activity of CNFs toward NADH [40]. A study of the direct electrochemistry of NADH at a CNF-modified carbon electrode demonstrated a decrease of the oxidation potential of NADH by more than 300 mV when compared with a bare glassy-carbon electrode.

Hao et al. utilized the chemical stability and mechanical strengths of CNFs to design a biocompatible architecture for attaching and cytosensing human K562 cells on an electrode [41]. The impedance of the electronic transduction of the K562 cells was found to be related to the amount of the adhered cells through electrochemical impedance spectroscopy and cyclic voltammetry, which results in a highly sensitive impedance sensor for K562 cells.

These results confirmed that CNFs provide a promising material for assembling electrochemical sensors and biosensors.

4.5 INTEGRATION WITH SENSOR ARCHITECTURE

Fletcher et al. proposed a CMOS chip, which can be used for integration with VACNF array [42]. A fixed potential is required at the VACNF array for amperometric test. Redox/oxidation reactions take place at a fixed potential depending on the target analyte. The proposed chip contained 8 selectable voltage reference circuits and a 3 × 8 decoder to select different voltage reference circuits for VACNF electrode array. Ko et al. also proposed a fabrication method to transfer VACNF on a flexible polycarbonate surface [43]. In this process, a polycarbonate film covers the VACNF grown on a silicon substrate and annealed at a temperature of 190°C for 2 h and with a pressure of 2–3 N/cm^{-2}.

The VACNF partially embed into the softened polycarbonate substrate at elevated temperature. The cured substrate is then peeled off along with the VACNF embedded on it. The weak interconnection of the VACNF with silicon substrate facilitates the peel-off process. The signal conditioning circuit for VACNF array requires measurement of the current produced from oxidation/redox reactions at the VACNF surface. A typical low-current measuring ammeter can be added to the VACNF sensors to act as the measuring unit for the purpose.

4.6 SILICON COMPATIBILITY

CNF growth process requires elevated temperature in the range of 600°C–700°C [15]. This high growth temperature is a barrier to the growth of the nanofibers on a commercial CMOS chip, which is a low-temperature process. The functionality of complementary MOS degrades around 400°C. One alternative way to integrate VACNF with low-temperature electronics is to transfer the grown VACNF on a metal pad, which is patterned in standard CMOS process. Such transfer process has recently been verified where flexible arrays of nanofiber were grown in a Si substrate in high-temperature PECVD process and then transferred to a metal pad array [41]. The process requires several steps. The VACNF array grown in PECVD process was partially buried in an epoxy (SU8), and the VACNF-embedded SU8 was subsequently peeled off from the Si substrate for transfer. The peeled layer was then placed on prepatterned metal pads, and to mate the two structures, a patterned silver-loaded epoxy was used. A recent study by Yang et al. [16] shows that in fact a semiconducting SiC layer forms beneath the VACNF during the growth process. The layer results a Schottky-barrier junction between the semiconducting layer and the nanofiber. Current–voltage characteristics for the junction also support the study and show a rectifying behavior of nanofibers grown in a DC PECVD process.

4.7 ASSOCIATED CHALLENGES

The weak bonding between CNF and silicon substrate is one of the major challenges for the implementation of a robust nanofiber-based electrode platform as the bonding plays an important role in terms of electrode stability. Repeated washing and electrochemical measurements can pull out fibers in an amperometric test. To circumvent the weak bonding issue, one of the ways reported in literature is to coat nanofiber forest with SiO_2 and SU8 [42]. However, this always comes with reduced sensing efficiency of the fiber forest. Other challenges include poor stability of nanofiber-based electrode, electrode storage requirements, noise in nanofiber electrode, integration with standard CMOS circuit, and lowering of the cost associated with nanofiber fabrication.

The stability of CNF electrode is still a subject of further research. Due to their hydrophobic nature, the nanofibers react with airborne water vapor, and salt deposition may occur throughout the fiber surface. Storing of the nanofiber forest thus requires extra care. One option to preserve the fiber forest is to preserve the electrodes in buffer solution while they are not in use. Dry fiber forest tends to be brittle and can break during processing or testing. In cases when dry preservation of nanofiber electrode is the only option, soaking fiber forest in a buffer solution or deionized (DI) water is recommended before any electrochemical test. Furthermore, there are many acidic and basic groups throughout the nanofiber surface, which are highly reactive to processing steps and can cause a net acidic or basic nature of the dissolved solution. For CNF electrode, a buffer is thus needed in any electrochemical analysis to keep pH at a constant level. This may also limit the usage of bare nanofiber in any implantable system.

CNF tips have the highest reactivity compared to other carbon nanomaterials [42]. For high-resolution sensing, signal interferences from neighboring fiber forests become an unavoidable event. One way to get around is to cover the forest surface with an insulating layer and expose only the fiber tips for electrochemical reactions. Again, CNF growth process requires elevated temperature in the range of 600°C–700°C. This requires specialized process techniques for nanofiber growth and necessitates costly infrastructure. The temperature range is not at all suitable for direct integration of nanofiber with standard CMOS chips. One alternative may be the use of RF energy in RF PECVD process to elevate temperature only at the spot where nanofiber is grown on a CMOS chip. Also for lowering the fabrication cost, a low-cost SPUN CNF growth process has been reported in literature [44].

4.8 CONCLUSIONS

CNFs represent an effective platform for the realization of biosensors. Due to the presence of abundant number of defect sites composed of exposed carboxyl groups, CNFs allow successful immobilization of the biological elements on their surfaces and provide better electrical response compared to other carbon nanomaterials. By plugging enzymes with the metal electrode through the CNFs by enzyme wiring technique, CNF-based biosensors provide an effective way to transfer electrons from the electrode to the electrochemical reaction centers. These sensors demonstrate good dynamic range and manifest high selectivity toward analytes in the presence of interfering electroactive compounds. Mediator- or membrane-free operation facilitates potential application of these biosensors in healthcare, environmental monitoring, scientific experiments, etc. Since CNFs can be grown on silicon-compatible substrates, these sensors can potentially be integrated with existing sensor technologies for the development of fully integrated analytical instruments based on lab-on-a-chip platform. However, successful commercial realization of the CNF biosensors will require further research to overcome number of challenges associated with the material properties, growth technologies, functionalization of the target biomolecules, and packaging strategies.

REFERENCES

1. B. R. Eggins, *Chemical Sensors and Biosensors*, Wiley, New York, 2002.
2. S. V. Dzyadevych, V. N. Arkhypova, A. P. Soldatkina, A. V. El'skaya, C. Martelet, N. Jaffrezic-Renault, Amperometric enzyme biosensors: Past, present and future, *BioMedical Engineering and Research (IRBM)*, 29(2–3), 171–180, 2008.
3. C. J. Slevin, P. R. Unwin, Microelectrochemical measurements at expanding droplets (MEMED): Mass-transport characterization and assessment of amperometric and potentiometric electrodes as concentration boundary layer probes of liquid/liquid interfaces, *Langmuir*, 15(21), 7361–7371, 1999.
4. S. J. Sadeghi, Amperometric biosensors, *Encyclopedia of Biophysics*, ed.: G. C. K. Roberts, Springer-Verlag, Berlin, Germany, pp. 61–67, 2013.
5. F. Mizutani, S. Yabuki, Y. Sato, T. Sawaguchi, S. Iijima, Amperometric determination of pyruvate, phosphate and urea using enzyme electrodes based on pyruvate oxidase-containing poly(vinyl alcohol)/polyion complex-bilayer membrane, *Electrochimica Acta*, 45(18), 2945–2952, 2000.
6. A. E. G. Cass, G. Davis, G. D. Francis, H. A. O. Hill, W. J. Aston, I. J. Higgins, E. V. Plotkin, L. D. L. Scott, A. P. F. Turner, Ferrocene-mediated enzyme electrode for amperometric determination of glucose, *Analytical Chemistry*, 56(4), 667–671, 1984.
7. N. J. Forrow, G. S. Sanghera, S. J. Walters, The influence of structure in the reaction of electrochemically generated ferrocenium derivatives with reduced glucose oxidase, *Journal of the Chemical Society: Dalton Transactions*, (16), 3187–3194, 2002.
8. A. Chaubey, B. D. Malhotra, Mediated biosensors, *Biosensors and Bioelectronics*, 17(6–7), 441–456, 2002.
9. G. M. Cote, R. M. Lec, M. V. Pishko, Emerging biomedical sensing technologies and their applications, *IEEE Sensors Journal*, 3(3), 251–266, 2003.
10. M. Zayats, E. Katz, R. Baron, I. Willner, Reconstitution of apo-glucose dehydrogenase on pyrroloquinolinequinone-functionalized Au nanoparticles yields an electrically contacted biocatalyst, *Journal of American Chemical Society*, 127(35), 12400–12406, 2005.
11. S. Boussaad, N. J. Tao, T. Hopson, L. A. Nagahara, In situ detection of cytochrome c adsorption with single walled carbon nanotube device, *Chemical Communication*, 13, 1502–1503, 2003.
12. P. G. He, L. M. Dai, Aligned carbon nanotube–DNA electrochemical sensors, *Chemical Communication*, 3, 348–349, 2004.

13. D. Lee, J. Lee, J. Kim, J. Kim, H. B. Na, B. Kim, C.-H. Shin, J. H. Kwak, A. Dohnalkova, J. W. Grate, T. Hyeon, H.-S. Kim, Simple fabrication of a highly sensitive and fast glucose biosensor using enzymes immobilized in mesocellular carbon foam, *Advanced Materials*, 17(23), 2828–2833, 2005.

14. O. Niwa, Electroanalytical chemistry with carbon film electrodes and micro and nano-structured carbon film-based electrodes, *Bulletin of the Chemical Society of Japan*, 78(4), 555–571, 2005.

15. A. V. Melechko, V. I. Merkulov, T. E. McKnight, M. A. Guillorn, K. L. Klein, D. H. Lowndes, M. L. Simpson, Vertically aligned carbon nanofibers and related structures: Controlled synthesis and directed assembly, *Journal of Applied Physics*, 97(4), 041301, 2005.

16. X. Yang, M. A. Guillorn, D. Austin, A. V. Melechko, H. Cui, H. M. Meyer III, V. I. Merkulov, J. B. O. Caughman, D. H. Lowndes, M. L. Simpson, Fabrication and characterization of carbon nanofiber-based vertically integrated Schottky barrier junction diodes, *Nano Letters*, 3(12), 1751–1755, 2003.

17. Q. Zhao, Z. Gan, Q. Zhuang, Electrochemical sensors based on carbon nanotubes, *Electroanalysis*, 14(23), 1609–1613, 2002.

18. S.-U. Kim, K.-H. Lee, Carbon nanofiber composites for the electrodes of electrochemical capacitors, *Chemical Physics Letters*, 400, 253–257, 2004.

19. J. Jang, J. Bae, M. Choi, S.-H. Yoon, Fabrication and characterization of polyaniline coated carbon nanofiber for supercapacitor, *Carbon*, 43(13), 2730–2736, 2005.

20. Y.-L. Yao, K.-K. Shiu, A mediator-free bienzymaticamperometric biosensor based on horseradish peroxidase and glucose oxidase immobilized on carbon nanotube modified electrode, *Electroanalysis*, 20(19), 2090–2095, 2008.

21. J. Park, S. Kwon, S. I. Jun, T. E. Mcknight, A. V. Melechko, M. L. Simpson, M. Dhindsa, J. Heikenfeld, P. D. Rack, Active-matrix microelectrode arrays integrated with vertically aligned carbon nanofibers, *IEEE Electron Device Letters*, 30(3), 254–257, 2009.

22. A. V. Melechko, R. Desikan, T. E. McKnight, K. L. Klein, P. D. Rack, Synthesis of vertically aligned carbon nanofibres for interfacing with live systems, *Journal of Physics D: Applied Physics*, 42(19), 193001, 2009.

23. J. Wang, Y. Lin, Functionalized carbon nanotubes and nanofibers for biosensing applications, *Trends in Analytical Chemistry*, 27(7), 619–626, 2008.

24. C. S. Lee, S. E. Baker, M. S. Marcus, W. S. Yang, M. A. Eriksson, R. J. Hamers, Electrically addressable biomolecular functionalization of carbon nanotube and carbon nanofiber electrodes, *Nano Letters*, 4(9), 1713–1716, 2004.

25. R. A. Marcus, On the theory of electron–transfer reactions. VI. Unified treatment for homogeneous and electrode reactions, *Journal of Chemical Physics*, 43(2), 679–701, 1965.

26. S. E. Baker, K. Y. Tse, E. Hindin, B. M. Nichols, T. L. Clare, R. J. Hamers, Covalent functionalization for biomolecular recognition on vertically aligned carbon nanofibers, *Chemistry of Materials*, 17(20), 4971–4978, 2005.

27. T. Strother, T. Knickerbocker, J. N. Russell, J. E. Butler, Jr., L. M. Smith, R. J. Hamers, Photochemical functionalization of diamond films, *Langmuir*, 18(4), 968–971, 2002.

28. T. L. Lasseter, W. Cai, R. J. Hamers, Frequency-dependent electrical detection of protein binding events, *Analyst*, 129, 3–8, 2004.

29. W. S. Yang, O. Auciello, J. E. Butler, W. Cai, J. A. Carlisle, J. E. Gerbi, D. M. Gruen et al., DNA-modified nanocrystalline diamond thin-films as stable, biologically active substrates, *Nature Materials*, 1, 253–257, 2002.

30. P. Allongue, M. Delamar, B. Desbat, O. Fagebaume, R. Hitmi, J. Pinson, J.-M. Savéant, Covalent modification of carbon surfaces by aryl radicals generated from the electrochemical reduction of diazonium salts, *Journal of American Chemical Society*, 119, 201–207, 1997.

31. T. E. McKnight, C. Peeraphatdit, S. W. Jones, J. D. Fowlkes, B. L. Fletcher, K. L. Klein, A. V. Melechko, M. J. Doktycz, M. L. Simpson, Site-specific biochemical functionalization along the height of vertically aligned carbon nanofiber arrays, *Chemistry of Materials*, 18(14), 3203–3211, 2006.

32. B. L. Fletcher, T. E. McKnight, A. V. Melechko, M. L. Simpson, M. J. Doktycz, Biochemical functionalization of vertically aligned carbon nanofibres, *Nanotechnology*, 17(8), 2032–3039, 2006.

33. R. J. Chen, Y. Zhang, D. Wang, H. Dai, Noncovalent sidewall functionalization of single-walled carbon nanotubes for protein immobilization, *Journal of American Chemical Society*, 123(16), 3838–3839, 2001.

34. V. Vamvakaki, K. Tsagaraki, N. Chaniotakis, Carbon nanofiber-based glucose biosensor, *Analytical Chemistry*, 78(15), 5538–5542, 2006.

35. L. Wu, X. Zhang, H. Ju, Amperometric glucose sensor based on catalytic reduction of dissolved oxygen at soluble carbon nanofiber, *Biosensors and Bioelectronics*, 23(4), 479–484, 2007.

36. A. B. Islam, F. S. Tulip, S. K. Islam, T. Rahman, K. C. MacArthur, A mediator free amperometric bienzymatic glucose biosensor using vertically aligned carbon nanofibers (VACNFs), *IEEE Sensors Journal*, 11(11), 2798–2804, 2011.

37. K. C. MacArthur, K. A. A. Mamun, F. S. Tulip, N. McFarlane, S. K. Islam, Fabrication and characterization of vertically aligned carbon nanofiber as a biosensor platform for hypoglycemia, *Lester Eastman Conference of High Performance Devices (LEC 2012)*, August 7–9, Brown University, Providence, RI, 2012.

38. M. L. Weeks, T. Rahman, P. D. Frymier, S. K. Islam, T. E. McKnight, A reagentless enzymatic amperometric biosensor using vertically aligned carbon nanofibers (VACNF), *Sensors and Actuators B*, 133(1), 53–59, 2008.

39. Z. Li, X. Cui, J. Zheng, Q. Wang, Y. Lin, Effects of microstructure of carbon nanofibers for amperometric detection of hydrogen peroxide, *Analytica Chimica Acta*, 597(2), 238–244, 2007.

40. A. Arvinte, F. Valentini, A. Radoi, F. Arduini, E. Tamburri, L. Rotariu, G. Palleschi, C. Bala, The NADH electrochemical detection performed at carbon nanofibers modified glassy carbon electrode, *Electroanalysis*, 19(14), 1455–1459, 2007.

41. C. Hao, L. Ding, X. Zhang, H. Ju, Biocompatible conductive architecture of carbon nanofiber-doped chitosan prepared with controllable electrodeposition for cytosensing, *Analytical Chemistry*, 79(12), 4442–4447, 2007.

42. B. L. Fletcher, T. E. McKnight, A. V. Melechko, D. K. Hensley, D. K. Thomas, M. N. Ericson, M. L. Simpson, Transfer of flexible arrays of vertically aligned carbon nanofiber electrodes to temperature-sensitive substrates, *Advanced Materials*, 18(13), 1689–1694, 2006.

43. H. Ko, Z. Zhang, J. C. Ho, K. Takei, R. Kapadia, Y.-L. Chueh, W. Cao, B. A. Cruden, A. Javey, Flexible carbon nanofiber connectors with anisotropic adhesion properties, *Small*, 6(1), 22–26, 2010.

44. C. Kim, Y.-J. Kim, Y.-A. Kim, Fabrication and structural characterization of electro-spun polybenzimidazol-derived carbon nanofiber by graphitization, *Solid State Communications*, 132(8), 567–571, 2004.

5 Translating Sensor Technology into the Medical Device Environment

Robert D. Black

CONTENTS

5.1 INTRODUCTION

The translation of a sensor technology into clinical medicine is a complex and multifaceted undertaking [1–4]. In addition to challenges in the discovery phase, there are regulatory, reimbursement, and physician acceptance hurdles to clear. The latter challenges are often not considered in research-oriented journal articles, and authors of such articles run the risk of reporting on developments that may ultimately be sterile in a translational sense. As research funding sources become increasingly challenging to obtain, work that purports to support new capabilities in human medicine must be complete in the sense of being able to clear all hurdles that will be arrayed against it. This is not to say that there isn't a vital role for pure research projects, but work that is *justified* by an implied payback in advancing human health must evolve with a full recognition and acceptance of what that entails. Some questions that should be asked, even at the earliest stages of development, are the following:

- Will the product fit into existing reimbursement codes?
- Will the product be approved by the Food and Drug Administration (FDA) when a tractable clinical trial is completed?
- If the device is an implantable sensor, how is it implanted and can it be included with an existing procedure?

- Does the device fit with existing medical training and practices, and if not, what is the learning curve for acceptance?
- If the device involves sensor data, specifically, how will that data be used to influence patient care?
- Can the device be manufactured in a cost-effective way?
- Does the implanted device meet the *burden of inconvenience* test, meaning that the information provided is unique and valuable and can't be obtained by less-invasive means (e.g., a blood test)?

Starting with a review of some of the essential considerations involved in the regulatory process, several examples of sensor-based technologies that are now in wide use with patients will be noted. Finally, some future trends that rely on sensor-based devices, as projected by the FDA, will be mentioned. In the end, this article will have succeeded if it provides the reader with a better appreciation of the multifaceted environment in which discoveries that seek to be translated into human medicine must navigate.

5.2 FDA PROCESS FOR MEDICAL DEVICES

In the United States, FDA oversees commercialization of medical devices [5]. In Europe, for example, the Medical Device Directive (MDD) leads to harmonization of standards, but each country applies the MDD for consistency with national goals. Therefore, generalizing the regulations surrounding the commercialization of devices is a nation-by-nation process, and herein, we will focus on the particular requirements of the FDA. The rules established for the FDA are based on laws contained in the Code of Federal Regulations (CFR).

5.2.1 How the Device Is to Be Used

When speaking of regulatory control of medical devices, one must first define both the intended use and indication for use. Though similar, the concepts differ and the latter is in some sense a subset of the former. Intended use refers to the overall goal in using the device and can be fairly generic (21 CFR 801.4). The indication(s) for use (e.g., 21 CFR 814.20(b)(3)(i)) refers to "a general description of the disease or condition the device will diagnose, treat, prevent, cure, or mitigate, including a description of the patient population for which the device is intended." Therefore, a given device can in principle have many indications for use, and each new use must be evaluated in terms of whether it creates new questions of safety and efficacy, and if so, the device maker will need to provide data establishing safety and efficacy (possibly clinical data). Developers of medical sensors, therefore, must navigate this preliminary classification question and delineate clearly the role of the sensor and how it will affect patient health.

5.2.2 Safety and Efficacy

The essence of what the FDA regulates is the safety of the device: does it have potential to harm the patient directly or through information it provides that affects patient care, and does the device accomplish its indication for use? A common misconception is that a product that *doesn't touch the patient* is not a medical device. But even software that records patient medical records, for example, is a subject to regulation (clearly, if the information provided by the software is faulty, that could affect patient care and well-being). Even if a sensor system does not perform a direct diagnostic or therapeutic function, it may still be classified as a medical device if, again, it has the potential to alter patient care. An element of a telemetry system, were it to malfunction, could provide erroneous data to a physician and lead to improper interventions being taken with a patient. Sensors in medical devices must have rigorous calibration records demonstrating the accuracy of the sensor for the intended use for the duration of a diagnostic or therapeutic procedure. If, during development, a sensor designed for a medical application is not deemed to be accurate enough, then a secondary method of calibration would have to be established. For example, continuous glucose monitors (CGMs) rely on calibration by established glucometers.

5.2.3 Establishing a Predicate

Although most new medical devices have novel features, it is often the case that they share an indication for use with an established commercial device: a predicate device. Identifying a predicate device provides a distinct advantage when pursuing device commercialization since the burden of proof is lessened. In effect, one can take advantage of an existing track record of safety and efficacy. But how close must such a predicate be? There is no clear-cut means of making such a determination, and the device maker must consult with the FDA so as to work out a development plan that will be accepted when the device is proffered for regulatory scrutiny. This is the most challenging concept for inventors and researchers seeking to introduce sensors into the medical environment. Whereas novelty is essential for the process of invention and seeking patent protection, it works against the concept of predicate-use scenarios. Researchers are rewarded for devising new and different sensors that can, in theory, be useful in advancing medicine. Too often, the promise, sometimes hype, is not commensurate with the likelihood of translation into medical practice. False hope and optimism can be generated when the stakeholders in an invention have not paid adequate attention to the entire chain of events that must be addressed when moving from the lab to the clinic. One often sees cautionary comments, indicating that many years separate a discovery from implementation in the clinic, but interested parties are left wondering, *if not now, when?* Premature announcements of potential medical significance, frankly, serve no purpose and should be eschewed.

An important caveat when speaking of predicate devices is to establish the risk classification of the predicate. There are two overlapping risk assessment categories, the first being whether the device is significant or nonsignificant risk (SR or NSR). Here, risk covers both risk to the patient and the operator of the device. Additionally, devices are categorized into three classes. Class I devices are products that are not for a substantive use in preventing impairment of human health and that do not present a potentially unreasonable risk of patient injury. Examination gloves are Class I devices. Class II devices include higher-technology products that do not by themselves maintain life, but nonetheless can have an important impact on patient care. Magnetic resonance imaging (MRI) machines are Class II devices. Class III devices are used in supporting or sustaining human life or are substantially important in preventing impairment of human health. Novel implants, like neurostimulators, are in this category. Obviously, Class III devices are SR. Class II devices can be SR or NSR. Class I devices are NSR.

5.2.4 Regulatory Pathway

Focusing on medical sensors, there are two primary potential procedural pathways for gaining regulatory certification. The first is commonly called a 510(k) and the second a premarket approval (PMA). Devices are *cleared* in the 510(k) pathway and *approved* via the PMA pathway. There is a third pathway, referred to as *De Novo*, that can serve as an intermediate route as discussed in the following. Understanding how to navigate the approval pathway is a primary undertaking in the translational process. A 510(k) device is one that has a demonstrable predicate, is Class I or II, and may be SR or NSR. If a device is determined to be SR, then a formal application, termed an investigational device exemption (IDE), must be submitted to the FDA before clinical data are acquired in support of a device application. In the IDE, the sponsor lays out the reasoning behind a proposed clinical investigation, summarizes laboratory or animal data, etc. The aim is to explain to the FDA how the proposed approach will answer questions germane to the clearance process. Until the IDE is approved, no human research is permitted. If a device is NSR, approval for clinical work may be obtained through an Institutional Review Board (IRB), which exists within many hospitals and may also be separate, private organizations. Determining *whether* a sensor-based device clearance requires human test data is another important consideration when evaluating clinical potential.

Frequently, a medical sensor would be expected to fall into the 510(k) category, but what about a sensor type that truly is novel as applied to medical science? What about a sensor device that has the attributes of a Class II device, is NSR, and yet has no direct predicate in the market? With only two pathways available, 510(k) and PMA, such a device would have to be regulated through the PMA route and be Class III, even though it may be extremely safe and effective. In an effort to resolve this logical impasse, the *De Novo* pathway was initiated by the FDA. A *De Novo* device is one that is viewed to be low risk and yet does not have an indication for use that matches currently marketed products. Effectively, the device becomes its own predicate...*de novo*. The process of obtaining *De Novo* approval is comparable to that for a 510(k) device, but the FDA may invoke an advisory panel review, which adds time to the process.

Devices that are both high risk and nonsubstantially equivalent to a predicate device will be regulated via the PMA pathway. A PMA application requires compilation and presentation of considerable amounts of information. This includes a complete description of the device and components, photos and engineering diagrams, a detailed description of the methods, facilities and controls used to manufacture the device, the proposed labeling and advertising literature, training materials, software documentation, biocompatibility information, and references to applicable standards, also a summary of all clinical, animal, and bench data. An advisory panel is often assembled to provide an external review of the device, and an inspection of manufacturing facilities will be arranged. Finally, after PMA approval, the FDA maintains market surveillance rights as the product is commercialized. The PMA process can take years and cost millions or tens-of-millions of dollars. This is a substantial hurdle for novel, untried technology, such as a new sensor system, to clear, and the process focuses attention on the market potential of the device.

5.2.5 Independent Device Testing

Since any new medical device raises questions of safety, it is essential to have independent bodies test and verify compliance with safety standards. For example, with respect to biocompatibility, a device may need to undergo toxicology, sensitization, and irritation testing if it is to be implanted or even just contact the skin. There are several laboratories that perform these testing services. For a medical sensor that must operate in a biological environment, additional testing of its functional state during exposure to that environment must be included. Devices must also adhere to a set of technical standards set forth in the International Electrotechnical Commission (IEC) 60601, and the device maker must show empirical evidence that the device has passed the safety and functional tests set forth in this standard. The so-called third edition rules under this standard outline significant new rules for home-use devices. Thus, medical sensors in home-use products face additional testing in order to demonstrate compliance. The time and cost involved in biocompatibility and IEC testing can also be significant.

5.2.6 Some Specific Examples of Sensor Characteristics That Evoke Regulatory Decision Points

1. *Implanted sensor breakage or migration*: There is strong interest in diagnostic sensors that can reside in or near bodily organs and provide feedback on medical conditions (e.g., blood pressure in the heart or serum glucose levels in diabetics). In addition to the question of

how to place such sensors, that is, the surgical approach, the issues of how to keep such sensors in the desired location and avoid damage to the sensor that could expose the patient to risk must be addressed. Typically, sensors will have some means of retention associated with the outer package. This could take the form of a feature that allows for suturing or some sort of adhesive coating that works against sensor slippage. If the sensor is near the skin surface, will it be subjected to forces that could potentially fracture the device? If the sensor does move away from the point of placement, might it inflict damage on nearby tissues or organs? For instance, a sensor that resides in the heart or nearby large vessels must be evaluated for its potential to create an embolism should it come free. If the sensor components include potentially toxic components, testing should account for the possibility of breakage and leakage in the patient.

2. *Biocompatibility*: In addition to the physical concerns stated earlier for implanted sensors, biological compatibility must also be established. The general requirements are summarized in the ISO-10993 standards. For implanted devices, cytotoxicity (toxic effects on cells), irritation, and sensitization are routinely studied. Additionally, based on the component materials, questions about carcinogenicity and genotoxicity may arise. These latter tests can extend the time horizon for biocompatibility testing in general and should be factored into the timeline of the translational plan. The use of novel materials that have not been utilized in already-marketed medical devices can be particularly challenging since regulatory agencies are unable to derive guidance from previous studies in gauging new risks that novel materials may present. Extensive in vitro and animal testing will be expected for new implantable substances.

3. *Biofouling*: Separate from the effects of an implanted sensor on the body are the effects residing on the body may have on the sensor. For implanted medical sensors, key to the success is the need to ensure reliable performance when exposed to living tissue and fluids. The term biofouling has gained common usage as a description of the buildup of proteins and fibroblasts that accompany the body's immune reaction against foreign objects. Effectively, the body attempts to wall off the foreign material, and the growth layer is typically 100 μm thick or more. This layer is problematic for any device that seeks to sample serum or blood, for example, since it acts as a diffusion barrier. Implanted glucose sensors provide a case-in-point.

4. *Electromagnetic interference*: Electronically active medical devices must work in the expected range of environments to which they will be exposed and also not interfere with other electronic systems in the healthcare setting. The rules differ based on whether the device is intended for home use or hospital use. Both IEC (e.g., IEC 60601-1-2) and Federal Communications Commission (FCC) requirements must be met. Typically, testing by a third party is required to demonstrate that a medical device complies with existing industry and local regulations. With the ubiquity of wireless communication devices operating in the ~2.4 GHz band, sensor systems that rely on wireless connections must avoid packet collisions that could corrupt data flow. The wireless medical telemetry service was established by the FCC in an attempt to provide a less-crowded frequency space for medical devices, but the advantages of accessing the wide array of existing commercial products operating in the WiFi and Bluetooth bands mean that the issue has not gone away.

5. *Telemetry*: The growth of sensors systems for portable monitoring has prompted the FDA to issue new guidance (e.g., wireless medical telemetry risks and recommendations). In addition to the EMI concerns noted earlier, the security of patient data is a significant area of regulatory scrutiny. The Health Insurance Portability and Accountability Act (HIPPA) includes a set of directives that put strict limits on the sharing of patient data. In this context, even the initials of the patient, for instance, cannot be transferred in transmissions

that may be subject to unintended viewing by unauthorized persons. Uninformed device developers have encountered significant hurdles when attempting to commercialize *apps* that cross the boundary into the definition of a medical device because the app collects and transmits sensitive patient data.

6. *Clinical utility*: Strictly speaking, the FDA evaluates medical devices for safety and efficacy. Does the device perform as stated and is it safe for the intended use? A separate question is whether the device performs a necessary or useful medical function. This is a question faced often by medical sensor technology, and although the FDA does not formally comment on *commercial potential*, it is common to have a requirement that the device must demonstrate utility, most commonly as the result of a well-run clinical study. Just as with drug studies, devices must show superior performance to a placebo in order to gain regulatory clearance. Additionally, even if the device performs a useful function, is there an existing, easier means for obtaining the same information? A good rule of thumb for implanted sensors, for instance, is that if you can acquire the same data by taking a blood sample, the implanted sensor won't gain traction. Even in cases such as the development around implanted glucose sensors, thus far the older and accepted glucometers that rely on a blood draw retain a controlling role in the management of diabetic patients. Clinical utility is perhaps the single most underanalyzed aspect of the translational process (Figure 5.1).

What follows is a short list of case examples demonstrating FDA pathway, design considerations, and the process of translation in general.

FIGURE 5.1 The interplay of essential questions that should be addressed even in the earliest stages of medical device/sensor development.

5.3 GLUCOSE SENSING

There are roughly 800,000 type 1 diabetics in the United States and a portion of the estimated 26 million type 2 diabetics who require regular insulin infusions. Insulin, working with serotonin, enables cells to transport glucose through the cell membrane to meet metabolic demand. An insulin injection provides a bolus to the patient, and this is suboptimal in terms of proper serum glucose maintenance. There are current developments aimed at alternative means for introducing insulin, such as inhalers, but the outcome is the same: a large pulse of insulin, which is not how the body can best utilize it. The development of implanted insulin pumps was viewed as a significant advance, as they enable effectively *on demand* supplies of insulin, providing a more natural delivery means. Missing still is a means for feedback that adjusts insulin release based on serum glucose levels. Ergo the vigorous pursuit of an implantable glucose sensor (IGS) that would allow for a closed-loop system with feedback: a true artificial pancreas using only electronic devices.

Most IGS devices that have been proposed envision relatively simple placements under the skin. Yet ultimate removal of and longevity of the sensor are obvious concerns. In terms of *fit* within existing medical practice, an IGS would find fairly ready acceptance. Still, the change in diabetic management that such a device would imply is not to be underestimated. The question of cost-effective manufacturing depends almost entirely on the longevity of the device within the body. Finally, the *burden of inconvenience* test in the case of an IGS would initially seem to be clear-cut. A device that frees a patient from multiple daily blood draws and enables consistent and automated insulin management would be a true advance indeed. But as the various challenges earlier that face a successful IGS are weighed in comparison with cheap, accurate, and reliable modern glucometers (a *simple blood test*), it is easier to understand why it doesn't yet exist in medical practice.

What about FDA regulation of an IGS? There is little question that such devices would face a PMA pathway once sufficient clinical data could be collected. The complexity of the software needed to enable automated means for controlling an insulin pump would draw substantial scrutiny. The inherent risk involved in controlling insulin levels is so high, that the device manufacturer will face a very rigorous clinical trial and approval process. And the FDA reserves the right to perform postmarket surveillance on PMA devices as a means for checking on the device's performance in the field. It is unlikely that any manufacturer, even a large public company, would undertake that level of burden lightly. To highlight the challenge involved, the current so-called CGMs, which are percutaneous sensors that stay in place for 3–7 days, must be independently calibrated using conventional glucometers, and the FDA does not allow them to be used without such calibration. An IGS that still requires frequent calibration with a conventional glucometer would be underwhelming, to say the least, and thus the challenge of FDA approval is significant for these devices.

CGM has dominated the recent literature. Though not strictly based on implantable sensors, in the sense of a true IGS, these devices have nonetheless established themselves as *the best alternative* in pursuit of the goal of continuous feedback to an insulin pump for adaptive insulin regulation. Several clinical trials [6–10] have established the utility of this approach, and the FDA has cleared devices for this purpose. Investigators have started to examine the additional potential benefits of tighter glucose control for patients with comorbid disease. Hermanides et al. [11] used CGM in a study of hyperglycemic myocardial infarction patients, concluding that: "Although a promising tool for in-hospital hyperglycemia therapy, (CGM) needs improvement before continuing to large-scale randomized controlled trials." Klupa et al. [12] evaluated the apparent benefit of CGM for cystic fibrosis patients. Development on new formats for percutaneous glucose sensors continues (e.g., [13–17]).

There is little doubt that CGM enabled by percutaneous sensors has gained ground and produced real benefits for many diabetics. Gough et al. [18] point out that none of these devices has been cleared by the FDA as a primary standard, and they must still be calibrated using conventional glucometers. Using a porcine model, Gough et al. implanted devices using glucose oxidase chemistry and found: "(i) acceptable long-term biocompatibility, assessed after 18-month implant periods; (ii) immobilized enzyme life exceeding 1 year; (iii) battery life exceeding 1 year; (iv) electronic circuitry reliability and telemetry performance; (v) sensor mechanical robustness including long-term maintenance of hermeticity; (vi) stability of the electrochemical detector structure; and (vii) acceptability and tolerance of the animals to the implanted device." This study may be viewed as a proof-of-concept demonstration that the salient technical challenges facing fully IGSs can be met. In addition to amperometric approaches, the use of fluorescent reporters has been explored with some success in laboratory and animal studies [19–26], but translational work is not as advanced.

5.4   CAPSULE ENDOSCOPY

Capsule endoscopy is the term applied to a device with a miniature camera that can be swallowed, allowing for filming inside the gastrointestinal (GI) tract. Though not an implantable sensor, most of the same concerns about biocompatibility, hermeticity, and data collection obtain. The leader in this field is Given Imaging (www.givenimaging.com/en-int/Innovative-Solutions/Capsule-Endoscopy/Pages/default.aspx), an entity that successfully traversed the route from small start-up to public company. The device, called *Pill Cam*, has been in common use for many years and thus no discrete listing of clinical studies using it is tractable (a Pubmed search on capsule endoscopy will be rewarding to the reader). The *Pill Cam* device makes use of a charge-coupled device (CCD) camera and light-emitting diode (LED) light sources. The initial medical need it addressed was the ability to image parts of the small intestine not accessible to other endoscopic tools. The device sends images taken at a preset frequency to a recorder worn by the patient during the transit time of the capsule through his or her body. The physician is then able to review these images in a movie-loop format with some location/position data in addition to anatomical landmark identification. *Pill Cam* received 510(k) clearance with a relatively small human clinical trial. There is no question that the short duration of time in the body as well as the ability to encase the entire device in a biocompatible plastic simplified this process. An interesting question is whether the device would be viewed as substantially equivalent today as it was a decade ago. At very least, it is an interesting case study of the interplay of risk, efficacy, and predicate weighting. Reimbursement for capsule endoscopy was not immediately forthcoming; however, medical demand blossomed, and physician adoption forcefully pushed aside any question that the device would be broadly covered. This product had to, essentially, define a new subfield of gastroenterology, and it had to gain favor with GI professional societies. But it started with a product that is comparably easy to manufacture and deploy and thus serves as a useful case study for those interested in translational medical devices development.

5.5   COCHLEAR IMPLANT

The cochlear implant is a sensor-based technology that evolved over decades, and it was eventually approved as a PMA device. It is an interesting example of how sometimes ideas must wait for technical evolution before becoming practical in medicine. Wilson and Dorman [27] provided an excellent history of the development of the cochlear implant, which the authors rightly called the most successful current neural prosthesis. It is illustrative to review some of the points made by the authors, as they have general applicability. They note that in the early 1980s, many knowledgeable people believed that cochlear implants would provide only modest awareness of ambient sounds and that speech recognition was a highly unlikely outcome. It is fair to say that the evolution in electronics from that time made the available toolkit much more complete for both compact spectral analysis chips and electrode design. Viewed as an opportunity for product development in the early 1980s, even by a large company, one would have concluded without doubt that the task was impossible.

Here is a case where dogged researchers and unrelated consumer and business electronics developments altered the landscape drastically by the mid-1990s. This is an important point, for the intent of the current review is *not* to suggest that some lines of research should be abandoned (e.g., the silicon retina), only that translational goals must be near-term (years, not decades) for any sensible advancements in actual patient-worthy device development. Though cardiac pacing had been around for some time when cochlear implants became a serious possibility, the field of neuromodulation was relatively new, and one may appropriately classify cochlear implants as neuromodulators. One other lesson to emphasize from Wilson and Dorman is that there may be unexpected resistance to adoption of a new implant technology. In the case of the cochlear implant, it was the deaf community itself. Though by no means universal, there was nonetheless some pushback on the idea that deaf children (for whom the success with early intervention is the greatest) must regain hearing to be productive people and citizens. The historical struggle of deaf people to gain respect and proper recognition could understandably lead to a perception that cochlear implants were meant as salvations of some sort, even if this was not at all in the minds of the developers of the technology. Today, there is a generally balanced view that cochlear implant recipients should not be stigmatized, one way or the other and that, at best, the implants are still tools to assist the user, not unlike the corrective action of glasses or hearing aids.

5.6   CURRENT TRENDS IN FDA REGULATIONS

A presentation by Dr. William Maisel, Deputy Center Director for Science and Chief Scientist at the Center for Devices and Radiological Health (CDRH) at MEDCON 2011, provides insight into areas where the FDA expects growth (CDRH is the branch of the FDA that regulates medical devices). His list of emerging trends:

- Robotics
- Miniaturized devices (nanotechnology)
- Combination products
- Sophisticated, computer-related technologies
- Organ replacements and assist devices
- Personalized medicine
- Wireless systems
- Home use

includes several areas wherein sensors, either alone or in a system, will be relevant. He also defined the CDRH innovation initiative:

- Facilitate the development and regulatory evaluation of innovative medical devices.
- Strengthen the US research infrastructure and promote high-quality regulatory science.
- Prepare for and respond to transformative innovative technologies and scientific breakthroughs.

This summary parallels a report (*Future Trends in Medical Device Technologies: A Ten-Year Forecast*: www.fda.gov/downloads/AboutFDA/CentersOffices/CDRH/CDRHReports/UCM238527.pdf) by CDRH in which 15 non-FDA medical device experts were polled about technology developments in the device field. In particular, the committee highlighted photonic and acoustic devices as being *invasiveness-reducing* and sensor technologies in general for detection, diagnosis, and monitoring.

One of the frequent mistakes made by companies during the development of new devices is to avoid talking to the FDA early in the process. This doesn't mean calling the FDA to ask frequent, minor questions, but it is important to talk about strategy before going too far down the development path. For example, a primary determination for devices is whether they present an SR or NSR to the patient. If a device

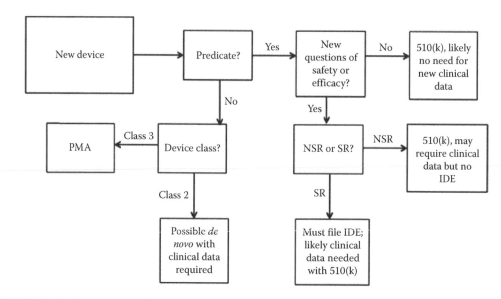

FIGURE 5.2 Representative flowchart for assessing likely disposition of a new medical device. (*Note*: this diagram should not be viewed as being absolute, and the device sponsor should consult with the FDA in order to ascertain the appropriate procedural pathway.)

is found to be SR, then its use in patients must wait until an IDE is approved. If however an IRB agrees that a device is NSR, then clinical studies can proceed under IRB approval only. Most implanted sensor devices will be considered SR, and therefore an IDE approval must precede clinical use.

The new guidelines on the *De Novo* regulatory structure will have a significant potential impact on implanted devices. As explained earlier, this category is meant to deal with devices that don't have obvious predicates (substantial equivalence) but nonetheless are not Class III devices (presenting a major risk to health and well-being of the patient). The ability to claim substantial equivalence for novel devices has become more limited in the last several years, especially in the use of combination predicates (using two or more separate existing devices to match to the new product). Without a strong *De Novo* mechanism, therefore, devices that should rightly have Class II designation are forced into a PMA approval path. Though the *De Novo* system has been around for a number of years, it has been used very infrequently, mostly because of residual confusion as to when it should be applied and a degree of uneven responsiveness within the FDA. This has led to a significant movement of early device sales outside the United States, where often devices that would be Class II in the United States are granted market clearance (e.g., via the CE mark). The US medical device community is looking hopefully at a reenergized *De Novo* process to provide more tractable early access to US patients (Figure 5.2).

5.7 SUMMARY

The concept of personalized medicine has not come to pass as rapidly as expected based on the number of encouraging, perhaps overly naïve, press releases describing new patient-specific devices, often based on advanced sensor technology. However, drawing from the lesson of cochlear implants, there are several factors that come to bear: the maturity of the technical approach, the translatability to the clinic, and the degree to which physicians and patients accept new technology. Several decades of work and refinement were needed before cochlear implants became practical. By comparison, the Pill Cam was adopted fairly rapidly since the technology, small imaging arrays and transmitters, had undergone significant refinement in the commercial electronics sector. Additionally, minimally invasive imaging of the small intestine was an application in search of a solution, and thus physician acceptance was rapid. Finally, the story of continuous glucose monitoring is illustrative in that the *ultimate* sensor system, one that would be fully implantable, proved to be too challenging, and an intermediate approach, limited-duration percutaneous sensor implants, allowed manufacturers, patients, and physicians to provide greater control and convenience sooner. Sensor researchers who wish to impact medicine need to think expansively and consider the trajectory of their work beyond the laboratory/discovery phase. It is important to manage not only the expectations of future patients but also their own predictions and motivations as they embark on what can be vital and fulfilling work.

REFERENCES

1. Bergmann, J. H. M. et al. (2012). Wearable and implantable sensors: The patient's perspective. *Sensors* **12**, 16695–16709.
2. Black, R. D. (2011). Recent advances in translational work on implantable sensors. *IEEE Sensors Journal* **11**(12), 3171–3182.
3. Ledet, E. H. et al. (2012). Implantable sensor technology: From research to clinical practice. *Journal of the American Academy of Orthopaedic Surgeons* **20**(6), 383–392.
4. Inmann, A. and Hodgins, D. eds. (2013). *Implantable Sensor Systems for Medical Applications*. Vol. 52. Philadelphia, PA, Woodhead Pub.
5. Smith, J. J. and Henderson, J. A. (2008). FDA regulation of implantable sensors: Demonstrating safety and effectiveness for marketing in the U.S. *IEEE Sensors Journal* **8**(1), 52–56.
6. Tamborlane, W. V. et al. (2008). Continuous glucose monitoring and intensive treatment of type 1 diabetes. *New England Journal of Medicine* **359**, 1464–1476.
7. Raccah, D. et al. (2009). Incremental value of continuous glucose monitoring when starting pump therapy in patients with poorly controlled type 1 diabetes: The RealTrend study. *Diabetes Care* **32**, 2245–2250.
8. Conget, I. et al. (2011). The SWITCH study (sensing with insulin pump therapy to control HbA(1c)): Design and methods of a randomized controlled crossover trial on sensor-augmented insulin pump efficacy in type 1 diabetes suboptimally controlled with pump therapy. *Diabetes Technology and Therapeutics* **13**, 49–54.
9. Nishida, K. et al. (2009). What is artificial endocrine pancreas? Mechanism and history. *World Journal of Gastroenterology* **15**, 4105–4110.
10. Rubin, R. R. et al. (2011). Crossing the technology divide: Practical strategies for transitioning patients from multiple daily insulin injections to sensor-augmented pump therapy. *The Diabetes Educator* **37**(Suppl 1), 5S–18S; quiz 19S–20S.
11. Hermanides, J. et al. (2010). Sensor-augmented insulin pump therapy to treat hyperglycemia at the coronary care unit: A randomized clinical pilot trial. *Diabetes Technology and Therapeutics* **12**, 537–542.
12. Klupa, T. et al. (2008). Use of sensor-augmented insulin pump in patient with diabetes and cystic fibrosis: Evidence for improvement in metabolic control. *Diabetes Technology and Therapeutics* **10**, 46–49.
13. Castle, J. R. and Ward, W. K. (2010). Amperometric glucose sensors: Sources of error and potential benefit of redundancy. *Journal of Diabetes Science and Technology* **4**, 221–225.
14. Takaoka, H. and Yasuzawa, M. (2010). Fabrication of an implantable fine needle-type glucose sensor using gamma-polyglutamic acid. *Analytical Science* **26**, 551–555.
15. Patel, J. N. et al. (2008). Flexible glucose sensor utilizing multilayer PDMS process. *Conference Proceedings, IEEE Engineering in Medicine and Biology Society* **2008**, 5749–5752.
16. Qiang, L. et al. (2011). Edge-plane microwire electrodes for highly sensitive H(2)O(2) and glucose detection. *Biosensors and Bioelectronics* **26**, 3755–3760.
17. Yehezkeli, O. et al. (2009). Integrated oligoaniline-cross-linked composites of Au nanoparticles/glucose oxidase electrodes: A generic paradigm for electrically contacted enzyme systems. *Chemistry* **15**, 2674–2679.
18. Gough, D. A. et al. (2010). Function of an implanted tissue glucose sensor for more than 1 year in animals. *Science Translational Medicine* **2**, 42ra53.
19. Stein, E. W. et al. (2008). Microscale enzymatic optical biosensors using mass transport limiting nanofilms. 2. Response modulation by varying analyte transport properties. *Analaytical Chemistry* **80**, 1408–1417.

20. Long, R. and McShane, M. (2010). Three-dimensional, multiwavelength Monte Carlo simulations of dermally implantable luminescent sensors. *Journal of Biomedical Optics* **15**, 027011.
21. Singh, S. and McShane, M. (2010). Enhancing the longevity of microparticle-based glucose sensors towards 1 month continuous operation. *Biosensors and Bioelectronics* **25**, 1075–1081.
22. Singh, S. and McShane, M. (2011). Role of porosity in tuning the response range of microsphere-based glucose sensors. *Biosensors and Bioelectronics* **26**, 2478–2483.
23. Chaudhary, A. et al. (2010). Glucose response of dissolved-core alginate microspheres: Towards a continuous glucose biosensor. *Analyst* **135**, 2620–2628.
24. Jayant, R. D. et al. (2011). In vitro and in vivo evaluation of anti-inflammatory agents using nanoengineered alginate carriers: Towards localized implant inflammation suppression. *International Journal of Pharmaceutics* **403**, 268–275.
25. Valdastri, P. et al. (2011). Wireless implantable electronic platform for chronic fluorescent-based biosensors. *IEEE Transactions of Biomedical Engineering* **58**, 1846–1854.
26. Veetil, J. V. et al. (2010). A glucose sensor protein for continuous glucose monitoring. *Biosensors and Bioelectronics* **26**, 1650–1655.
27. Wilson, B. S. and Dorman, M. F. (2008). Interfacing sensors with the nervous system: Lessons from the development and success of the cochlear implant. *IEEE Sensors Journal* **8**, 131–147.

Part II

Chemical and Environmental Sensors

6 Multiregion Surface Plasmon Resonance Fiber-Optic Sensors

Kent B. Pfeifer and Steven M. Thornberg

CONTENTS

6.1   INTRODUCTION

The first report of the effect of surface plasmon resonance (SPR) was made by Wood in 1902 when he noticed the absence of light in narrow spectral bands from a diffraction grating. He also reported in 1935 that these *anomalies*, as he called them, had never yet been properly modeled [1,2]. It was subsequently left to others to derive a model for the origin of the observed effect [3,4]. The significant advance in SPR research that transformed it from an *anomaly* to a valuable analytical technique was the recognition by Kretschmann and Raether that a thin metallic film supported by a dielectric substrate could have surface plasmon waves excited on the ambient exposed metallic surface when optical excitation was from the substrate–metal interface. Thus, the probe light beam would be incident on one side of the metal film, and chemistry could be performed on the other making a useful sensor system [5]. The following discussion is based on our *IEEE Sensors Journal* paper from 2010 and follows a similar development [6].

SPR spectroscopy has now been applied to a number of analytical problems due to its high sensitivity to variations in the electronic nature of a surface. In particular, reports of sensing of chemicals and of biological samples using functionalized surfaces are widespread in the literature [7–13]. The majority of examples employ the open beam optical arrangement known as a Kretschmann configuration that has a single angle of incidence on a metallic layer and a single angle of exit to an optical spectrometer [14]. The SPR is supported in a thin metallic layer and the chemistry of interest takes place on the side of the metal layer opposite the incident and reflected light. In practice, this thin metallic layer is usually deposited on the diagonal facet of a right-angle prism allowing the thin film access to the chemical system while having structural support for the film provided by the prism.

6.2 PLANAR SPR THEORY OVERVIEW

There are a number of excellent references that detail the mathematical development of the conditions for SPR in a planar geometry such as the Kretschmann configuration, and that development will not be repeated in detail here except to summarize the results and to clarify the origin of the important equations of SPR [15–18].

If we consider the geometry shown in Figure 6.1, a wave equation can be solved at the boundary ($x = 0$) for surface waves propagating in the z-direction and decaying exponentially as a function of the x-direction. We will demonstrate that a decaying exponential solution can be derived where the sign convention for the x-direction of positive implies above the material interface and negative implies below the material interface as indicated in Figure 6.1. This assumes that there is no net charge in the system and no external currents. This wave equation (also known as a Helmholtz equation) is formulated for the electric field $\vec{E}$ and the magnetic field $\vec{H}$ as follows:

$$\nabla^2 \vec{E} - \varepsilon\mu \frac{\partial^2 \vec{E}}{\partial t^2} = 0$$

$$\nabla^2 \vec{H} - \varepsilon\mu \frac{\partial^2 \vec{H}}{\partial t^2} = 0$$

(6.1)

In Equation 6.1, ε and μ are the electric permittivity and magnetic permeability of the materials, and t is the time. The general solution to these equations for a surface plasmon wave traveling in the z-direction and confined to the surface of the metal dielectric interface, assuming that they are separable functions, is traveling waves of the following form (β_z is the propagation constant, $\vec{r}$ is the location of the wave, and ω is the natural frequency):

$$\vec{E}(\vec{r}) = \vec{E}(x) e^{j(\beta_z z - \omega t)}$$

$$\vec{H}(\vec{r}) = \vec{H}(x) e^{j(\beta_z z - \omega t)}$$

(6.2)

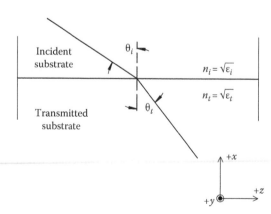

FIGURE 6.1 Diagram illustrating a two-material interface that can support SPR with materials of dielectric constants ε_i and ε_t.

Equation 6.2, substituted into Equation 6.1, leads to the following forms of the wave equation for transverse electric (TE) and transverse magnetic (TM) modes, respectively. A similar result occurs for the TM mode:

$$\frac{1}{\vec{E}(x)}\frac{\partial^2 \vec{E}(x)}{\partial x^2} + \frac{1}{e^{j\beta_z z}}\frac{\partial^2 e^{j\beta_z z}}{\partial z^2} + \varepsilon\mu\omega^2 = 0$$

$$\frac{1}{\vec{H}(x)}\frac{\partial^2 \vec{H}(x)}{\partial x^2} + \frac{1}{e^{j\beta_z z}}\frac{\partial^2 e^{j\beta_z z}}{\partial z^2} + \varepsilon\mu\omega^2 = 0$$

(6.3)

As indicated earlier, Equation 6.3 is a separable partial differential equation that we can solve for the separation constants in the usual way as $\beta_x^2 - \beta_z^2 = 0$ [19,20]. Since we are only concerned with the bounded modes, we solve Equation 6.3 in the y-dimension as Equation 6.4 for the TE mode and similarly for the TM mode (Equation 6.5), where β_x is the separation constant in the partial differential equation:

$$\frac{\partial^2 E_y(x)}{\partial x^2} + \left(\varepsilon\mu\omega^2 - \beta_x^2\right)E_y(x) = 0$$

(6.4)

$$\frac{\partial^2 H_y(x)}{\partial x^2} + \left(\varepsilon\mu\omega^2 - \beta_x^2\right)H_y(x) = 0$$

(6.5)

Equations 6.4 and 6.5 are now ordinary differential equations; the solutions of which, when combined with Equation 6.2, yield

$$E_y(x,y,z,t) = E_{y0}e^{\pm j\sqrt{\left(\varepsilon\mu\omega^2 - \beta_x^2\right)}x}e^{j(\beta_z z - \omega t)}$$

$$H_y(x,y,z,t) = H_{y0}e^{\pm j\sqrt{\left(\varepsilon\mu\omega^2 - \beta_x^2\right)}x}e^{j(\beta_z z - \omega t)}$$

(6.6)

The derivation of Equation 6.1 from Maxwell's equation implies that the wave equation is a consequence of Maxwell's formulas; however, solutions to the wave equation are not necessarily solutions to Maxwell's equations. Thus, the components of the H vector for the TE case and the E vector for the TM case must be found by applying Maxwell's equations to the solutions in Equation 6.6 [21]. In Equation 6.6, the sign in front of the x-dependent exponential is positive for negative x values and is negative for positive x in order to keep the function bounded in x.

Thus, the following field components are found for the TE case:

$$\vec{\nabla}\times\vec{E}_y(x,y,z,t) = \begin{pmatrix} -\dfrac{\partial E_y(x,z,t)}{\partial z} \\ 0 \\ \dfrac{\partial E_y(x,z,t)}{\partial x} \end{pmatrix} = -\mu\begin{pmatrix} \dfrac{\partial H_x}{\partial t} \\ 0 \\ \dfrac{\partial H_z}{\partial t} \end{pmatrix}$$

(6.7)

$$H_x = \int \frac{j\beta_z}{\mu}E_y(x,z,t)dt = \frac{-\beta_z}{\omega\mu}E_y(x,z,t)$$

$$H_z = \mp j\int \frac{\sqrt{\left(\varepsilon\mu\omega^2 - \beta_x^2\right)}}{\mu}E_y(x,z,t)dt = \frac{\pm\sqrt{\left(\varepsilon\mu\omega^2 - \beta_x^2\right)}}{\omega\mu}E_y(x,z,t)$$

Similarly, for the TM case, we get the following:

$$\vec{\nabla} \times \vec{H}_y(x,y,z,t) = \begin{pmatrix} -\dfrac{\partial H_y(x,z,t)}{\partial z} \\ 0 \\ \dfrac{\partial H_y(x,z,t)}{\partial x} \end{pmatrix} = \varepsilon \begin{pmatrix} \dfrac{\partial E_x}{\partial t} \\ 0 \\ \dfrac{\partial E_z}{\partial t} \end{pmatrix}$$

(6.8)

$$E_x = -\int \frac{j\beta_z}{\varepsilon} H_y(x,z,t)\,dt = \frac{\beta_z}{\omega\varepsilon} H_y(x,z,t)$$

$$E_z = \pm j \int \frac{\sqrt{(\omega^2 \varepsilon\mu - \beta_x^2)}}{\varepsilon} H_y(x,z,t)\,dt = \frac{\mp\sqrt{(\omega^2 \varepsilon\mu - \beta_x^2)}}{\omega\varepsilon} H_y(x,z,t)$$

By applying the continuous tangential component boundary conditions $H_{zi} = H_{zt}$, $E_{zi} = E_{zt}$, $H_{yi} = H_{yt}$, and $E_{yi} = E_{yt}$ at $x = 0$, where i implies the incident medium and t implies the transmitted medium, we can write the following set of simultaneous equations:

$$\frac{\sqrt{(\varepsilon_i \mu \omega^2 - \beta_x^2)}}{\omega\mu} E_{yi}(x) - \frac{\sqrt{(\varepsilon_t \mu \omega^2 - \beta_x^2)}}{\omega\mu} E_{yt}(x) = 0$$

(6.9)

$$E_{yi}(x) - E_{yt}(x) = 0$$

$$\frac{\sqrt{(\varepsilon_i \mu \omega^2 - \beta_x^2)}}{\omega\varepsilon_i} H_{yi}(x) - \frac{\sqrt{(\varepsilon_t \mu \omega^2 - \beta_x^2)}}{\omega\varepsilon_t} H_{yt}(x) = 0$$

(6.10)

$$H_{yi}(x) - H_{yt}(x) = 0$$

In order to solve for the eigenvalue β_x, we must set the determinate of Equations 6.9 and 6.10 equal to zero. The solution of Equation 6.9 leads to the conclusion that $\varepsilon_i = \varepsilon_t$. This is a nonsensical result and implies that there are no bounded modes for the TE case, which has been predicted and demonstrated experimentally [22].

The interesting result comes from Equation 6.10, which leads to a propagation constant for the surface plasmon waves at the surface (Equation 6.11). Since we have a nonzero propagation constant for the x-direction, we conclude that SPR can only be excited by light polarized in the TM mode:

$$\frac{\varepsilon_i \mu_0 \omega^2 - \beta_x^2}{\varepsilon_i^2} = \frac{\varepsilon_t \mu_0 \omega^2 - \beta_x^2}{\varepsilon_t^2}$$

$$\beta_x = \omega \sqrt{\frac{\mu_0 \varepsilon_i \varepsilon_t}{\varepsilon_i + \varepsilon_t}}$$

(6.11)

Thus, we have shown that there exists the potential for a bound electromagnetic mode that is confined to the surface according to Equation 6.6. It has a propagation constant that is dependent on the material properties of the two substrates according to Equation 6.11 and can only be excited by TM polarized light.

Equation 6.11 demonstrates that the surface plasmon wave will have a propagation constant that is complex since the transmitting material is a metal. Examination of Equation 6.6 illustrates the relationship between the propagation constant and the material properties of the system. If the term $\sqrt{\varepsilon_i \mu \omega^2 - \beta_x^2}$ is complex, then the wave is an exponentially decreasing function of x implying a confined mode at the interface between the two materials. This occurs when $|\varepsilon_i| > \varepsilon_j$. This is the surface plasmon mode. Similarly, if $\sqrt{\varepsilon_i \mu \omega^2 - \beta_x^2}$ is real, then the wave is an evanescent mode into the metal which is a nonpropagating mode [23].

In order to use SPR as a sensor, a three-layer medium is necessary where the incident medium (dielectric) is labeled subscript 3, the metallic medium is labeled subscript 1, and the surrounding media outside the metal is labeled subscript 2. The reflectivity of the three-layer structure can be found by calculating the film scattering matrix (S) as follows [24]:

$$S = \frac{1}{\tau_{31}}\begin{pmatrix} 1 & \rho_{31} \\ \rho_{31} & 1 \end{pmatrix}\begin{pmatrix} e^{-j\gamma_1} & 0 \\ 0 & e^{j\gamma_1} \end{pmatrix}\frac{1}{\tau_{12}}\begin{pmatrix} 1 & \rho_{12} \\ \rho_{12} & 1 \end{pmatrix} = \frac{1}{\tau_{31}\tau_{12}}\begin{pmatrix} e^{-j\gamma_1} + \rho_{12}\rho_{31}e^{j\beta_1} & \rho_{12}e^{-j\gamma_1} + \rho_{31}e^{j\gamma_1} \\ \rho_{12}e^{j\gamma_1} + \rho_{31}e^{-j\beta_1} & e^{j\gamma_1} + \rho_{12}\rho_{31}e^{-j\gamma_1} \end{pmatrix}$$

(6.12)

In Equation 6.12, ρ_{ij} and τ_{ij} are the TM Fresnel amplitude reflection coefficient and amplitude transmission coefficients, respectively, for the transition from layer i to layer j. For completeness, these equations are reproduced here where ε_j is the permittivity of the layer material, θ_i and θ_j are the incident and transmitted angles respectively d is the thickness of the metal film, and λ is the wavelength of the incident light [25–27]:

$$\tau_{ij} = \frac{2\sqrt{\varepsilon_i}\cos\theta_i}{\sqrt{\varepsilon_j}\cos\theta_i + \sqrt{\varepsilon_i}\cos\theta_j}$$

$$\rho_{ij} = \frac{\sqrt{\varepsilon_j}\cos\theta_i - \sqrt{\varepsilon_i}\cos\theta_j}{\sqrt{\varepsilon_j}\cos\theta_i + \sqrt{\varepsilon_i}\cos\theta_j}$$

(6.13)

$$\gamma_j = \frac{2\pi}{\lambda}\sqrt{\varepsilon_j}d\cos\theta_j$$

(6.14)

The reflectance (R) is then found as the following:

$$R_{31} = \left|\frac{S_{12}}{S_{22}}\right|^2 = \left|\frac{\rho_{31} + \rho_{12}e^{-j2\gamma_1}}{1 + \rho_{12}\rho_{31}e^{-j2\gamma_1}}\right|^2$$

(6.15)

Finally, the resonance condition is found by recognizing that the component of the propagation constant vector in the incident medium that is parallel to the surface must be equal to the surface plasmon propagation constant found in Equation 6.11. Thus, the resonance condition for a planar surface plasmon geometry is the following [28]:

$$\frac{\omega n_i}{c}\sin\theta_i = \omega\sqrt{\frac{\mu_0\varepsilon_1\varepsilon_2}{\varepsilon_1 + \varepsilon_2}}$$

(6.16)

$$\sin\theta_i = \frac{c}{n_i}\sqrt{\frac{\mu_0\varepsilon_1\varepsilon_2}{\varepsilon_1 + \varepsilon_2}}$$

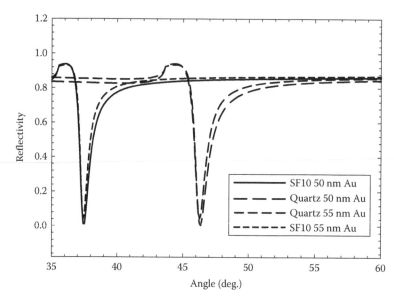

FIGURE 6.2 Plot of the normalized reflectivity from a Kretschmann-configured SPR experiment with two different thicknesses of Au (n_{Au} = 0.1726 + j3.4218) and two different dielectric substrates (n_{SF10} = 1.723 and n_{quartz} = 1.515). Note that the position of the resonance is strongly influenced by the thickness of the Au film but is not strongly dependent on the index of refraction of the dielectric material.

In the aforementioned development, we have derived the basic notions of SPR in a planar system. Namely, that the angle of resonance is related to the material properties of the system but is independent of the frequency of excitation except the variation of permittivity due to dispersion (Equation 6.16). Thus, any shift in the resonance angle is a function of the dispersion of the light in the medium only and not the excitation frequency. Second, the propagation constant of surface plasmons is found as in Equation 6.11 and is only derived from TM polarized electromagnetic energy implying that TE mode light will not lead to SPR. This has been extensively confirmed using the model developed earlier for a planar system. Finally, the reflectivity for a three-layer system can be found from Equation 6.15.

Figure 6.2 is a plot of the modeled response for a three-layer SPR experiment in the Kretschmann geometry for two different substrate indices of refraction and two different thicknesses of a Au film with a complex index of refraction as noted. These conditions result in a narrow resonance peak that is very sensitive to the thickness of the Au layer. Thus, for SPR measurements in a laboratory setting where the geometry of the optical system can be conveniently constructed, the Kretschmann configuration is optimum due to the narrow resonances measured.

6.3 OPTICAL FIBER RATIONALE

However, for monitoring of sealed systems in field environments where internal volume and external access is limited by the general function of the system, the Kretschmann geometry is often difficult or impossible to implement. For such systems, optical-fiber-based SPR systems have been applied. An optical-fiber-based SPR sensor is normally made from a single fiber that has its cladding etched down to the optical core of the fiber over a known section of length. The exposed core is then coated with an SPR-supporting metal and the sensor is used in a single-pass configuration [29–32]. A fundamental difference between an optical fiber and the common Kretschmann configuration is that in a multimode optical fiber, the allowed incident angles are a continuum from the critical angle of the fiber as set by the numerical

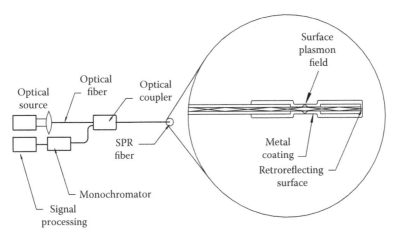

FIGURE 6.3 Schematic diagram of the SPR sensor showing the optical system. Broadband light is injected into the fiber that is then sent to the SPR end via a three-port coupler. The light is passed through the SPR section of the fiber and reflected back by a retroreflecting end back through the coupler and is detected by the monochromator.

aperture to $\pi/2$ from normal. As has been demonstrated both experimentally and theoretically, this has the effect of significantly broadening the SPR peak [30,31].

A single-fiber version of the fiber-optic SPR sensor that employs a retroreflecting metal film on the end of a multimode fiber similar in geometry (Figure 6.3) to the intensity-based *micromirror* sensors documented in the literature has been demonstrated [33–36]. While the geometry is similar to the previous work, the previous work only sampled the reflectance of the reflecting mirror on the fiber. In our system, light is injected into the fiber that then travels by way of the optical coupler to the SPR-coated fiber where it interacts with the sensing film. The light is then retroreflected back through the fiber to the coupler where half returns to the light source and is lost and the other half is directed via the second leg of the coupler to the monochromator where it is analyzed. The reflecting end serves the purpose of returning the light back through the fiber to the monochromator but does not participate in the sensing since the film is optically thick (100 nm) and each ray only interacts with the end film once as compared to multiple times for the axial coatings. In addition, the end is made from an inert noble metal such as Au or overcoated with a sealant. The retroreflecting coating was chosen to be optically thick to minimize any signal due to a chemical change in reflectivity of the retroreflector. The calculated skin depth for bulk Au at 550 nm is less than 10 nm indicating that a 100 nm film is at least a factor of 10 greater. In addition, Butler et al. [35] reported minimal reflectivity changes due to chemical reactions on the surface of optically thick Au films deposited on optical fibers.

Since the fiber is single ended, it becomes practical to use this sensor to monitor the atmospheres of complex subsystems for contamination and aging effects. For example, deterioration of some important compounds found in packaging can lead to production of sulfur compounds that can corrode connectors and lead to electronic failures. In addition, batteries, transformers, and thermal degradation of packaging can evolve H_2 leading to metallic embrittlement, water formation, and explosive environments. Further, leakage of seals can allow moisture to infiltrate and condense on critical components contributing to corrosion and other failure mechanisms.

6.4 FIBER-OPTIC SPR THEORY

A numerical model was used to determine the approximate thickness for the SPR films required. These calculations were accomplished using a 600 μm core fused quartz fiber and assuming that the index of refraction was constant over the spectral range of 250–1000 nm at a value of $n_q \cong 1.46$

and a numerical aperture of $NA = 0.22$. Comparison to the fiber data sheet confirms the validity of this constant assumption to within 3% over this wavelength range. These quantities allow the calculation of the critical angle in the fiber that establishes the lower limit of integration in the model as follows [37]:

$$\theta_c = \frac{\pi}{2} - \sin^{-1}\left(\frac{NA}{n_q}\right)$$

(6.17)

For a single-pass system with no retroreflector, the reflectance (R) as a function of wavelength (λ) of the SPR surface is given where $p(\theta)$ is the incident power on the film at the angle θ, $r^N(\theta)$ is the reflectance for each individual ray at the angle θ, N is the number of reflections from the surface as determined by the incident angle and the length of the SPR-supporting metal (L), and λ is the wavelength of the light:

$$R(\lambda) = \frac{\int_{\theta_c}^{\pi/2} p(\theta) r^N(\theta) d\theta}{\int_{\theta_c}^{\pi/2} p(\theta) d\theta}$$

(6.18)

$$N = \frac{L}{d_2 \tan(\theta)}$$

A complete description of the method is outlined in detail in Xu et al. [30]. In Equation 6.18, the argument of the numerator integral is the power that remains after multiple reflections of any single TM ray at any single angle θ. Recognizing that this is a continuum of rays, integration was performed on the guided rays from the critical angle θ_c below which the light will be lost to fiber cladding leakage to an angle of incidence of $\pi/2$ (grazing incidence with the core cladding interface). A Lambertian model was employed for our light source to model the tungsten-halogen lamp used to illuminate our sample fibers. A Lambertian source has the property of having a uniform radiance that is independent of the angle into which the radiation is directed [38]. Thus, the optical power can be expressed as the function [30]

$$p(\theta) = p_o n_q^2 \sin\theta \cos\theta$$

(6.19)

In Equation 6.19, p_o is the nominal power from the light source. Since the power function appears in both the numerator and the denominator of Equation 6.18, the absolute magnitude of the power is normalized leaving only the ratio of the reflected light to the incident light or the reflectance of the fiber-film system.

A fundamental concept of SPR is that SPR can only be varied for the TM polarized light as a function of angle; thus, the transmission through a coated section of fiber must be expressed as

$$T(\lambda) = \frac{\Phi_{TE} + R(\theta)\Phi_{TM}}{\Phi_{total}}$$

$$= \frac{1}{2} + \frac{R(\theta)}{2}$$

(6.20)

In Equation 6.20, Φ is the transmitted optical power in each of the TM and TE modes. Light is assumed to be evenly distributed between the two modes resulting in half the transmission coming from the TE mode that is undisturbed by SPR and the rest coming from the total reflection from the coating divided by two since only half of the original incident light can participate in SPR.

A double-pass optical configuration was employed as shown in Figure 6.3, which requires a metallic film to be deposited on the end facet of the fiber to retroreflect the light back toward the detector. Thus, an additional correction to $T(\lambda)$ must be made to compensate for the spectral influence of the retroreflector on the incident light. Thus, again assuming that half the light is TE mode and does not contribute to SPR but is modified by the retroreflector according to Fresnel's equations and the other half of the light is modified both by the SPR and the retroreflecting surface, Equations 6.18 and 6.20 can be modified as follows [39]:

$$T(\lambda) = \frac{1}{2} \frac{\int_{\theta_c}^{\pi/2} p(\theta) R_{TE}(\theta) d\theta}{\int_{\theta_c}^{\pi/2} p(\theta) d\theta} + \frac{1}{2} \frac{\int_{\theta_c}^{\pi/2} p(\theta) R_{TM}(\theta) d\theta}{\int_{\theta_c}^{\pi/2} p(\theta) d\theta} \tag{6.21}$$

where R_{TE} is defined as [40]

$$R_{TE}(\theta) = \left(\frac{n_2 \cos\theta_i - n_{rr} \cos\theta_t}{n_2 \cos\theta_i + n_{rr} \cos\theta_t} \right)^2 \times \left(\frac{n_2 \cos\theta - n_1 \cos\theta'}{n_2 \cos\theta + n_1 \cos\theta'} \right)^{4N} \tag{6.22}$$

and R_{TM} is defined as

$$R_{TM}(\theta) = \left(\frac{n_{rr} \cos\theta_i - n_2 \cos\theta_t}{n_{rr} \cos\theta_i + n_2 \cos\theta_t} \right)^2 r^{2N}(\theta) \tag{6.23}$$

The angles in Equations 6.22 and 6.23 are defined in Figure 6.4 and Equation 6.24. In Equation 6.22, the left-hand term is Fresnel's equation for light polarized such that the electric field is perpendicular to the plane of incidence or TE as defined earlier for the single reflection off of the retroreflector. The right-hand term is the same Fresnel equation configured for the multiple reflections off of the fiber core to SPR film interface again in TE mode. Note this term is raised to the *4N* rather than *2N* power because the length of the SPR-supporting metal is twice as long due to the double-pass nature of the fiber architecture and there are *N* individual reflections in each direction. The right-hand term in Equation 6.23 is the SPR term derived by Xu et al. (Equation 6.25) and the left-hand term is Fresnel's equation for reflection off of the retroreflector for TM mode light as a function of incident angle [30]. This term is raised to the *2N* rather than *N* power as in Equation 6.18 again because of the double-pass nature of the fiber architecture. The relationships between the various angles and the integration angle are as follows:

$$\theta_i = \frac{\pi}{2} - \theta$$

$$\theta' = \sin^{-1}\left(\frac{n_2}{n_1} \sin\theta \right) \tag{6.24}$$

$$\theta_t = \sin^{-1}\left(\frac{n_2}{n_{rr}} \cos\theta \right)$$

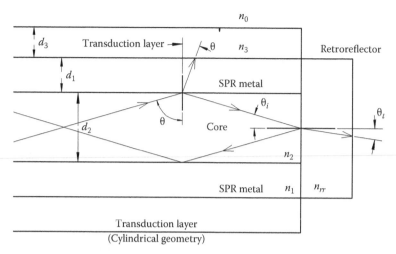

FIGURE 6.4 Diagram of geometry used to derive equations where the subscript 0 represents the outside atmosphere, 1 is the SPR-supporting metal layer, 2 is the core, and 3 is an optional second transduction layer. (Used from Pfeifer, K.B. and Thornberg, S.M., *IEEE Sensors J.*, 10(8), 1360, 2010. With permission.)

The reflectance function due to SPR ($r(\theta)$) is given by the following [30]:

$$r\left(\theta\right) = \left|\frac{r_{21} + r_{130}e^{2ik_1d_1}}{1 + r_{21}r_{130}e^{2ik_1d_1}}\right|^2 \tag{6.25}$$

where

$$r_{130} = \frac{z_{10} - iz_{43}\tan\left(k_3d_3\right)}{n_{10} - in_{43}\tan\left(k_3d_3\right)}$$

$$r_{12} = \frac{z_{21}}{n_{21}}$$

$$z_{l,m} = k_l\varepsilon_m - k_m\varepsilon_l$$

$$n_{l,m} = k_l\varepsilon_m + k_m\varepsilon_l \tag{6.26}$$

$$k_l = \left[\varepsilon_l\left(\frac{2\pi}{\lambda}\right)^2 - \left(\frac{2\pi}{\lambda}n_2\sin\theta\right)^2\right]^{1/2}, \quad l = 1,2,3$$

$$\varepsilon_4 = \frac{\varepsilon_0\varepsilon_1}{\varepsilon_3}, \quad k_4 = \frac{k_0k_1}{k_3}$$

and the subscripts represent the layers as defined in Figure 6.4.

6.5 MODELING RESULTS

A Mathcad* program was constructed to solve for the transmission through the optical system and vary the thickness parameters of the fibers based on the earlier formalism. Results were applied to choosing various thicknesses of metal films in order to locate the SPR minimum and estimate the

---

* PTC, 140 Kendrick St., Needham, MA 02494, United States.

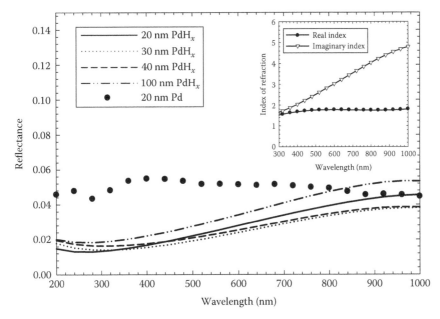

FIGURE 6.5 Plot of calculated reflectance for several thicknesses of Pd film exposed to H_2 to form PdH_x as calculated and compared to a measured unexposed 12.5 mm length, 20 nm thick Pd film (●) in the SPR geometry. Data indicate that good candidates for fabrication and testing due to their broad absorption resonance being predicted in the 250–400 nm range are films with thicknesses of 20–40 nm. The real and imaginary parts of the index of refraction were obtained from the literature for PdH_x and are plotted in the inset graph [41,42]. (Used from Pfeifer, K.B. and Thornberg, S.M., *IEEE Sensors J.*, 10(8), 1360, 2010. With permission.)

response to various exposures. The results from the models, in general, do not exactly describe the measured behavior of the films tested; however, the general trends are described and employed to direct the sensor design.

Figure 6.5 is a plot of the calculated response of Pd to exposure to H_2 as given by the model and literature values of refractive index [41,42]. The values of refractive index for PdH_x were the values reported by von Rottkay et al. [42] and are for thin films of Pd exposed to an atmosphere of 10^5 Pa of H_2. This concentration is significantly higher than any measurements that were made on our fiber structures; however, they represent an upward bound on the refractive index values as a function of hydride state. These simulations indicated that significant variations between pure Pd and PdH_x should be expected in the 200–400 nm range, which enables the detection of H_2. Calculations were done with 600 μm fiber with a numerical aperture of $NA = 0.22$ and various thicknesses of PdH_x. These types of calculations were employed to direct the choice of metal thicknesses for the experimental studies but do not exactly model the position of the minimum response. In the example of Figure 6.5, various thicknesses of films were simulated in order to locate thicknesses that had substantial resonance peaks in the visible spectrum and could be easily fabricated. Similar studies were performed on Au/SiO_2 and Ag films.

6.6 EXPERIMENTAL

6.6.1 PREPARATION OF THE FIBER

Following the ray optics model described by Xu et al. [30], the metal thickness was calculated for monitoring of H_2 using deposited Pd films, H_2S using Ag films, and moisture using a cover layer of SiO_2 on a Au film. In addition, the core diameter of the optical fiber was chosen using this theoretical model.

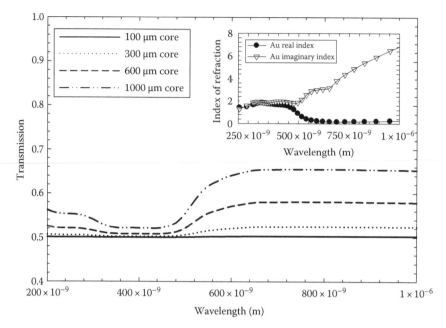

FIGURE 6.6 Plot of optical transmission for a 20 nm Au-coated fiber as a function of wavelength for several different core diameters using the theory of Xu et al. [30]. Plot illustrates that for SPR, the larger diameter gives a higher signal-to-noise ratio SPR peak allowing better wavelength discrimination. (Used from Pfeifer, K.B. and Thornberg, S.M., *IEEE Sensors J.*, 10(8), 1360, 2010. With permission.)

Figure 6.6 is a simulation for a Au film on a fiber of various core diameters from 100 to 1000 μm with the real and complex index of refraction components of Au plotted as a function of wavelength in the inset. The trend is for higher throughput and deeper SPR minima as the core diameter becomes larger. However, in practice, 600 μm couplers are the largest conveniently available, and therefore, fused quartz fiber of this diameter was chosen (Ocean Optics Fiber-600-UV*) for the experiment.

Sensing is accomplished by first etching off the cladding material and then coating a cylindrical section of the fiber with an SPR-supporting metal film as illustrated in Figure 6.3. This geometry has the distinct advantage of doubling the sensitivity per unit length of the SPR film since the light travels through the sensitive section twice. In principle, multiple regions could be etched to the core and coated with different metals to vary the location of the SPR peak and allow sensing of several compounds using the same probe fiber.

Based on the results from the numerical simulations, a series of 600 μm core fibers were fabricated and coated with SPR-supporting metals. The general process was to first cleave sections of optical fiber approximately 20 cm in length and then pyrolyze the last ~3 cm of polyimide coating to allow the etching chemical access to the fused silica. The pyrolysis was accomplished by inserting the optical fiber in an alumina cylinder with a coil of nichrome wire wrapped around the cylinder and then using a power supply to heat the nichrome to incandescence. The fiber was then cleaned and etched by dipping the last 1.25 cm into a solution of 5 g NH_4HF_2 dissolved in 50 mL deionized H_2O. The initial outside diameter of the fiber is 660 μm and nominal core diameter is 600 μm; thus, the fibers were etched until they measured an outside diameter on the order of 590 nm to guarantee that the cladding had been completely removed and the core material alone was exposed. The measurement was accomplished using a digital micrometer. The etching of the cladding required time intervals on the order of 30 min, but the etching was stopped based on the micrometer measurement rather

---

* Ocean Optics, 830 Douglas Ave., Dunedin, FL 34698, United States.

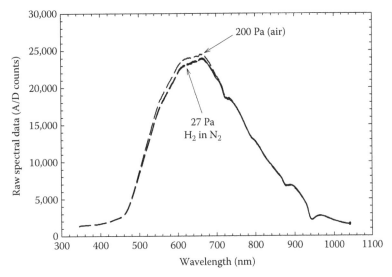

FIGURE 6.7 Plot of the raw spectral response from a 30 nm Pd-coated fiber under conditions of near vacuum pumped from an air background and exposed to ~27 Pa of H_2. (Used from K.B. Pfeifer and Thornberg, S.M., *IEEE Sensors J.*, 10(8), 1360, 2010. With permission.)

than the elapsed time. In addition, scanning electron microscope measurements of the surface revealed that variations in diameter of <250 nm were observed over the etched lengths.

Next, the fibers were rinsed in deionized water and then placed in a vacuum compatible rotating fixture. The fibers were then placed in an electron beam deposition system and pumped down to a base pressure of below 100 μPa. Next, they were coated with the respective SPR metal at rates of approximately 1 Å/s for Pd and Ag and a rate of 4 Å/s for Au. The fibers were rotated at 3 RPM with a substrate temperature of ~24°C until the quartz crystal thickness monitor indicated that the desired thickness was achieved. Finally, the fibers were masked from the source except for the cleaved fiber end and the retroreflecting coating was applied. In all cases, the retroreflecting coating is a 100 nm Au film deposited using the same parameters as the SPR Au films. The thicknesses were verified using a stylus profilometer and found to be within ±20% of the expected values.

The responses of the fibers were then measured using the system depicted in Figure 6.3 and the data were analyzed. The system of Figure 6.3 consists of a tungsten-halogen light source (Ocean Optics LS-1) connected to a three-port optical coupler (Ocean Optics 600 μm bifurcated fiber assembly). The spectrometer is an Ocean Optics USB-2000 system interfaced to a PC for data collection. An example of the raw data is shown in Figure 6.7, which illustrates the change in the spectral response as a function of exposure. The film in Figure 6.7 is a 30 nm Pd film deposited on ~1.25 cm of exposed fiber core. The plot shows the spectrum of an unexposed film in vacuum (200 Pa air) and the spectrum of a film hydrated in a partial pressure of ~27 Pa of H_2 in N_2. Since this is a large concentration, the effect is visible in the raw spectral data. However, in most cases, the effect is small and must be extracted using numerical methods written in MATLAB.®*

6.6.2 MATLAB® ALGORITHMS

The MATLAB algorithm employed takes the spectrum from the unexposed fiber and averages several scans. This becomes the reference or background spectrum. All the scans are then normalized such that the maximum value is fixed at a value of unity. By doing this, effects such as light source aging, connector loss, and other drift mechanisms in the optical system are eliminated. This is key

---

* The MathWorks, Inc., 3 Apple Hill Drive, Natick, MA 01760–2098, United States.

to application of this technique for long-term unpowered applications where it is neither practical nor desirable to continuously measure the spectrum. The rest of the data is then normalized to the beginning value and plotted as a function of time to look for trends in the spectral response of the fiber.

6.6.3 Gas Sample Generation

Vapor generation in the experiment was accomplished by constructing a crossed tubing arrangement with the fiber inserted into one leg and sealed using a solid Teflon compression ferrule. The ferrule was drilled to accommodate the fiber using a wire drill that was slightly larger than the fiber diameter. By placing the fiber in the ferrule and compressing the ferrule into the fitting, a gas-tight seal was achieved. The other legs of the cross arrangement were connected to (1) a venturi pump for evacuation of the volume, (2) a thermocouple gauge for measurement of the pressures, and (3) a length of tubing connected to a sample gas source via a sealed bellows valve.

The measurements were made by first evacuating the volume and then back filling the volume with standard gas mixtures while monitoring the partial pressure of the volume with the thermocouple gauge. Our standards were 1.00% H_2 in N_2 and 1.09 ppm H_2S in N_2. Moisture samples were generated in the laboratory using a Thunder Scientific Model 3900 two-pressure, two-temperature, low-humidity generator* rather than a gas bottle and the output was connected in a flow mode past the fiber end.

6.7 RESULTS

6.7.1 Pd/H_2

Typical data for the Pd fiber is plotted in Figure 6.8, which illustrates the response of a 40 nm Pd film with a 100 nm Au retroreflecting film exposed to various partial pressures of H_2 in N_2. The data at each wavelength has been normalized and then compared to its initial value. The data illustrate that at partial pressures below 27 Pa, the response of the sensor is linear and the sign of the response is

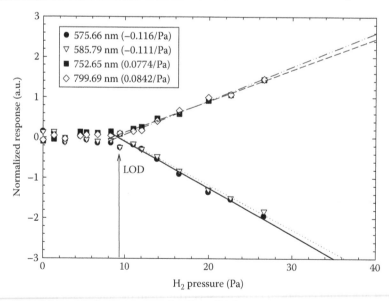

FIGURE 6.8   Plot of the normalized response of a 40 nm Pd film with a 100 nm Au retroreflector as a function of H_2 partial pressure at several wavelengths indicating that the sign of the response is wavelength dependent. Limit of detection is on the order of 9 Pa of H_2. (Used from Pfeifer, K.B. and Thornberg, S.M., *IEEE Sensors J.*, 10(8), 1360, 2010. With permission.)

---

* Thunder Scientific Corporation, 623 Wyoming Blvd. SE, Albuquerque, NM 87123-3198, United States.

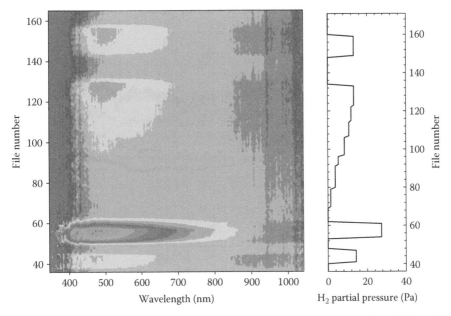

FIGURE 6.9 Plot of sensor response as a function of wavelength and file number plotted along with the H_2 partial pressure. Data illustrate that the sensor is reversible and has a strong dependence on wavelength. Each separate file was acquired on a 30 s interval. (Used from Pfeifer, K.B. and Thornberg, S.M., *IEEE Sensors J.*, 10(8), 1360, 2010. With permission.)

wavelength dependent. A linear response to H_2 at low concentrations has been reported previously for an SPR experiment in a Kretschmann configuration with a 633 nm light source [43]. It has been demonstrated that the linear trend continues over a sampling of wavelengths, but the absolute magnitude changes as a function of wavelength as well as the sign of the response, making it suitable for multivariate spectral analysis. As illustrated in the graph, changes are no longer observable as a function of concentration below about 9 Pa corresponding to a signal-to-noise ratio of unity in the system. The predominant noise sources in the system are from the spectrometer. The spectrometer was set to average 20 scans with an integration interval of 100 ms.

Palladium hydride (PdH_υ) forms for low partial pressures of H_2 in the α-phase. Transformation from the α-phase to the of β-phase of PdH_υ occurs on exposure to higher concentrations of H_2 (2000–2700 Pa) where $\upsilon \sim 0.03$ and is undesirable since it causes a 3.5% increase in the lattice spacing and will result in a mechanical failure of the Pd film on the fiber [44]. Our application is concerned with trace concentrations of H_2 well below the β-phase; hence, concentrations below 27 Pa of H_2 partial pressure were tested. PdH_υ formation is linear with partial pressure in this range at a constant temperature [45].

Figure 6.9 is a plot of the measured response of a 40 nm Pd film exposed to H_2 plotted as a function of wavelength and as a function of file number. Each file was acquired on a 30 s interval implying that the entire scan occurred over approximately 74 min. Results demonstrate that the measurement is strongly a function of wavelength with the maximum response wavelength in the same general wavelength region as was calculated using the model of Figure 6.5. In addition, the response is reversible and indicates the feasibility of using this technique for measurement of H_2 in a sealed environment.

6.7.2 Ag/H₂S

Similar measurements have been conducted using Ag films exposed to 0.04 Pa H_2S in N_2 for extended intervals of 7 days that demonstrate a nonreversible SPR response at wavelengths starting around 450 nm. This is about 100 nm shorter wavelength than the Pd examples. These experiments

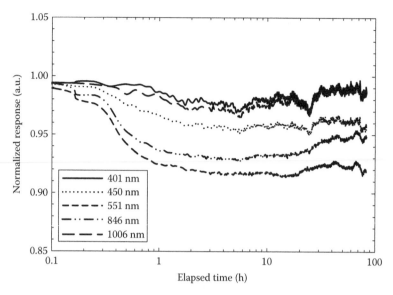

FIGURE 6.10 Plot of the normalized response for 401, 450, 551, 846, and 1006 nm for a 40 nm Ag SPR film on the fiber exposed to 0.04 Pa of H_2S in N_2. Data illustrate that response is complete after approximately 10 h of exposure and is wavelength dependent. (Used from Pfeifer, K.B. and Thornberg, S.M., *IEEE Sensors J.*, 10(8), 1360, 2010. With permission.)

were performed on the same 600 μm fiber coated with similar thicknesses of Ag. Figure 6.10 is a plot of the response of five separate wavelengths as a function of time on exposure to 0.04 Pa H_2S in N_2. Data illustrate that the response is nominally complete after approximately 10 h and is strongly wavelength dependent as expected. Data at 401 and 1006 nm illustrate very little response, while the response in the 450–550 nm region of the spectrum is significantly larger. In the figure, H_2S is introduced approximately 0.3 h after beginning the data collection. We believe that the majority of this signal is due to SPR interaction with the coatings on the walls of the fiber and not due to the interaction with the sulfur bonding to the retroreflecting Au coating on the end of the fiber. This is because the 100 nm thick Au coating is optically thick at visible wavelengths and the light would not interact with the exposed surface of the Au film [35].

6.7.3 SiO₂/H₂O

An additional system consisting of 20 nm of Au deposited on a fiber with a 20 nm cover layer of SiO_2 was used to monitor moisture changes in atmospheres with dew/frost points of 10°C to −70°C in N_2. The experimental results at around 550–600 nm show the most dramatic change in this moisture range and indicate that good discrimination between the spectra at −70°C and −10°C allows SPR measurements of moisture ingress in sealed systems using this SPR geometry. Example data illustrating the increase in dew point from −10°C to +10°C are shown in Figure 6.11.

Figure 6.11 is the normalized response at five different wavelengths from 415 to 701 nm and illustrates that the films are sensitive to moisture at all the wavelengths plotted, but the sensitivity is greater in the range of 455–557 nm.

6.8 MULTICOATING SPR FIBER

The ability to measure several contaminants of interest and to discriminate between them by monitoring a series of key wavelengths implies that a single fiber with multiple regions etched and coated with different metals could be used to produce a multicomponent sensor (Figure 6.12). Such a sensor

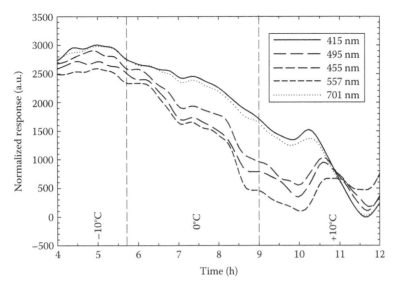

FIGURE 6.11 Plot of response of a 20 nm Au film overcoated with a 20 nm SiO$_2$ film for moisture monitoring. Data illustrate the normalized response of five different wavelengths as a function of dew/frost point from −10°C to +10°C at atmospheric pressure. Data illustrate that wavelengths over a wideband respond to changes in moisture level but wavelengths around between 455 and 557 nm respond with greater sensitivity than wavelengths away from the center of the band (415 and 701 nm). Each step of the moisture generator was programmed to remain constant for 4.5 h. As is observed in the 0°C zone, the sensor response was slower than this and therefore was not stable when the generator changed to a dew point of +10°C resulting in the constant downward drift of the response. (Used from Pfeifer, K.B. and Thornberg, S.M., *IEEE Sensors J.*, 10(8), 1360, 2010. With permission.)

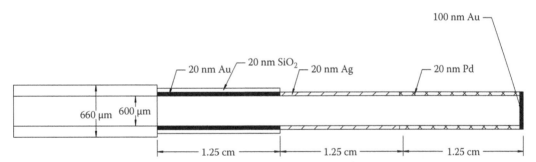

FIGURE 6.12 Diagram of three-section SPR sensor showing Au, Ag, and Pd regions on a 600 μm optical fiber core. (Used from Pfeifer, K.B. and Thornberg, S.M., *IEEE Sens. J.*, 10(8), 1360, 2010. With permission.)

has been demonstrated that was coated with a 12.5 mm length, 20 nm Pd film followed by a 12.5 mm length 20 nm Ag film followed by a 12.5 mm length 20 nm Au/20 nm SiO$_2$ film stack for detection of H$_2$, sulfur compounds, and moisture, respectively (Figure 6.12). Data from these tests were first acquired using the unexposed fiber to detect 27 Pa H$_2$ (Figure 6.13a) and then exposed to 0.04 Pa of H$_2$S (Figure 6.13b). The Ag section of the fiber was then monitored until the nonreversible reaction was complete. The fiber was then reexposed to 27 Pa H$_2$ with the permanent H$_2$S/Ag response normalized into the data to determine if the sulfur atmosphere poisoned the Pd film (Figure 6.13c). The data indicate that the resonance peak that appears on exposure to H$_2$ is essentially unchanged compared to Figure 6.13a in shape and magnitude after the exposure of the Ag film to H$_2$S.

Multiple measurements of the response of the Pd fiber as a function of H$_2$ partial pressure were conducted both prior to H$_2$S exposure and then again post-H$_2$S exposure. These measurements consisted of stepping the concentration exposed to the fiber to various partial pressures between 0 Pa

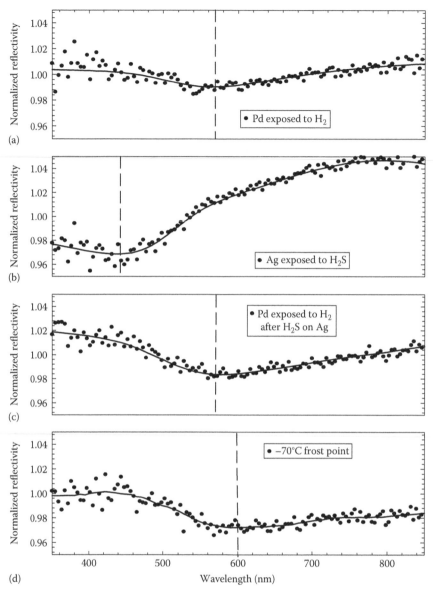

FIGURE 6.13 Plots showing the response to (a) 27 Pa H_2 in Pd, (b) 0.4 Pa H_2S on Ag, (c) 27 Pa H_2 in Pd, and (d) the response of the fiber when dried from 0°C dew point to −70°C frost point on a single SPR fiber. The time progression is from top to bottom and the data are renormalized after the H_2S on Ag exposure since that change is not reversible. Thus, (c) and (d) are normalized to (b). The SPR wavelengths are separated with the minimum for Ag at approximately 450 nm, the minimum for Pd at about 550 nm, and the minimum for Au at about 600 nm. The dots are the raw data from the spectrometer and the solid lines are running averages of the data. The vertical dashed lines locate the minimum of the averaged data. (Used from Pfeifer, K.B. and Thornberg, S.M., *IEEE Sensors J.*, 10(8), 1360, 2010. With permission.)

and 27 Pa of H_2 in N_2 and then back to a vacuum baseline. The response of the fiber after exposure to the H_2S at four discrete wavelengths (547.7, 574.2, 586.4, and 751.3 nm) agrees to within 30% with the pre-H_2S exposures. In addition, the response time prior to H_2S exposure for a 27–0 Pa H_2 concentration change is on the order of 2.5 min. The response time postexposure to H_2S is very similar indicating that the response time is also essentially unchanged.

It should be noted that Figure 6.13a and b are normalized to the unexposed spectrum. Figure 6.13c is renormalized after the H_2S exposure and thus includes the spectral features introduced on the exposure of the Ag film to H_2S. Since the two metals have nonoverlapping SPRs, the responses to the two analytes can be separately resolved. The resulting response indicated that low background levels of sulfur exposure for several days do not appreciably degrade the response of the Pd section to H_2. In all cases, a horizontal line at 1 would represent the reflectivity of the unexposed, normalized optical spectrum.

Figure 6.13d is a plot of the response of the SiO_2/Au section of the fiber to drying from a frost point of $0°C$ to $-70°C$. Again, the response is with respect to normalization of the response at $0°C$. This data illustrates that the moisture sensing section of the fiber is not poisoned by the H_2S and that monitoring of multiple components in the system is feasible by judicious choice of wavelengths in the SPR spectrum. The wavelength of maximum SPR response is similar in both the Pd and SiO_2/Au cases indicating that multiple wavelengths would need to be monitored simultaneously and analyzed with multivariate spectral analysis to separate H_2O and H_2. Binary or ternary mixtures have not yet been tested using this approach.

6.9 CONCLUSIONS

We have developed the background mathematics that predicts SPR in a planar geometry. These models were then expanded to a single-ended cylindrical geometry that can be made into a practical single-ended fiber-optic probe. Mathematical models to simulate the response of a fiber-optic-based SPR sensor using a single fiber with a retroreflective coating have then been demonstrated. Several SPR-supporting metals were then modeled and used to design sensors for testing including H_2 sensing using Pd films, H_2S sensing using Ag films, and H_2O sensing using SiO_2 films on a Au film. Data illustrate that different metals allow variation in the spectral response to various analytes allowing the manufacture of a three-region SPR sensor that is sensitive to all three analytes using the single fiber and experimental system. It has been demonstrated that the H_2 response is unchanged after an extended interval of exposure to H_2S and that moisture can be detected at low frost points ($-70°C$) after repeated exposure to H_2 and prolonged exposure to H_2S. Thus, a simplified fiber-optic SPR geometry using a single-ended fiber has been demonstrated and shown to have multiple responses to several chemical species that can be present due to aging effects in sealed systems.

Sandia National Laboratories is a multiprogram laboratory managed and operated by Sandia Corporation, a wholly owned subsidiary of Lockheed Martin Corporation, for the US Department of Energy's National Nuclear Security Administration under contract DE-AC04-94AL85000.

REFERENCES

1. R. W. Wood, On a remarkable case of uneven distribution of light in a diffraction grating spectrum, *Philosophical Magazine*, 4(19–24), 396–402, 1902.
2. R. W. Wood, Anomalous diffraction gratings, *Physical Review*, 48, 928–937, 1935.
3. U. Fano, The theory of anomalous diffraction grating and of quasi-stationary waves on metallic surfaces (Sommerfeld's waves), *Journal of Optical Society of America*, 31, 213–222, 1941.
4. R. H. Ritchie, Plasma losses by fast electrons in thin films, *Physical Review*, 196(5), 874–881, 1957.
5. E. Kretschmann and H. Raether, Radiative decay of non-radiative surface plasmons excited by light, *Zeitschrift für Naturforschung*, 23a, 2135, 1968.
6. K. B. Pfeifer and S. M. Thornberg, Surface plasmon sensing of gas phase contaminants using a single-ended multiregion optical fiber, *IEEE Sensors Journal*, 10(8), 1360, 2010.
7. S. Ekgasit, C. Thammacharoen, F. Yu, and W. Knoll, Influence of the metal film thickness on the sensitivity of surface plasmon resonance biosensors, *Applied Spectroscopy*, 59(5), 661–667, 2005.
8. I. Stemmler, A. Brecht, and G. Gauglitz, Compact surface plasmon resonance-transducers with spectral readout for biosensing applications, *Sensors and Actuators B: Chemical*, 54, 98–105, 1999.

9. P. T. Leung, D. Pollard-Knight, G. P. Malan, and M. F. Finlan, Modeling of particle-enhanced sensitivity of the surface-plasmon-resonance biosensor, *Sensors and Actuators B: Chemical*, 22, 175–180, 1994.

10. C. M. Pettit and D. Roy, Surface plasmon resonance as a time–resolved probe of structural changes in molecular films: Consideration of correlating resonance shifts with adsorbate layer parameters, *Analyst*, 132, 524–535, 2007.

11. A. Ikehata, K. Ohara, and Y. Ozaki, Direct determination of the experimentally observed penetration depth of the evanescent field via near-infrared absorption enhanced by the off-resonance of surface plasmons, *Applied Spectroscopy*, 62(5), 512–516, 2008.

12. S. A. Love, B. J. Marquis, and C. L. Haynes, Recent advances in nanomaterials plasmonics: Fundamental studies and applications, *Applied Spectroscopy*, 62(12), 346A–362A, 2008.

13. W. Yuan, H. P. Ho, C. L. Wong, S. K. Kong, and C. Lin, Surface plasmon resonance biosensor incorporated in a Michelson interferometer with enhanced sensitivity, *IEEE Sensors Journal*, 7, 70–73, 2007.

14. E. Kretschmann, Decay of non-radiative surface plasmons into light on rough silver films. Comparison of experimental and theoretical results, *Optics Communications*, 6(2), 185–187, 1972.

15. H. Raether, *Surface Plasmons on Smooth and Rough Surfaces and on Gratings*. Springer-Verlag, Berlin, Germany, pp. 118–120, 1988.

16. J. Homola, Electromagnetic theory of surface plasmons, in *Surface Plasmon Resonance Based Sensors*, J. Homola, ed., Springer-Verlag, Berlin, Germany, pp. 1–10, 2006.

17. M. Yamamoto, Surface plasmon resonance (SPR) theory: Tutorial, *Review of Polarography*, 48(3), 209–237, 2002.

18. A. K. Sharma, R. Jha, and B. D. Gupta, Fiber-optic sensors based on surface plasmon resonance: A comprehensive review, *IEEE Sensors Journal*, 7(8), 1118–1129, 2007.

19. J. D. Jackson, *Classical Electrodynamics*, 2nd edn., John Wiley & Sons, New York, pp. 68–71, 1975.

20. S. Ramo, J. R. Whinnery, and T. Van Duzer, *Fields and Waves in Communications Electronics*, 3rd edn., John Wiley & Sons, New York, pp. 385–387, 1994.

21. J. R. Reitz, F. J. Milford, and R. W. Christy, *Foundations of Electromagnetic Theory*, Addison-Wesley Publishing Company, Reading, MA, pp. 341, 1980.

22. A. D. Boardman, Hydrodynamic theory of plasmon-polaritions on plane surfaces, in *Electromagnetic Surface Modes*, A. D. Boardman, ed., John Wiley & Sons, Chichester, U.K., p. 17, 1982.

23. M. J. Adams, *Án Introduction to Optical Waveguides*, John Wiley & Sons, Chichester, U.K., pp. 64–67, 1981.

24. M. V. Klein and T. E. Furtak, *Optics*, 2nd edn., John Wiley & Sons, New York, pp. 295–300, 1986.

25. M. V. Klein and T. E. Furtak, *Optics*, 2nd edn., John Wiley & Sons, New York, pp. 76–80, 1986.

26. E. Hecht and A. Zajac, *Optics*, Addison-Wesley, Menlo Park, CA, pp. 72–75, 1979.

27. C. A. Balanis, *Advanced Engineering Electromagnetics*, John Wiley & Sons, New York, p. 191, 1989.

28. H. Raether, *Surface Plasmons on Smooth and Rough Surfaces and on Gratings*, Springer-Verlag, Berlin, p. 11, 1988.

29. X. Bévenot, A. Trouillet, C. Veillas, H. Gagnaire, and M. Clément, Surface plasmon resonance hydrogen sensor using an optical fibre, *Measurement Science and Technology*, 13, 118–124, 2002.

30. Y. Xu, N. B. Jones, J. C. Fothergill, and C. D. Hanning, Analytical estimates of the characteristics of surface plasmon resonance fiber-optic sensors, *Journal of Modern Optics*, 47(6), 1099–1110, 2000.

31. A. K. Sharma and B. D. Gupta, Absorption-based fiber optic surface plasmon resonance sensor: A theoretical evaluation, *Sensors and Actuators B: Chemical*, 100, 423–431, 2004.

32. M. Mitsushio, K. Miyashita, and M. Higo, Sensor properties and surface characterization of the metal-deposited SPR optical fiber sensor with Au, Ag, Cu, and Al, *Sensors and Actuators A: Physical*, 125, 296–303, 2006.

33. M. A. Butler and A. J. Ricco, Reflectivity changes of optically-thin nickel films exposed to oxygen, *Sensors and Actuators*, 19, 249–257, 1989.

34. M. A. Butler and A. J. Ricco, Chemisorption-induced reflectivity changes in optically thin silver films, *Applied Physics Letters*, 53(16), 1471–1473, 1988.

35. M. A. Butler, A. J. Ricco, and R. J. Baughman, Hg adsorption on optically thin Au films, *Journal of Applied Physics*, 67(9), 4320–4326, 1990.

36. K. B. Pfeifer, R. L. Jarecki, and T. J. Dalton, Fiber-optic polymer residue monitor, *Proceedings of the SPIE*, Vol. 3539, Boston, MA, pp. 36–44, 1998.

37. B. E. A. Saleh and M. C. Teich, *Fundamentals of Photonics*, John Wiley & Sons, Inc., New York, p. 18, 1991.

38. W. L. Wolfe, *Introduction to Radiometry*, SPIE Optical Engineering Press, Bellingham, WA, p. 17, 1998.

39. K. B. Pfeifer, S. M. Thornberg, M. I. White, and A. N. Rumpf, Sandia National Laboratories, Albuquerque, NM, Report SAND2009–6096, p. 8, 2009.
40. E. Hecht and A. Zajac, *Optics*, Addison Wesley, Reading, MA, p. 74, 1979.
41. M. A. Ordal, L. L. Long, R. J. Bell, S. E. Bell, R. R. Bell, R. W. Alexander, and C. A. Ward, Optical properties of the metals Al, Co, Cu, Au, Fe, Pb, Ni, Pd, Pt, Ag, Ti, and W in the infrared and far infrared, *Applied Optics*, 22(7), 1099–1119, 1983.
42. K. V. Rottkay, M. Rubin, and P. A. Duine, Refractive index changes of Pd-coated magnesium lanthanide switchable mirrors upon hydrogen insertion, *Journal of Applied Physics*, 85(1), 408–413, 1999.
43. B. Chadwick, J. Tann, M. Brungs, and M. Gal, A hydrogen sensor based on the optical generation of surface plasmons in a palladium alloy, *Sensors and Actuators B: Chemical*, 17, 215–220, 1994.
44. M. A. Butler, Optical fiber hydrogen sensor, *Applied Physics Letters*, 45(10), 1007–1009, 1984.
45. R. R. J. Maier, B. J. S. Jones, J. S. Barton, S. McCulloch, T. Allsop, J. D. C. Jones, and I. Bennion, Fibre optics in palladium-based, hydrogen sensing, *Journal of Optics A: Pure and Applied Optics*, 9, S45–S59, 2007.

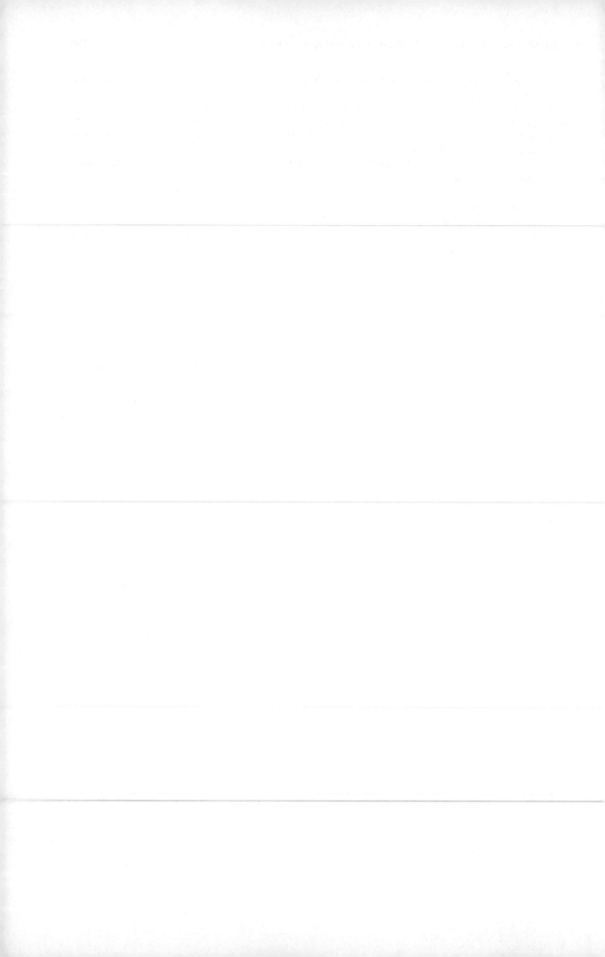

7 Active Core Optical Fiber Chemical Sensors and Applications

Shiquan Tao

CONTENTS

7.1 PRINCIPLE OF AC-OFCS AND EW-OFCS

An optical fiber chemical sensor (OFCS) detects the existence of and measures the concentration of a compound in a sample through detecting the interaction of the compound with light propagating inside an optical fiber.[1,2] Depending on the location at which the interaction occurs, OFCS can be divided into two classes: active core OFCS (AC-OFCS)[3,4] and evanescent wave OFCS (EW-OFCS).[5-7] The principle of AC-OFCS and EW-OFCS, which detect the analyte's optical absorption as sensing signals, are diagrammatically shown in Figure 7.1. In an AC-OFCS, the interaction of an analyte compound with the light occurs inside an optical fiber core, while in an EW-OFCS, the interaction of an analyte compound with the light occurs in the cladding layer of an optical fiber. A light beam traveling down an optical fiber can be scattered or absorbed by a compound existing inside the fiber core or inside the cladding. The light propagating in an optical fiber can also excite a compound in the fiber to a higher energy level and causes the emission of fluorescence (FL). All these interactions can be used in designing OFCS. Therefore, analytical spectroscopic techniques, such as ultraviolet/visible (UV/Vis) absorption spectrometry, near-infrared (NIR) and mid-infrared (IR) absorption spectrometry, Raman scattering spectrometry, FL spectrometry, have been used in OFCS design.[1-11] In addition, the existence of an analyte compound in the fiber core or cladding can change the refractive index of the core or cladding materials, which causes the change of light intensity guided through the fiber. This phenomenon has also been used in designing OFCS.[12-14] The characteristics, including sensitivity, precision, reversibility, response time, and selectivity, of an OFCS are decided by the properties of the compound to be detected, the chemical reactions involved in the sensing process, the analyte/light interaction used for the detection, the location of analyte/light interaction occurs, and the microstructure of the optical fiber core and the cladding.

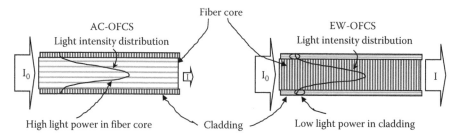

FIGURE 7.1 Diagrammatic graphs show the principles of AC-OFCS and EW-OFCS. The horizontal-lined parts in the fibers are where an analyte (or reaction product(s) of an analyte with a sensing reagent) interacts with light in the optical fiber. The vertical-lined parts in the fibers are not involved in the sensing, but only for light-guiding. Optical absorption spectrometry is used as an example in the graphs. I_0 is the light intensity injected to the fiber, and I is the intensity of light transmitted out of the fiber. In AC-OFCS, most of the light are absorbed, while in EW-OFCS only small part of the light is absorbed.

7.1.1 OFCS WITH AN OPTICAL FIBER CORE AS A TRANSDUCER

When a light beam is injected into an optical fiber, the light travels down the fiber through a series of total internal reflections (TIR) at the interface of the optical fiber core and cladding layer. If an analyte exists inside the fiber core, the analyte molecules can interact with light propagating inside the fiber core. In this case, the optical fiber core also acts as an optical cell for detecting analyte/light interaction. Theories established in conventional analytical optical spectroscopy can be used to describe the interaction of an analyte with light inside the fiber core. For example, the absorption of light at a specific wavelength by an analyte inside the fiber core can be described by using the Lambert–Beer law[15]:

$$A = \mathrm{Log}\left(\frac{1}{T}\right) = \varepsilon C L \tag{7.1}$$

where
   A is absorbance
   T is the transmittance
   ε is the absorption coefficient
   C is the concentration of the analyte inside the fiber core
   L is the length of the analyte/light interaction, which is decided by following equation[4]:

$$L = \frac{\ell}{\left(1 - \sin^2 \theta\right)^{1/2}} \tag{7.2}$$

where
   ℓ is the length of the optical fiber transducer
   θ is the incident angle of light beam to the optical fiber

Similarly, a molecule inside an optical fiber core can also be excited by the light guided inside the fiber and emits FL. If the concentration of fluorescent molecule inside the fiber core is very low, the intensity of FL can be expressed as[16]

$$F = K\phi I_0 \varepsilon C L \tag{7.3}$$

where
   K is a constant
   ϕ is the quantum efficiency of the fluorescent molecule
   I_0 is the incident light intensity

The interaction of light guided in an optical fiber core with an analyte or reaction product(s) of an analyte with a sensing reagent inside the optical fiber core is the basis of AC-OFCS using porous solid optical fibers (PSOF), liquid core waveguides (LCW), or hollow waveguides (HWG) as transducers. In optical fiber industry, the interactions of light guided in an optical fiber with the fiber core materials have also been used for fiber quality control, optical fiber amplifier design, etc. For example, laser-induced optical fiber FL spectroscopy can be used to monitor possible contamination of fiber core material.[17] Optical fiber core Raman scattering is another technique having been used to monitor the quality of optical fiber cable.[18] The in-fiber amplification technologies significantly improved optical fiber communication capacity. Present in-fiber amplification technologies used in optical fiber communication industry are based on NIR laser-induced FL of erbium ions doped in a silica fiber core.[19,20] Fiber core optical absorption techniques have also been proposed for monitoring high-energy radiation (x-ray, γ-ray) and ionization particles (α-particle, neutron) for applications in nuclear facilities. The irradiation of silica optical fibers by the high-energy radiation and ionization particles breaks down chemical bounds, which causes the generation of radicals, nonbinding oxygen species inside the fiber core. The formed radical species can be detected through monitoring the absorption spectrum of the optical fiber core.[21]

7.1.2 OFCS WITH TAILORED CLADDING AS A TRANSDUCER

When a light beam is guided through an optical fiber, a standing wave, called EW, is formed at each point of TIRs. The existence of EW distributes part of the light power into the cladding layer adjunct to the fiber core. If a compound inside the cladding layer can interact with the electromagnetic wave distributed in the cladding, this compound can be sensed through detecting the interaction of the compound with EW in the cladding. The interactions of a compound in the cladding with EW can be optical absorption, FL emission, or scattering. Almost all of the analytical optical spectroscopic techniques used in conventional analytical spectroscopy can be used for detecting the interaction of an analyte with the EW in the cladding layer. Taking optical absorption as an example, the EW absorption by a compound in the cladding layer can be expressed as follows[4]:

$$A_{EW} = Log\left(\frac{1}{T}\right) = \gamma\varepsilon C\left[\frac{d_p\ell n_2\sin\theta}{a\left(n_1^2 - n_2^2\sin^2\theta\right)^{1/2}}\right] \tag{7.4}$$

where

γ is the ratio of light power distributed in the cladding over total light power guided through the optical fiber

n_1 and n_2 are the refractive index of the fiber core and cladding materials

ε is the absorption coefficient of the analyte, and $d_p\ell n_2\sin\theta \big/ a\left(n_1^2 - n_2^2\sin^2\theta\right)^{1/2}$ is the absorption path length, which equals the penetration depth (d_p) times the number of TIRs. The number of TIR is calculated as $\ell n_2\sin\theta \big/ a\left(n_1^2 - n_2^2\sin^2\theta\right)^{1/2}$, as light travels a length of ℓ inside an optical fiber (diameter equals a).

Similarly, EW excited FL,[22–24] EW Raman scattering,[25,26] and EW scattering[27] can be used in designing OFCS to detect the interaction of an analyte in the cladding with the EW penetrated into the cladding.

7.2   COMPARISON OF AC-OFCS AND EW-OFCS

Comparing Equation 7.1 with Equation 7.4, it is clear that the sensitivity of an optical fiber core absorption-based chemical sensor is much higher than that of an EW absorption-based chemical sensor. Two factors, the light intensity ($I_{cladding} = \gamma I_{total}$, γ value is usually smaller than 0.05 for multimode optical fibers) and the interaction path length ($d_p \ell n_2 \sin\theta \big/ a \left(n_1^2 - n_2^2 \sin^2\theta \right)^{1/2}$, d_p is only in µm level), limit the sensitivity of an EW absorption-based sensor. For example, G. L. Klunder et al. calculated the absorption path length of an EW-based OFCS using a $\ell = 12$ m optical fiber as a transducer to be only 3 mm.[28] In FL spectroscopy and scattering spectroscopy, the intensity of emitted FL light and scattered light is proportional to the intensity of excitation light and the path length of the interaction. Therefore, the sensitivity of an AC-OFCS using FL emission or Raman scattering as a sensing mechanism is also higher than that of EW excited FL sensor or EW scattering sensor.

Most reported OFCS are based on optical fiber EW spectrometry.[1,2,8–11] An EW-OFCS can use conventional optical fibers made for communication industry to construct the sensor. These fibers are inexpensive and easy to handle and are compatible with all kinds of tools and instruments developed for the communication industry. Therefore, these EW-OFCS are easy to construct and low cost. On the other hand, commercially available optical fibers used by communication industry are solid fibers. It is almost impossible to introduce a sample into fiber core of such a fiber to detect the interaction of an analyte in the sample with light guided inside the fiber core. Therefore, in order to make an AC-OFCS, a specially tailored optical fiber core has to be developed. This fiber core should be able to guide light and more importantly allows the introduction of samples into the fiber core. Presently, three type special optical fibers, PSOF, LCW, and HWG, have been used in designing AC-OFCS. Some very sensitive chemical sensors have been developed using these special optical fibers. Following are examples of such AC-OFCS.

7.3   AC-OFCS AND APPLICATIONS

7.3.1   AC-OFCS USING TAILOR-MADE PSOF AS TRANSDUCERS

Several techniques have been reported for making PSOF. Shahriari et al.[29] reported a high-temperature glass fiber pulling process followed by wet chemical etching to make a porous glass optical fiber. A sodium borate-doped glass rod is first made by normal high-temperature glass-making process. The glass rod is then pulled by a fiber-pulling device to make a glass optical fiber. The glass optical fiber is then soaked in a hot concentrated hydrochloric acid solution for more than 12 h to etch out the sodium borate. A porous glass optical fiber was obtained after the etching process. This porous optical fiber can then be soaked into a solution of a chemical agent to impregnate the chemical agent into the fiber core. A short piece of this reagent-impregnated porous glass optical fiber can be used as a transducer of an AC-OFCS. An optical fiber ammonia sensor has been developed by using a short piece of such a porous glass fiber impregnated with a pH indicator as a transducer. This sensor can detect trace ammonia in gas samples with a detection limit of 0.7 ppm.

A significant progress in designing AC-OFCS using PSOF is reported by Tao et al.[3,4] They reported a wet chemical process to make porous silica optical fibers. The starting material for making such porous silica fibers is an ester of silicic acid. The ester is first hydrolyzed in the presence of a mineral acid as a catalyst. The obtained silicic acid is not stable. They dehydrate through hydroxide condensation to form silicon dioxide nanoparticles, which distributed in water to form a colloidal solution (called silica sol solution). The formed sol solution is then mixed with a solution of a sensing agent. The mixed solution is injected into a tube of small diameter and kept in the tube at room temperature until the solution inside the tube is gelatinized. During the gelatinizing process, part of the solvent separates out of the gel and forms a film between the formed silica gel and the wall of the tube. The liquid solvent film prevents the contact of silica gel with the tube and avoids adhesion of the gel to the tube wall. As gelatinizing proceeds, part of the solvent permeates out of the tube and the formed silica gel shrinks. Finally, a silica gel monolith in the shape of a fiber with diameter smaller than that of the tube is formed inside the tube. This gel fiber can be pushed out of the tube by injecting a liquid (water or other liquid) through the tube from one end. The gel fiber just pushed out of the tube is rigid but gradually hardens after exposed to air. Finally, a sensing agent-doped porous silica optical fiber is obtained after the fiber is hardened. This process for making porous silica optical fibers is simple and low cost. The porous silica optical fiber is made from a wet chemical process, and any chemical or biochemical agent, which can be dissolved in water or an appropriate organic solvent, can be doped into the fiber for designing sensors for different application purposes.

Figure 7.2 is a scanning electron microscope (SEM) imaging of such an as-made porous silica optical fiber. The rough structure on the surface of the optical fiber was believed to be originated from the inner surface microstructure of the Tygon tube, which was used as the model in making the porous silica optical fiber. This rough structure scatters light out of the fiber. A chemical polishing procedure by using a hydrofluoric acid solution was developed to remove the surface structure and improve the fiber's light-guiding efficiency. The SEM has also been used to study the inner microstructure of the porous silica optical fiber. However, it was found that the pore size inside the fiber is smaller than the resolution limit (2.5 nm) of the used SEM. The small pore size inside the porous silica optical fiber is one of the reasons the fiber can efficiently guide light, because scattering loss is very low with such small pores. However, the porous silica optical fibers have a high loss level in transmitting UV light below 350 nm.

Several AC-OFCS using reagent-doped porous silica optical fibers have been reported.[3,4,30,31] Figure 7.3 shows the transducer structure of such an AC-OFCS. A $CoCl_2$-doped porous silica optical fiber has been used to design a moisture sensor having a structure similar to that shown in Figure 7.3.[3,4] This sensor is based on the reversible reaction of $CoCl_2$ with water vapor to form $CoCl_2(H_2O)_x$ complex inside the porous silica optical fiber core when it is exposed to a water vapor containing gas sample. The concentration of formed $CoCl_2(H_2O)_x$ complex is in equilibrium with the concentration of water vapor in the gas sample. The formation of $CoCl_2(H_2O)_x$ complex reduces $CoCl_2$ concentration in the porous silica optical fiber. $CoCl_2$ absorbs light at 632 nm. Therefore, water vapor concentration in a gas sample can be detected through detecting the fiber's optical absorbance signal by using a 632 nm diode laser as a light source. This sensor has high sensitivity

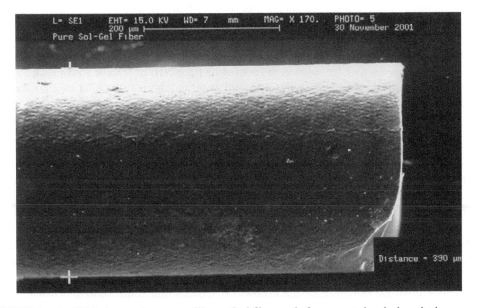

FIGURE 7.2   An SEM picture of a porous silica optical fiber made from a wet chemical method.

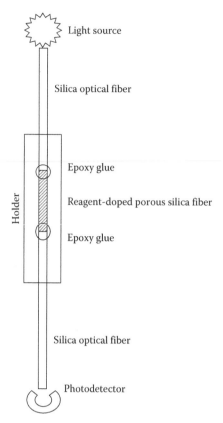

FIGURE 7.3   A diagrammatic graph shows the structure of an AC-OFCS using a reagent-doped porous silica optical fiber as a transducer.

and can easily detect water vapor in air to ppm level. However, the sensor has a slow response for monitoring water vapor in air at room temperature because the diffusion of water molecules into the porous structure inside the fiber takes a long time at ambient temperature.

An AC-OFCS using a $CuCl_2$-doped porous silica optical fiber has been reported for monitoring trace NH_3 in high-temperature gases, such as stack gas in coal-fired power plant.[30] In recent years, ammonia has been added to the exhaust gas of some combustion systems in order to reduce NO_x emission. Therefore, a sensor for continuous monitoring trace NH_3 in exhaust gas of such combustion system is significant for process control as well as for monitoring air pollutant emission. The AC-OFCS for NH_3 monitoring is based on a simple reversible chemical reaction:

$$CuCl_2 + NH_3 \longleftrightarrow Cu(NH_3)_x Cl_2$$

The formed $Cu(NH_3)_x^{2+}$ complexes inside the porous silica optical fiber is in equilibrium with the concentration of NH_3 in the gas sample. The formed $Cu(NH_3)_x^{2+}$ absorbs light with peak absorption wavelength at around 550 nm. Therefore, the concentration of NH_3 in gas phase can be monitored through monitoring the fiber optic absorption signal at around 550 nm. The sensor was tested for monitoring trace ammonia in air gas sample at a temperature of 450°C. Figure 7.4 is the recorded absorption spectra of the AC-OFCS exposed to high-temperature air samples containing NH_3 of different concentration. With the increase of NH_3 concentration in air sample, the absorbance at 540 nm increased. The recorded absorbance at 540 nm has a linear relationship with the NH_3 concentration in air gas sample. Figure 7.5 shows the sensor's time response to the change of NH_3 concentration in an air gas sample. This result shows that the sensor is reversible and can be used

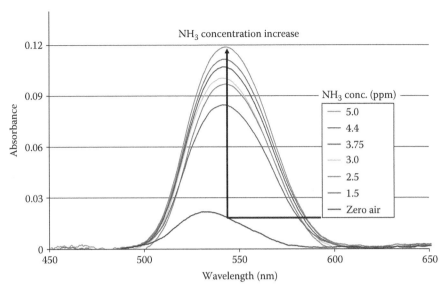

FIGURE 7.4 The recorded UV/Vis absorption spectra of an AC-OFCS using a $CuCl_2$-doped porous silica optical fiber (fiber length, 3 cm) as a transducer exposed to high-temperature air samples containing ammonia of different concentrations.

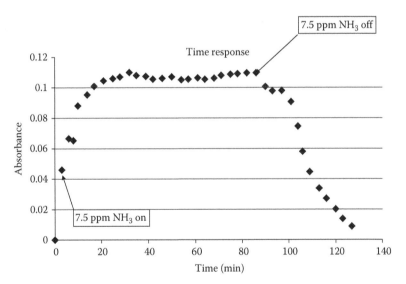

FIGURE 7.5 Time response of the $CuCl_2$-doped porous silica optical fiber ammonia sensor alternatively exposed to an ammonia-containing air sample and a blank air sample.

for continuous monitoring trace NH_3 in a gas sample. The response time is < 20 min. It has to be indicated that the response time of such a sensor can be much shorter when a porous silica optical fiber of small diameter will be used to design the sensor.

AC-OFCS using polydimethylsiloxane (PDMS) polymer optical fibers have also been reported. PDMS has been used in manufacturing organic polymer optical fibers as well as a cladding material for conventional optical fibers that are used by communication industry. PDMS is a hydrophobic polymer and can be made by polymerizing dimethylsiloxane with some cross-linking agents in the existence of special catalysts. The reagent kits for making this polymer are available from commercial sources. Klunder et al. made PDMS fibers by using silicone sealing products, such as RTV-732 and RTV-from department stores. They investigated the application of these polymer fibers for sensing trichloroethene (TCE) in environmental samples.[28,32] TCE absorbs NIR light with peak absorption wavelength at around 1.64 μm. A PDMS optical fiber of 10 mm length was connected with conventional optical fibers as a transducer. This sensor was reported for detecting TCE in water samples down to 1.1 ppm.

7.3.2   AC-OFCS Using LCW as Transducers

LCW is an optical fiber using a liquid confined inside a tube (capillary) or another liquid sheath as light-guiding medium. LCW development was originally targeted for delivering high intensive laser energy. Liquid materials having high refractive index and polymers of low refractive index have been studied for constructing LCW.[33–36] For example, carbon disulfide (CS_2) is an organic solvent used in organic chemistry. The refractive index of this solvent is 1.63, which is higher than the refractive index of glass and fused silica. Therefore, a glass capillary filled with CS_2 can be used as an LCW. The CS_2 LCW has been used as a sample cell for optical spectroscopic detection of chemical compounds extracted into CS_2.[37] However, due to the compound's strong unpleasant odor, the application of CS_2-based LCW in optical spectroscopy as a sample cell is not widely accepted.

The most significant progress in LCW and the application of LCW in AC-OFCS is the development of special amorphous polyfluoropolymers.[38–46] These polymers are transparent materials and do not absorb light of broad wavelength range, from UV to mid-IR. The refractive index of these fluoropolymers is in the range from 1.29 to 1.31, which is lower than the refractive index of water (1.33) and most organic solvents.[47] Therefore, a tube or a capillary made from one of these polymer materials filled with water can act as an optical fiber with water or an organic solvent inside the tube as a light-guiding medium. In addition, these fluoropolymers are chemically stable and do not react with normal chemical reagent. Therefore, an LCW with water or an organic solvent filled fluoropolymer capillary is a perfect long path sample cell for optical spectroscopic detection/monitoring of a chemical/biochemical species in the water or organic solution. Moreover, the fluoropolymers are also highly permeable to many gas molecules.[48,49] Molecules in a gas sample can diffuse into the solution inside the tube and interact with light guided through the solution. This makes the fluoropolymer-based LCW also useful in gas sensing.[38,50,51] The LCW presently most widely used are made from Teflon® AF amorphous fluoropolymer resin developed by DuPont™.[52] Two fluoropolymer resins, Teflon® AF 1600 and Teflon® AF 2400, are available from DuPont™. The resin can be made to the form of tube or capillary to form LCW. The resin can also be coated on the inner surface of a silica capillary to form an LCW.[39] Presently, Teflon® AF capillary products are available from commercial sources.

An LCW can be used as an optical cell in optical spectrometry. Compared with traditional optical spectrometers, which have optical cells with path length in centimeter range, an LCW optical cell can be as long as meters to hundreds of meters. In optical spectrometric methods, the sensitivity of the methods is usually proportional to the path length of the optical sample cell. Therefore, the application of LCW in optical spectrometry can significantly improve sensitivity. This is demonstrated in several LCW-based sensors.

An LCW-based Cr(VI) sensor has been reported.[45] This simple sensor uses a 2 m LCW as a transducer, a UV light-emitting diode (LED, 375 nm peak wavelength) as a light source, and a photodiode as a photodetector. Light from the UV LED is coupled into the LCW by using a short piece of conventional silica optical fiber and a three-way connector as shown in Figure 7.6. Light guided through the LCW is delivered to the photodiode via another piece of conventional optical fiber. A water sample is delivered into the LCW with a pump. This sensor monitors the optical absorption signal by chromate ion itself, and, therefore, no chemical reagent is needed. Figure 7.7 shows the time response of this sensor when water samples having Cr(VI) of different concentrations were pumped through the LCW. As demonstrated in the test results, this simple structured sensor can detect Cr(VI) in water sample with a detection limit of 0.10 ng/mL. Cr(VI) has been broadly used as a chemical reagent, and inappropriate discharge of Cr(VI)-containing waste caused

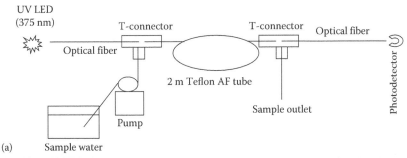

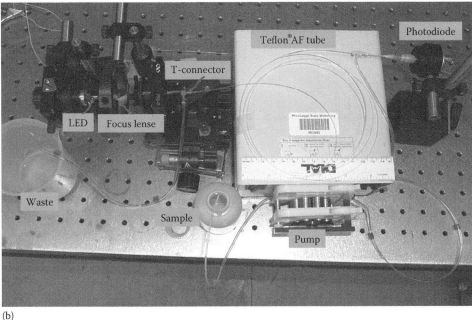

FIGURE 7.6 (a) A diagrammatic structure of an AC-OFCS using an LCW for monitoring trace Cr(VI) in water samples and (b) a picture of a laboratory-made LCW Cr(VI) sensor for testing the sensing principle.

water contamination worldwide. Because Cr(VI) is cancerogenic, the existence of Cr(VI) in drinking water is a serious concern today. This simple and easy to use sensor was proposed for testing Cr(VI) level in tap water at home.

Another interesting work involving LCW AC-OFCS is the detection of mercury atoms in water by using LCW atomic absorption spectrometry.[46] In traditional analytical chemistry, atomic absorption spectrometry is used to detect the interaction of free atoms in gas phase with light. Conventional samples (liquid or solid) must be atomized to generate free atoms in high-temperature flames or plasmas. Mercury is a special element. Mercury ions in a sample can be reduced to elementary mercury in an aqueous solution. The elementary mercury can be purged out of the aqueous solution by burbling the solution with an inert gas. Mercury in the gas phase exists as mercury atoms, which can be detected by atomic absorption spectrometry. However, what is the existing form of mercury after chemical reduction in aqueous solution is unclear. Tao et al. investigated the state of reduced mercury in water by using LCW optical absorption spectrometry. Mercury in ionic form (Hg^{2+}) was reduced by mixing a $HgCl_2$ solution with the solution of a strong reducing agent ($NaBH_4$). The mixed solution was pumped into a 1.6 m LCW. Light from a UV light source was coupled into the LCW. Light emerged from the LCW was coupled into an optical fiber compatible UV/Vis spectrometer with a structure similar to that shown in Figure 7.6. The recorded absorption spectra are shown in Figure 7.8. These results indicate that the reduced mercury in the aqueous solution absorbs light with peak absorption

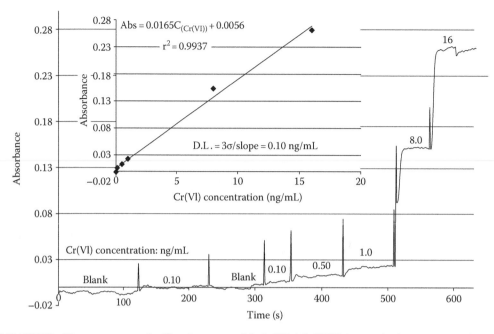

FIGURE 7.7   Time response and calibration curve of the LCW AC-OFCS for monitoring water samples containing Cr(VI) of different concentrations. The sensor achieved a 0.10 ng/mL detection limit, which is much lower than the detection limit achieved by using expensive instruments, such as AAS and ICP-AES.

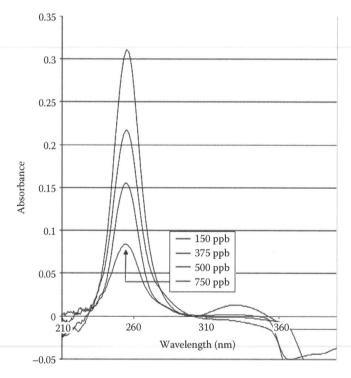

FIGURE 7.8   Atomic absorption spectra of mercury atoms in water recorded by using an LCW as a long pathlength optical cell. The half-height width of the absorption spectra is around 20 nm, which is thousands of times broader than atomic absorption spectrum of mercury atoms in gas phase. The mercury atoms in water encounter high-frequency collisions, which broadens the absorption spectrum according to quantum mechanical principles.

wavelength exactly the same as that of atomic absorption spectrum by mercury atoms in gas phase. It can be concluded from these results that the reduced mercury exists in aqueous solutions as mercury atoms. However, the mercury atoms in aqueous solutions are hydrated and encounter high-frequency collisions from molecules surrounding the atoms. This is reflected in the width of the atomic absorption spectrum in Figure 7.8. The half-height width of atomic absorption spectrum by gas phase mercury atoms is around 10 pico-meter, while the half-height width of the recorded atomic absorption spectrum by mercury atoms in the aqueous solution is more than 20 nano-meter. The LCW atomic absorption spectrometric method can also be used for monitoring mercury in water samples.

LCW can also be used as sample cells for FL spectrometry. A compound inside the solution filled in a Teflon® AF capillary can be perpendicularly excited. The emitted FL light are collected by the LCW and guided to the capillary's end, which is connected to an optical fiber compatible spectrometer or a photodetector via a conventional optical fiber. For example, a Teflon® AF 2400 tube has been helically wounded round a quartz tube.[40] Two UV lamps (370 nm) were inserted into the quartz tube. Quinine sulfate and chlorophyll a in the solution filled inside the LCW has been excited by the UV light from the UV lamps. The emitted FL photons are detected with optical fiber compatible spectrometer. This sensor can easily detect the analytes to sub-ppb concentration. LCW FL spectrometry using a UV LED as an excitation source has also been used in detecting atmospheric formaldehyde.[43] This fieldable device can be used to detect formaldehyde in air sample in sub-ppb level. A Teflon® 2400 capillary has been used as both a separation column and a light-guiding device. Proteins in a sample filled inside the LCW are separated by isoelectric focusing technique. A laser beam has been injected into the LCW to excite the separated proteins. A CCD imager was used to detect the distribution of FL from the separated proteins. When compared to a commercially available instrument with UV detection, the separation efficiency and peak capacity were similar, while the detection sensitivity was enhanced by 3–5 orders of magnitude.[53]

Laser-induced Raman spectrometry can be considered as a convenient detection technique, because the excitation is not wavelength dependent, and a laser can excite many compounds. However, Raman spectrometry has an inherited low sensitivity compared with FL spectrometry. The application of a long path-length sample cell in laser-induced Raman spectrometry can be a solution for improving sensitivity of Raman spectrometry. LCWs have been investigated as sample cells in laser-induced Raman spectrometry, and significant sensitivity improvement has been achieved.[54,55] LCW as sample cell for laser-induced Raman is especially attractive as a detection technique for separation technologies, such as HPLC and CE, because Raman spectrometry can be considered as a general detection technique and is well suited for detecting multiple analytes from separation eluents.

The gas permeation property of Teflon® AF material makes LCW very useful for gas sensing. Teflon® AF materials are highly permeable to H_2O, CO_2, O_3, H_2, and N_2. LCW gas sensors can be developed either by directly monitoring the intrinsic optical property of the gas compound or by monitoring reaction products of gas compounds permeated into the solution with sensing reagents in the solution filled inside the LCW. Tao and Le reported an LCW ozone sensor for monitoring water ozonation process.[56] Water ozonation is presently used in industry for water sanitation. A sensor with capability of continuous monitoring ozone concentration in water is significant in the ozonation process control. The principle of the reported sensor is diagrammatically shown in Figure 7.9. This sensor is simple structured and consists of a Teflon® AF 2400 capillary filled with pure water, a UV light source (254 nm), and a photodetector. The pure water-filled Teflon® AF 2400 capillary is deployed into a water sample. Ozone in the water sample permeated through the Teflon® AF 2400 tube and dissolved in the pure water filled in the capillary. Light from the UV light source guided through the LCW is absorbed by ozone molecules dissolved in water, which is monitored with a photodetector as a sensing signal. This sensor is reversible with response time shorter than 4 min. It can be used for monitoring trace ozone in water to sub-ppb level. An LCW-based sensor for monitoring trace SO_2 in gas and liquid samples has also been reported.[57] Similar to the ozone sensor, this sensor also

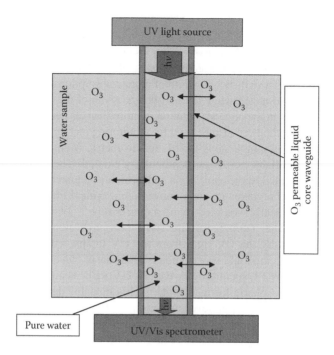

FIGURE 7.9 A diagrammatic graph shows the principle of an LCW AC-OFCS for monitoring trace ozone in water samples. The LCW acts as a light-guiding device as well as a permeation separator, which prevents interference species in water sample from entering the LCW.

uses a pure water-filled Teflon® AF 2400 capillary as a transducer. The intrinsic absorption signal by SO_2 molecule permeated to water filled in the Teflon® AF capillary at 254 nm has been monitored as a sensing signal. LCW gas sensors have also been reported for monitoring CO_2.[58] In such a sensor, a pH-indicating reagent is dissolved in water filled in a Teflon® AF 2400 capillary. Carbon dioxide permeated into the solution filled inside the capillary changes the pH and thus changes the color of the solution. This color change is monitored by using fiber optic UV/Vis absorption spectrometry. This sensor can be used for continuous monitoring CO_2 in air or for industrial process control.

7.3.3   AC-OFCS Using HWG as Transducer

HWG is originally developed for the purpose of delivering high-energy laser power, such as high inten-sive light from CO_2 lasers and Er:YAG lasers.[59–61] An HWG uses a gas (air or specially prepared gas) filled in a special capillary or tube as light-guiding media. Compared with solid and liquid, it is much easier to obtain high-purity gases, which do not absorb laser light of specific wavelength. Therefore, there is much less chance of fiber damage when an HWG is used for guiding high intensive laser power. Depending on the refractive index of capillary wall material, HWG can be grouped into two categories: (1) those whose inner wall materials have refractive indices greater than one (leaky guide) and (2) those whose inner wall materials have refractive index less than one. The HWG with n < 1 wall material is similar to normal optical fibers in that the refractive index of fiber core is higher than that of cladding. Hollow sapphire fiber operating at 10.6 μm (n = 0.67) is an example of this class HWG.[51] Presently, the most popular HWG is a leak guide having a dielectric silver iodide (AgI) film deposited on the surface of a silver film, which is coated on the inner surface of a glass tube or capillary.[61–64] The smooth surface of glass tube reduced the possible scattering loss. With the present technique, AgI film-based HWG can work in the wavelength range from 2.9 to 16 μm with loss level at <1 dB/m.[65]

Photonic band-gap HWG is a recent development of optical fibers. Light guiding in this fiber is based on Bragg grating. The transmittance of a photonic band-gap hollow optical fiber depends on wavelength. Light of specific wavelength ranges can be guided through such a fiber, while light of other wavelengths will have a high loss level. The transmitted wavelength range of such a fiber can be tuned through controlling the diameter of holes surrounding the central light-guiding hole.[66]

Early HWG gas sensors use leaky guides or dielectric AgI membrane-coated capillaries as transducers.[59–64] In recent development, photonic band-gap HWG gas sensors have been reported for monitoring organic compounds (ethane, dichloromethane, trichloromethane, ethyl chloride, etc.).[67–69] A high-resolution quantum cascade laser has been used as a light source in photonic band-gap HWG gas sensors.[66] The stable quantum cascade laser can provide stable light beams of precisely tuned wavelength. These sensors do not require an IR monochromator and largely simplify the sensor structure. FTIR and mid-IR optical spectroscopy using cascade lasers as light sources are powerful for identifying/detecting individual organic compounds, and they have been proposed as detectors for gas chromatography. However, the sensitivity of these mid-IR spectrometric methods is very limited for gas detection. Although HWG can be used for increasing the path length of interaction, the reported detection limit of these sensors is still limited in ppm level. As detection techniques, these detectors are not as good as broadly used mass spectrometers.

In viewing the three type AC-OFCS, LCW can be used for monitoring both liquid and gas samples and can achieve high sensitivity. PSOF-based AC-OFCS are especially attractive in monitoring high-temperature gas samples. However, PSOF-based sensor's response can be too slow for many industrial process control applications if operated at ambient temperature. HWG-based AC-OFCS have the potential to achieve high sensitivity and fast response detection if high-resolution lasers are available as light sources.

REFERENCES

1. R. Narayanaswamy and O. S. Wolfbeis, eds., *Optical Sensors for Industrial, Environmental and Clinical Applications*, Springer-Verlag, Berlin, Germany, 2003.
2. K. T. V. Grattan and B. T. Meggitt, eds., *Optical Fiber Sensor Technology: Vol. 4, Chemical and Environmental Sensing*, Kluwer Academic Publishers, Boston, MA, 1999.
3. S. Tao, C. B. Winstead, J. P. Singh, and R. Jindal, *Opt. Lett.*, **27**, 1382 (2002).
4. S. Tao, C. B. Winstead, R. Jindal, and J. P. Singh, *IEEE Sens. J.*, **4**, 322 (2004).
5. B. D. MacCraith, *Sens. Actuators,* **B11**, 29 (1993).
6. G. Stewarg and W. Johnstone, *Optic. Fiber Sens.*, **3**, 69 (1996).
7. W. Jin, H. L. Ho, G. Stewart, and B. Culshaw, *Trends Anal. Spectrosc.*, **4**, 155 (2002).
8. V. Ruddy, B. D. MacCraith, and J. A. Murphy, *J. Appl. Phys.*, **67**, 8070 (1990).
9. O. S. Wolfbeis, *Anal. Chem.*, **76**, 3269 (2004).
10. O. S. Wolfbeis, *Anal. Chem.*, **74**, 2663 (2002).
11. L. Su, T. H. Lee, and S. R. Elliott, *Opt. Lett.*, **34**, 2685 (2009).
12. M. Chomat, D. Berkova, V. Matejec, I. Kasik, G. Kuncova, and M. Hayer, *Sens. Actuat.*, **B87**, 258 (2002).
13. R. G. Heideman, Rob P. H. Kooyman, J. Greve, and Bert S. F. Altenburg, *Appl. Opt.*, **30**, 1474 (1991).
14. S. Korposh, S. W. James, S. W. Lee, S. Topliss, S. C. Cheung, W. J. Batty, and R. P. Tatam, *Opt. Expr.*, **18**, 13227 (2010).
15. D. A. Skoog, D. M. West, and F. J. Holler, eds., *Fundamentals of Analytical Chemistry*, 7th edn., Sauders College Publishing, New York, 1996.
16. J. R. Lakowicz, ed., *Topics in Fluorescence Spectroscopy, Vol. 4: Probe Design and Chemical sensing*, Plenum Press, New York, 1991.
17. W. J. Miniscalco and B. A. Thompson, *Mater. Res. Soc. Symp. Proc.*, **88** (*Opt. Fiber Mater. Prop.*), 127 (1987).
18. B. E. Hubbard, N. I. Agladze, J. J. Tu, and A. J. Sievers, *Physica B: Conden. Matter.*, **316–317**, 531 (2002).
19. J. M. P. Delavaux and J. A. Nagel, *J. Lightwave Technol.*, **13**, 703 (1995).
20. M. Artiglia, P. Di Vita, and M. Potenza, *Opt. Quantum Electron.*, **26**, 585 (1994).
21. D. W. Cooke, B. L. Bennett, and E. H. Farnum, *J. Nucl. Mater.* **232**, 214 (1996).
22. B. D. MacCraith, V. Ruddy, C. Potter, B. O'Kelly, and J. F. McGilp, *Electron. Lett.*, **27**, 1247 (1991).
23. L. C. Shriver-Lake, K. A. Breslin, P. T. Charles, D. W. Conrad, J. P. Golden, and F. S. Ligler, *Anal. Chem.*, **67**, 2431 (1995).

24. B. D. MacCraith, C. M. McDonagh, G. O'Keeffe, E. T. Keyes, J. G. Vos, B. O'Kelly, and J. F. McGilp, *Analyst*, **118**, 385 (1993).
25. B. Mizaikoff, M. Karlowatz, and M. Kraft, *Proc. SPIE-Int. Soc. Optic. Eng.*, **4202**, 263 (2001).
26. J. Baldwin, N. Schuehler, I. S. Butler, and M. P. Andrews, *Langmuir*, **12**, 6389 (1996).
27. L. Xu, J. C. Fanguy, K. Soni, and S. Tao, *Opt. Lett.*, **29**, 1191 (2004).
28. G. L. Klunder and R. E. Russo, *Appl. Spectrosc.*, **49**, 379 (1995).
29. M. R. Shahriari, Q. Zhou, and G. H. Sigel, *Opt. Lett.*, **13**, 407 (1988).
30. S. Tao, J. C. Fanguy, and T. V. S. Sarma, *IEEE Sens. J.*, **8**, 2000 (2008).
31. T. V. S. Sarma and S. Tao, *Sens. Actuat.*, **B127**, 471 (2007).
32. G. L. Klunder, R. J. Silva, and R. E. Russo, *Proc. SPIE-Int Soc. Optic Eng.*, **2068** (*Chem. Biochem. Environ. Fiber Sens. V*), 186 (1994).
33. J. Stone, *Appl. Phys. Lett.*, **20**, 239 (1972).
34. G. J. Oglivie, R. J. Esdaile, and G. P. Kidd, *Electron. Lett.*, **8**, 533 (1972).
35. I. Pinnau and L. G. Toy, *J. Mater. Sci.*, **109**, 125 (1996).
36. R. Altkorn, I. Koev, R. P. Van Duyne, and M. Litorja, *Appl. Opt.*, **36**, 8992 (1997).
37. K. Fujiwara and K. Fuwa, *Anal. Chem.*, **57**, 1012 (1985).
38. M. Belz, P. Dress, A. Sukhitskiy, and S. Y. Liu, *Proc. SPIE-Int. Soc. Optic. Eng.*, **3856** (Internal Standardization and Calibration Architectures for Chemical Sensors), 271 (1999).
39. M. Holtz, P. K. Dasgupta, and G. Zhang, *Anal. Chem.*, **71**, 2934 (1999).
40. J. Li, P. K. Dasgupta, and G. Zhang, *Talanta*, **50**, 617 (1999).
41. R. Altkorn, I. Koev, and M. J. Pelletier, *Appl. Spectrosc.*, **53**, 1169 (1999).
42. B. J. Marquardt, P. G. Vahey, R. E. Synovec, and L. W. Burgess, *Anal. Chem.*, **71**, 4808 (1999).
43. J. Li and P. K. Dasgupta, *Anal. Chem.*, **72**, 5338 (2000).
44. P. K. Dasgupta, G. Zhang, J. Li, C. B. Boring, S. Jambunathan, and R. Al-Horr, *Anal. Chem.*, **71**, 1400 (1999).
45. S. Tao, C. B. Winstead, H. Xia, and K. Soni, *J. Environ. Monit.*, **4**, 815 (2002).
46. S. Tao, S. Gong, L. Xu, and J. C. Fanguy, *Analyst*, **129**, 342 (2004).
47. R. C. Weast, M. J. Astle, and W. H. Beyer, eds., *CRC Handbook of Chemistry and Physics*, 65th edn., CRC Press, Inc., Boca Raton, FL, 1984.
48. A. Yu. Alentiev, Yu. P. Yampolskii, V. P. Shantarovich, S. M. Nemser, and N. A. Plate, *J. Mater. Sci.*, **126**, 123 (1997).
49. P. R. Resnick and W. H. Buck, *Fluoropolymers*, **2**, 25 (1999).
50. M. R. Milani and P. K. Dasgupta, *Anal. Chim. Acta*, **431**, 169 (2001).
51. Z. A. Wang, W. J. Cai, Y. Wang, and B. L. Upchurch, *Anal. Chem.*, **84**, 73 (2003).
52. Teflon® AF amorphous fluoroplastics. http://www2.dupont.com/Teflon_Industrial/en_US/products/product_by_name/teflon_af/
53. Z. Liu and J. Pawliszyn, *Anal. Chem.*, **75**, 4887 (2003).
54. M. J. Pelletier and R. Altkorn, *Anal. Chem.*, **73**, 1393 (2001).
55. M. Holtz, P. K. Dasgupta, and G. Zhang, *Anal. Chem.*, **71**, 2934 (1999).
56. L. Le and S. Tao, *Analyst*, **136**, 3335 (2011).
57. S. Gong, J. C. Fanguy, and S. Tao, *226th ACS National Meeting Paper Abstract*, New York, September 7–11, 2003, ANYL-111.
58. Z. A. Wang, Y. C. Wang, W. J. Cai, and S. Y. Liu, *Talanta*, **57**, 69 (2002).
59. J. A. Harrington, *Fiber Integrated Opt.*, **19**, 211 (2000).
60. R. K. Nubling and J. A. Harrington, *Appl. Opt.*, **34**, 372 (1996).
61. R. L. Kozodoy, A. T. Pagkalinawan, and J. A. Harrington, *Appl. Opt.*, **35**, 1077 (1996).
62. M. Alaluf, J. Dror, R. Dahan, and N. Croitoru, *J. Appl. Phys.*, **72**, 3878 (1992).
63. Y. Matsuura, T. Abel, and J. A. Harrington, *Appl. Opt.*, **34**, 6842 (1995).
64. C. D. Rabii and J. A. Harrington, *Appl. Opt.*, **35**, 6249 (1996).
65. Polymicro hollow silica waveguide HSW, in Polymicro Technologies™ optical fibers. http://www.molex.com/molex/products/group?channel=products&key=polymicro
66. J. C. Knight, *Nature*, **424**, 847 (2003).
67. O. Frazao, J. L. Santos, F. M. Araujo, and L. A. Ferreira, *Laser Photon. Rev.* **2**, 449 (2008).
68. C. Charlton, *Appl. Phys. Lett.*, **86**, 194102 (2005).
69. N. Gayraud, L. W. Kornaszewski, J. M. Stone, J. C. Knight, D. T. Reid, D. P. Hand, and W. N. MacPherson, *Appl. Opt.*, **47**, 1269 (2008).

8 All-Polymer Flexural Plate Wave Devices

Christoph Sielmann, John Berring, Suresha Mahadeva,
John Robert Busch, Konrad Walus, and Boris Stoeber

CONTENTS

8.1 INTRODUCTION

8.1.1 MOTIVATION

Versatile sensors and actuators continue to gain traction in a variety of industries, with new applications appearing as manufacturing costs decrease. This chapter focuses on a subset of sensors, comprised of polymer materials, that use flexural plate waves (FPW) traveling across a substrate as part of the mechanism for detecting changes to the sensor's environment. The acoustic waves are generated and measured using interdigitated transducers (IDTs) printed directly on the polymer substrate utilizing inkjet printing to pattern conductive, organic, polymer ink. The substrate is piezoelectric polyvinylidene fluoride (PVDF), a highly chemically resistant polymer that can be purchased in large, thin sheets. As the waves propagate along the substrate, they travel through a sensing medium that changes in properties based on the concentration of an environmental measurand. Sensors that detect mass changes in the sensing layer are known as gravimetric sensors.

A strong motivation for fabricating sensors from low-cost materials, such as polymers, is the creation of new applications that can leverage low-cost, lower-quality sensors. These applications, including some material testing, biomedical testing, and testing hazardous environment sensing, require disposable sensors that can be inexpensively procured and replaced. The low-cost of production, mass reproducibility, biocompatibility, and chemical resilience of all-polymer sensors can encourage new commercial applications in sensing by providing disposable gravimetric sensors.

PVDF has captured significant attention as a material substrate for sensing platforms [1–7]. These sensing applications typically leverage the piezoelectric properties of the material to generate or receive acoustic waveforms. The waveforms interact with another material acoustically coupled to the PVDF to provide a mechanism for gravimetric sensing [5,8–10], nondestructive testing (NDT) [2], biosensing [1,4], or curvature sensing [8]. The growing popularity of PVDF is restricted by some fundamental limitations in the material, specifically a low electromechanical coupling coefficient causing poor energy transfer between electrical and acoustic waves and a high mechanical attenuation coefficient that limits the ability for high-frequency waves to traverse through the material.

Low cost, chemical resiliency, and ease of processing are sufficiently intriguing for continued efforts in exploring this material for its potential in sensing applications. The focus of this chapter is the fundamental analysis of the suitability of PVDF for general acoustic sensing applications utilizing FPWs and organic polymer, printed electrodes. Choosing organic electrodes compounds on the benefits of PVDF by using a low-cost, biocompatible electrode material, while leveraging FPWs rather than surface acoustic or bulk waves, further simplifies the design by allowing large and easy to fabricate electrodes, low operating frequencies, and high sensitivity.

8.1.2 ACOUSTIC WAVE MICROSENSORS

Acoustic wave microsensors are sensing devices that consist of a vibrating substrate, an acoustic actuation mechanism, an acoustic detection mechanism, and, optionally, a coupling to another material that changes mechanical characteristics in response to external stimuli. An acoustic wave is generated within the substrate, either through optical, thermal, electrical (piezoelectric), dielectric, or similar actuation [11]. The waves travel outward from the generation region along the substrate and couple through a sensing material. The sensing material is specially chosen to vary in mechanical properties such as stiffness, density, attenuation, or speed of sound as a known function of a change in the environment, such as temperature, gas concentration, particle density, or fluid viscosity. The changing properties of the sensing material perturb the traveling wave, by modifying the frequency or speed of this wave that is then picked up at an acoustic receiver. An example of such a device is a multisensor matrix device produced by Cai et al. for volatile organic compound (VOC) detection [10].

Acoustic wave microsensors typically operate using either bulk acoustic waves (BAW), which consist of thickness shear (TSM) and acoustic plate (APM) modes, surface acoustic waves (SAWs), and FPWs [12,13]. The acoustic mode selected depends on the properties of the substrate material, purpose of the sensor, and mechanical structure of the acoustic actuation and detection mechanisms. FPWs and SAWs can both be actuated and detected through piezoelectric excitation using IDTs as shown in Figure 8.1. SAWs are generated when the substrate thickness is much greater than the wavelength of the traveling wave. When the wavelength of the wave is less than the thickness of the substrate, FPWs are generated [13].

8.1.3 GRAVIMETRIC SENSING

Acoustic gravimetric sensing involves passing acoustic waves through a material whose density changes as a function of an external measurand. The change in density varies the mass of the sensing material, which can affect the resonance of the entire structure, such as in BAW sensing, or change the velocity of a traveling wave propagating through the material, such as in the case of FPW and SAW sensors [14]. When applied to FPW VOC sensing, a polymer layer is deposited on either the top side or bottom side of the sensor substrate between the acoustic actuator and receiver. The polymer is chosen to be selectively

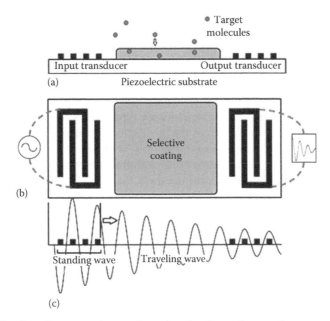

FIGURE 8.1 (a) Gravimetric gas sensing configuration showing polymer substrate, transducers, and molecules absorbing in polymer sensing layer. (b) Top view showing transducers, selecting coating, and interface wiring. (c) Acoustic standing wave generated over electrical input transducer and traveling wave propagating to the electrical output transducer. Practical devices consist of many more fingers in the IDTs.

sorptive to a specific or small range of VOCs. For low VOC concentrations, there is a roughly linear relationship between the concentration of the VOC in the surrounding environment and the concentration of the VOC in the sense polymer. The ratio between the VOC concentration in the polymer and the VOC concentration in the environment is often characterized by the so-called partition coefficient.

The change in density of the sensing layer varies the linear and rotational inertia of the material as the wave travels through it. The linear inertia, and rotational inertia in some cases, influences the phase velocity of the traveling wave. For an open-loop actuation/detection system, the change in wave velocity can be measured as a change in multiple FPW characteristics as described in Table 8.1.

TABLE 8.1
Perturbative Wave Modes in FPW Gravimetric Sensing

FPW Characteristic	Sensing Material	Perturbation Mechanism
Frequency	Above or below electrodes	For a constant wavelength and varying wave velocity as the wave travels through the electrodes, the resonance frequency of the structure will change.
Phase/delay	Above, below, or between electrodes	As the wave velocity changes, a given deflection in the traveling wave will take more or less time to reach the detector.
Amplitude	Above, below, or between electrodes	Wave amplitude can be affected by many factors, including how well tuned the sensor is to the resonance frequency of the device. The sorption of an analyte such as a VOC can also change the attenuation properties of a sensing material, thereby affecting the amplitude of the received waveform.

Source: Sielmann, C. et al., Implications of a low stiffness substrate in lamb wave gas sensing applications, in *2012 IEEE Sensors*, October 28–31, 2012, Taipei, Taiwan, pp. 1–4.

8.2 FLEXURAL PLATE WAVES

8.2.1 BACKGROUND

Lamb waves are traveling waves that propagate along a thin plate. Lamb waves comprise of symmetric and antisymmetric waves, with the latter also being known as FPWs [11]. For a given wavelength and an infinitely thin plate, the symmetric and antisymmetric waves correspond to different frequencies with the symmetric wave frequency typically higher. As the plate thickens such that its thickness exceeds the wavelength of the device, the symmetric and antisymmetric modes merge together to form a SAW with a unique frequency or phase velocity, as shown in Figure 8.2. FPWs have been examined for use in some sensing platforms, but compensating for the temperature sensitivity of the sensor often complicates the design [8,10].

The phase velocity of an FPW

$$v_{pa} = \sqrt{\frac{T + \beta^2 D}{M + \beta^2 H}} \qquad (8.1)$$

as determined by Wenzel [11] depends on the stress, T, stiffness, D, linear inertia, M, and rotational inertia, H. The stress is modeled as

$$T = \int_{-d}^{0} \tau_{yy}(\xi)d\xi \qquad (8.2)$$

where
  $\tau_{yy}(\xi)$ is the stress through the thickness
  d, of the sensor, perpendicular to the direction of wave propagation

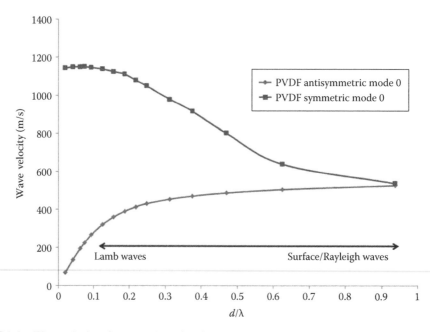

FIGURE 8.2 Wave velocity of symmetric and antisymmetric Lamb waves as a function of PVDF thickness to acoustic wave wavelength ratio from a finite element simulation under no stress, $T = 0$ MPa.

The stiffness, D, is given by

$$D = \int_{x=-d}^{x=0} E'(\xi)(\xi - x_0)^2 \, d\xi \tag{8.3}$$

where the location of zero stress during flexure in the sensor is indicated by x_0. The adjusted Young's modulus is described using

$$E'_n = \frac{E_n}{(1 - \sigma_n^2)} \tag{8.4}$$

where
   σ is the Poisson ratio of material n
   E_n is the corresponding Young's modulus

The linear inertia term, in (kg/m^2), is a combination of the linear inertia terms of both the PVDF substrate (M_{PVDF}) and polymer/analyte layer (M_A); that is,

$$M = M_{PVDF} + M_A \tag{8.5}$$

The rotational inertia, H, can be neglected in many practical sensor designs where material thickness is an order of magnitude less than the wavelength, as the linear inertia term is much larger [15]. Finally, the adjusted wavenumber, β, of the traveling wave can be determined from the wavelength λ of the transducer fingers.

$$\beta = 2\pi/\lambda$$

8.2.2 SENSITIVITY

It can be seen from Equation 8.1 that varying the linear inertia, stress, or stiffness of the material through which the FPW propagates will cause a corresponding change in phase velocity. In strictly gravimetric sensing, M_A varies as a function of absorbed or adsorbed analyte, causing the change in the phase velocity. The change in phase velocity due to a change in mass per unit area can be expressed by the mass sensitivity given by [14]

$$S_M = \frac{1}{\Delta m} \frac{\Delta v_{pa}}{v_{pa}} = \frac{-1}{2M} = \frac{1}{\Delta m} \frac{\Delta f}{f_0} \tag{8.6}$$

where
   Δm is the change in mass per unit area due to the sorption of a target analyte
   Δf represents the corresponding change in frequency
   f_0 is the initial resonance frequency of the sensor

The mass sensitivity of a gravimetric FPW sensor therefore depends only on the total mass per unit area of the sensor, M. This simple dependence is very advantageous for FPW sensors over other acoustic wave sensors, such as SAW sensors, which have a mass sensitivity that is dependent on the square of the resonance frequency, necessitating high frequency electronics for high-sensitivity operation [14].

8.2.3 Low-Stiffness Substrate Effects

Recent publications have suggested that all-polymer FPW sensors may have applications in stiffness and stress sensing in addition to gravimetric sensing. The sensitivity of FPWs to the stress and stiffness suggests that variations in stress and stiffness can also modulate the phase velocity of the traveling wave. According to Sielmann et al., a simplified total sensitivity can be derived theoretically using the following assumptions [16]:

- The sensing material has the same Young's modulus and Poisson ratio as the substrate material (PVDF).
- The sensing material coats the entire PVDF substrate.
- At the zero state (no analyte present in sensing layer), there is no stress in either the sensing layer or PVDF substrate.
- All materials are linearly elastic and isotropic.
- Changes in Young's modulus due to sorption are neglected.
- Swelling of the sensing material occurs uniformly in all directions.
- The sensing layer is much thinner than the substrate.

Under these assumptions, a new expression for sensitivity due to changes in stress/tension, stiffness, and mass is

$$S_{TDM} = \frac{1}{\Delta m}\frac{\Delta v_{pa}}{v_{pa}} = -\frac{1}{2M} + \frac{3s_{3D}}{2d_0}\left(1 - \frac{\lambda^2(1+\sigma)}{\pi^2 d_0^2}\right) = \frac{1}{\Delta m}\frac{\Delta f}{f_0} \qquad (8.7)$$

where
   d_0 is the total initial thickness of the device
   s_{3D} is the ratio of the sensing layer swelling along one of three axes per unit change in mass

The new expression for sensitivity consists of two terms: a term describing the mass sensitivity, which is identical to Equation 8.6, and a new stiffness/stress sensitivity term. The stiffness/stress term can be further broken down into a stiffness term, inversely proportional to initial material thickness, and a stress term, inversely proportional to the cube of the initial material thickness. The implication is that improvements in sensitivity can be found by reducing device thickness. If the sensor stress is not negligible, then an increase in substrate stress causes a decrease in stress and stiffness sensitivity, allowing for selective sensitivity to either mass effects or stiffness/stress effects [16].

   As the sensing material swells due to mass absorption, the stiffness term causes a positive shift in frequency. During absorption, the polymer also swells parallel to the substrate, causing stress within the sensing material and the substrate, leading to a negative shift in frequency. The acknowledgement that changes in stress and stiffness can play significant roles in sensor response creates opportunities for new designs that can measure multiple material parameters concurrently but also introduces new challenges associated with signal cross talk and interference [6].

8.2.4 Limitations in Sensing

Although using FPWs creates new opportunities in stiffness and stress sensing, lower-frequency operation, and low-cost fabrication with inexpensive materials, there is a corresponding high sensitivity to temperature and aging effects [15]. Methods for compensating for the sensitivity to temperature have been investigated, but this challenge requires further work before a reliable solution is available [5,9,10].

The significant sensitivity of FPW devices to temperature is due to the impact of changing substrate stress on wave velocity [16]. Small changes in substrate temperature cause expansion and contraction of the substrate, which, when the substrate is bound within a frame, lead to variations in substrate stress. The first method for managing temperature effects involves performing measurements within an enclosure designed to stabilize the temperature of the sensor and carrier gas. An example of such a system is a Neslab refrigerated circulating water bath encompassing a foam sensing chamber [9]. The gas travels through a long coil immersed within the bath to stabilize the carrier gas temperature prior to exposure to the sensor to aid in establishing temperature equilibrium. The enclosure also aids sensing by reducing electromagnetic noise from the environment.

In scenarios where a temperature stabilizing enclosure is not feasible, particularly if cost is a concern, then a sensitive temperature monitoring sensor can provide compensation information to the data acquisition system. The use of this technique requires profiling each sensor's performance at different temperatures, adding many steps to the calibration process. The calibration curve can also be very difficult to fit, is nonlinear, and shifts due to effects of aging. Using a temperature sensor to compensate the FPW sensor also limits the accuracy of the FPW sensor based on the accuracy, repeatability, and sensitivity of the temperature sensor [11].

A third method for reducing the effects of temperature involves the use of a differential measurement. A differential measurement subtracts a waveform influenced by environmental effects and the measurand from a reference that is only affected by the environmental effects. A differential configuration is proposed by Sielmann et al. [15]. A single IDT generates acoustic waveforms that propagate outward in both directions. One receiver is coated with a sensing layer that is exposed to the carrier gas. The other receiver is coated with the same sensing layer on the opposite side of the substrate, which is not exposed to the gas. Consequently, the influences of temperature affect both channels identically, but changes in analyte concentration influence only one of the channels. This method is also proposed to overcome effects of aging, such as polymer creep and relaxation.

PVDF also provides a significant challenge to overcome in sensor attenuation and low electromechanical coupling coefficient. High attenuation causes acoustic wave amplitudes to decay rapidly, reducing the energy transferred from the actuator to the receiver as well as the sensor signal-to-noise ratio (SNR). The low signal strength produced by the acoustic wave at the receiver has been experimentally demonstrated to be equal to or less than the electromagnetic cross talk caused by the electrical actuation signal, requiring either discontinuous excitation of the sensor or extensive electrode shielding. The high attenuation also causes a low-quality factor of the device, detrimentally reducing the precision at which phase and frequency can be measured [15].

8.3 FABRICATION

8.3.1 Materials, Preparation, and Characterization

PVDF is a semicrystalline polymer that is approximately 50% amorphous, in which fluorinated hydrocarbon units form the unit cell as shown in Figure 8.3. It exists in a combination of four crystalline phases: α, β, γ, and δ. Among them, the α-phase is the most common form and is composed of polymer chains that occur in nonpolar trans–gauche–trans–gauche (TGTG) conformation. While α-phase PVDF is nonpolar and non-piezoelectric, β-phase PVDF is highly oriented with polymer chains in *all-trans* zigzag conformation, such that β-phase PVDF can show piezo- and pyroelectric properties [17].

The preparation of piezoelectric β-phase PVDF is well documented and involves multistep processing: (1) Preparation of nonpolar, non-piezoelectric PVDF film (α-phase PVDF). (2) Stretching at high temperature: this process ensures phase transformation of PVDF from α-phase to β-phase. It involves drawing PVDF film mechanically to 4x their original length within an oven at 80°C, at a rate of approximately 3–4 mm/min. Mechanical deformation of the PVDF resulted

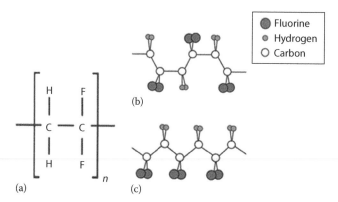

FIGURE 8.3   (a) Chemical structure of PVDF and structure of (b) α-PVDF and (c) β-PVDF.

in necking at the center of sample along with a decrease in their thickness. This process caused the spherulitic structure of the polymer chains to be torn down concurrent with the formation of a microfibrillar structure, which led to the transformation of PVDF from α- to β-phase via molecular chain alignment [18,19]. Figure 8.3 shows the different molecule conformations of α- and β-phase PVDF. (3) *Electric/corona poling*: the PVDF molecular chains are composed of hydrogen and fluorine atoms that act as dipoles [20–22]. Although stretching causes α-phase PVDF to be transformed into β-phase, these dipoles are randomly arranged within it as shown in Figure 8.4, which renders the film non-piezoelectric. Corona poling is a well-known method of aligning the dipoles within β-phase PVDF by exposing the material to a strong electric field at an elevated temperature [23].

A corona poling setup as shown in Figure 8.5, in general, consists of a poling dome above the polymer substrate that is placed on a ground electrode, while the poling dome contains one or several needles (referred to as corona needles) suspended within it. As a high voltage V_N is applied to these needles, the air within the dome ionizes and the electric potential gradient between the needles and the ground plate results in an ion flux directed at the sample. The ions create a charge layer on the surface of the sample, generating a poling field between the sample top surface and the ground plate located beneath. A conducting grid is affixed above the PVDF and held at a constant voltage V_G that controls the potential at the surface of the film. This grid ensures that the charged ions are evenly distributed on the sample surface. The high electric field within the sample causes dipole alignment resulting in a net polarization [20–22]; the mobility of the molecules during this process is enhanced at elevated temperature. The net polarization of PVDF shown in Figure 8.4 causes the polymer chains to collectively exhibit piezoelectric behavior.

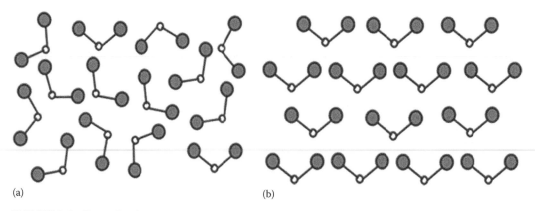

FIGURE 8.4   Example of the alignment of diploes within β-phase PVDF. (a) Random dipole arrangement and (b) aligned dipole arrangement.

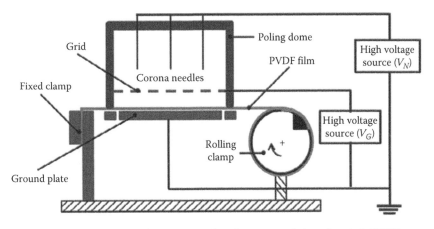

FIGURE 8.5 Schematic of corona poling setup employed to prepared piezoelectric β-PVDF.

The effectiveness of the PVDF preparation process can be assessed by characterizing the PVDF sheet. Available tools include x-ray diffraction (XRD), Fourier transform infrared (FTIR) spectroscopy, and piezoelectric constant d_{33} measurement. All these methods are briefly discussed in the following.

XRD is a powerful nondestructive technique for characterizing crystalline materials; it provides details of material structures, phases, crystal orientation, and other structural details such as crystal defects, grain size, and crystallinity. The crystalline atoms in a material cause a beam of x-rays to diffract into various specific directions; the angle and intensities of these diffracted beams represent the fingerprint of atomic arrangements in a given material. X-ray patterns of α-PVDF and β-PVDF are shown in Figure 8.6a. The pattern for α-PVDF shows reflection peaks at 17.72°, 18.44°, and 20.08°, which represents the semicrystalline structure, mainly composed of α-phase and no peaks corresponding to polar β-crystallites. Stretching and poling of PVDF resulted in appearance of reflections at 20.84° and 36.6° corresponding to β-phase, while reflections corresponding to the α-phase are absent, demonstrating evidence of α- to β-phase transition of PVDF [18].

FTIR is a useful method to study the structural changes in materials and any change in the characteristics of the peak conformations of molecules at the molecular level. When IR radiation is passed through a sample, some of the IR radiation is absorbed by the material and some of it is passed through (transmitted). The resulting spectrum represents molecular absorption and transmission, which can be used to determine information about the molecular structure of the sample. The FTIR spectra of raw and stretched, poled PVDF in the 600–1500 cm^{-1} region are shown in Figure 8.6. α-PVDF exhibits the

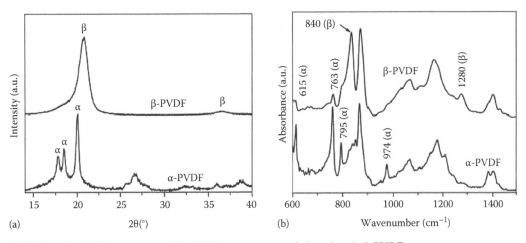

FIGURE 8.6 (a) XRD pattern and (b) FTIR spectroscopy of piezoelectric β-PVDF.

characteristic absorption peaks at 615, 763, 795, and 974 cm^{-1} [18,23,24] and no peaks corresponding to β-phase. After stretching and poling, most of the α-phase peaks have disappeared and new peaks emerge at 840 and 1280 cm^{-1} corresponding to β-phase [18,23–25]. By measuring the absorbance at the characteristic absorbance bands of α-phase (763 cm^{-1}) and β-phase (840 cm^{-1}), β-phase content of the processed PVDF is estimated and is found to 83.3 ± 7%, meaning that the material contains no ternary phase after being processed.

The piezoelectric charge constant, d_{33}, was determined by measuring the charge induced by the PVDF sheet. In this investigation, a 6 mm × 30 mm sample is placed between two electrodes and is subjected to compressive load (0.5 to 3 N to 0.5 N in steps of 0.5 N) to measure the charge generated by the piezoelectric β-PVDF on the electrodes. The piezoelectric charge constant

$$d_{33} = \frac{\text{Induced charge}}{\text{Applied load}} \tag{8.8}$$

of the PVDF sheet was measured at $d_{33} = 34.3 \pm 7.2$ pC/N under the following poling conditions: corona needle voltage $V_N = 15$ kV, grid voltage $V_G = 2$ kV, poling temperature $T_p = 80°C$, and poling time $t_p = 45$ min, showing good agreement with data reported in literature [26,27].

8.3.2 FRAMING

As discussed in Section 8.1, the tension in the PVDF films is an important factor in determining the acoustic wave speed. For accurate results, the tension in the film should be static and known. In Sielmann et al., the PVDF films were clamped into stainless steel frames while under tension [5]. This not only ensured a constant tension in the film but also allowed for easy handling and inkjet printing.

8.3.3 INKJET PRINTING

Inkjet printing is a low-cost method of depositing a wide variety of materials in arbitrary patterns. Several conductive inks are readily available, though most require curing once they have been printed. Poly(3,4-ethylenedioxythiophene)–poly(styrenesulfonate) (PEDOT:PSS) is a polymer-based conductive ink and does not require curing to exhibit conductive behavior. Inkjet printing with PEDOT:PSS offers several advantages for this application since it is relatively low cost and it is a room temperature process, minimizing the possibility of depolarizing the PVDF films.

Various properties of PEDOT:PSS thin films have been extensively studied [28]. Findings by Kim et al. demonstrated that the electrical properties of PEDOT:PSS can be enhanced with the use of solvents such as dimethylsiloxane [29]. The surface wetting properties of PEDOT:PSS inks, as printed with inkjet nozzles, on PVDF have also been studied [30]. It was found that, using an inkjet nozzle with a 20 μm orifice, the minimum achievable PEDOT:PSS track width was 55 μm. Further work has been done concerning the practical aspects of inkjet printing IDTs of PEDOT:PSS on PVDF for the application of FPW sensing [26]. This work demonstrated that printing uniform and repeatable PEDOT:PSS tracks of 100 μm width on PVDF is possible using the multilayer printing process illustrated with an example in Figure 8.7.

Inkjet printing can also be used to pattern various sensing materials on a completed FPW device. The use of FPW devices for VOC sensing, for example, requires the application of a polymer sensing layer and utilizing polymers that are selectively sorptive to certain VOCs. Many of these polymer layers can be dissolved in the appropriate solvents to produce a solution that can be inkjet printed. This method allows the deposition of polymer layers in controlled patterns and densities. In the work by Sielmann et al., inkjet printing was used to deposit layers of polyvinyl acetate (PVA) on an FPW device to characterize the response of the device to mass loading [5]. A solution of 2%wt, 85,000–124,000 MW PVA in distilled water was used.

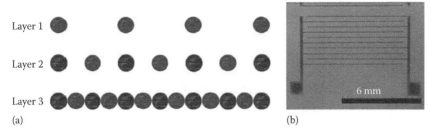

FIGURE 8.7 Schematic of multilayer printing process (a) and image of a printed PEDOT:PSS IDT on PVDF with 100 μm track widths and λ = 800 (b).

8.3.4 FPW DEVICE TESTING

During the development of PVDF FPW devices, it is important to characterize the efficacy of the various steps outlined earlier. Some of the tools that can be used to study these include laser Doppler vibrometry (LDV) and electrical characterization. Each of these methods is summarized in the following with further information available through various publications [26,31].

Scanning LDV is useful for visualizing the out-of-plane vibrations of the surface of the PVDF films resulting from the electromechanical actuation using the printed IDTs. The propagation of acoustic waves along the substrate depends on the geometry of the IDTs, the electromechanical coupling to the PVDF substrate, and the mechanical properties of the substrate. Visualizations of these acoustic waves, as well as velocity, displacement, and frequency spectrum data, can provide much insight into the operation of a FPW device, including information about the acoustic wave speed, fundamental frequency, and other dominant modes.

Through the electromechanical coupling of the IDTs, an electrical signal can be converted to an acoustic signal, which propagates through the PVDF substrate and is converted back to electrical energy at a receiving IDT. By applying appropriate electrical signals and measuring the resulting electrical output, the frequency, phase, and amplitude can be determined. The measured phase is representative of the time taken for the acoustic wave to reach the receiving IDT after it leaves the generating IDT and relates to the wave speed in the substrate. These data can be digitally processed and logged to determine the response of the sensors to various stimuli. This measurement setup allows for accurate measurement of the device parameters without the need for specialized equipment such as an LDV.

8.4 FPW DEVICE PERFORMANCE

The PVDF FPW device is unique among guided acoustic sensors in that it allows the detection of polymer sensing layer stress and stiffness, along with mass loading. As such, recent investigations have focused on the detection through these different modes. The ability to measure simultaneous changes in sensing layer density, thickness, and Young's modulus was demonstrated in a series of mass-loading and gas exposure tests. In the former, successive layers of polymer were applied to a sensor and the shift in frequency was measured [5]. In the latter, sensing layers of varying thickness were exposed to different concentrations of an analyte. The shift in frequency and phase driven by analyte absorption was then measured [6,15]. These tests involved 18 μm thick, 800 μm wavelength sensors under 125 MPa of in-plane tension. One set of transmitting IDTs was applied to each device. Two receiving IDTs were patterned on either side of the transmitting transducer such that a differential signal could be measured. For each investigation, results were compared with models developed using the relationships described in Section 8.2.

A schematic of the setup used to interrogate the sensors is shown in Figure 8.8. An excitation centered at 200 kHz is used to induce acoustic waves at the transmitting IDTs. The signal is detected at the receiving IDTs and fed to a 10× to 50× gain amplifier before being passed to an ADC and DSP.

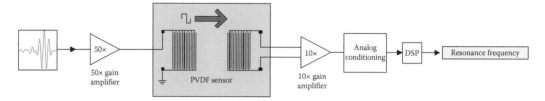

FIGURE 8.8   Schematic of the PVDF FPW sensor test setup.

8.4.1 MASS LOADING

Common FPW acoustic devices leverage gravimetric sensing to detect the presence of an analyte. As described by Equation 8.1, phase velocity and frequency of a zeroth order antisymmetric plate wave are inversely proportional to the mass density of the medium. An increasing concentration of analyte in a sensing layer adjacent to the substrate should result in a decreasing resonant frequency. Given the relative softness of the PVDF substrate, however, this trend does not always hold. A recent publication details an investigation into the mass-loading performance of an all-polymer FPW device [5]. In this work, successively thicker layers of polyvinyl acetate film were applied to a sensor and the resulting shift in resonant frequency was measured. The frequency vs. mass-loading plot produced demonstrates both sensitivity to mass changes and stiffness changes.

Prior to testing, the device was interrogated using LDV in order to confirm the existence of planar Lamb waves and measure the resonant frequency. Samples were fabricated with 20–125 µm thick substrates and wavelengths of 800 µm. While being imaged by the LDV, each device was excited with a $150V_{pp}$ periodic chirp. Figure 8.9 shows the frequency response of a 125 µm thick sensor. Antisymmetric waves were observed here with a resonant frequency of 335 kHz.

Mass-loading tests were carried out by drop casting successively thicker layers of polyvinyl acetate (PVA) over half of the sensor and measuring the resulting change in frequency on one channel. The signal from the unloaded reference channel was used to compensate for any variation caused by the changing environment. Figure 8.10 shows the frequency response of a device as a function of time and PVA mass loading.

For the addition of thin layers of PVA, the frequency drops, indicating that the device is acting as a gravimetric sensor. Above a certain mass loading, the frequency starts to increase as more PVA is added, indicating that the device is responding to changes in sensing layer stiffness. Due to the low Young's modulus of the substrate, the addition of PVA changes the overall stiffness of the propagation medium significantly, along with the mass loading.

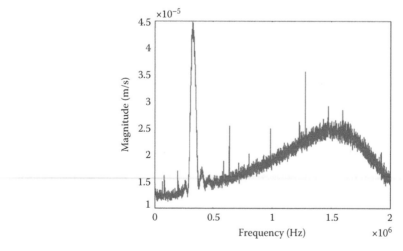

FIGURE 8.9   Average out-of-plane velocity of a 125 µm thick PVDF FPW sensor measured via LDV.

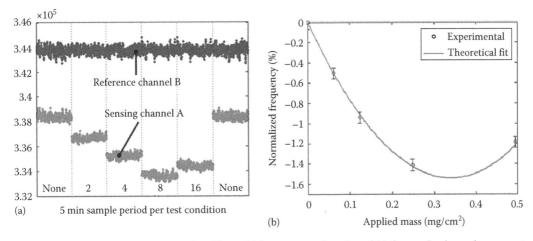

FIGURE 8.10   Frequency response of an 18 μm thick sensor as a function of (a) time and polymer layer count and (b) applied mass loading.

Equation 8.1 was used to predict the response given the added mass and thickness of the polymer layer. In the linear region of the curve near the zero applied mass, the predicted mass sensitivity was -156 cm^2/g while the measured sensitivity was -153 cm^2/g, showing good agreement with expectations.

8.4.2   Gas Sensing

As the sensing layer of a polymer FPW device is exposed to an analyte that will dissolve it, the velocity of the acoustic wave will shift due to mass loading, softening, and expansion. The concentration of that gas may then be calculated using Equation 8.1.

Gas sensing has been demonstrated in two recent publications [6,15]. In these investigations, a device was exposed to air with water vapor ranging from 10% to 60% RH and the resulting frequency and phase shift was measured. The sensing channel was coated in 1.9–12.6 μm of PVA while the reference channel was coated in a 1.9 μm film of PVA. Water vapor was generated and measured using an Owlstone OVG-4 gas calibration system. Samples were exposed at room temperature and pressure.

Figure 8.11 shows the results from one set of tests. As water vapor concentration increases, the phase of the signal decreases, along with frequency (not shown). Given knowledge of the partition coefficient for water vapor and PVA, it is possible to predict the relative humidity given the phase shift and frequency shift it induces.

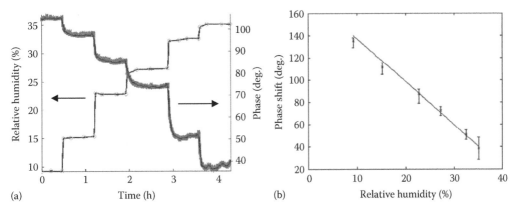

FIGURE 8.11   (a) Phase response of a PVA-coated FPW sensor to varying concentrations of water vapor. The sensor is 18 μm thick with a 6.6 μm thick sensing layer. The phase measured corresponds to a signal phase shift measured at the receiver. (b) Phase shift vs. RH of the PVA-coated sensor.

8.4.3 POLYMER CHARACTERIZATION

PVDF FPW devices have the unique property of being sensitive to stiffness and tension changes in the sensing layer as shown in Section 8.2. In a recent publication [6], it has been demonstrated that this can be exploited to characterize the mechanical properties of a thin polymer film as a function of an analyte concentration.

As a polymer absorbs a solvent, it will gain mass, swell, soften, and relax. The combined effect of these changes on the resonant frequency of a polymer acoustic sensor will produce a unique frequency vs. analyte concentration curve for a given polymer–solvent pair and film geometry. This curve may be fit to Equation 8.1 to determine the Young's modulus, density, thickness, and tension of the film as a function of solvent concentration.

In [6], PVA films of three different thicknesses are exposed to varying concentrations of water vapor and the resulting frequency shifts were fit such that the physical data of the film as a function of humidity could be extracted. The results were compared with material characteristics obtained from literature and experiment. PVA density, Young's modulus, and thickness as a function of water concentration were obtained from past experiments, while partition coefficient was measured directly by weighing a sample of PVA exposed to a constant humidity over a long period of time.

Figure 8.12 shows the frequency vs. concentration curves for the device tested in [6]. As relative humidity increases, the frequency falls suggesting that rising mass density, falling Young's modulus, and falling tension are driving the resonant shift. For thicker PVA layers, the unexposed resonant frequency increases significantly. This is caused by the thickness-driven increase in stiffness of the sensing layer.

The predicted and measured film properties were compared in a dry state at 0% RH and a saturated state at 52% RH. With the exception of Young's modulus, the measured and predicted dry characteristics of the film agreed within 3%. The error increases significantly for saturated values of mass and thickness. This variation is attributed to differences in hydrolyzation, density, and molecular mass of the PVA used in the investigation and that referenced for comparison.

8.4.4 LIMITATIONS IN FPW DEVICE PERFORMANCE

Performance of PVDF-based FPW devices is primarily limited by the strong attenuation of acoustic waves and low electromechanical coupling coefficient in PVDF. As discussed in Section 8.2.4, this damping leads to low SNR, which results in a reduced frequency resolution. The decimated acoustic amplitude also contributes to applied polymer film thickness limitations. Thicker polymer films adjacent to the substrate result in greater attenuation. For an 18 μm PVDF substrate coated with PVA, the maximum sensing layer thickness was about 15 μm.

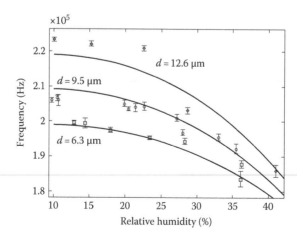

FIGURE 8.12 Frequency response of a PVA-coated 18 μm thick, 800 μm wavelength FPW sensor to varying RH concentrations given different sensing layer thicknesses, d. The solid curves correspond to fit results.

When designing a device to act as a gas detector, the properties of the sensing layer also play a role in defining the sensitivity of the device. A polymer with a high partition coefficient for a given solvent is necessary to detect small quantities of vapor. If possible, the polymer swelling to mass change ratio, s_{3D}, and thickness must also be selected to allow maximum mass sensitivity or thickness sensitivity. s_{3D} is an empirical constant that describes the amount that the polymer expands per unit added mass. As shown in Equation 8.7, it affects the stiffness sensitivity of a device. A thinner film with a smaller s_{3D} will result in greater mass sensitivity than a thicker film with a greater swelling to mass ratio, for example. Choosing a polymer with a well-characterized s_{3D} will allow devices to be tailored to primarily detect mass or stiffness changes.

Performance is also limited by environmental characteristics. In order to maintain the piezo-electricity of the substrate, the device must be kept well below the poling temperature of PVDF, 80°C. Similarly, in order to account for pyroelectric signal generation and thermal mechanically induced stress variation in the substrate, a reference channel must be used in tandem with the sensing channel.

8.5 CONCLUSIONS

Although acoustic sensing using FPWs and an all-polymer, printed device has been demonstrated for sensing applications, the low signal strength and corresponding poor sensitivity of the sensors indicate that further research is required to overcome the inherent material limitations in polymer substrates such as PVDF. Innovation in the area, including the use of reference electrode channels, concurrent sampling of multiple sensor measurands, inkjet and screen printing of organic conductive polymers, and discontinuous excitation and sampling demonstrate ongoing efforts to resolve these limitations.

Two advancements that can lead to a significant improvement in performance are a reduction in substrate thickness and a reduction in substrate stress. Reducing the former increases the mass sensitivity and signal strength of the sensor, whereas reducing the latter increases the stress and stiffness sensitivity and lowers the operating frequency. As PVDF and similar piezoelectric polymers are improved and micron-thick piezoelectric films become possible, the benefits to mass, stiffness, and stress sensing in all-polymer FPW devices can be substantial.

Further improvements in the SNR and sensitivity of the sensing platform can lead to many new and interesting applications. The use of FPWs allows for the application of the sensing polymer on the underside of the sensor, and the chemical resiliency of PVDF permits the detection of potentially hazardous and corrosive chemicals without interfering with sensor electrodes. Low cost of the sensor creates opportunities in biological sensing where the sensor is disposed of after the test to avoid contamination. The sensitivity of the sensor substrate to mass, stress, and stiffness can enable thorough examination of the drying or hardening behavior of polymers, elastomers, epoxies, and other substances that undergo structural changes at room temperature. The combination of these attributes has the potential to address needs within the oil and gas and chemical manufacturing sectors.

REFERENCES

1. P. Inacio, J. N. Marat Mendes, and C. J. Dias, Development of a biosensor based on a piezoelectric film, in *11th International Symposium on Electrets, 2002. ISE 11. Proceedings*, 2002, pp. 287–290.
2. L. F. Brown and J. L. Mason, Disposable PVDF ultrasonic transducers for nondestructive testing applications, *IEEE Transactions on Ultrasonics, Ferroelectrics and Frequency Control*, 43(4), 560–568, 1996.
3. R. S. C. Monkhouse, P. D. Wilcox, and P. Cawley, Flexible interdigital PVDF transducers for the generation of Lamb waves in structures, *Ultrasonics*, 35(7), 489–498, 1997.
4. P. W. Walton, M. R. O'Flaherty, M. E. Butler, and P. Compton, Gravimetric biosensors based on acoustic waves in thin polymer films, *Biosensors and Bioelectronics*, 8(9–10), 401–407, 1993.
5. C. Sielmann, J. R. Busch, B. Stoeber, and K. Walus, Inkjet printed all-polymer flexural plate wave sensors, *IEEE Sensors Journal*, 13(10), 4005–4013, 2013.

6. C. Sielmann, J. Berring, K. Walus, and B. Stoeber, Application of an all-polymer flexural plate wave sensor to polymer/solvent material characterization, in *2012 IEEE Sensors*, Taipei, Taiwan, 2012, pp. 1–4.

7. D. M. G. Preethichandra and K. Kaneto, SAW sensor network fabricated on a polyvinylidine difluoride (PVDF) substrate for dynamic surface profile sensing, *IEEE Sensors Journal*, 7(5), 646–649, 2007.

8. S. W. Wenzel and R. M. White, Flexural plate-wave gravimetric chemical sensor, *Sensors and Actuators A: Physical*, 22(1–3), 700–703, 1989.

9. J. W. Grate, S. W. Wenzel, and R. M. White, Flexural plate wave devices for chemical analysis, *Analytical Chemistry*, 63(15), 1552–1561, 1991.

10. Q.-Y. Cai, J. Park, D. Heldsinger, M.-D. Hsieh, and E. T. Zellers, Vapor recognition with an integrated array of polymer-coated flexural plate wave sensors, *Sensors and Actuators B: Chemical*, 62(2), 121–130, 2000.

11. S. W. Wenzel, Applications of ultrasonic lamb waves, Doctoral dissertation, Berkeley, CA, 1992.

12. J. W. Grate, S. J. Martin, and R. M. White, Acoustic wave microsensors, *Analytical Chemistry*, 65(21), 940A–948A, 1993.

13. J. D. N. Cheeke, *Fundamentals and Applications of Ultrasonic Waves*, 1st edn. CRC Press, Boca Raton, FL, 2002.

14. S. W. Wenzel and R. M. White, Analytic comparison of the sensitivities of bulk-wave, surface-wave, and flexural plate-wave ultrasonic gravimetric sensors, *Applied Physics Letters*, 54(20), 1976–1978, 1989.

15. C. Sielmann, Design and performance of all-polymer acoustic sensors, The University of British Columbia, Vancouver, British Columbia, Canada, 2012.

16. C. Sielmann, B. Stoeber, and K. Walus, Implications of a low stiffness substrate in lamb wave gas sensing applications, in *2012 IEEE Sensors*, Taipei, Taiwan, 2012, pp. 1–4.

17. T. R. Dargaville, M. C. Celina, J. M. Elliott, P. M. Chaplya, G. D. Jones, D. M. Mowery, R. A. Assink, R. L. Clough, and J. W. Martin, Characterization, performance and optimization of PVDF as a piezoelectric film for advanced space mirror concepts, Sandia Labs Report (SAND2005-6846), Sandia Labs, Albuquerque, NM, 2005.

18. A. Salimi and A. A. Yousefi, Analysis method: FTIR studies of β-phase crystal formation in stretched PVDF films, *Polymer Testing*, 22(6), 699–704, 2003.

19. R. Hasegawa, Y. Takahashi, Y. Chatani, and H. Tadokoro, Crystal structures of three crystalline forms of poly(vinylidene fluoride), *Polymer Journal*, 3(5), 600–610, 1972.

20. T. Furukawa, Ferroelectric properties of vinylidene fluoride copolymers, *Phase Transitions*, 18(3–4), 143–211, 1989.

21. X. Yang, X. Kong, S. Tan, G. Li, W. Ling, and E. Zhou, Spatially-confined crystallization of poly(vinylidene fluoride), *Polymer International*, 49(11), 1525–1528, 2000.

22. D. M. Esterly, Manufacturing of poly(vinylidene fluoride) and evaluation of its mechanical properties, Master's thesis, Virginia Tech, Blacksburg, VA, 2002.

23. J. Hong, J. Chen, X. Li, and A. Ye, Effects of the bias-controlled grid on performances of the corona poling system for electro-optic polymers, *International Journal of Modern Physics B*, 19(14), 2205–2211, 2005.

24. R. Gregorio, Determination of the α, β, and γ crystalline phases of poly(vinylidene fluoride) films prepared at different conditions, *Journal of Applied Polymer Science*, 100(4), 3272–3279, 2006.

25. S. Lanceros-Méndez, J. F. Mano, A. M. Costa, and V. H. Schmidt, FTIR and DSC studies of mechanically deformed β-PVDF films, *Journal of Macromolecular Science, Part B*, 40(3–4), 517–527, 2001.

26. J. R. Busch, All polymer flexural plate wave sensors, The University of British Columbia, Vancouver, British Columbia, Canada, 2011.

27. S. K. Mahadeva, J. Berring, K. Walus, and B. Stoeber, Effect of poling time and grid voltage on phase transition and piezoelectricity of poly(vinyledene fluoride) thin films using corona poling, *Journal of Physics D: Applied Physics*, 46(28), 285305, 2013.

28. D. J. Lipomi, J. A. Lee, M. Vosgueritchian, B. C.-K. Tee, J. A. Bolander, and Z. Bao, Electronic properties of transparent conductive films of PEDOT:PSS on stretchable substrates, *Chemistry of Materials*, 24(2), 373–382, 2012.

29. J. Y. Kim, J. H. Jung, D. E. Lee, and J. Joo, Enhancement of electrical conductivity of poly (3,4-ethylenedioxythiophene)/poly (4-styrenesulfonate) by a change of solvents, *Synthetic Metals*, 126, (2–3), 311–316, 2002.

30. G. Man, Towards all-polymer surface acoustic wave chemical sensors for air quality monitoring, The University of British Columbia, Vancouver, British Columbia, Canada, 2009.

31. J. R. Busch, C. Sielmann, G. Man, D. Tsan, K. Walus, and B. Stoeber, Inkjet printed all-polymer flexural plate wave sensors, Presented at the *25th IEEE International Conference on Micro Electro Mechanical Systems*, Paris, France, 2012.

9 Whispering Gallery Microcavity Sensing

Serge Vincent, Xuan Du, and Tao Lu

CONTENTS

9.1 INTRODUCTION

The whispering gallery at Saint Paul's Cathedral, London, is known to be able to convey whispers tens of meters away along the gallery wall. The mechanism that triggers these effects, later discovered by scientists [1], is due to the fact that the closed gallery wall effectively forms a waveguide where a whisper can deliver sound energy over long distances without significant loss. Ever since the first scientific interpretation of the whispering gallery, researchers began to speculate over its optical counterpart. The first theory-based investigation of optical whispering gallery microresonators (WGMs) was published as early as 1939 [2], while the experimental demonstration was pioneered by several research groups in the late 1970s to early 1980s [3–5]. Currently, WGMs are the subject of many applications ranging from nonlinear optics, low-pump-power narrow-linewidth lasers, nanodetection, to biosensing. Starting from a review on the basic properties of a whispering gallery microcavity, we will investigate microresonators from a biodetection technology perspective. Various types of WGM cavities and their fabrication procedures will be reviewed in the following section, while their ultrahigh sensitivity in a subset of the previous applications will be highlighted.

9.2 BASICS OF WHISPERING GALLERY MICRORESONATORS

The operational principles of a WGM can be explained in terms of geometrical optics to some extent. Consider a general whispering gallery mode cavity, as illustrated in Figure 9.1a, with refractive index n_1 and a surrounding medium with refractive index n_2. When a light ray is continuously fed into and propagating inside the cavity from a plane of origin at $\phi = 0°$, it will bounce back at the edge of the cavity due to reflection. In particular, if the ray strikes the edge at an angle larger than the critical angle defined by Snell's law

$$\phi_c = \sin^{-1}\left(\frac{n_2}{n_1}\right) \tag{9.1}$$

total internal reflection will occur, and thus the ray will be fully contained within the cavity without any loss due to refraction. The ray will continually leap forward at the cavity edge, building a strong resonance field if the optical path the photon travels coincides with an integer multiple of its wavelength. A more quantitative description of the whispering gallery cavity requires the adoption of Maxwell's equations:

$$\vec{\nabla} \times \vec{E}(\vec{r},t) = -\frac{\partial \vec{B}(\vec{r},t)}{\partial t}$$

$$\vec{\nabla} \times \vec{H}(\vec{r},t) = \frac{\partial \vec{D}(\vec{r},t)}{\partial t} + \vec{j}_f(\vec{r},t) \tag{9.2}$$

$$\vec{\nabla} \cdot \vec{D}(\vec{r},t) = \rho_f(\vec{r},t)$$

$$\vec{\nabla} \cdot \vec{B}(\vec{r},t) = 0$$

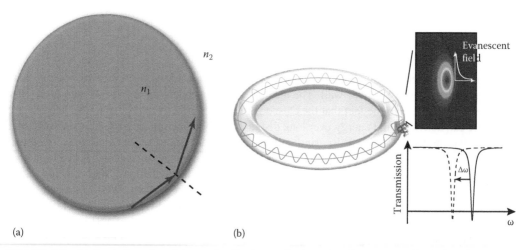

(a)   (b)

FIGURE 9.1 (a) Depiction of total internal reflection at the boundary between a spherical cavity and its surrounding environment, as described by ray optics. (b) Whispering gallery mode resonance of a glass microtoroid, consisting of an energy exchange within the evanescent field by which the adsorbed biosample is polarized and an increase in optical path length (i.e., along the circular trajectory) takes place. Adjacent to this is an illustration of the associated resonance frequency shift $\Delta\omega$, resulting in the displacement of a transmission dip.

where, for the cavities under study, we assume that the free current density $\vec{j}_f(\vec{r}, t)$ and free charge density $\rho_f(\vec{r}, t)$ vanish. The electrical field $\vec{E}(\vec{r}, t)$, electrical displacement $\vec{D}(\vec{r}, t)$, magnetic field $\vec{B}(\vec{r}, t)$, and magnetic field strength $\vec{H}(\vec{r}, t)$ are related to the material permittivity $\varepsilon = \varepsilon_r \varepsilon_0$ and permeability $\mu = \mu_r \mu_0$ according to

$$\vec{D}(\vec{r}) = \epsilon(\vec{r}, \omega_0)\vec{E}(\vec{r})$$

$$\vec{B}(\vec{r}) = \mu(\vec{r}, \omega_0)\vec{H}(\vec{r})$$

(9.3)

Here, we assume the electromagnetic field is monochromatic with optical angular frequency ω_0 such that

$$\left\{\vec{D}(\vec{r},t), \vec{E}(\vec{r},t), \vec{B}(\vec{r},t), \vec{H}(\vec{r},t)\right\}^T = \left\{\vec{D}(\vec{r}), \vec{E}(\vec{r}), \vec{B}(\vec{r}), \vec{H}(\vec{r})\right\}^T e^{j\omega_0 t}$$

(9.4)

By taking the curl of the first two Maxwell's equations and applying a vector identity, we obtain the Helmholtz equation that governs the wave propagation behavior in a whispering gallery cavity:

$$\left[\frac{\partial^2}{\partial\rho^2} + \frac{1}{\rho}\frac{\partial}{\partial\rho} + \frac{1}{\rho^2}\frac{\partial^2}{\partial\phi^2} + \frac{\partial^2}{\partial z^2} + k_0^2 \tilde{n}^2(\vec{r})\right]\left\{\begin{matrix} \vec{E}(\vec{r}) \\ \vec{H}(\vec{r}) \end{matrix}\right\} = \vec{0}$$

(9.5)

For convenience, we select a cylindrical coordinate system that is concentric with the whispering gallery microcavity. A vacuum wave number $k_0 = 2\pi/\lambda_0 = \omega_0/c$ and a complex refractive index $\tilde{n}$ derived from the relative permittivity $\varepsilon_r = \tilde{n}^2$ are adopted. Note that light absorption from a material is quantified by the imaginary part of the refractive index. Also, we assume the materials that form the cavity and its surrounding environment are nonmagnetic (i.e., $\mu = \mu_0$). For an ideal whispering gallery cavity, the refractive index is independent of the azimuth (i.e., $\tilde{n}(\vec{r}) = \tilde{n}(\rho, z)$) and so we may separate the transverse and azimuthal dependence of $\vec{E}$:

$$\left\{\begin{matrix} \vec{E}(\vec{r}) \\ \vec{H}(\vec{r}) \end{matrix}\right\} = \left\{\begin{matrix} \vec{E}(\rho, z) \\ \vec{H}(\rho, z) \end{matrix}\right\} G(\phi)$$

(9.6)

As a result, the azimuth-dependent term becomes

$$\frac{\partial^2}{\partial\phi^2} G(\phi) = -m^2 G(\phi)$$

(9.7)

where $m = m_r + jm_i$ is, in general, a complex constant with real and imaginary components m_r and m_i such that

$$\left\{\begin{matrix} \vec{E}(\vec{r}) \\ \vec{H}(\vec{r}) \end{matrix}\right\} = \left\{\begin{matrix} \vec{E}(\rho, z) \\ \vec{H}(\rho, z) \end{matrix}\right\} e^{-m_i\phi} e^{jm_r\phi}$$

(9.8)

Evidently, m_i determines the loss of light energy along the propagation path in the azimuthal direction. The transverse field components can be separated to construct the mode equations:

$$\left[\rho^2\frac{\partial^2}{\partial\rho^2}+\rho\frac{\partial}{\partial\rho}+\rho^2\frac{\partial^2}{\partial z^2}+k_0^2\rho^2\tilde{n}^2(\rho,z)\right]\left\{\begin{matrix}\vec{E}(\rho,z)\\\vec{H}(\rho,z)\end{matrix}\right\}=m^2\left\{\begin{matrix}\vec{E}(\rho,z)\\\vec{H}(\rho,z)\end{matrix}\right\} \tag{9.9}$$

Similar to the mode equations in the case of a straight waveguide, the second equation determines a set of nonzero orthogonal field patterns in the absence of external sources, known as the whispering gallery modes. By discretizing the mode equation into an eigenvalue form, one may accumulate a set of eigenvalues m^2 with corresponding eigenvectors $\{\hat{e}(\rho, z), \hat{h}(\rho, z)\}^T$ representing the whispering gallery modes. For simplicity, we assume the modal power is normalized to 1 W at the cavity cross section:

$$\iint d\rho dz[\hat{e}^*(\rho,z)\times\hat{h}(\rho,z)]=1\text{ W} \tag{9.10}$$

Furthermore, we assume $P_{in} = a \cdot a^*$ watts of light is delivered to the cavity mode as a continuous wave mode at $\phi = 0$ so that one may obtain the accumulated EM field components $\vec{E}_T$ and $\vec{H}_T$ as

$$\left\{\begin{matrix}\vec{E}_T(\rho,z)\\\vec{H}_T(\rho,z)\end{matrix}\right\}=a\sum_{p=0}^{\infty}e^{jp(2\pi m)}\left\{\begin{matrix}\hat{e}(\rho,z)\\\hat{h}(\rho,z)\end{matrix}\right\}=\frac{a}{1-e^{-2\pi m_i}e^{j2\pi m_r}}\left\{\begin{matrix}\hat{e}(\rho,z)\\\hat{h}(\rho,z)\end{matrix}\right\} \tag{9.11}$$

and the intracavity power P_{cav} as

$$P_{cav}=\frac{P_{in}}{1+e^{-4\pi m_i}-2e^{-2\pi m_i}\cos(2\pi m_r)} \tag{9.12}$$

Evidently, for a fixed input power, the intracavity power is maximized when m_r is an integer. Consequently, we define the light wavelength that leads to m_r being an integer as the cavity resonance wavelength λ_r, since the cavity is in a state of resonance under this condition. Employing the definition of quality factor Q of a resonator, we arrive at

$$Q=2\pi\frac{\text{Energy stored in the cavity}}{\text{Energy dissipated per cycle}}=\frac{m_r}{2m_i} \tag{9.13}$$

Illustrated in Equations 9.8 and 9.9, the loss of an ideal whispering gallery microcavity consists of two mechanisms. One is due to the nonzero imaginary part of the refractive index or the material absorption from the cavity and its surrounding medium. Its contribution to the quality factor is known as Q_{abs}. Another mechanism is due to the first-order dissipative term in Equation 9.9, which leads to a radiation loss-related quality factor contribution Q_{rad}. In reality, other factors such as Rayleigh scattering arising from the surface roughness of the cavity and moisture-derived water layers at the cavity surface may introduce additional loss. Their contributions to quality factor are named Q_{scatt} and Q_{surf}. Consequently, an intrinsic quality factor Q_0 of a whispering gallery microcavity is defined as the quality factor arising from the overall losses intrinsic to the cavity, expressed as

$$\frac{1}{Q_0}=\frac{1}{Q_{rad}}+\frac{1}{Q_{scat}}+\frac{1}{Q_{surf}}+\frac{1}{Q_{abs}} \tag{9.14}$$

Conventionally, light is delivered to or extracted from the cavity through a coupling component such as a tapered waveguide [6] or microprism. Since the extraction of light from the cavity can be regarded as a loss mechanism, one defines a coupling or extrinsic quality factor Q_c as to characterize the portion of energy lost from the cavity. Therefore, the overall quality factor Q_T follows

$$\frac{1}{Q_T} = \frac{1}{Q_0} + \frac{1}{Q_c} \qquad (9.15)$$

Assuming light with total power P_i and optical angular frequency ω_0 is carried from, for example, a tapered waveguide to the cavity, we obtain an intracavity power of

$$P_{cav} = \frac{\dfrac{\omega_0}{\tau_0 Q_c}}{\dfrac{\omega_0^2}{4}\left(\dfrac{1}{Q_c} + \dfrac{1}{Q_0}\right)^2 + \Delta\omega^2} P_i \qquad (9.16)$$

where
   τ_0 is the photon round-trip time
   $\Delta\omega = \omega_0 - \omega_r$ is the frequency detune

The intracavity power is maximized when the cavity is in resonance and critical coupling $Q_c = Q_0$ occurs, wherein the intracavity power becomes

$$\frac{P_{cav}}{P_i} = \frac{Q_0}{2\pi m_r} \qquad (9.17)$$

A typical silica microcavity immersed in water has an intrinsic quality factor as high as 1.8×10^9 at 633 nm as derived from material absorption coefficient of fused silica and water [7,8], for a diameter around 100 μm and a mode order $m_r = 707$; therefore, an intracavity power as high as 4.1 W can be reached for such a structure when probed by a typical external cavity laser source with 10 mW of power. Given that a whispering gallery mode has a tight mode field diameter on the order of 1–2 μm (1.5 μm wide, 3 μm high, and edge at $1/e$ of the maximum intensity as illustrated in Figure 9.1b), a maximum intracavity intensity of 2.6×10^{12} W/m^2 can be acquired. Such a high intensity along with an ultra-narrow cavity resonance linewidth enables WGMs to become powerful tools to detect nanoscale biological samples adsorbing to the surface of a cavity.

9.3 MICROCAVITY MATERIALS

Numerous categories of small-mode-volume, ultrahigh-Q resonant cavities exist. An optical resonator can be classified by the materials it is composed of and its geometry, both impacting the lower limit to its sensing resolution. Benefiting from its ultralow loss, fused or thermally grown silica (SiO_2) is extensively selected as a microcavity material. Silicon, on the other hand, is an alternate material that has garnered continual interest from the scientific community given that it is compatible with silicon photonics. Silicon nitride (SiN), whose high nonlinear properties have remained a focus of research into nonlinear optics, is also used in fabricating whispering gallery microcavities. Electro-optical crystals, such as lithium niobate ($LiNbO_3$), have been explored for fabricating potentially electrically tunable microcavities. The III–V semiconductors, such as gallium arsenide (GaAs), have also been used in the fabrication of functionalized laser microcavities. It is important, however, to recognize that the record quality factors of 6.3×10^{10} for microcavities were obtained from cavities made of calcium fluoride (CaF_2) crystals [9].

9.4 MICROCAVITY STRUCTURES

Some of the most prominent ultrahigh-Q microcavities that have been researched, which support whispering gallery modes, include microspheres, microdisks, microtoroids, ring resonators, bottleneck resonators, microbubbles, double disks, and liquid-core optical ring resonators or LCORRs (Figure 9.2).

9.4.1 SILICA MICROSPHERE

The preliminary procedure for creating SiO_2 microspheres consists of stripping a commercial optical fiber, followed by etching with a buffered hydrofluoric acid (HF) solution for size control (hence, shrinking the mode volume). A focused CO_2 laser module's laser pulse, with an emitted light wavelength of roughly 10.6 μm that corresponds to high absorption by silica, can then be used to melt the tip of a fiber and thus form a microsphere—a process that is referred to as *reflow*. Other reflow procedures, such as melting the tip with a hydrogen–oxygen flame, have also been elaborated. While cooling, the surface of the microresonator becomes naturally smooth due to surface tension. Up to this date, the record quality factor nearest to theoretical limits has been reported to be $0.8 \pm 0.1 \times 10^{10}$ at a wavelength close to 633 nm [10]. This quality factor in the near-infrared regime, however, dropped to 10^8 within hours after fabrication due to moisture forming a thin water layer, resulting in substantial absorption of light.

9.4.2 MICRODISK

Functional microdisk WGMs often require more elaborate lithography and etching steps, yet they are advantageous due to their compatibility with integrated circuits. A typical silica microdisk is fabricated through the following procedure:

1. An oxide is thermally grown on a silicon wafer. Alternatively, the silica thin film can be fabricated through other standard processes such as flame hydrolysis deposition (FHD), chemical vapor deposition (CVD), plasma-enhanced CVD (PECVD), or the sol-gel method.
2. A layer of photoresist is then coated on the wafer by a spin coater.
3. A soft baking process takes place.
4. Patterning processes are executed to transfer the microdisk image from a chromium mask to the wafer through ultraviolet (UV) light exposure.
5. The patterned wafer is hard baked to smooth the edge of the disk.
6. After removing the unexposed photoresist for a film development process, the sample is immersed in buffered HF as to remove the unprotected silica and form a microdisk.
7. A xenon difluoride isotropic etching procedure is then applied to create the silicon pillar that is to support the fabricated disk.

Due to the smooth wedge created in the fabrication process, a decent quality factor in the mid-10^7 range can be obtained [11]. With procedural modifications, such as e-beam lithography or the use of a stepper to attain finer resolution, microcavities of this kind with Q-factors of 875 million can be made [12]. Microdisks made from other materials, such as silicon, have been demonstrated where a quality factor as high as 10^6 was obtainable [13].

9.4.3 SILICA MICROTOROID

Microtoroids are frequently fabricated by reflowing a microdisk with a CO_2 laser source to boost quality factors by an order of magnitude, to 5×10^8 [14]. As one of the only on-chip ultrahigh-Q microcavities demonstrated by the time of its invention, silica microtoroids have been extensively used in the study of nonlinear optics [15], frequency microcomb generation [16], cavity optomechanics [17], cavity quantum electrodynamics [18], low threshold power, narrow-linewidth laser sources [19], as well as ultrasensitive biosensors [20].

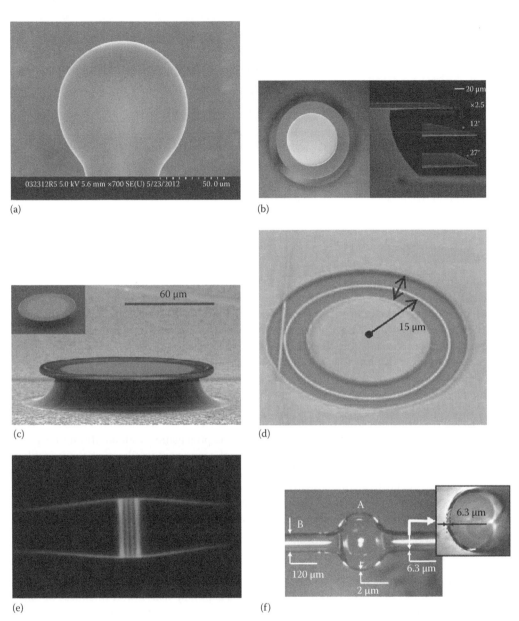

FIGURE 9.2 Assortment of microcavities, with quality factors Q extracted from the literature. The entries from top to bottom, left to right, are (a) a typical silica microsphere – $Q \approx 1010$ at 633 nm, in air [10]; (b) wedge resonator – $Q \approx 109$ within the 1500 nm band, in air [12]. (Reprinted from Lee, H. et al., *Nat. Photon.*, 6, 370, 2012.); (c) silica microtoroid – $Q \approx 108$ within the 1500 nm band, in air [14]. (Reprinted from Armani, D. et al., *Nature*, 421, 926, 2003.); (d) silicon ring resonator – $Q \approx 105$ at 650 nm, in water [21]. (Reprinted from Iqbal, M. et al., *IEEE J. Sel. Top. Quantum Electron.*, 16, 655, 2010.); (e) bottleneck microresonator – $Q \approx 10^8$ at 850 nm, in air [22]. (Reprinted from Pöllinger, M. et al., *Phys. Rev. Lett.*, 103, 053901, 2009.); (f) silica microbubble resonator – $Q \approx 10^7$ at 1550 nm, in air [23,27]. (Reprinted from Sumetsky, M. et al., *Opt. Lett.*, 35, 899, 2010.)

(*continued*)

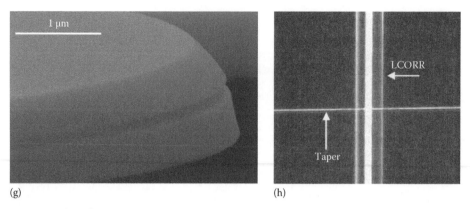

(g)   (h)

FIGURE 9.2 (continued) Assortment of microcavities, with quality factors Q extracted from the literature. The entries from top to bottom, left to right, are (g) double-disk microresonator – $Q \approx 10^8$ at 1518.17 nm, in air [24]. (Reprinted from Lin, Q. et al., *Phys. Rev. Lett.*, **103**, 103601, 2009.); and (h) LCORR – $Q \approx 10^6$ at 980 nm, containing solution [25]. (Reprinted from White, I.M. et al., *Opt. Lett.*, **31**, 1320, 2006.)

9.4.4 DOUBLE-DISK MICRORESONATOR

Double-disk cavity geometries, composed of two silica disks separated by a gap distance on the order of nanometers, exhibit strong dynamical backaction effects surpassing many comparable, optomechanical devices [24]. These resonators are formed from selective plasma etching of a silicon substrate and a sacrificial amorphous silicon layer sandwiched between two silica layers sitting above the substrate. With particular relevance to optomechanics, double disks allow for high-Q-factor WGM feedback combined with large gradient-like per-photon force.

9.4.5 SILICON RING RESONATOR

Silicon ring resonators provide an avenue for single-mode propagation with an affinity for manufacturing and very low loss in signal power. As a critical component in silicon photonics, utilization of silicon ring resonators as an optical filter, high-speed modulator, as well as biosensor has been studied. Large-scale arrays of these microresonators can be made via cost-effective, deep UV lithography methods, the products of which researchers indicated had highly uniform biosensing data with impressive dynamic range and minimal crosstalk [26].

9.4.6 LIQUID-CORE OPTICAL RING RESONATOR

Functioning in a similar fashion as that of silica microbubbles, an LCORR carries aqueous samples internally. An LCORR is essentially a fused silica capillary with a wall thickness of only a few microns with an inner polymer coating. It is shaped by stretching the central region of a capillary under heating until a desired outer radius is achieved, by which point etching determines the wall thickness. The resonator's Q-factor is slightly above that of standard ring resonators, despite the persisting lack of imposed constraints on the whispering gallery mode in the longitudinal direction [25].

9.4.7 BOTTLENECK MICRORESONATOR

An exotic type of microcavity is the bottleneck resonator, which is fabricated by a heat-and-pull process and which hosts whispering gallery modes that are displaced toward and away from the bottlenecks. The resonator structure is useful in customizing mode patterns, as the axial eigenmodes confined by an effective harmonic potential are tuned by the curvature profile alone [22]. Recently, sensing exploiting optomechanical properties of a whispering gallery microcavity has been demonstrated on a bottleneck resonator.

9.4.8　Silica Microbubble Resonator

Silica microbubbles [23,27] are another type of microcavity that is blown from silica tube-drawn microcapillaries. This design befits microfluidic integration by staging the transport of liquid and nanoparticle specimens within the internal confines of the silica tube and microbubbles. Moreover, such microcavities have favorable coupling geometries and insertion losses, as the former is streamlined and the latter has been affirmed to be small. A microbubble has the potential to be configured as a gas sensor, considering that the gas under test can be contained inside the bubble enclosure.

9.5　REACTIVE SENSING

The early exploration of reactive sensing dates back to 1995 [28], where a prediction of subatomic sensitivity with the technique irrevocably excited active research in the field. At present, much research has proven that reactive sensing using a high-Q microcavity can be a vivid technique for high-resolution measurements. In addition, prominent works have rapidly pushed the sensitivity toward theoretical predictions.

Ultrahigh-Q WGMs have the capability of trapping photons within an orbit for a markedly large number of round trips. This provides a window for biosensing where optical properties can observably change. In particular, light can circulate over hundreds of thousands of times if energy losses due to material absorption and scattering at the external interface are minimized. A dielectric particle within a WGM's evanescent field produces a photon resonant wavelength shift, given that there is mode energy spent toward polarization in order to satisfy conservation laws.

Due to the proportionality of a particle's polarizability to its volume [29], excess polarizability of adsorbed bioagents can be related to reactive shifts. Unlike refractometric sensors [30], a modality can be chosen as to include microresonator outer layers that are functionalized for analyte binding. With a perturbative approach, one may obtain the resonance wavelength shift $\Delta\lambda_{res}$ according to a volume integral over the cavity [31,32]:

$$\frac{\Delta\lambda_{res}}{\lambda_{res}} = \frac{\int dV \Delta\varepsilon \, |E|^2}{4 \int dV \varepsilon \, |E|^2} \tag{9.18}$$

where $\Delta\varepsilon$ is the permittivity difference between the pre- and post-adsorption states. For a biolayer-coated silica microsphere where resonance conditions are met (i.e., $\ell\lambda_{Res}/n = 2\pi R$), the resonant wavelength shift $\Delta\lambda_{Res}$ caused by binding to a WGM can be expressed as [33]

$$\frac{\Delta\lambda_{Res}}{\lambda_{Res}} = \frac{\alpha_{Exc}\sigma}{\varepsilon_o \left(n_{Sph}^2 - n_{Ext}^2\right) R} \tag{9.19}$$

where
　α_{Exc} is the excess polarizability of a biomolecule (e.g., protein)
　σ is the biomolecules' associated mean surface density
　n_{Sph} and n_{Ext} are, respectively, the refractive indices of the silica microsphere and external medium
　R is the resonator's radius

The role of photon lifetime is not explicitly shown in this equation, although it certainly arises once the limit of detection σ_{LOD} of the system is considered [34]:

$$\sigma_{LOD} = \frac{\varepsilon_o \left(\dfrac{\Delta\lambda_{Res}}{\delta\lambda_{Res}} \right) \left(n_{Sph}^2 - n_{Ext}^2 \right) R}{\alpha_{Exc} Q} \tag{9.20}$$

Most importantly, the overall Q-factor ($\propto$ mean photon lifetime) of the resonator is fundamentally linked to the resonance linewidth $\delta\lambda_{Res}$ via $Q = \lambda_{Res}/\delta\lambda_{Res}$. Mathematically, Q can also be related to L_{eff} (generally on the order of cm) for a photon wavelength λ using the formula $L_{eff} = Q\lambda/2\pi n$ [10,35].

Great care is needed in controlling the influence of obscuring thermal or mechanical effects, as correlated noise threatens to lower the signal-to-noise ratio and limit of detection—two critical aspects of sensitivity. Such influences clearly produce accountable background noise, such as ambient temperature fluctuations that vary a resonator material's refractive index n and the path length. That is, a positive thermal expansion term dL/dT and thermo-optic coefficient dn/dT will serve in raising $\Delta\lambda_{Res}$ through

$$\Delta\lambda_{Res} = \lambda_{Res}\Delta T \left(\frac{1}{n}\frac{dn}{dT} + \frac{1}{L}\frac{dL}{dT} \right) \tag{9.21}$$

where L often represents a circumference-like path that the electromagnetic wave travels.

In an experiment setting, an optical fiber can be efficiently coupled to a WGM structure and a connected narrow-linewidth laser source can be swept over a specified wavelength range, as depicted in Figure 9.3 [36]. The light transmitted past a microresonator can then be monitored with a photodetector–oscilloscope pair as to track the expected shifting of resonance dips. Commonly, target biological substances can be directly transported to the sensor surface via microfluidic delivery. This can consist of a fiber taper fixed to a glass slide via UV adhesion with a silicon microfluidic cell molded as to contain a cell accommodating the fiber, a WGM, and buffer/biomolecule channels as well as a drainage channel [37]. Syringe pumps are necessary in regulating flow rates, while ensuring that realistic detection times and low fluid cross contamination are maintained.

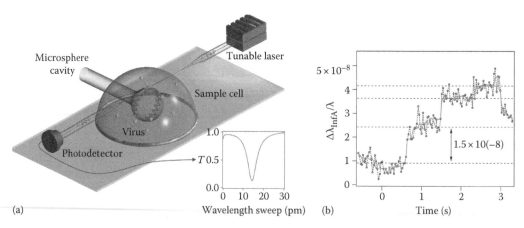

FIGURE 9.3 (a) Microsphere WGM-based label-free biosensing configuration, entailing standard coupling techniques and the injection of a biosample solution in a cell initially filled with phosphate-buffered saline. (b) A wavelength shift response instigated by the presence of a biosample (i.e., influenza A virus) for a 39 mm radius microsphere and distributed feedback laser with a nominal wavelength of 763 nm. (Reprinted from Vollmer, F. et al., *Proc. Natl. Acad. Sci.*, 105, 20701, 2008.)

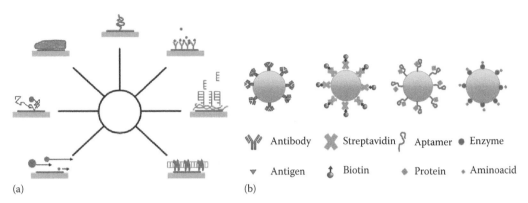

(a) (b)

FIGURE 9.4 (a) Assortment of biological materials, in varying configurations, that can be analyzed with a WGM biosensor. (Reprinted from Vollmer, F. and Arnold, S., *Nat. Methods*, 5, 595, 2008.) (b) Biosensing WGM with biological adlayers. The middle row displays primary ligands/receptors, while the bottom row displays primary analytes [39]. (Reprinted from Righini, G.C. et al., *Riv. Nuovo Cimento*, 34, 476, 2011.)

To lessen the factor of temperature fluctuations, a thermoelectric stage can be used. Increasingly sophisticated arrangements, however, such as LCORRs, can facilitate integration by situating resonator-sample interactions within a liquid core region. This arrangement has been reported to support whispering gallery modes with limited impact on Q and sensitivity, additionally providing multiplexing functionalities by simultaneous excitation of multiple modes [38].

Functionalization of the sensor surface is conventionally realized by silanizing silica, followed by biotin or streptavidin layer formation. Each has distinct advantages: biotin has high specificity and leads to further ligand or receptor functionalization [40], while streptavidin has extraordinarily high affinity for biotin and biotinylated molecules [41]. Further feasible recognition elements can be viewed in Figure 9.4.

9.6 REFERENCE INTERFEROMETER SENSING

Although subatomic sensitivity has been predicted, relatively large biosamples such as the influenza A virion have been detected with reactive sensing at a signal-to-noise ratio close to unity [36]. The hurdle standing in the way of driving the sensitivity towards theoretical limitations is the probe laser jitter noise. For a typical external cavity laser, this unwanted variation in the range of 10 MHz will establish a noise floor on the sensitivity such that the minimum detectable reactive optical frequency shift is at the same level. In [42], the jitter noise was mitigated by a low jitter noise laser built on a periodically poled lithium niobate (PPLN) crystal. Further, the previously referenced real-time reference interferometry can be utilized to significantly improve overall stability and performance of a WGM sensing methodology. Other researchers have aimed to confront the same challenge by adopting feedback mechanisms [43]; however, such implementations are either expensive or difficult to configure. On the other hand, a Mach–Zehnder interferometer, which is both thermally and mechanically stabilized, may be placed in parallel with a fiber taper-microcavity system so that it can record laser jitter noise information that is shared by the sensing signal. The laser jitter, responsible for bringing about uncertainty in pinpointing the location of transmission dips, can be cancelled by the data measured by the reference interferometer.

Upon calibration, the observed sensitivity of a microcavity sensing device can be increased, and hence, the error contributions that sensing schemes normally fall prey to can be suppressed. Due to the nature of its implementation, this technique permits any microcavity structure to be used granted that a sufficient fraction of the energy input from the exciting laser source is diverted to the interferometer. An example of reference interferometer biosensing includes the recent detection of influenza A virus binding to a microtoroid resonator with a reported signal-to-noise ratio of 38:1, wherein a 1 pM solution of the biomolecules diluted in a saline solution was tested [20]. Record sensitivity was also demonstrated by detecting a 12.5 nm radius polystyrene bead binding to the surface of the microtoroid.

9.7 SPLIT FREQUENCY SENSING

In ultrahigh-Q microcavities, Rayleigh backscattered light from the forward propagating mode in a WGM may overcome the round-trip loss and form a backward propagating mode with power equal to that of the forward propagating one. The two counterpropagating modes will form a pair of standing wave modes, also known as a spatially separated sine and cosine mode. The by-product of this, arising from the originally twofold degenerate modes with identical resonance frequencies and transverse field distributions, is resonances occurring as doublets [44]. The relative frequency difference between the doublets is related to the backscattering strength. Upon adsorption of a subwavelength nanoparticle or biomolecule, the twin modes will confront the light scattering center and cause a change in the frequency splitting. Similar to reactive sensing, shifts of the split frequency can be estimated from perturbation theory:

$$\frac{\Delta\lambda_{res}}{\lambda_{res}} = \frac{\int dV \Delta\varepsilon \, |E|^2}{2 \int dV\varepsilon \, |E|^2} \tag{9.22}$$

Therefore, frequency splitting can be employed to monitor biosamples in aqueous media. First discovered in early 2008, a split frequency change in the presence of high-concentration protein molecule binding at the surface of an ultrahigh-Q microtoroid in an aqueous environment [45] was observed. Substantial progress has been made to achieve detection comparable to the aforementioned reference interferometry technique for a single polystyrene nanobead with a 12.5 nm radius [20,46]. The effects in combination with resonance shifting can improve the signal-to-noise ratio, as can be seen in Figure 9.5. Due to the differential nature of the detection scheme, common mode noise (e.g., cavity resonance shifting caused by temperature fluctuations) will be shared between the two whispering gallery modes and so can be suppressed.

The highest quality factor for a microcavity in an aqueous environment is on the order of 10^8 at a wavelength of 670 nm, setting the linewidth of a cavity resonance to several MHz. It is known that the resonance frequency shift can, at this stage, be best resolved at a fraction of the resonance linewidth. The next anticipated advancement is to make use of an active microresonator—a microresonator that has optical gain, accordingly obtaining a narrower linewidth. The fundamental quantum limit for the linewidth of a resonator's lasing mode, titled the Schawlow–Townes linewidth [47], is

$$\Delta\nu_{laser} = \frac{\pi h\nu(\Delta\nu_c)^2}{P_{out}} \tag{9.23}$$

where
$\Delta\nu_c$ is the spectrum linewidth of the passive cavity resonance
$h\nu$ is the resonance photon energy ($h = 6.63 \times 10^{-34}$ J · s is the Planck constant and ν is the resonance frequency)
P_{out} is the power of the lasing mode

If a microcavity operates above the lasing threshold, the linewidth of the lasing line will be likely narrower than the $\Delta\nu_c$ term. In the past, a fundamental linewidth of 3 Hz has been reported on a

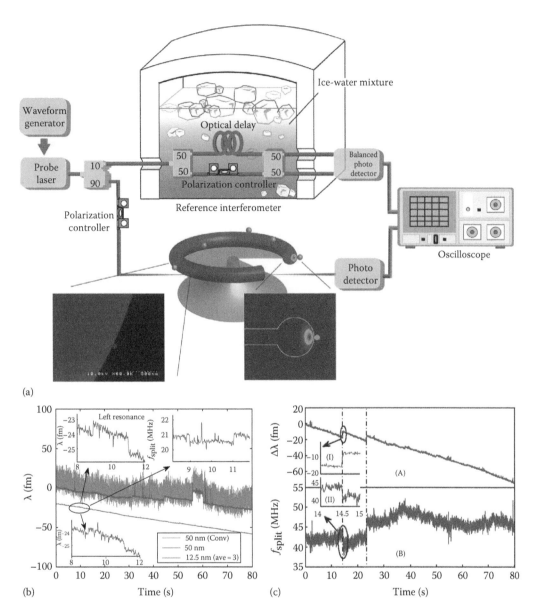

(a)

(b)

(c)

FIGURE 9.5 (a) Diagram of an experimental setup for nanodetection that incorporates reference interferometry. (b) Plot of resonance shift steps for detected polystyrene nanobeads of a 50 nm radius using a conventional method (i.e., noise-ridden curve) and the above reference interferometer system, as well as accompanying data for a 12.5 nm radius that includes mode-splitting descriptors. Note that the microcavity that produced these is a microtoroid. (Reprinted from Lu, T. et al., *Proc. Natl. Acad. Sci.*, 108, 5976, 2011.) (c) Plot of resonance wavelength and split frequency shift, where a comparable resonator to that of (a) was doused with a 1 pM influenza A virus solution. (Reprinted from Lu, T. et al., Nano-sensing with a reference interferometer, in *Frontiers in Optics 2010/Laser Science XXVI*, Optical Society of America, 2010, p. PDPB5.)

silica microtoroid platform [48]. In particular, a microcavity laser will display a split frequency tone in its lasing modes due to the exact principle applicable to its passive counterpart. The removal of the common mode noise leaves a split frequency tone that has a linewidth mostly limited by the fundamental linewidth. An ultrahigh-Q cavity laser operating in an aqueous environment is thus expected to reach a sensitivity orders of magnitude higher. Inspired by the findings of [45], research has been carried out to fabricate an ytterbium-doped microtoroid laser operating in an aqueous environment [49]. Sensing with this laser (exhibiting frequency splitting) was demonstrated in [50–52] with orders of magnitude improvement in the sensitivity. Research in this topic has been similarly carried out by other groups as well [53–55].

9.8 PLASMONIC ENHANCEMENT

To assess resonance shifts of a WGM sensing platform with plasmon coupling (Figure 9.6), one can once again adopt a first-order perturbation theory analysis approach. In this way, particle

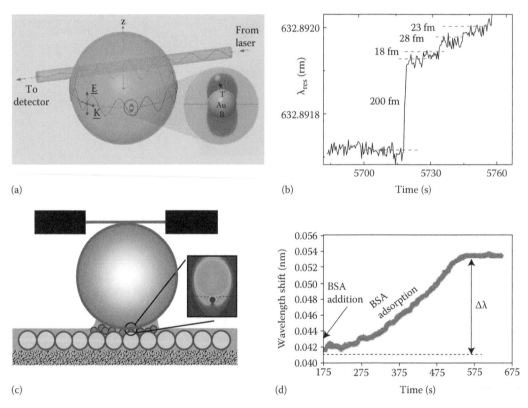

FIGURE 9.6 (a) Photonic–plasmonic mode coupling involving the interaction between a WGM and an equator-bound gold nanoparticle's induced dipole plasmon resonance mode. Emerging hot spots exert an attractive force on nearby nanoparticles, ushering forth (b) accentuated resonance wavelength shift steps for the microcavity [59]. (Reprinted from Shopova, S.I. et al., *Appl. Phys. Lett.*, 98, 243104, 2011.) (c) Similar drawing of a sensing arrangement as that of (a), showing a gold nanoparticle layer (adsorbed to a porous membrane) coupling to a WGM to form a hybrid resonance. In the original publication from which this figure was derived [60], bovine serum albumin proteins (small darker circles) were added within the hot-spots as to bring about a (d) real-time observable increase in the resonance wavelength. (Modified with permission from Santiago-Cordoba, M.A. et al., *J. Biophoton.*, 5, 629, 2012; Santiago-Cordoba, M.A. et al., *Appl. Phys. Lett.*, 9, 073701, 2011.)

adsorption-induced changes at a plasmonic hot-spot positioned at $\vec{r}_{AP}$ can be described in terms of the local field intensity [32]:

$$\frac{\Delta\lambda_{Res}}{\lambda_{Res}} \cong \frac{\alpha_{Exc} \mid \vec{E}(\vec{r}_{AP}) \mid^2}{2\varepsilon_o \int_V \varepsilon_r(\vec{r}) \mid \vec{E}(\vec{r}) \mid^2 dV} \tag{9.24}$$

The denominator (excluding the permittivity of free space term, ε_o) contains the mode volume integral of the energy density. It is clear that if the field $\vec{E}$ is amplified in the binding site region of interest, a larger sensitivity can be achieved. For hybrid photonic–plasmonic whispering gallery mode resonators such as a silica microsphere with bound gold nanoparticle antennas [56], the necessary field enhancements can be attained without substantial Q degradation. As opposed to the goal of surface-enhanced Raman spectroscopy (SERS), which aims to enhance the scattered far-field intensity, hybrid photonic–plasmonic WGM biosensors aim to solely increase the near-field signal. This implies that slight detuning of the WGM resonance wavelength from the plasmon resonance aids in minimizing scattering losses, therefore boosting the device's sensitivity. Ideal coupling with these features should, in principle, amount to enhancements in the range of over three orders of magnitude [57]. It is important to note that the remarkably high sensitivity improvements for this scheme have also been shown to apply to cavity quantum electrodynamics [58].

9.9 OPTOMECHANICAL SENSING

When a photon travels at the edge of a whispering gallery microcavity, its momentum continuously changes direction along the trajectory. The alterations in the photon momentum exert an optical force strong enough to push the cavity outward, which in turn will reduce the strength of the force along with the intracavity power as it drives the cavity away from resonance. Eventually, the cavity tensile force dominates over the diminishing optical force, and so it drags the cavity backwards towards the resonance point. The combination of the alternating optical and tensile forces forms restoring forces similar to those existing in a plain spring oscillator in a plain, loaded spring oscillator, initiating periodic expansion and shrinking of the cavity. This phenomenon is known as the optomechanical oscillation existing in ultrahigh-Q cavities, such as a silica microtoroid [17,61–63]. The eigenfrequency of the cavity mechanical oscillation is within the range of radio frequencies and is determined by the mechanical structure of the cavity. Therefore, fluctuations in the surrounding medium due to the binding of biomolecules at the cavity surface will introduce a sudden shift in mechanical frequency. Similar to reactive sensing, monitoring the change of cavity mechanical frequency provides another mechanism for biosensing. In recent publications [64,65], sensing with optomechanics has been demonstrated by injecting water–glucose solutions with different concentrations in a hollow-bubble resonator as seen in Figure 9.7 [66].

9.10 SENSING WITH SECOND-HARMONIC GENERATION

Another relevant topic concerning whispering gallery mode resonators with optical gain is second-harmonic light generation. Quasi-phase matching has been proposed as mechanism for second-harmonic generation (SHG), which is in turn made possible through periodically modulating a resonator's nonlinear susceptibility [67]. SHG has been experimentally demonstrated using spherical WGMs dip coated in a crystal violet solution, wherein strips between the two poles on the WGM surface are selectively exposed to a 10 kV electron beam, as outlined in Figure 9.8 [68].

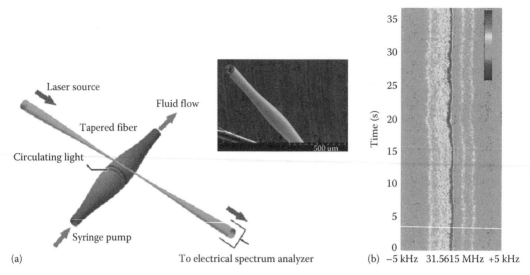

FIGURE 9.7 Images of light coupling to a microfluidic optomechanical resonator and a stability test spectrogram, pertaining to a 31 MHz vibration with 112 Hz standard deviation and 35 s period for a water-encapsulating bubble. (Modified from Kim, K.H. et al., Observation of optically excited mechanical vibrations in a fluid containing microresonator, in *Lasers and Electro-Optics (CLEO), 2012 Conference on*, May 6–11, 2012, San Jose, CA, pp. 1–2.)

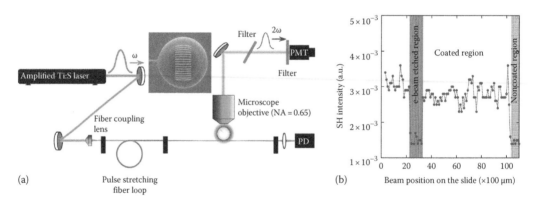

FIGURE 9.8 (a) Second-harmonic generation measurement set up for a microsphere WGM with a periodic molecular pattern on its surface (where NA = numerical aperture, PD = photodetector, PMT = photomultiplier tube). (b) Second-harmonic generation data for a flat silica substrate partially covered by a nonlinear monolayer, with an e-beam-exposed region. (From Dominguez-Juarez, J. et al., *Nat. Commun.*, 2, 3, 2011.)

9.11 CONCLUSION AND FUTURE RESEARCH

In this chapter, we provided a brief overview on some fundamental concepts regarding whispering gallery microcavities, materials used for cavity fabrication, as well as various types of cavities under investigation. Scenarios for utilizing fastidious microcavity properties to detect biomolecules down to the nanoscale were presented.

Thanks to their intrinsic detection sensitivity and amalgamation with plasmonic structures, whispering gallery microcavities are promptly becoming a contender for the preeminent sensor platform. One major challenge that has yet to be met, however, is the detection specificity or ability to identify the specific kind of biomolecules that are detected. Studies on protein identification through spectroscopic enhancement utilizing the high-intensity feature of microcavities should be a logical direction for future research. In addition, low-cost, efficient biosample delivery through microfluidic channels is among the inevitable developments for whispering gallery microcavity sensors that will render them practical.

REFERENCES

1. L. Rayleigh, The problem of the whispering gallery, *Philosophical Magazine* **20**, 1001–1004 (1910).
2. R. D. Richtmyer, Dielectric resonators, *Journal of Applied Physics* **10**, 391 (1939).
3. A. Ashkin and J. M. Dziedzic, Observation of resonances in the radiation pressure on dielectric spheres, *Physical Review Letters* **38**, 1351–1355 (1977).
4. H. E. Benner, P. W. Barber, J. F. Owen, and B. K. Chang, Observation of structure resonances in the fluorescence spectra from microspheres, *Physical Review Letters* **44**, 475–478 (1980).
5. V. Braginsky, M. Gorodetsky, and V. Ilchenko, Quality-factor and nonlinear properties of optical whispering-gallery modes, *Physics Letters A* **137**, 393–397 (1989).
6. M. Cai, O. Painter, and K. J. Vahala, Observation of critical coupling in a fiber taper to a silica-microsphere whispering-gallery mode system, *Physical Review Letters* **85**, 74–77 (2000).
7. I. H. Malitson, Interspecimen comparison of the refractive index of fused silica, *Journal of the Optical Society of America* **55**, 1205–1208 (1965).
8. G. M. Hale and M. R. Querry, Optical constants of water in the 200-nm to 200-μm wavelength region, *Applied Optics* **12**, 555–563 (1973).
9. I. S. Grudinin, V. S. Ilchenko, and L. Maleki, Ultrahigh optical Q factors of crystalline resonators in the linear regime, *Physical Review A* **74**, 063806 (2006).
10. M. L. Gorodetsky, A. A. Savchenkov, and V. S. Ilchenko, Ultimate Q of optical microsphere resonators, *Optics Letters* **21**, 453–455 (1996).
11. T. Kippenberg and K. Vahala, Demonstration of high-Q microdisk resonators: Fabrication and nonlinear properties, in *Lasers and Electro-Optics, 2007. CLEO 2007. Conference on*, May 6–11, 2007, Baltimore, MD, pp. 1–2.
12. H. Lee, T. Chen, J. Li, K. Y. Yang, S. Jeon, O. Painter, and K. J. Vahala, Chemically etched ultrahigh-Q wedge-resonator on a silicon chip, *Nature Photonics* **6**, 369–373 (2012).
13. Q. Lin, T. J. Johnson, C. P. Michael, and O. Painter, Adiabatic self-tuning in a silicon microdisk optical resonator, *Optics Express* **16**, 14801–14811 (2008).
14. D. Armani, T. Kippenberg, S. Spillane, and K. Vahala, Ultra-high-Q toroid microcavity on a chip, *Nature* **421**, 925–928 (2003).
15. T. J. Kippenberg, S. Spillane, and K. J. Vahala, Kerr-nonlinearity optical parametric oscillation in an ultrahigh-Q toroid microcavity, *Physical Review Letters* **93**, 083904 (2004).
16. P. Del'Haye, A. Schliesser, O. Arcizet, T. Wilken, R. Holzwarth, and T. J. Kippenberg, Optical frequency comb generation from a monolithic microresonator, *Nature* **450**, 1214–1217 (2007).
17. T. J. Kippenberg and K. J. Vahala, Cavity opto-mechanics, *Optics Express* **15**, 17172–17205 (2007).
18. S. Spillane, T. J. Kippenberg, O. J. Painter, and K. J. Vahala, Ideality in a fiber-taper-coupled microresonator system for application to cavity quantum electrodynamics, *Physical Review Letters* **91**, 043902 (2003).
19. S. M. Spillane, T. J. Kippenberg, and K. J. Vahala, Ultralow-threshold Raman laser using a spherical dielectric microcavity, *Nature* **415**, 621–623 (2002).
20. T. Lu, H. Lee, T. Chen, S. Herchak, J.-H. Kim, S. E. Fraser, R. C. Flagan, and K. Vahala, High sensitivity nanoparticle detection using optical microcavities, *Proceedings of the National Academy of Sciences* **108**, 5976–5979 (2011).
21. Q. Lin, J. Rosenberg, X. Jiang, K. J. Vahala, and O. Painter, Mechanical oscillation and cooling actuated by the optical gradient force, *Physical Review Letters* **103**, 103601 (2009).
22. M. Iqbal, M. Gleeson, B. Spaugh, F. Tybor, W. Gunn, M. Hochberg, T. Baehr-Jones, R. Bailey, and L. Gunn, Label-free biosensor arrays based on silicon ring resonators and high-speed optical scanning instrumentation, *Selected Topics in Quantum Electronics, IEEE Journal of* **16**, 654–661 (2010).
23. I. M. White, H. Oveys, and X. Fan, Liquid-core optical ring-resonator sensors, *Optics Letters* **31**, 1319–1321 (2006).
24. M. Pöllinger, D. O'Shea, F. Warken, and A. Rauschenbeutel, Ultrahigh-Q tunable whispering-gallery-mode microresonator, *Physical Review Letters* **103**, 053901 (2009).
25. M. Sumetsky, Y. Dulashko, and R. S. Windeler, Optical microbubble resonator, *Optics Letters* **35**, 898–900 (2010).

26. S. Berneschi, D. Farnesi, F. Cosi, G. N. Conti, S. Pelli, G. C. Righini, and S. Soria, High Q microbubble resonators fabricated by arc discharge, *Optics Letters* **36**, 3521–3523 (2011).

27. A. Nitkowski, A. Baeumner, and M. Lipson, On-chip spectrophotometry for bioanalysis using microring resonators, *Biomedical Optics Express* **2**, 271–277 (2011).

28. A. Serpenguzel, S. Arnold, and G. Griffel, Excitation of resonances of microspheres on an optical fiber, *Optics Letters* **20**, 654–656 (1995).

29. F. Vollmer and S. Arnold, Whispering-gallery-mode biosensing: Label-free detection down to single molecules, *Nature Methods* **5**, 591–596 (2008).

30. N. M. Hanumegowda, C. J. Stica, B. C. Patel, I. White, and X. Fan, Refractometric sensors based on microsphere resonators, *Applied Physics Letters* **87**, 201107 (2005).

31. G. Griffel, S. Arnold, D. Taskent, A. Serpengzel, J. Connolly, and N. Morris, Morphology-dependent resonances of a microsphere-optical fiber system, *Optics Letters* **21**, 695–697 (1996).

32. I. Teraoka and S. Arnold, Perturbation approach to resonance shifts of whispering-gallery modes in a dielectric microsphere as a probe of a surrounding medium, *Journal of the Optical Society of America B* **20**, 1937–1946 (2003).

33. S. Arnold, M. Khoshsima, I. Teraoka, S. Holler, and F. Vollmer, Shift of whispering-gallery modes in microspheres by protein adsorption, *Optics Letters* **28**, 272–274 (2003).

34. S. Arnold, R. Ramjit, D. Keng, V. Kolchenko, and I. Teraoka, Microparticle photophysics illuminates viral bio-sensing, *Faraday Discussions* **137**, 65–83 (2008).

35. R. K. Chang and A. J. Campillo, eds., *Optical Processes in Microcavities*, vol. 3 of Advanced Series in Applied Physics. World Scientific Publishing Co., Singapore, 1996.

36. F. Vollmer, S. Arnold, and D. Keng, Single virus detection from the reactive shift of a whispering-gallery mode, *Proceedings of the National Academy of Sciences* **105**, 20701–20704 (2008).

37. D. Keng, S. R. McAnanama, I. Teraoka, and S. Arnold, Resonance fluctuations of a whispering gallery mode biosensor by particles undergoing Brownian motion, *Applied Physics Letters* **91**, 103902 (2007).

38. G. C. Righini, Y. Dumeige, P. Féron, M. Ferrari, G. N. Conti, D. Ristic, and S. Soria, Whispering gallery mode microresonators: Fundamentals and applications, *Rivista Del Nuovo Cimento* **34**, 435–487 (2011).

39. I. M. White, H. Oveys, X. Fan, T. L. Smith, and J. Zhang, Integrated multiplexed biosensors based on liquid core optical ring resonators and antiresonant reflecting optical waveguides, *Applied Physics Letters* **89**, 191106 (2006).

40. F. Vollmer, S. Arnold, D. Braun, I. Teraoka, and A. Libchaber, Multiplexed DNA quantification by spectroscopic shift of two microsphere cavities, *Biophysical Journal* **85**, 1974–1979 (2003).

41. Y. Lin, V. Ilchenko, J. Nadeau, and L. Maleki, Biochemical detection with optical whispering-gallery resonators, *Biophysical Journal* **6452**, 64520U (2007).

42. S. Shopova, R. Rajmangal, Y. Nishida, and S. Arnold, Ultrasensitive nanoparticle detection using a portable whispering gallery mode biosensor driven by a periodically poled lithium-niobate frequency doubled distributed feedback laser, *Review of Scientific Instruments* **81**, 103110 (2010).

43. J. H. Chow, M. A. Taylor, T. T.-Y. Lam, J. Knittel, J. D. Sawtell-Rickson, D. A. Shaddock, M. B. Gray, D. E. McClelland, and W. P. Bowen, Critical coupling control of a microresonator by laser amplitude modulation, *Optics Express* **20**, 12622–12630 (2012).

44. M. L. Gorodetsky, A. D. Pryamikov, and V. S. Ilchenko, Rayleigh scattering in high-Q microspheres, *Journal of Optical Society of America B* **17**, 1051–1057 (2000).

45. T. Lu, T.-T. J. Su, K. J. Vahala, and S. Fraser, Split frequency sensing methods and systems, Patent pending. p. 20100085573 (2009). Preliminary filing in October 2008.

46. T. Lu, H. Lee, T. Chen, S. Herchak, J.-H. Kim, and K. Vahala, Nano-sensing with a reference interferometer, in *Frontiers in Optics 2010/Laser Science XXVI* (Optical Society of America, Rochester, NY, 2010), p. PDPB5.

47. A. Schawlow and C. H. Townes, Infrared and optical masers, *Physical Review* **112**, 1940–1949 (1958).

48. T. Lu, L. Yang, T. Carmon, and B. Min, A narrow-linewidth on-chip toroid Raman laser, *IEEE Journal of Quantum Electronics* **47**, 320–326 (2011).

49. E. P. Ostby and K. J. Vahala, Yb-doped glass microcavity laser operation in water, *Optics Letters* **34**, 1153–1155 (2009).

50. T. Lu, H. Lee, T. Chen, and S. Herchak, An ultra-narrow-linewidth microlaser for nanosensing, in *Conference on Lasers and Electro-Optics 2012* (Optical Society of America, 2012), p. CTu2L.6.

51. T. Lu, H. Lee, T. Chen, and S. Herchak, Single molecule detection with an Yb-doped microlaser, in *CLEO: 2013* (Optical Society of America, San Jose, CA, 2013), p. CM2H.4.

52. T. Lu, H. Lee, T. Chen, and S. Herchak, Fast nano particle and single molecule detection with an ultra-narrow-linewidth microlaser, under revision (2013).

53. J. Knittel, T. G. McRae, K. H. Lee, and W. P. Bowen, Interferometric detection of mode splitting for whispering gallery mode biosensors, *Applied Physics Letters* **97**, 1–3 (2010).

54. J. Zhu, S. Ozdemir, Y. Xiao, L. Li, L. He, D. Chen, and L. Yang, On-chip single nanoparticle detection and sizing by mode splitting in an ultrahigh-Q microresonator, *Nature Photonics* **4**, 46–49 (2010).

55. L. He, S. K. Ozdemir, J. Zhu, W. Kim, and L. Yang, Detecting single viruses and nanoparticles using whispering gallery microlasers, *Nature Nanotechnology* **6**, 428–432 (2011).

56. M. A. Santiago-Cordoba, M. Cetinkaya, F. Boriskina, Svetlana V. Vollmer, and M. C. Demirel, Ultrasensitive detection of a protein by optical trapping in a photonic-plasmonic microcavity, *Journal of Biophotonics* **5**, 629–638 (2012).

57. J. D. Swaim, J. Knittel, and W. P. Bowen, Detection limits in whispering gallery biosensors with plasmonic enhancement, *Applied Physics Letters* **99**, 243109 (2011).

58. Y.-F. Xiao, Y.-C. Liu, B.-B. Li, Y.-L. Chen, Y. Li, and Q. Gong, Strongly enhanced light-matter interaction in a hybrid photonic-plasmonic resonator, *Physical Review A* **85**, 031805 (2012).

59. H. Rokhsari, T. Kippenberg, T. Carmon, and K. Vahala, Radiation-pressure-driven micro-mechanical oscillator, *Optics Express* **13**, 5293–5301 (2005).

60. T. J. Kippenberg, H. Rokhsari, T. Carmon, A. Scherer, and K. J. Vahala, Analysis of radiation-pressure induced mechanical oscillation of an optical microcavity, *Physical Review Letters* **95**, 033901 (2005).

61. T. Carmon, H. Rokhsari, L. Yang, T. Kippenberg, and K. Vahala, Temporal behavior of radiation-pressure-induced vibrations of an optical microcavity phonon mode, *Physical Review Letters* **94**, 223902 (2005).

62. S. I. Shopova, R. Rajmangal, S. Holler, and S. Arnold, Plasmonic enhancement of a whispering-gallery-mode biosensor for single nanoparticle detection, *Applied Physics Letters* **98**, 243104 (2011).

63. M. A. Santiago-Cordoba, S. V. Boriskina, F. Vollmer, and M. C. Demirel, Nanoparticle-based protein detection by optical shift of a resonant microcavity, *Applied Physics Letters* **9**, 073701 (2011).

64. K. H. Kim, G. Bahl, W. Lee, J. Liu, M. Tomes, X. Fan, and T. Carmon, Cavity optomechanics on a microfluidic resonator with water and viscous liquids, http://arxiv.org/abs/1205.5477v2 (2012).

65. G. Bahl, X. Fan, and T. Carmon, Acoustic whispering-gallery modes in optomechanical shells, *New Journal of Physics* **14**, 115026 (2012).

66. K. H. Kim, G. Bahl, W. Lee, J. Liu, M. Tomes, X. Fan, and T. Carmon, Observation of optically excited mechanical vibrations in a fluid containing microresonator, in *Lasers and Electro-Optics (CLEO), 2012 Conference on*, May 6–11, 2012, San Jose, CA, pp. 1–2.

67. G. Kozyreff, J. L. Dominguez-Juarez, and J. Martorell, Nonlinear optics in spheres: From second harmonic scattering to quasi-phase matched generation in whispering gallery modes, *Lasers and Photonics Reviews* **5**, 737–749 (2011).

68. J. Dominguez-Juarez, G. Kozyreff, and J. Martorell, Whispering gallery microresonators for second harmonic light generation from a low number of small molecules, *Nature Communications* **2**, 1–8 (2011).

10 Coupled Chemical Reactions in Dynamic Nanometric Confinement

III. Electronic Characterization of Ag₂O Membranes within Etched Tracks and of Their Precursor Structures

Dietmar Fink, W.R. Fahrner, K. Hoppe,
G. Muñoz Hernandez, H. García Arellano,
A. Kiv, J. Vacik, and L. Alfonta

CONTENTS

10.1 INTRODUCTION

Since about half a century, it is known that the transition of swift heavy ions through thin polymer foils leaves a trail of radiochemical and structural damage behind—the so-called latent tracks—that can easily be removed by adequate etchants, thus creating nanopores, the so-called etched tracks [1]. With a previously developed strategy, it is also possible to create nanopores with central membranes embedded therein, thus separating the pores into two independent individual adjacent compartments (part I of this series [2]). The membranes can consist of, for example, Ag₂O [2] or other materials such as LiF, CaO, or BaCO₃ (to be published). In the same way as normal etched tracks can be filled with various materials of interest for electronics [3–9], medicine [10], or biosensing [11–19], also these nanopores with central membranes can be used for various applications. That will be treated in the next paper of this series.

As to our knowledge, thorough examinations are missing not only on the electronic characterization of such membrane-containing nanopores but also on their precursor structures; we decided to fill this gap with the present work.

10.2 EXPERIMENTAL

10.2.1 FORMATION OF ETCHED TRACKS WITH EMBEDDED Ag_2O MEMBRANES

The formation of etched tracks with embedded Ag_2O membranes requires two steps that we denoted as the pre-etching and the membrane formation step. In the pre-etching step, 12 μm thick polyethylene terephthalate foils had been irradiated by Kr ions at 250 MeV energy at the Joint Institute for Nuclear Research (JINR), Dubna, at fluences of 1×10^5, 5×10^7, and 1×10^9 cm^{-2}. One cm^2 large pieces of these foils were inserted into the center of a measuring chamber (made after the example of Ref. [20]) with two adjacent compartments and then etched with 9 M KOH from both sides at ambient (~25°C) temperature [2]. The etching was interrupted shortly before the emerging conical nanopores could merge, by removing the KOH etchant and washing the foil thoroughly.

Thereafter, in the subsequent membrane formation step, 1 M $AgNO_3$ solution was inserted on one side—let us call this the left side—and 1 M KOH was inserted on the other side—the right side—of the membrane. Thus the etching continued at a slower speed from side B only, until etchant breakthrough occurred. As reported in [2], simultaneously, AgOH is formed at the point of etched track intersection, which readily transforms to Ag_2O. As these silver compounds are insoluble in water, they will precipitate within the etched track at its narrowest place and thus form a plug or a membrane. The fact that Ag_2O is rather impermeable for both ions and electrons (except at very high applied electric field strengths or frequencies) gives us a tool for the detection of the Ag_2O membrane formation, by observing a current reduction to near zero (*quiet phases*) when these membranes have been formed.

For comparison of the foils with embedded membranes in the etched tracks, also, foils with etched tracks without membranes were produced under nearly identical conditions. In this case, etching was stopped very soon (i.e., within 1–2 min) after etchant breakthrough, so that the etched tracks were very narrow (with radii of about 10–20 nm, as estimated from the transmitted currents).

10.2.2 ELECTRONIC CHARACTERIZATIONS

Both the control of the track etching and membrane formation processes and the subsequent membrane characterization are performed electrically, by applying a voltage across the measuring chamber (including both the foil and the electrolytes) and determining the passing currents. This was achieved by means of a Velleman PCSGU250 pulse generator/oscilloscope combination. In this work, we restrict to the low-frequency results.

During the pre-etching stage, a sinusoidal ac voltage U of 5 $V_{peak-peak}$ at a frequency of ~0.5 Hz was applied through Ag electrodes to the system. Both the voltage and the corresponding electrical current were continuously measured in the *transient recording* mode of that equipment (settings: dc, 0.3 V/div for measuring the applied voltage and 10 mV/div for measuring the current via a 1 MΩ probe resistance; time resolution, 0.1 s/div) as a function of time. This measurement served only for control that the etching proceeded reliably; in the standard case, no current signal emerged before etchant breakthrough. Deviations from this behavior indicated erroneous production situations (such as the leakage of etchant due to incomplete sealing of the reaction chamber with eventual subsequent shortcuts), after which the experiment had to be repeated.

During the membrane formation stage, the ac voltage setting was reduced to U = 1 V$_{p-p}$, to minimize the influence of the voltage on the membrane formation, and the measurement took place in the ac mode to filter the background currents away that emerge from the chemical potential differences between both compartments of the reaction chamber. As described in [2], at the moment of etchant breakthrough from one track side to the other, a strongly spiky current emerges that vanishes rather abruptly when stable membranes have been formed. These *quiet phases* serve as a fingerprint for the Ag$_2$O membrane formation. Therefore, whenever such a quiet phase could be identified unambiguously, the corrosion was stopped.

For the characterization of the final Ag$_2$O membrane-containing samples embedded in some electrolytes, recordings took place either in the *oscilloscope*, the *circuit analyzer*, or in the *spectrum analyzer* mode. Furthermore, the samples were inserted into the small test circuit of Figure 10.1, to describe them by a set of matrix elements—the four-pole parameters—according to the electronic network theory. In the *oscilloscope* mode, the time dependence of both the applied voltage U(t) and the recorded current I(t) was determined, as well as their I(U) correlation; deviations in time of both the U(t) and I(t) curves from each other reveal possible phase shifts. In the *circuit analyzer* mode, Bode plots were recorded, that is, the frequency dependence of the logarithm of the inverse impedance (in [Vrms]* or [db]) of the {electrolyte/(foil with membrane)/electrolyte} system. In the *spectrum analyzer* mode, the Fourier spectra of the recorded current signals were determined, which describe the impact of an eventual frequency filtering onto the frequency distribution transmitted by the experimental system. Usually, 5 V$_{p-p}$ was applied.

The knowledge of the sample's complex four-pole parameters h$_{ij}$ (i, j = 1,2) according to the electronic network theory enables one to predict the behavior of any electronic circuit with these elements, which will become important for future industrial applications of such structures as biological sensors. Here, h$_{11}$ is the short-circuit input impedance, h$_{12}$ describes the open-circuit reverse voltage transfer ratio, h$_{21}$ is the short-circuit forward transfer ratio, and h$_{22}$ is the open-circuit output admittance. These parameters can be obtained for the samples in their various production stages at different frequencies from the voltages U$_0$, U$_1$, and U$_2$ and the corresponding phase shifts φ_0, φ_1, and φ_2 as determined according to Figure 10.1, via a small MATLAB® program. Though the samples considered here are in fact dipoles, they have been incorporated in the conventional four-pole scheme (as shown in Figure 10.1), for the sake of higher generality. This means, whenever one might add in a later stage one or two more poles to create more advanced track-based samples, this approach can be readily overtaken. The frequencies applied in this work range from 0.5 Hz to 10 kHz. Furthermore, the parameters z$_{11}$ were derived from these data that describe the open-circuit input resistance, that is, the actual ohmic resistance of the given samples.

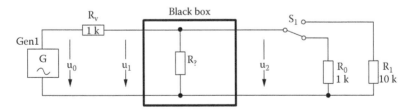

FIGURE 10.1 Measuring circuit for determining the two-port parameters of the examined samples. G: frequency voltage generator; black box: measuring chamber for foil etching with unknown resistance R$_?$; R$_v$: input resistance, R$_0$; R$_1$: load resistance; S$_1$: switch to change between different loads. The resistances R$_v$, R$_0$, and R$_1$ have to be adjusted approximately to the size of the unknown resistance R$_?$; in the given example, R$_v$ = R$_0$ = 1 kΩ and R$_1$ = 10 kΩ, to probe the h$_{ij}$ parameters for a low-resistance sample of a few kΩ or so. The voltages and phases determined at the positions u$_0$, u$_1$, and u$_2$ serve as input for a MATLAB® program from which the complex quadripole parameters h$_{ij}$ are derived.

---

* The electrotechnical unit V$_{rms}$ is defined as the square root of the mean of the squares of the values for the one time period of the sine wave; V$_{rms}$ = 0.3535 * V$_{peak-peak}$.

10.3   ELECTRONIC CHARACTERIZATION OF THE FORMATION PROCESS OF Ag$_2$O MEMBRANES WITHIN ETCHED TRACKS IN PET FOILS: RESULTS AND DISCUSSION

10.3.1   CURRENT/VOLTAGE SPECTROMETRY

Figure 10.2 shows the time dependence of the current I(t) through the experimental vessel in different configurations, as compared with the applied alternating voltage U(t), and additionally the I/U correlations, all as a function of the applied frequency υ. Depending on the studied system and the frequency, the I(t) curves show at least two different behaviors: (1) coincidence of the I(t) and U(t) positions, that is, zero phase shifts and (2) I(t) running ahead of U(t), that is, negative phase shifts. Eventually, also, I(t) might follow U(t) in time (which would signify positive phase shifts), but this appears to stem from instrumental artifacts and therefore is discarded here. The first two cases are indicative of dominant (1) ohmic and (2) capacitive behavior, respectively; the last case of positive phase shifts usually describes an inductive behavior that is thought to be negligible for etched ion tracks.

Whereas the first case stems from the ohmic resistance of the narrow electrolyte-filled tracks, the second one is a consequence of the nanometric topological obstacles in the current paths, such as given in very narrow intrinsic or radiation-induced pores in both the pristine and ion-irradiated polymer, or by very small etched track diameters, or by membranes within the etched tracks. All these lead to a transient pileup of charges in the current paths, that is, to an increasingly capacitive behavior. The fact that even the relatively small overall pore volumes available here already give rise to a measurable effect at all is ascribed to the very high dielectric coefficient of water.

In some cases (pristine polymers and polymers with latent tracks at very low frequencies; see Figure 10.2c through f), the combination of the obstacle's very high resistances and sufficiently high applied voltages leads to current spike emission, as described in detail in recent work [21]. In the most pronounced cases (Figure 10.2d and f), exactly four main spikes can be distinguished in the I/V diagrams, which is consistent with earlier theoretical predictions [21].

Figure 10.3 compiles the results extracted from Figure 10.2. The combination of both the I(t) and U(t) curves to only one curve, I(U), implies the loss of the information about the phase shift, as the different rotation directions in the I(U) diagram describing the transition from capacitive to inductive behavior cannot be recognized from the graphs. The water-filled measuring chamber alone shows a constant purely ohmic conductivity for frequencies up to some 5–10 kHz as in this frequency range, the H$^+$ and OH$^-$ ions can easily follow the applied alternating electric field (Figure 10.3, circles). Deviations occur at higher frequencies due to instrumental reasons. Therefore, we discard this region in this work.

In Figure 10.3a, the differences I_{pp} between the maximum and minimum current amplitudes (i.e., I_{max}–I_{min}) are plotted vs. the frequency, which are indicative for the system's impedances, and in Figure 10.3b, the corresponding phase shifts are plotted that give a clue for the dominant nature of the impedances at the given frequencies. If, for contrast, the thicknesses I_{mm} of the I(U) elipsoides (i.e., the differences between the maximum and minimum elipsoide current values at zero voltage: $I_{max}(U = 0 \text{ V})$–$I_{min}(U = 0 \text{ V})$) of Figure 10.2 are evaluated, one arrives at Figure 10.3c. These I_{mm} values reflect the system's nonohmic components, however without the information whether they describe a capacitive or inductive behavior. As this lack could yield misleading interpretations, it usually has to be taken with caution. However, in the given case, we know from the I(t) and U(t) curves of Figure 10.2 that the behavior is limited to ohmic and capacitive cases.

Upon insertion of a pristine polymer foil into the electrolyte-filled measuring chamber (Figure 10.3, light grey triangles), the overall low-frequency current drops considerably as there exist only very few continuous water-filled nanopores in the polymer that allow for ohmic charge transport across the foil. However, the insulating polymer itself enables some capacitive charge transport as the system, that is, the (electrolyte–polymer–electrolyte) system acts as a capacitor. This can be clearly seen from the strongly negative phase shifts at low frequencies, Figure 10.3b. With increasing frequency, the phase shift increases until it becomes zero, indicating ohmic behavior.

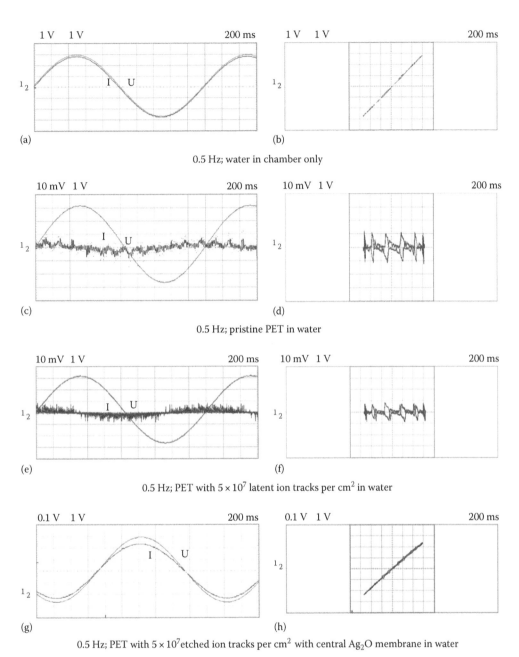

FIGURE 10.2   Typical examples of the measured I(t) and U(t) curves (a,c,e,g,i,k,m,o; left side) and I(U) characteristics (b,d,f,h,j,l,n,p; right side) for the precursor structures and the final membrane-containing track-based structures, as a function of the applied frequencies of 0.5 Hz (a–h) NS 500 Hz (i–p). Cases: a,b,i,j—water in chamber only; c.d,k,l—pristine PET in water; e,f,m,n—PET with 5×10^7 latent ion tracks per cm^2 in water; g,h,o,p—PET with 5×10^7 etched ion tracks per cm^2 with central Ag_2O membrane in water. Scales of the U(t) and I(t) curves: measuring times—200 ms/division in all cases; applied voltages—1 V/div in all cases; measured currents—1 µA/div (a,i), 10 nA/div (c,e), 100 nA/div (g), and 300 nA/div (k,m,o). Scales of the I(U) plots: voltages—1 V/div in all cases; currents—1 µA/div (b,j), 10 nA/div (d,f), 100 nA/div (h), and 300 nA/div (l,n,p). In the I(U) plots, the x-axis always denotes the applied voltage and the y-axis the measured current.

(continued)

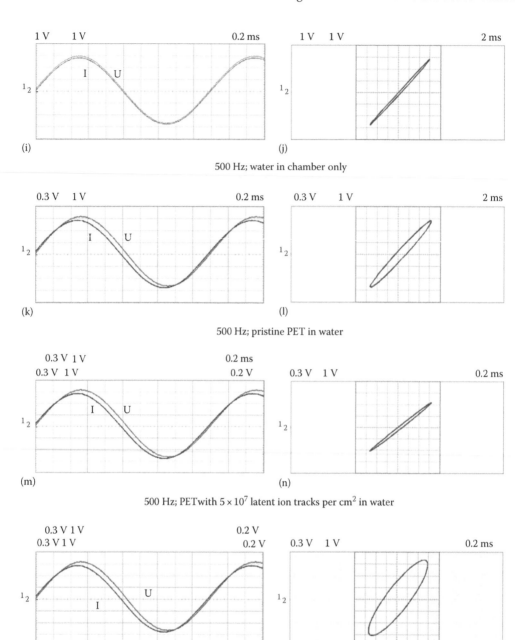

FIGURE 10.2 (continued)   Typical examples of the measured I(t) and U(t) curves (a,c,e,g,i,k,m,o; left side) and I(U) characteristics (b,d,f,h,j,l,n,p; right side) for the precursor structures and the final membrane-containing track-based structures, as a function of the applied frequencies of 0.5 Hz (a–h) NS 500 Hz (i–p). Cases: a,b,i,j—water in chamber only; c.d,k,l—pristine PET in water; e,f,m,n—PET with 5×10^7 latent ion tracks per cm^2 in water; g,h,o,p—PET with 5×10^7 etched ion tracks per cm^2 with central Ag_2O membrane in water. Scales of the U(t) and I(t) curves: measuring times—200 ms/division in all cases; applied voltages—1 V/div in all cases; measured currents—1 µA/div (a,i), 10 nA/div (c,e), 100 nA/div (g), and 300 nA/div (k,m,o). Scales of the I(U) plots: voltages—1 V/div in all cases; currents—1 µA/div (b,j), 10 nA/div (d,f), 100 nA/div (h), and 300 nA/div (l,n,p). In the I(U) plots, the x-axis always denotes the applied voltage and the y-axis the measured current.

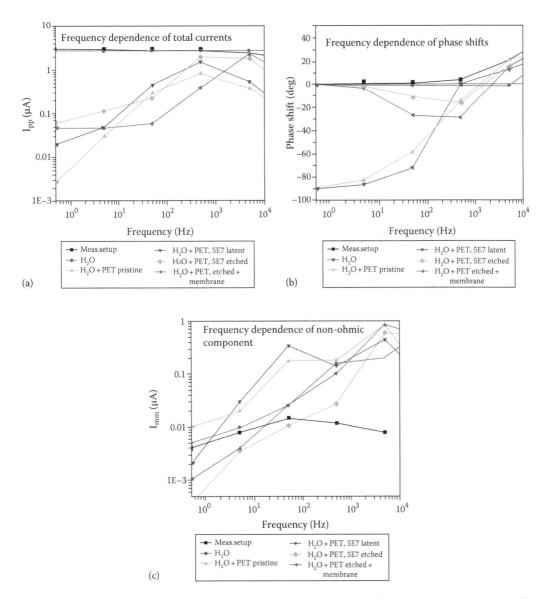

FIGURE 10.3 Evaluation of the frequency-dependent results from Figure 10.2: (a) current peak–peak amplitudes I$_{pp}$ (reflecting the magnitude of the impedances) and (b) the corresponding phase shifts (indicating the dominant ohmic, capacitive, or inductive nature of the currents), as derived from the I(t) and U(t) curves. In (c), the current amplitudes I$_{mm}$ between the minimum and maximum thickness of the I(U) ellipsoides are plotted. Values are shown for water only (circles), the PET foils with latent ion tracks in water (light grey triangles), the PET foils with etched ion tracks in water (standing blue triangles), and the etched PET foil with embedded Ag$_2$O membranes in water (dark grey triangles pointing to the left). The Ag$_2$O membranes studied here represent a relatively late formation stage (i.e., after passing ~10 *quiet phases*) and therefore are considered to be somewhat broad and large.

Replacing now these pristine polymer foils by ion-irradiated but not etched ones yields only slight changes (Figure 10.3, inverted dark blue triangle). The fact that most currents slightly increase signifies that some additional free radiation-induced volume is formed. Etching the latent tracks regularly from both sides up to etchant breakthrough (Figure 10.3, standing light blue squares) leads to some additional water incorporation along the tracks. Hence, the overall available free volume increases so that the low-frequency overall current also increases due to an emerging ohmic component. Consequently, the previous capacitive component decreases due to partial shortcut via the new conductive connection.

One might expect that the irradiation fluence should have a considerable influence on the magnitude of the results. This is, however, not the case. The reason is that upon impact of highly energetic heavy ions into polymers, strong pressure waves—up to 400 MPa [22]—are emitted from the centers of the ion tracks that compress the surrounding material efficiently—the higher the fluence, the greater the polymer's compaction (thus, the polymer's density may increase by up to 20%). Also, cross-linking contributes to this effect. This leads to a profound reduction of the polymer's intrinsic free volume due to its plastic deformation and hence also of the amount of electrolyte incorporation and the etching rate. This means that the expected increase in total volume by chain scissioning and degassing of light radiochemical products is more or less compensated by compaction effects. Consequently, the differences between low and high fluences (here: spanning a factor of 10^4) of the irradiated polymers lead to only marginal differences in the polymer's free volume [22] and hence also in the electronic effects (not shown here).

When, for comparison, in the last stage insulating Ag_2O membranes are embedded in the etched tracks (Figure 10.3, with dark grey triangles pointing left), the medium- and higher-frequency conductivities are lowered due to the new obstacles in the current paths, and the overall current maximum shifts to somewhat higher frequencies (typically a few kHz). Most remarkable is the membrane-induced change of the phase shift, Figure 10.3b.

At the end of this paragraph, let us consider the ohmic-type frequency region around 1 kHz in more detail. In the case of electrolyte-filled etched ion tracks in the polymer foil, the total ohmic resistance R_t of the track-containing polymer foil is here equal to the geometric mean of both the electrolyte resistance R_e in the tracks and the neighboring polymer resistance R_p: $R_t = 1/(1/R_e + 1/R_p)$ (neglecting here the resistance of the water outside the polymer foil). However, in the case that additional Ag_2O membranes with resistance R_{ag} exist within the tracks, one has to replace in the earlier equation the ohmic track resistance R_e by $(R_e + R_{ag})$. Assuming that the frequency dependence of the Ag_2O membrane resistance shows qualitatively the same tendency as the polymer's frequency dependence (i.e., high resistance at low frequencies and decreasing resistance at higher frequencies), then the shift of the observed transmitted current maximum in the presence of Ag_2O membranes towards higher frequencies can be understood.

The reproducibility of the measurements was estimated by comparing the results of two independent measuring series, performed under rather similar conditions (not shown here). Though the derived tendencies are the same in both cases, they differ from each other qualitatively somewhat in their current magnitudes, essentially due to different resistivities of the used highly pure water and different membrane geometries.

In conclusion, each of the membrane production steps shows pronounced peculiarities that distinguish it from the other ones. Specifically, Ag_2O membranes embedded in etched tracks exhibit unique fingerprints in the I(t) and I(U) plots, as well as for the conductive and the capacitive signals, so that this can be used for their unambiguous identification. The prepared foils with membranes are stable in their characteristics for at least several weeks at ambient temperature.

10.3.2 Bode Plots

All Bode plots performed for the same foils as described earlier exhibit the same qualitative behavior even after different manipulation stages, insofar as the measured impedances increase with increasing frequency up to some maximum (Figure 10.4). This also coincides qualitatively with what has been mentioned previously for proton microbeam-irradiated PET foils [23]. The recorded Bode plots reveal somewhat similar tendencies as found earlier for the I(U) plots (Figure 10.4). However, for contrast to the considerable changes found there during the different production stages, the Bode plots differ amazingly little from each other, indicating that the use of the logarithm of the inverse impedance in the Bode plots largely masks the effects under study here and hence is not a good parameter in our case.

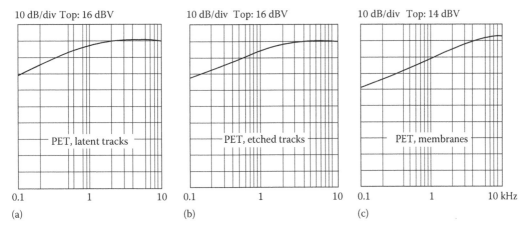

10 dB/div Top: 16 dBV   10 dB/div Top: 16 dBV   10 dB/div Top: 14 dBV

PET, latent tracks   PET, etched tracks   PET, membranes

0.1   1   10   0.1   1   10   0.1   1   10 kHz

(a)   (b)   (c)

FIGURE 10.4   Bode plots for the different foil manipulation stages: (a) as implanted but not etched PET foil, embedded in deionized water, (b) after etching and 5× cleaning with water, and (c) after producing the Ag$_2$O membrane and washing the latter in water. The x-axes—log (frequency [Hz]) from 100 Hz to 10 kHz; the y-axes—10 db/division, with the top denoting 16, 16, and 14 dbV, respectively. The Ag$_2$O membranes studied here represent a relatively late formation stage (i.e., after passing ~10 *quiet phases*) and thus are considered to be somewhat broad and large.

It appears that in the low-frequency region, the impedance is largely dominated by the polymer, and in the medium frequency range, both the polymer and electrolyte (here: water) play a role. A slight shift of the Bode plot maximum to higher frequencies and the pronounced decrease in the lower-frequency region may be indicative for the formed Ag$_2$O membrane within the etched tracks, as derived from the comparison of Figure 10.4b with c.

Another consequence of these findings is that polymer foils with etched tracks only as well as foils with etched tracks with embedded membranes act as frequency filters that suppress lower frequencies to some extent. It appears that the low-frequency suppression becomes somewhat more pronounced by the insertion of especially thin membranes into the etched tracks.

10.3.3   Fourier Spectra

Fourier spectra show all frequencies from which a certain oscillation pattern is composed. In an ideal passive electronic structure, the Fourier spectra of both the applied voltage and the current transmitted through some electronic structure should be identical. However, any frequency dependence of the resistance of such a structure will change the Fourier spectra. Therefore, Fourier spectrometry can be used as a very sensitive tool to detect such frequency dependences. In our case, the distortions of Fourier spectra may reveal changes of the electronic properties of both the electrolyte and the nanostructures within the polymer foil.

Figure 10.5 shows some Fourier spectra of the samples described earlier, after different manipulations. It was advantageous in this case that we had taken a rather simple sinusoidal frequency generator that exhibited still some (~15) higher and lower less pronounced computer-generated side frequency lines. These side frequencies (Figure 10.5a) served as probes to indicate the frequency-dependent current modifications of the samples during their various production stages, resulting from the previously shown nonlinearity of the impedance's frequency dependence. Indeed, it is seen that essentially higher frequency peaks are virtually lost after the manipulations (Figure 10.5b and c). This is the obvious consequence of the system's action as a band-pass. There are hints that also longer aging might eventually induce some changes in the Fourier spectra. Also, the change of the applied electrolyte (e.g., replacing H$_2$O by KOH) has some impact on both the main and side frequency amplitudes and the peak's spacings from each other.

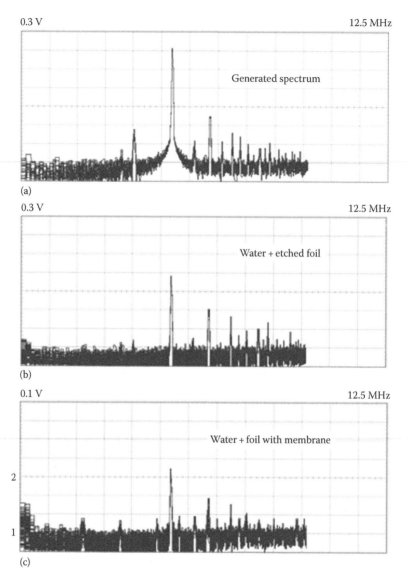

0.3 V   12.5 MHz

Generated spectrum

(a)

0.3 V   12.5 MHz

Water + etched foil

(b)

0.1 V   12.5 MHz

Water + foil with membrane

2

1

(c)

FIGURE 10.5 Comparison of the Fourier spectrum of (a) a computer-generated sinusoidal voltage waveform with poor quality with (b,c) the currents measured across the measuring chamber filled with deionized water and a PET foil on both sides; (b) chamber with irradiated and etched PET foil in water; (c) chamber with etched PET foil with embedded Ag membrane in water. The x-axes: linear, measured spectra extending from 0 to 12.5 MHz; the y-axes: frequency peak abundances in arbitrary units.

10.3.4 FOUR-POLE PARAMETERS

From the measured voltages U_0, U_1, and U_2 and the corresponding phase shifts φ_0, φ_1, and φ_2 as determined according to Figure 10.1 for the different frequencies of the samples in their various production stages, the corresponding four-pole parameters h_{11}, h_{12}, h_{21}, and h_{22} and the open-circuit input resistance z_{11} as well as the corresponding phases are derived, Figure 10.6. If the h parameters are multiplied with the cosine of the corresponding given angle, one obtains the resistive contribution, that is, the real part of the impedance; if this value is multiplied with the sine of the same angle, one obtains its capacitive contribution X_c. From this, the equivalent capacity of the samples follows to $C = 1/(2 * \pi * \nu * X_c)$.

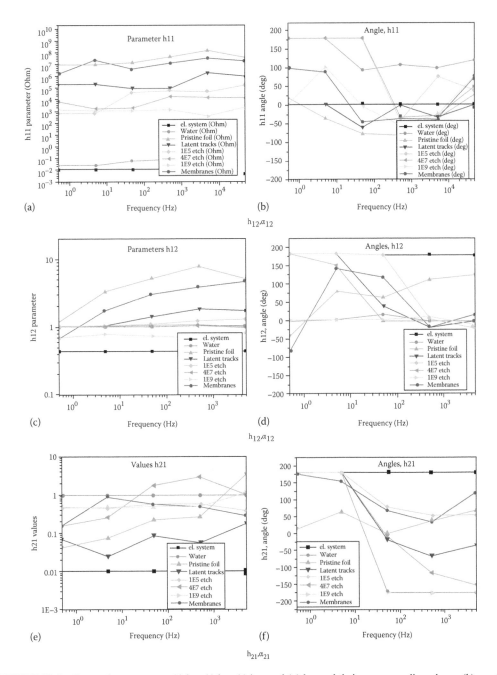

FIGURE 10.6 Four-pole parameters (a) h$_{11}$, (c) h$_{12}$, (e) h$_{21}$, and (g) h$_{22}$ and their corresponding phases (b) α$_{11}$, (d) α$_{12}$, (f) α$_{21}$, and (h) α$_{22}$; the last two graphs show (i) the parameter z$_{11}$, describing the open-circuit input resistance and (j) its corresponding phase β$_{11}$. The values are given for the different components of the measuring system and for the different stages of sample preparation: black rectangle—electronic system (consisting of the frequency generator, the measuring chamber, the electronic circuit with a fixed 10 kΩ resistance, and the oscilloscope) only; red circle—measuring chamber filled with double deionized water; blue standing triangle—polymer foil separating the water-filled chamber into two compartments; green falling triangle—polymer foil with 1 × 10^5 cm^{-2} etched tracks (etching from both sides) within water; violet triangle pointing towards left—polymer foil with 4 × 10^7 cm^{-2} etched tracks within water; olive triangle pointing towards right—polymer foil with 1 × 10^9 cm^{-2} etched tracks within water; dark blue tilted square—polymer foil with Ag$_2$O membranes embedded within etched tracks within water.

(continued)

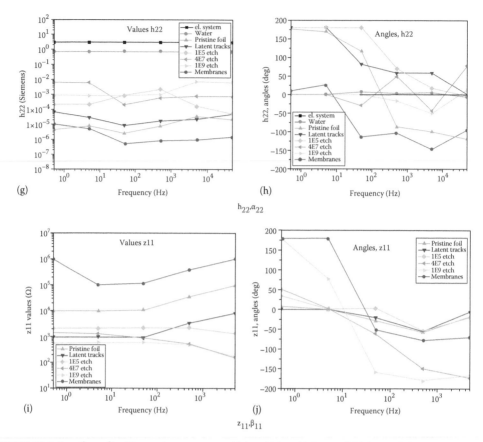

FIGURE 10.6 (continued) Four-pole parameters (a) h_{11}, (c) h_{12}, (e) h_{21}, and (g) h_{22} and their corresponding phases (b) α_{11}, (d) α_{12}, (f) α_{21}, and (h) α_{22}; the last two graphs show (i) the parameter z_{11}, describing the open-circuit input resistance and (j) its corresponding phase β_{11}. The values are given for the different components of the measuring system and for the different stages of sample preparation: black rectangle—electronic system (consisting of the frequency generator, the measuring chamber, the electronic circuit with a fixed 10 kΩ resistance, and the oscilloscope) only; red circle—measuring chamber filled with double deionized water; blue standing triangle—polymer foil separating the water-filled chamber into two compartments; green falling triangle—polymer foil with 1×10^5 cm^{-2} etched tracks (etching from both sides) within water; violet triangle pointing towards left—polymer foil with 4×10^7 cm^{-2} etched tracks within water; olive triangle pointing towards right—polymer foil with 1×10^9 cm^{-2} etched tracks within water; dark blue tilted square—polymer foil with Ag$_2$O membranes embedded within etched tracks within water.

The measuring system itself does not show any frequency dependence, thus indicating the general reliability of the obtained data in the used frequency interval. It is interesting that in general, the h parameters are much less frequency dependent than the current/voltage parameters are. The different production steps show up directly in the graphs of Figure 10.6.

For example, for the parameter h_{11} (Figure 10.6a), one can see that, whereas the parameter value for the electronic measuring system alone is lowest ($\sim 10^{-2}$ Ω), it rises when the chamber is filled with water, but nevertheless is still low ($\sim 10^{-1}$ Ω). When a pristine foil is inserted, h_{11} rises strongly up to ~ 10–100 MΩ, reflecting the still high resistance of the insulating foil immersed in water. If, for contrast, an ion-irradiated but unetched foil is inserted, h_{11} decreases to about 0.1–1 MΩ, as the latent tracks therein enable more water uptake. Ion track etching decreases the h_{11} values further, progressively with increasing track density in the foil down to about 1 kΩ, for the same reason. It is surprising that the h_{11} values of the foil with our lowest track density (1×10^5 cm^{-2}) are in a similar order than those of the foil with latent tracks. This reflects the rather narrow diameter of our etched tracks of typically ~10–20 nm. The insertion of membranes into the etched tracks leads to a dramatic jump of the h$_{11}$ values towards some 10 MΩ. Unfortunately, the results for the corresponding angles are less clear-cut, though at least some rough tendencies can be derived (Figure 10.6b), so that we do not discuss them here.

It appears that the h$_{11}$ parameters change stronger than the current/voltage results when going from one system to another. Similar tendencies can also be found in the graphs of the other quadripole parameters. Whereas the water-filled measuring chamber as well as the polymer foils with large etched track densities exhibits h$_{12}$ values around unity, pristine foils and foils with membranes reach higher h$_{12}$ values up to 10 or so. Most of the corresponding h$_{12}$ angles follow the common tendency of decreasing from about +180° towards zero at around 1–10 Hz. An exception is the pristine polymer foil that reaches values of around 100° or so for frequencies around 1 Hz to 10 kHz.

The measured h$_{21}$ parameters show a less clear tendency. Whereas water and polymer foils with many water-containing etched tracks exhibit h$_{21}$ values around unity, water-deficient polymer foils exhibit lower h$_{21}$ parameters. Interestingly, most h$_{21}$ values increase gradually with frequency. The corresponding h$_{21}$ angles show a characteristic drop from about +180° towards −180° in the frequency interval from ~10 Hz to ~10 kHz for water and strongly water-containing transparent polymer foils, but show angles between about +50° and −50° for less water-transparent structures.

With decreasing water content, the h$_{22}$ parameters roughly decrease from about unity (for water only) towards ~$10^{-5, \ldots, 6}$ S. The corresponding angles are around zero for water and for PET foils with high track densities for all frequencies, but they fall from about +180° towards angles between zero and −100° or so for pristine foils and foils with few etched ion tracks. Polymer foils with embedded membranes exhibit the lowest angles with typically −100° to −150°.

Finally, the z$_{11}$ parameters are low for transparent foils with high water content (around 10^2–10^4 Ω), but very high (around 10^4–10^6 Ω) for less-transparent foils (pristine foils and foils with embedded membranes). Samples with high z$_{11}$ values exhibit an increasing tendency, and those with low z$_{11}$ values exhibit a decreasing tendency with increasing frequency. All corresponding angles show a pronounced minimum between ~10 Hz and the kHz range, with transparent water-enriched polymer foils reaching angles as low as −180°, whereas the less transparent foils reach minimum angles around −50° only.

10.4 SUMMARY

A peculiarity of experiments on coupled chemical reactions in dynamic nanometric confinement is the transient formation of stable, unsoluble, and impermeable membranes of the precipitating material (here: Ag$_2$O) within the etched tracks, which separate the latter into two compartments. Especially at sufficiently low etching speed, membranes of good quality may emerge. Closer details of the membrane growth have been elaborated in the previous work [2,24].

These Ag$_2$O precipitations as well as their precursor structures are studied here in the low-frequency range by analyzing the corresponding current/voltage characteristics as a function of the applied frequency and the Bode plots and by observing the distortion of input frequency spectra by the membrane structures, as revealed by Fourier spectral analysis. Furthermore the four-pole parameters of these structures were derived to enable their easy future use for the design of electronic circuits.

Though the membranes embedded in the etched tracks guarantee a tight separation of the two etched track sides from each other—both for ions, biomolecules, and electrical signals at low frequencies—information exchange between the two sides can readily be accomplished by turning to higher frequencies—typically in the kHz regime.

This signifies that the formation of membranes embedded in etched tracks enables the filtering of low frequencies, depending on the membrane's properties. This makes these structures somewhat inert against hostile electrosmog stemming from low-frequency household, transport, and industrial noise. In a forthcoming paper of this series, the use of these structures for the production of a new type of robust biosensors will be reported.

ACKNOWLEDGMENT

D.F. is grateful to the Universidad Autónoma Metropolitana-Cuajimalpa, Mexico City, for the guest professorship in the frame of the Cathedra *Roberto Quintero Ramírez*, the Ben Gurion University of the Negev (BGU), Beer Sheva, and both the Grant Agency of the Czech Republic (P108-12G-108) and the Nuclear Physics Institute, Řež near Prague, for support. We are especially obliged to Profs. S. Revah, R. Quintero R., and M. Sales Cruz for their continuous help, encouragement, and discussions and for providing us with adequate working facilities. We are further obliged to Dr. P. Apel from JINR Dubna, Russia, for providing us with the ion-irradiated polymer foils. L.A. would like to acknowledge the Edmond J. Safra Center for the Design and Engineering of Functional Biopolymers at BGU.

REFERENCES

1. R. L. Fleischer, P. B. Price, R. M. Walker, *Nuclear Tracks in Solids: Principles and Applications.* University of California, Berkeley, CA, (1975).
2. G. Muñoz Hernandezab, S. A. Cruz, R. Quintero, D. Fink, L. Alfonta, Y. Mandabi, A. Kiv, J. Vacik, Coupled chemical reactions in dynamic nanometric confinement: Ag_2O formation during ion track etching. *Radiation Effects and Defects in Solids.* (2013), in print.
3. D. Fink, P. Yu. Apel, R. H. Iyer, Chapter II. 5, Ion track applications. Fink, D. ed., *Transport Processes in Ion Irradiated Polymers*, Springer Series in Materials Science, Vol. 65, pp. 269, 300, Springer-Verlag, Berlin, Germany, (2004), and references therein.
4. A. Biswas, D. K. Avasthi, B. K. Singh, S. Lotha, J. P. Singh, D. Fink, B. K. Yadav, B. Bhattacharya, S. K. Bose, Resonant tunneling in single quantum well heterostructure junction of electrodeposited metal semiconductor nanostructures using nuclear track filters. *Nuclear Instruments and Methods in Physics Research Section B.* 151 (1999) 84–88.
5. L. Piraux, J. M. George, J. F. Despres, C. Leroy, E. Ferain, R. Legras, K. Ounadjela, A. Fert, Giant magnetoresistance in magnetic multilayered nanowires. *Applied Physics Letters.* 65(19) (1994) 2484–2486.
6. M. Lindeberg, L. Gravier, J. P. Ansermet, K. Hjort, Processing magnetic field sensors based on magnetoresistive ion track defined nanowire cluster links, *Proceedings of the Workshop on European Network on Ion Track Technology*, Caen, France, (February 24–26, 2002).
7. K. Hjort, The European network on ion track technology. Presented at the *Fifth International Symposium on Swift Heavy Ions in Matter*, Giordano Naxos, Italy, (May 22–25, 2002) (unpublished).
8. D. Fink, A. Petrov, K. Hoppe, W. R. Fahrner, Characterization of "TEMPOS": A new Tunable Electronic Material with Pores in Oxide on Silicon. *Proceedings of 2003 MRS Fall Meeting. Vol. 792-Symposium R–Radiation Effects and Ion Beam Processing of Materials*, Boston, MA, (December 1–5, 2003).
9. D. Fink, A. Petrov, H. Hoppe, A. G. Ulyashin, R. M. Papaleo, A. Berdinsky, W. R. Fahrner, Etched ion tracks in silicon oxide and silicon oxynitride as charge injection channels for novel electronic structures. *Nuclear Instruments and Methods in Physics Research Section B.* 218 (2004) 355.
10. M. Tamada, M. Yoshida, M. Asano, H. Omichi, R. Katakai, R. Spohr, J. Vetter. Thermo-response of ion track pores in copolymer films of methacryloyl-L-alaninemethylester and diethyleneglycol-bis-allylcarbonate (CR-39). *Polymer.* 33(15) (1992) 3169–3172.
11. C. G. J. Koopal, M. C. Feiters, R. J. M. Nolte, B. de Ruiter, R. B. M. Schasfoort, Glucose sensor utilizing polypyrrole incorporated in track-etch membranes as the mediator. *Biosensors and Bioelectronics.* 7 (1992) 461–471; S. Kuwabata, C. R. Martin, Mechanism of the amperometric response of a proposed glucose sensor based on a polypyrrole-tubule-impregnated membrane. *Analytical Chemistry.* 66 (1994) 2757–2762.
12. Z. Siwy, L. Trofin, P. Kohl, L. A. Baker, C. R. Martin, C. Trautmann, Protein biosensors based on biofunctionalized conical gold nanotubes. *Journal of the American Chemical Society.* 127 (2005) 5000–5001.
13. Z. S. Siwy, C. C. Harrell, E. Heins, C. R. Martin, B. Schiedt, C. Trautmann, L. Trofin, A. Polman, Nanopores as ion-current rectifiers and protein sensors. Presented at the *Sixth International Conference on Swift Heavy Ions in Matter*, Aschaffenburg, Germany, (May 28–31, 2005) (unpublished).
14. L. Alfonta, O. Bukelman, A. Chandra, W. R. Fahrner, D. Fink, D. Fuks, V. Golovanov et al., Strategies towards advanced ion track-based biosensors. *Radiation Effects and Defects in Solids.* 164 (2013) 431–437.
15. C. R. Martin and Z. S. Siwy, Learning nature's way: Biosensing with synthetic nanopores. *Science.* 317 (2007) 331.

16. D. Fink, I. Klinkovich, O. Bukelman, R. S. Marks, A. Kiv, D. Fuks, W. R. Fahrner, L. Alfonta, Glucose determination using a re-usable ion track membrane sensor. *Biosensors and Bioelectronics*. 24 (2009) 2702–2706.

17. D. Fink, G. Muñoz Hernandezab, L. Alfonta, Highly sensitive ion track-based urea sensing with ion-irradiated polymer foils. *Nuclear Instruments and Methods in Physics Research Section B*. 273 (2012) 164–170.

18. D. Fink, G. Muñoz Hernandezab, J. Vacik, L. Alfonta, Pulsed biosensing. *IEEE Sensors Journal*. 11 (2011) 1084–1087.

19. Y. Mandabi, S. A. Carnally, D. Fink, L. Alfonta, Label free DNA detection using the narrow side of conical etched nanopores. *Biosensors and Bioelectronics*. (2013) in print.

20. M. Daub, I. Enculescu, R. Neumann, R. Spohr, Ni nanowires electrodeposited in single ion track templates. *Journal of Optoelectronics and Advanced Materials*. 7 (2005) 865–870.

21. D. Fink, S. Cruz, G. Muñoz Hernandezab, A. Kiv, Current spikes in polymeric latent and funnel-type ion tracks. *Radiation Effects and Defects in Solids*. 5 (2011) 373–388.

22. D. Fink, ed., *Fundamentals of Ion-Irradiated Polymers*, Vol. 63 of Springer Series in Materials Science, Springer-Verlag, Berlin, Germany, (2004), p. 190, 311, 312, 349, 357.

23. C. T. Souza, E. M. Stori, D. Fink, V. Vacík, V. Švorčík, R. M. Papaléo, L. Amaral, J. F. Dias, Electronic behavior of micro-structured polymer foils immersed in electrolyte. (2012) in print.

24. D. Fink, G. Muñoz Hernandezab, H. García Arellano, W. R. Fahrner, K. Hoppe, J. Vacik, Coupled chemical reactions in dynamic nanometric confinement: II Preparation conditions for Ag$_2$O membranes within etched tracks. To be published in 2013 in the same book.

11 Toward Unsupervised Smart Chemical Sensor Arrays

Leonardo Tomazeli Duarte and Christian Jutten

CONTENTS

11.1 INTRODUCTION

One of the major challenges in chemical analysis is how to deal with the lack of selectivity that is typical of chemical sensors [1]. Recently, much attention has been given to an approach brought from the field of signal processing, to deal with the interference problem related to chemical sensors. In this alternative approach, the sensing mechanism is based on an ensemble of sensors that are not necessarily highly selective with respect to a given analyte. The rationale behind such an approach, which is usually referred to as smart sensor arrays (SSAs), is that although the sensors within the array may respond to several chemical species, if there exists enough diversity between the sensors, then advanced signal processing methods can be applied in order to estimate the concentration—or detect the presence—of the chemical species under analysis.

Most of chemical SSAs are based on *supervised* signal processing methods, which require calibrating (or training) samples to adjust the parameters of the adopted signal processing method. The application of supervised methods in both quantitative and qualitative analyses has been proving quite successful in tasks such as odor and taste automatic recognition systems (electronic noses [2] and tongues [3], respectively). However, despite the success of SSAs based on supervised methods, this approach suffers from two important practical problems. Firstly, the acquisition of training samples is usually a costly and time-demanding task. Secondly, due to the drift in the response of chemical sensors, the calibration procedure must be performed every time the sensor array is used.

In view of the practical limitations associated with supervised SSAs, some researchers have been developing systems based on unsupervised (or blind) methods. The idea here is to adjust the SSA data processing stage by considering only the array responses and possibly some information on how the interference phenomenon takes place. As a result, the calibration stage may be eliminated or, at least, simplified in unsupervised solutions. When unsupervised systems for performing quantitative analysis are considered, the task at hand is to solve what the signal processing community calls the *blind source separation* (BSS) problem [4–6]. In fact, the goal of BSS is to retrieve a set of signals (sources) based only on the observation of a set of signals that correspond to mixed versions of the original sources. Therefore, in quantitative analysis via SSAs, the sources would correspond to the time variations of the concentrations of each chemical species, under analysis, whereas the mixed signals would be given by the array responses.

In this chapter, we discuss the main aspects underlying the application of BSS methods to the problem of quantitative analysis via SSAs. Our focus will be on sensor arrays composed of potentiometric electrodes. Concerning the chapter's organization, we first provide, in Section 11.2, a brief introduction to potentiometric sensors. In Section 11.3, we introduce the BSS problem. Then, in Section 11.4, we provide an overview on the existing results concerning the application of BSS methods to quantitative analysis via potentiometric sensor arrays. Section 11.5 is devoted to the discussion of some important practical aspects related to the use of BSS methods to performing chemical analyses. Finally, in Section 11.6, the chapter is closed with a concluding section.

11.2   SMART CHEMICAL SENSOR ARRAYS

11.2.1   Potentiometric Sensors

Potentiometric sensors can be defined as devices for which a variation of the concentration of a given chemical species induces a variation of its electrical potential. The best-known example of potentiometric sensor is the ion-selective electrode (ISE) [1,7,8,9]. This device is basically composed of an internal solution, an internal reference electrode, and sensitive membrane, as depicted in Figure 11.1. An electrical potential in this membrane comes from an electrochemical equilibrium and is directly related to the activity (which can be seen as the effective concentration) of a given target ion.

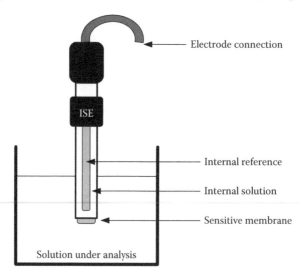

FIGURE 11.1   The ISE.

A well-known example of ISE is the glass electrode [1,8], which is used for measuring the pH of a given solution. Besides, one can find ISEs tailored to different ions such as ammonium, potassium, and sodium. These devices have been intensively used, for instance, in food and soil inspection, clinical analysis, and water quality monitoring. One of the reasons that explain the success of ISEs in such applications is the simplicity of this approach. Indeed, analyses through ISEs do not require sophisticated laboratory equipments and procedures and, thus, can be carried out in the field if necessary. Moreover, ISEs provide a very economical solution compared to other chemical sensing systems. Another popular example of potentiometric sensor is the ion-sensitive field-effect transistor (ISFET) [10]. In certain extent, the ISFET can be seen as a miniaturized version of the ISE, as the transduction mechanisms of both devices are essentially the same.

11.2.2 Selectivity Issues in Potentiometric Sensors

Electrodes such as ISEs and ISFETs suffer from an important drawback: these devices are not selective, as the generated potential usually depends on a given target ion but also on other interfering ions. This phenomenon may become important when the target ion and the interfering ones have similar physical and/or chemical properties [8]. In such cases, the measurements taken by the chemical sensor become uncertain when the concentrations of the interfering ions are high enough.

The interference phenomenon in potentiometric sensor can be modeled through the Nicolsky–Eisenman (NE) equation [1,8,9,11]. Assuming that s_i and s_j correspond, respectively, to the activity of the target ion and the activity of interfering ones, the response of a potentiometric sensor according to the NE equation is given by

$$x = e + d \log \left(s_i + \sum_{j, j \neq i} a_{i,j} s_j^{z_i/z_j} \right), \tag{11.1}$$

where
  e and d are constants that depend on some physical parameters
  z_i denotes the valence of the ith ion

The parameters a_{ij}, which are called selectivity coefficients, model the interference between the ions under analysis.

11.2.3 Chemical Sensor Arrays

The key concept underlying SSAs is diversity. Indeed, in an SSA, as illustrated in Figure 11.2, there is a data processing block that makes use of the diversity within the array to remove the effects caused by the interference. Since the core of SSAs lies at the data processing block, an interesting aspect of this strategy regards its flexibility: a same SSA can conduct different types of analyses through minor and often automatic changes. At some extent, the term *smart* makes reference to this adaptive character of SSA. Besides this advantage, other assets of SSAs include robustness and cost; indeed SSAs are usually composed of simple, and thus cheap, electrodes. There are also embedded SSAs, in which one can find a microcontroller.

The signal processing techniques used in SSAs can be firstly classified with respect to the paradigm adopted to adjust their parameters. If the method makes use of a set of training (or calibration) data, it is called *supervised* [12]. In supervised methods, the calibration data are considered in a stage (training stage) that precedes the effective use of the array. If, on the other hand, no calibration stage is considered, then the signal processing method is referred to as *unsupervised* or *blind*.

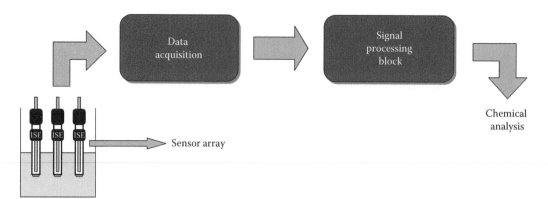

FIGURE 11.2 Smart chemical sensor array.

When one considers a supervised qualitative chemical analysis, then one ends up with a (supervised) pattern classification problem [13]. This problem can be tackled through machine learning techniques, such as multilayer perceptron (MLP) neural networks [14] and support vector machines (SVMs) [15]. Supervised quantitative analysis through sensor arrays can be formulated as a multivariate regression problem.

The majority of SSAs are based on a supervised paradigm. However, there is a growing interest in unsupervised methods. For instance, a number of works have considered chemical unsupervised qualitative analysis. In this case, since training points are not available, the classes that should be classified are not known beforehand. The resulting problem here is known as clustering analysis and can be approached via algorithms like k-means and self-organizing maps (SOMs) [15].

While the three signal processing tasks discussed so far (supervised and unsupervised qualitative analysis and supervised quantitative analysis) are now familiar to the community of chemical SSAs, the case of unsupervised quantitative chemical analysis is still in its beginning. As it will become clear later, this problem can be formulated as a BSS one.

11.3 BSS

In BSS [4], the goal is to retrieve a set of signals (sources) that were submitted to a mixing process. Since the mixing process is also assumed unknown, source separation is conducted by considering only observed signals (mixtures). BSS methods have been intensively used in biosignal processing, audio analysis, and image processing (see the textbooks [4–6] for more details.

Although the formulation of BSS is simple, its resolution became possible only after the introduction of a new learning paradigm in signal processing. Indeed, in a paper published in the 1980s by Hérault et al. [16], it was shown that the standard approach that was typically considered in statistical signal processing at that time—methods based on second-order statistics—could not solve the BSS problem. In the sequel, we will provide a mathematical formulation of the BSS problem and present the main strategies to solve this problem.

11.3.1 PROBLEM DESCRIPTION

Let us consider that vectors $s(t) = [s_1(t)\ s_2(t)\ \dots\ s_N(t)]^T$ and $x(t) = [x_1(t)\ x_2(t)\ \dots\ x_M(t)]^T$ represent the sources and the mixtures, respectively. Moreover, let the mixing process be represented by a mathematical mapping $F(\cdot)$, so the mixtures can be expressed as follows:

$$x(t) = F(s(t)).\tag{11.2}$$

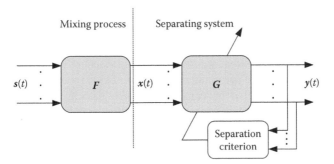

FIGURE 11.3 General scheme of the problem of BSS.

BSS methods aim at estimating the sources $s(t)$ based only on the mixtures $x(t)$, that is, without making use of any precise information about either the mixing mapping $F(\cdot)$ or the sources $s(t)$. It should be stressed here that, in the context of SSAs, the sources $s(t)$ correspond to the concentrations of the chemical species under analysis and the mixtures $x(t)$ to the signals recorded by the array.

In Figure 11.3, the problem and solution of BSS are illustrated. The basic idea is to define a separating system, which is represented by $G(\cdot)$ so the signals $y(t) = [y_1(t)\ y_2(t)\ \dots\ y_N(t)]^T$, given by

$$y(t) = G(x(t)) \tag{11.3}$$

are as close as possible to the actual sources $s(t)$. Since the sources are unknown, the fundamental question in this context is how to formulate a separation criterion that will guide adjustment of the parameters of the separating system $G(\cdot)$. In the sequel, we discuss some of the main strategies applied to accomplish this task. Henceforth, we assume that the number of mixtures and sources are the same ($N = M$).

11.3.2 Strategies to Perform BSS

The BSS problem is an ill-posed problem and, thus, cannot be solved if a minimum of prior information about the sources is not available. For example, let us consider that one knows that the sources present a given property. If this property is lost after the mixing process, a possible idea to build a separation criterion would be to search for the separating system $G(\cdot)$ that provides estimations $y(t)$ having the same property observed in $s(t)$. The idea is valid when the recovery of such original characteristic assures that $G(\cdot)$ provides a perfect inversion of the mixing mapping $F(\cdot)$. This is the case, for instance, in the context of independent component analysis (ICA), as will be discussed in the sequel.

11.3.2.1 ICA Methods

Initially, ICA methods were designed to deal with the situation in which the mixing process is linear and instantaneous, that is, without memory. In this case, it is possible to express the mixing process as follows:

$$x(t) = As(t), \tag{11.4}$$

where A corresponds to the mixing matrix. Typically, it is assumed that the number of sensors is equal to the number of sources, being A, a square matrix in this case. Therefore, in this situation, the separating system can also be defined as a square matrix, so the retrieved signals are given by

$$y(t) = Ws(t). \tag{11.5}$$

Ideally, the separating matrix should be given by $W = A^{-1}$.

In ICA, the sources are view as realizations of independent random variables. Since the independence property is lost after the linear mixing process, the basic idea in ICA is to adjust W so the signals $y(t)$ be again independent. Comon [17] showed that, for the situation in which at most one source follows a Gaussian distribution, the recovery of independent components $y(t)$ is possible if, and only if, $WA = PD$, where P is a permutation matrix and D a diagonal matrix. That is, recovering independent components assures that one can indeed recover the waveform of the original sources, but it is impossible to determine their exact order and their original scales. These limitations are usually referred to as the order and scale ambiguities and arise because statistical independence is a property that is invariant to order and scales.

The idea behind ICA can be implemented via an optimization problem, in which the cost function is related to a measure of statistical independence. Usually in the literature, this cost function is called contrast function. For instance, a measure that is commonly employed in the context of the information theory, the *mutual information* [18], can be defined as a contrast function. Indeed, the mutual information of a set of random variables is always greater or equal to zero and is zero if, and only if, these variable are mutually independent, which thus provide a natural measure of statistical independence.

One can find in the literature other approaches that also lead to simple ICA algorithms. Examples include the Infomax approach, cumulant-based methods, nonlinear decorrelation methods, and nonlinear PCA (see [4,6] for an introduction). An interesting point related to them is that all these approaches are somehow connected and can be described through a consistent theoretical framework [19], resulting thus in practical algorithms that are similar.

11.3.2.2 Bayesian Approach

Another approach that has been applied in the context of BSS is based on Bayesian estimation. Among the features found in this framework is the possibility of incorporating prior information that can be described in a probabilistic manner. Moreover, although Bayesian BSS methods usually consider, as prior information, that the sources are independent, this assumption is rather an instrumental one, as Bayesian methods do not rely on an independence recovery procedure. The development of Bayesian methods to BSS was addressed in a number of works (see Chapter 12 of [4], for an introduction), and one can find applications in, for instance, spectroscopy data analysis [20] and hyperspectral imaging [21].

In Bayesian BSS, instead of defining a separating system aimed at inverting the mixing process, one searches for a generative model that can correctly explain the observed data. To clarify this idea, let the $N \times T$ matrix X represent all the observations (T samples of N mixtures) of the problem. Moreover, let the matrix Θ represent all the unknown terms of our problem, that is, the sources and the coefficients of the mixing process. The key concept in the Bayesian approach is the posterior probability distribution, that is, the probability distribution of the unknown parameters Θ conditioned to the observed data X. This posterior distribution can be obtained via Bayes' rule as follows:

$$p\left(\Theta|X\right) = p\left(X|\Theta\right)\frac{p\left(\Theta\right)}{p\left(X\right)}. \tag{11.6}$$

In this expression, $p(X \mid \Theta)$ is the likelihood function and is directly related to the assumed mixing model. The term $p(\Theta)$ denotes the prior distribution and should be defined by taking into account the available information at hand. For example, if the sources are nonnegative [22], which is the case in chemical arrays, it is natural to consider prior distributions whose support takes only nonnegative values.

There are several possibilities to estimate Θ from in X in a Bayesian framework. For instance, a possible strategy is to find Θ that maximizes the posterior distribution. This strategy is known

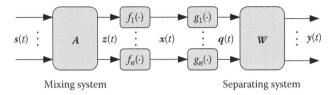

FIGURE 11.4 The PNL model (mixing and separating structures).

as maximum a posteriori (MAP) estimation. Another way to obtain estimation is based on the Bayesian minimum mean square error (MMSE) estimator. In this situation, the resulting estimator is obtained by taking the expected value of the posterior distribution $p(\Theta \mid X)$.

11.3.3 NONLINEAR MIXTURES

In some practical cases, the linear approximation does not give a good enough description of the interference model. For example, this is the case when dealing with potentiometric sensors (see Section 11.2.2). In such situations, the mixing process that takes place at the sensor array becomes nonlinear and, thus, must be tackled by nonlinear BSS models.

In nonlinear BSS, there are some problems that are not present in a linear context. The most important one is that ICA does not lead to source separation in a general nonlinear context. In order words, there are some mixing models for which retrieving independent components is not enough to retrieve the sources. In view of such difficulty, some researchers have been considering constrained classes of nonlinear model for which ICA is still valid [23]. For instance, as shown in [24,25], ICA-based solutions assure source separation in an important class of constrained models known as post-nonlinear (PNL) models.

The basic structure of a PNL is shown in Figure 11.4. The mixing process is composed of a mixing matrix A that is followed by component-wise functions, represented by the vector of functions $f(\cdot)$. The mixing model can be represented by $x(t) = f(As(t))$. As shown in Figure 11.4, the separating system in this case can be defined as $y(t) = Wg(x(t))$. Under the assumptions that the sources are independent and that the component-wise functions $f(\cdot)$ and $g(\cdot)$ are monotonic, the separating matrix W and the compensating functions $g(\cdot)$ can be adjusted by a independence recovery procedure [24].

11.4 APPLICATION OF BSS METHODS TO CHEMICAL SENSOR ARRAYS

11.4.1 FIRST RESULTS

The application of BSS methods to chemical sensor arrays is a relative new topic and was considered in a small number of papers. The first contributions in this area were made by Sergio Bermejo and collaborators [26]. In this work, linear source separation was employed to process the data obtained by an array of ISFETs, which, as discussed before, can be modeled by the NE equation. Hence, the mixing process that takes place in ISFETs is basically the same as that of the one found in ISEs.

As discussed before, one of the problems that one must deal with when applying source separation methods to potentiometric sensor arrays is related to the nonlinear nature of these sensors. In [26], the nonlinear terms present in the mixing process are compensated by using a small set of calibration points. Therefore, the solution proposed in [26] is not completely blind, since only the linear part of the mixing process is processed in a blind fashion.

The work presented in [27] investigated the application of the mutual information-based PNL method proposed in [24] to ISFET arrays. Indeed, if one assumes that the interference problem can be modeled by the NE equation and when all the ions under analysis have the same valences, the resulting mixing process becomes a special case of the PNL model, where the component-wise functions are given by

$$f_i(t) = e_i + d_i \log(t). \tag{11.7}$$

Ideally, the parameters d_i could be set by simply considering the theoretical Nernstian slope value predicted by the NE equation [1]. However, d_i usually deviates from this theoretical value and therefore should be estimated. The parameter e_i does not need to be estimated in a BSS framework, since it only introduces a scale gain and, as already discussed, BSS methods are not able to detect the original scales of the sources. The approach proposed in [27] was assessed by considering a set of synthetic data that were obtained from actual measurements.

11.4.2 ICA-BASED METHODS FOR THE CASE WITH DIFFERENT VALENCES

As pointed out by [27], when the valences of all ions under analysis are equal, the ratio z_i/z_j in Equation 11.1 takes one, and, as consequence, the mixing process in this case is given by a PNL model. When the valences are different, the resulting mixing model becomes tougher because nonlinearity (power term) appears inside the logarithm term.

In [28,29], we proposed algorithms suited to the situation in which there are power terms inside the logarithm function. In view of the complexity of the mixing resulting mixing model in this case, we consider some simplifying assumptions in a first moment. First, we assume that, although they are different, the valences of the ions under analysis are known in advance. Second, we consider that the parameters e_i and d_i are previously known. The estimation of d_i can be done, for instance, by the blind methods that will be discussed in Section 11.4.3. Moreover, even if e_i is not exactly known, we can still apply our method but the best we can do is to retrieve each source up to an unknown multiplicative gain. A last simplification is that we only considered the case with two sources (ions) and two mixtures (electrodes). This assumption is realistic in many practical situations where there is one interfering ion that is dominant, while the others can be neglected.

When the simplifications described previously are considered, the resulting mixing model can be written as

$$\begin{aligned} x_1 &= s_1 + a_{12}s_2^k \\ x_2 &= s_2 + a_{21}s_1^{1/k}. \end{aligned} \tag{11.8}$$

The term k corresponds to the valence ratio z_1/z_2. Our analysis assumes that k is always a natural number. This situation arises, for instance, in the analysis of Ca^{2+} and Na^+, since $k = 2$ in this case. It is worth noticing that since the sources are nonnegative in our problem, there is no risk of having complex-value numbers from the term $s_1^{1/k}$. Finally, as we are interested in an ICA solution, it is assumed that the sources are statistically independent.

A first problem that arises when dealing with the nonlinear mixing model (Equation 11.8) concerns the definition of a proper separating system. Differently from linear and even PNL systems, the mixing model (Equation 11.8) is not invertible, and, as a consequence, it is not possible to define a direct separating system. In order to overcome this problem, we defined in [28] a separating structure based on a recurrent system. This strategy, which was firstly adopted in the context of another class of nonlinear models known as linear–quadratic (LQ) models [30], is able to perform a sort of implicit inversion of the mixing model, given that, when the mixing parameters are known, the equilibrium points of the defined recurrent system correspond to the actual sources. Of course, these equilibrium points can be attained only if they are stable. As a consequence, for values of ionic activities and selectivity coefficients that lie outside the stability region, the adopted recurrent system cannot be used. In [28], we provide these stability regions.

Concerning the estimation of the parameters of the adopted recurrent separating system, we proposed in [28] an ICA method based on the minimization of a measure of nonlinear correlation. The idea was implemented by a considering a gradient descent approach. In particular, we studied some convergence issues related to the obtained gradient-based learning rule. In order to obtain a more robust learning algorithm to the recurrent separating system, we considered in [29] an algorithm based on the mutual information minimization. Our approach was based on the concept of the differential of the mutual information, which was proposed in [31]. More details can be found in [29].

To illustrate the performance of the method proposed in [29], we considered the problem of estimating the activities of Ca^{2+} and Na^+ ($k = 2$) by using two ISEs, being each one tailored to a different ion under analysis. The selectivity coefficients were obtained from [32] and the temporal evolution of the ionic activities (our sources) were artificially generated. In Figure 11.5, we consider the situation in which the activity of the ion Ca^{2+} lies in the interval $[10^{-4}; 10^{-3}]$ M, whereas the activity of Na^+ varies between $[10^{-4}; 10^{-1}]$ M, and we plot the sources (ionic activities), the mixtures (array responses), and the retrieved sources, estimated by the algorithm [29]. The total number of samples of the mixtures was 500. As shown in Figure 11.5, (1) there is indeed a mixing process taking place at the sensor array, and (2) our method was able to provide a very good estimation of the sources.

11.4.3 Use of Prior Information to Estimate the Electrode's Slope

The ICA methods discussed in the Section 11.4.2 were designed to deal with the model (Equation 11.8), and, thus, assumed that the slopes d_i were known in advance. In practice, though, it is necessary to estimate these slopes. In [33], we addressed this problem by considering the following additional assumption on the sources: there is, at least, a period of time where one, and only one, of the sources takes a constant value different from zero. In the context of chemical sources, this hypothesis seems valid in some applications as can be seen, for instance, in the waveforms presented in [34].

11.4.4 Bayesian Source Separation Applied to Chemical Sensor Arrays

As discussed in Section 11.3.2.2, another way to take advantage of prior information on the sources and mixing coefficients is to consider a Bayesian approach. One of the motivations behind this approach is related to the use of some prior information whose incorporation is easier in a probabilistic framework. For instance, in our problem, the fact that the sources are nonnegative can be easily taken into account by considering nonnegative priors.

In [35], we developed a Bayesian source separation method for mixtures generated by NE equation. The basis of our approach, which can be applied for both cases of equal and different valences, relies on the attribution of a nonnegative prior, the lognormal distribution, to the sources. Moreover, since the selectivity coefficients usually lie in the interval [0,1], we attributed to these parameters a uniform prior with support given by [0,1]. Concerning the implementation of the Bayesian estimation, we considered an approach based on Markov chain Monte Carlo methods (MCMCs). More details can be found in [35].

We tested our Bayesian method considering the set of real data available at the ISEA dataset (see Section 11.5.2). For instance, we show, in Figures 11.6 through 11.8, respectively, the sources, the mixtures, and the retrieved sources in a case in which the goal is to estimate the activities of K^+ and NH_4^+ by considering an SSA composed of one K^+-ISE and one NH_4^+-ISE. These results attest that the Bayesian algorithm was able to retrieve a good estimation of the sources even under adverse conditions, namely, reduced number of samples and sources with a high degree of correlation. Conversely, the application of an ICA-based PNL source separation method could not separate the sources in this case [35]. This limitation stems from the fact that the sources were not independent in this scenario, thus violating the essential assumption made in ICA.

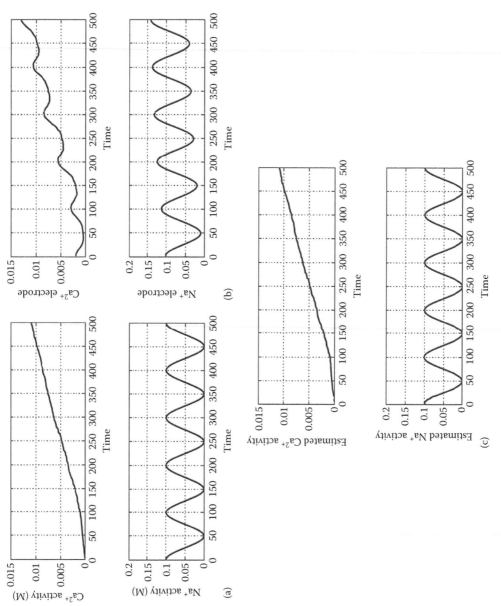

FIGURE 11.5 Example of mutual information minimization algorithm: (a) sources, (b) mixtures, and (c) retrieved sources after 600 iterations.

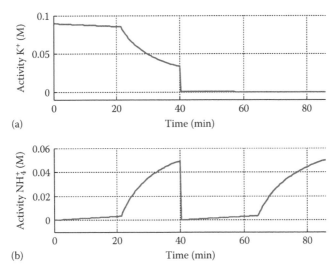

FIGURE 11.6 Application of the proposed Bayesian source separation approach to an array composed of (a) a K^+-ISE and (b) NH_4^+ ISE (sources).

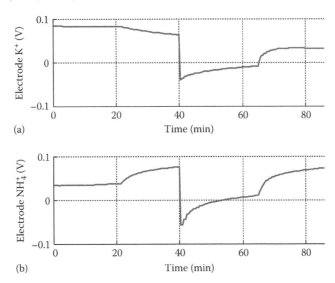

FIGURE 11.7 Application of the proposed Bayesian source separation approach to an array composed of (a) a K^+-ISE and (b) NH_4^+ ISE (mixtures).

11.5 PRACTICAL ISSUES

11.5.1 DEALING WITH THE SCALE AMBIGUITY

As mentioned before, in BSS, there is always a scale ambiguity, since usually separation criteria cannot recover the correct amplitude of the sources. While in many applications the scale ambiguity can be accepted, this limitation poses a problem in chemical sensing applications since the main goal is to retrieve the correct value of the concentration. Therefore, the source separation step must be followed by a postprocessing whose goal is to retrieve the correct scale. This additional stage requires, at least, two calibration points. Evidently, in view of this requirement, we may ask ourselves why not simply use the available calibration points for performing supervised processing. For example, we could define a separating system (inverse of NE model) adjusted with supervision. The key point here is that, as shown in [35] in a experiment with one K^+-ISE and one NH_4^+-ISE,

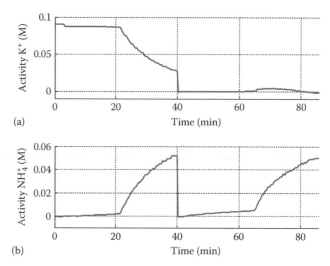

FIGURE 11.8 Application of the proposed Bayesian source separation approach to an array composed of (a) a K^+-ISE and (b) NH_4^+ ISE (estimated sources).

the supervised solution required at least 20 calibration points to provide a good estimation of the sources. Conversely, a good estimation was achieved by the Bayesian approach proposed in [35] followed by a postprocessing stage that made use of only three calibration points to retrieve the correct sources scale. From this difference, the benefits of unsupervised methods become clear: although it is not possible to completely operate without any calibration point, the number of calibration points needed in an unsupervised approach is rather small, which can be quite advantageous in practice.

11.5.2 Database ISEA

In the framework of our research on BSS-based chemical sensor arrays, we had some difficulty to find actual data that could be used to validate the developed methods. Motivated by this, in www.gipsa-lab.inpg.fr/isea, we provide a dataset (publicly available) acquired in a set of experiments with ISE arrays. These experiments are described in [36]. Basically, we considered three scenarios: (1) analysis of a solution containing K^+ and NH_4^+, (2) analysis of a solution containing K^+ and Na^+, and (3) analysis of a solution containing Na^+ and Ca^{2+}. These data can be used in the development of unsupervised signal processing methods as well as in the case of supervised solutions, since we also provide the original sources.

11.6 CONCLUSIONS

The goal of this chapter was to provide an overview on the application of BSS methods to potentiometric sensor arrays. The results are promising, since these methods may work even if only a reduced number of calibration points are available. Of course, in a practical context, this nice feature is quite useful; for instance, a less demanding calibration step may ease analysis in the field. Yet, despite the encouraging results obtained in this research, there are still many questions that should be investigated before envisaging the incorporation of source separation blocks into commercial chemical analyzers. These points include the development of methods for more than two sources and the search for more precise mixing models, which may eventually increase the estimation quality, thus permitting the use of source separation methods even in very-high-precision applications.

REFERENCES

1. P. Gründler, *Chemical Sensors: An Introduction for Scientists and Engineers*. Springer, Berlin, Germany, 2007.
2. H. Nagle, R. Gutierrez-Osuna, and S. S. Schiffman, The how and why of electronic noses, *IEEE Spectrum*, 35(9), 22–31, 1998.
3. Y. G. Vlasov, A. V. Legin, and A. M. Rudnitskaya, Electronic tongue: Chemical sensor systems for analysis of aquatic media, *Russian Journal of General Chemistry*, 78, 2532–2544, 2008.
4. P. Comon and C. Jutten (eds.), *Handbook of Blind Source Separation: Independent Component Analysis and Applications*. Academic Press, Oxford, U.K., 2010.
5. J. M. T. Romano, R. R. F. Attux, C. C. Cavalcante, and R. Suyama, *Unsupervised Signal Processing: Channel Equalization and Source Separation*. CRC Press, Boca Raton, FL, 2011.
6. A. Hyvärinen, J. Karhunen, and E. Oja, *Independent Component Analysis*. John Wiley & Sons, New York, 2001.
7. G. Bedoya, Nonlinear blind signal separation for chemical solid-state sensor arrays, PhD dissertation, Universitat Politecnica de Catalunya, Barcelona, Spain, 2006.
8. P. Fabry and J. Fouletier (eds.), *Microcapteurs chimiques et biologiques: Application en milieu liquide*. Lavoisier, France, 2003, in French.
9. E. Bakker, Electrochemical sensors, *Analytical Chemistry*, 76, 3285–3298, 2004.
10. P. Bergveld, Thirty years of ISFETOLOGY What happened in the past 30 years and what may happen in the next 30 years, *Sensors and Actuators B*, 88, 1–20, 2003.
11. E. Bakker and E. Pretsch, Modern potentiometry, *Angewandte Chemie International Edition*, 46, 5660–5668, 2007.
12. T. Hastie, R. Tibshirani, and J. Friedman, *The Elements of Statistical Learning*, 2nd edn. Springer, Berlin, Germany, 2009.
13. R. Blatt, A. Bonarini, E. Calabro, M. D. Torre, M. Matteucci, and U. Pastorino, Lung cancer identification by an electronic nose based on an array of mos sensors, in *Proceedings of International Joint Conference on Neural Networks (IJCNN)*, Orlando, FL, pp. 1423–1428, August 12–17, 2007.
14. M. Pardo and G. Sberveglieri, Classification of electronic nose data with support vector machines, *Sensors and Actuators B*, 107, 730–737, 2005.
15. R. O. Duda, P. E. Hart, and D. G. Stork, *Pattern Classification*, 2nd edn. Wiley-Interscience, New York, 2000.
16. J. Hérault, C. Jutten, and B. Ans, Détection de grandeurs primitives dans un message composite par une architecture de calcul neuromimétique en apprentissage non supervisé, in *Proceedings of the GRETSI*, pp. 1017–1022, 1985.
17. P. Comon, Independent component analysis, a new concept? *Signal Processing*, 36, 287–314, 1994.
18. T. M. Cover and J. A. Thomas, *Elements of Information Theory*, Wiley-Interscience, New York, 1991.
19. T.-W. Lee, M. Girolami, A. J. Bell, and T. J. Sejnowski, A unifying information-theoretic framework for independent component analysis, *Computers and Mathematics with Applications*, 39, 1–21, 2000.
20. S. Moussaoui, D. Brie, A. Mohammad-Djafari, and C. Carteret, Separation of non-negative mixture of non-negative sources using a Bayesian approach and MCMC sampling, *IEEE Transactions on Signal Processing*, 54, 4133–4145, 2006.
21. N. Dobigeon, S. Moussaoui, M. Coulon, J.-Y. Tourneret, and A. O. Hero, Joint Bayesian endmember extraction and linear unmixing for hyperspectral imagery, *IEEE Transactions on Signal Processing*, 57, 4355–4368, 2009.
22. A. Cichocki, R. Zdunek, A. H. Phan, and S. Amari, *Nonnegative Matrix and Tensor Factorizations: Applications to Exploratory Multiway Data Analysis and Blind Source Separation*. John Wiley & Sons, New York, 2009.
23. C. Jutten and J. Karhunen, Advances in blind source separation (BSS) and independent component analysis (ICA) for nonlinear mixtures, *International Journal of Neural Systems*, 14, 267–292, 2004.
24. A. Taleb and C. Jutten, Source separation in post-nonlinear mixtures, *IEEE Transactions on Signal Processing*, 47(10), 2807–2820, 1999.
25. S. Achard and C. Jutten, Identifiability of post-nonlinear mixtures, *IEEE Signal Processing Letters*, 12(5), 423–426, 2005.
26. S. Bermejo, C. Jutten, and J. Cabestany, ISFET source separation: Foundations and techniques, *Sensors and Actuators B*, 113, 222–233, 2006.

27. G. Bedoya, C. Jutten, S. Bermejo, and J. Cabestany, Improving semiconductor-based chemical sensor arrays using advanced algorithms for blind source separation, in *Proceedings of Sensors for Industry Conference (SIcon)*, New Orleans, LA, pp. 149–154, 2006.

28. L. T. Duarte and C. Jutten, Blind source separation of a class of nonlinear mixtures, in *Proceedings of the Seventh International Workshop on Independent Component Analysis and Signal Separation (ICA 2007)*, London, U.K., September 9–12, 2007.

29. L. T. Duarte and C. Jutten, A mutual information minimization approach for a class of nonlinear recurrent separating systems, in *Proceedings of the IEEE Workshop on Machine Learning for Signal Processing (MLSP)*, 2007.

30. S. Hosseini and Y. Deville, Blind separation of linear-quadratic mixtures of real sources using a recurrent structure, in *Proceedings of the Seventh International Work-Conference on Artificial and Natural Neural Networks (IWANN)*, Menorca, Spain, June 3–6, 2003.

31. M. Babaie-Zadeh, C. Jutten, and K. Nayebi, Differential of the mutual information, *IEEE Signal Processing Letters*, 11(1), 48–51, 2004.

32. Y. Umezawa, P. Bühlmann, K. Umezawa, K. Tohda, and S. Amemiya, Potentiometric selectivity coefficients of ion-selective electrodes, *Pure and Applied Chemistry*, 72, 1851–2082, 2000.

33. L. T. Duarte and C. Jutten, A nonlinear source separation approach to the Nicolsky-Eisenman model, in *Proceedings of the 16th European Signal Processing Conference (EUSIPCO 2008)*, Bucharest, Romania, August 27–31, 2008.

34. M. Gutiérrez, S. Alegret, R. Cáceres, J. Casadesús, and O. M. M. del Valle, Nutrient solution monitoring in greenhouse cultivation employing a potentiometric electronic tongue, *Journal of Agricultural and Food Chemistry*, 56, 1810–1817, 2008.

35. L. T. Duarte, C. Jutten, and S. Moussaoui, A Bayesian nonlinear source separation method for smart ion-selective electrode arrays, *IEEE Sensors Journal*, 9(12), 1763–1771, 2009.

36. L. T. Duarte, C. Jutten, P. Temple-Boyer, A. Benyahia, and J. Launay. A dataset for the design of smart ion-selective electrode arrays for quantitative analysis, *IEEE Sensors Journal*, 10(12), 1891–1892, 2010.

12 Zeolite Transformation Layers in Discriminating Metal Oxide Semiconductor Gas Sensors

Russell Binions

CONTENTS

12.1 INTRODUCTION

12.1.1 BRIEF INTRODUCTION TO MOS

According to band theory, in semiconductors, two distinct bands must exist: a conduction band and a valence band. The separation between these two bands is a function of energy, particularly the Fermi level, defined as the highest available electron energy levels at a temperature [1]. There are three main classes of material in band theory. Insulators have a large gap between the valence and conduction band (typically taken to be 10 eV or more), as such a lot of energy is required to pro-mote the electron in to the conduction band and so electronic conduction does not occur. The Fermi level is the highest occupied state at $T = 0$ [2]. Semiconductors have a sufficiently large energy gap (in the region of 0.5–5.0 eV) so that at energies below the Fermi level, conduction is not observed. Above the Fermi level, electrons can begin to occupy the conduction band, resulting in an increase in conductivity. Conductors are defined as having the Fermi level lying within the conduction band.

In metal oxide semiconductor (MOS) gas sensors, the target gas interacts with the surface of the metal oxide film, typically through reaction with surface-adsorbed oxygen ions. The electrons for such ionized oxygen species originate in the conduction band of the semiconductor and, on reaction with analyte gas, may be injected back into the material or further electrons withdrawn depending on the reaction that takes place resulting in a change in charge-carrier concentration of the material. This change in charge-carrier concentration serves to alter the conductivity (or resistivity) of the material. An *n*-type semiconductor is one where the majority charge carriers are electrons, and upon interaction with a reducing gas, an increase in conductivity occurs. Conversely, an oxidizing gas serves to deplete the sensing layer of charge-carrying electrons, resulting in a decrease in conductivity. A *p*-type semi-conductor is a material that conducts with positive holes being the majority charge carriers; hence, the opposite effects are observed with the material and showing an increase in conductivity in the presence of an oxidizing gas (where the gas has increased the number of positive holes). A resistance increase with a reducing gas is observed, where the negative charge introduced into the material reduces the positive (hole) charge-carrier concentration. A summary of the response is provided in Table 12.1 [3].

12.1.2 MODEL FOR GAS INTERACTION: *p*-TYPE SENSOR RESPONSE

A simple model for the response of a p-type sensor is demonstrated by Equations 12.1 and 12.2, showing the adsorption of an oxygen atom to the surface of the material, causing ionization of the atom and yielding a positive hole (p^+), demonstrated by Equation 12.1. The positive hole and the ion can then react with a reducing gas such as carbon monoxide, forming carbon dioxide (k_2), or be removed through interaction with each other (k_{-1}) [4] (Equation 12.2). The difference in charge-carrier concentration (in this case the positive hole) is manifest in a resistance change between the sensor's electrodes and read by the measurement circuitry.

$$\tfrac{1}{2}O_2 \underset{k_{-1}}{\overset{k_1}{\rightleftharpoons}} O^- + p^+ \tag{12.1}$$

$$p^+ + O^- + CO \xrightarrow{k_2} CO_2 \tag{12.2}$$

12.1.3 EQUIVALENT CIRCUIT MODEL

The equivalent circuit model, as described by Naisbitt et al. [5], is a refinement of the conventional response model (Equation 12.4) that is found to apply over a limited range of examples only, and that assumes the absorbed species on the surface of the metal oxide is the O^{2-} species, a species that is unlikely to be included as it is energetically unfavorable (Equation 12.3). Equation 12.4 describes

TABLE 12.1

Sign of Resistance Change (Increase or Decrease) to Change in Gas Atmosphere

Classification	Oxidizing Gases	Reducing Gases
n-type	Resistance increase	Resistance decrease
p-type	Resistance decrease	Resistance increase

a relationship where the change in resistance is proportional to the concentration of the gas (in this example, carbon monoxide [CO]) and a sensitivity parameter A (the sensitivity parameter is constant for a given material at a given temperature).

$$\tfrac{1}{2}O_2 \underset{k^{-3}}{\overset{k_3}{\rightleftharpoons}} O^{2-} + 2P^+ \tag{12.3}$$

$$\frac{R}{R_0} = 1 + A[CO] \tag{12.4}$$

Instead, Naisbitt et al. propose that Equation 12.1 is more likely, but to account for nonlinear responses, there must be other factors influencing the response. First, the assumption is made that the only parts of the material that exhibit a response to the target gas are areas where the gas can land and interact at the surface. Thus, the material is split into three regions (Figure 12.2): (1) the surface, (2) the bulk (inaccessible to the target gas), and (3) the neck or particle boundary (below this boundary, the material is no longer defined as a surface). The distance between the surface and the particle boundary is called the Debye length, the distance at which charge separation can occur. The model assumes that the gas sensitivities of the surface and particle boundary are the same.

The equation for the response in this model is found to be (12.5)

$$G_T = \gamma_{PB}\left(1 + A[CO]\right) + \frac{1}{\left[\left(1/\gamma_B\right) + \left(1/\gamma_S\left(1 + A[CO]\right)\right)\right]} \tag{12.5}$$

where G_T is the response ($G_T = R_T/R_{T,0}$), R_T is the total sensor resistance $R_{T,0}$ is the baseline resistance in clean, dry air and each $\gamma_x = R_{x,0}/R_{T,0}$ gives (x denotes particle boundary, PB; bulk, B; or surface, S. So γ is the ratio between the total baseline sensor resistance and the baseline resistance of X) [7].

This work shows that the response time is directly related to the grain size and the size of the particle boundary in the material. The response model is different for n-type and p-type semiconductors. In forming the baseline resistance, oxygen is adsorbed on to the surface and abstracts electrons from the material; hence, this process will determine R_0. The resistivity of the p-type decreases relative to the bulk and will increase for the n-type. The relative contributions from the three resistors in the model then differ. For very small grain sizes, the grain can be considered to contain no bulk area at all (so the whole grain is considered to contribute to the surface area); in this instance, the simpler model and Equation 12.4 are an adequate model for the response. If one considers the other extreme, where the grains are so large the contribution to the resistance or conductivity is negligible, the surface can be deemed to have a constant resistance. This model is expected to be generally applicable to both p- and n-type sensors.

12.1.4 SCREEN-PRINTING

Screen-printing is widely used in industry and the most widely used method for producing MOS gas sensors commercially [2,6]. Screen-printing involves pushing an ink through a porous layer or sheet, which has the geometry to match the substrate. The ink that is essentially the starting material is a viscous vehicle and is printed on the surface of the substrate (in this case, the electrode). Once the ink is on the surface, the print can be heated to remove the vehicle, leaving a solid material on the specific target area. Figure 12.1 indicates the general form of commercial MOS gas sensors.

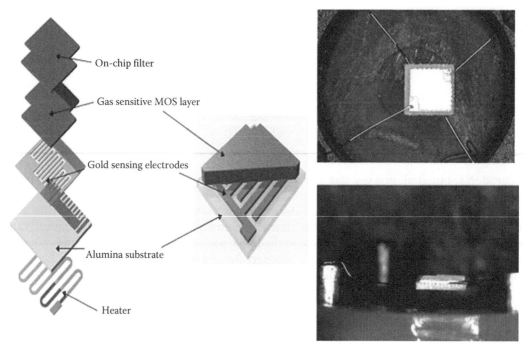

On-chip filter

Gas sensitive MOS layer

Gold sensing electrodes

Alumina substrate

Heater

FIGURE 12.1 Schematic of a gas-sensitive film on a sensor substrate (Diagram courtesy of Capteur Sensors and Analysers, Portsmouth, U.K.) and photographs of assembled device.

12.1.5 WHAT ARE ZEOLITES?

Zeolites are porous aluminosilicate cage-like structures that are able to accommodate cations such as Na^+, Mg^+, and Mg^{2+} and small molecules such as ethanol. These species are typically contained within a large 3D framework, with external pore sizes usually ranging from 4 to 12 Å (Figure 12.2). The oxygen atoms in the structure may be negatively charged due to substitution of silicon with aluminum; this charge can be used to counterbalance other species, most commonly cations such as Na^+ or H^+. Due to the pore structure and ability to host metal ions, zeolites can behave as selective catalysts and have been demonstrated to be capable of discriminating between small gaseous molecules on the basis of size and shape, allowing some gases into their structure, while preventing others from entering. Further to this, the zeolites can also behave in a chromatographic manner as each zeolite framework has specific diffusion characteristics.

Therefore, a mixture of gases may enter the zeolite pore, but because the zeolite framework has different binding strengths to the gaseous molecules, this leads to a difference in diffusion speed of the molecules through the zeolites. The zeolite may also be able to perform a catalytic transformation involving the target gas, which may lead to the production of one or more molecules that the sensor

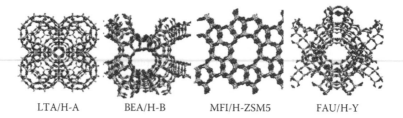

LTA/H-A   BEA/H-B   MFI/H-ZSM5   FAU/H-Y

FIGURE 12.2 Structural representations of various zeolite frameworks where black atoms represent silicon or aluminum and light gray atoms represent oxygen.

layer may be more or less sensitive too. In the ideal case, any catalytic reaction is specific to a particular target analyte and leads to the production of multiple species that the sensor element is more sensitive to, leading to a large enhancement in response signal for a given analyte with no chance of cross sensitivity.

12.1.6 OTHER GROUPS WORKING IN THIS AREA

A number of groups have published work in this area. This chapter will focus exclusively on work conducted in the authors research group, although interested readers are encouraged to investigate other groups work [4,5,7–18].

12.2 EXPERIMENTAL PREPARATION

12.2.1 MATERIALS PROCESSING

Chromium titanium oxide (CTO) was prepared as previously described [19] Cr_2O_3 (19.525 g, 0.13 mol, prepared by thermal decomposition of ammonium dichromate at 300°C) and was mixed with propan-2-ol (500 mL) for 10 min using a homogenizer (1000 rpm). Titanium isopropoxide (1.954 g, 6.6 mmol) and water (5 mL in 50 mL propan-2-ol) were added to the suspension over 15 min, and the mixture rotary evaporated in an ultrasonic bath. Tungsten oxide was sourced from New Metals Chemicals Limited and used as obtained.

The zeolites H-ZSM-5 and H-LTA were prepared by ion exchange of the Na zeolite form (obtained from Zeolyst) with a 1 M NH_4OH solution for 12 h at 60°C. The powder was then calcined in air at 100°C for 8 h and at 500°C for 12 h [19]. EDAX analysis was used to determine the zeolite Si/Al ratios and the extent of the ion exchange.

In preparation for screen-printing, inks were made by grinding the oxide and zeolite powders with an organic vehicle (Agmet ESL400) in a pestle and mortar followed by triple roll milling. The inks were then screen-printed onto strips of 3 × 3 mm alumina substrates with an integral platinum heater and gold electrode pattern [20]. Between each layer of print, the sensors were dried under an infrared lamp for 10 min.

The sensors consisted of metal oxide layers that were ~50 μm thick in total overlaid with layers of zeolite, again ~50 μm thick. The thickness was measured using a depth gauge at five points along the strip of substrates to give an average. The sensors were subsequently fired at 600°C in a Carbolite HTC1400 furnace for 4 h to burn off the organic vehicle.

Contacts to the devices were formed by spot welding 50 μm diameter platinum wire to pads of the track material in the corner of the sensor chip. The sensor heater was kept at constant temperature by incorporating it into constant resistance Wheatstone bridge. Gas sensing experiments were performed on a locally constructed test rig [21]. Test gas streams were prepared by dilution of synthetic air (79% nitrogen, 21% oxygen) containing ethanol (100 ppm), isopropyl alcohol (IPA) (100 ppm), NO_2 (10 ppm), or carbon monoxide (5000 ppm). The devices were operated at 350°C. Each device was duplicated and each experiment conducted several times to ensure repeatability.

12.2.2 MATERIALS CHARACTERIZATION

X-ray diffraction (XRD) was carried out on a Bruker D8 Discover diffractometer with a $CuK\alpha1/K\alpha2$ source λ = 1.5406 Å) and a GADDS detector. Energy dispersive x-ray spectroscopy (EDAX) was used to obtain compositions using an Oxford Instruments INCA energy system in conjunction with a Phillips XL30 environmental scanning electron microscope (SEM). X-ray photoelectron spectroscopy (XPS) spectra were taken using a VG Escalab 220 spectrometer. Top-down and cross-sectional SEM pictures were obtained using a JEOL 6301F SEM with an accelerating voltage of 15 kV. Gold sputter-coating of the SEM samples was carried out on an Edwards S105B sputter-coater. Triple

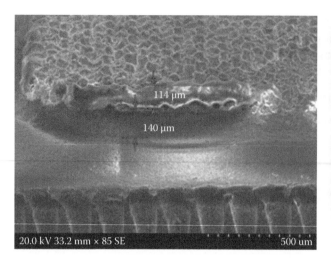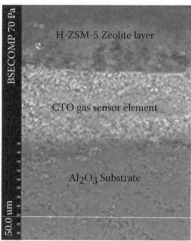

FIGURE 12.3   Representative SEM images of zeolite-overlayered MOS gas sensors.

roll milling of the screen-printing inks was carried out on a Pascall engineering triple roll milling machine and screen-printed using a DEK 1202. Welding of the sensors was done on a MacGregor DC601 parallel gap resistance welder.

The powders were analyzed using XRD. The CTO was isostructural with eskolaite (Cr_2O_3), with no other phases present, indicating that the titanium had been successfully substituted in place of chromium in the CTO structure. EDAX and XPS confirmed the powder composition of $Cr_{1.95}Ti_{0.05}O_3$. The XRD of the tungsten oxide indicated the expected WO_3 phase; no other phase was observed.

The zeolite powders were also examined by XRD before and after being incorporated into devices. In each case, the patterns were as expected, and no degradation of the zeolite crystallinity was seen after sensor construction and processing. The zeolite powders were also examined by EDAX to confirm that ion exchange had taken place. The results indicated that in the case of H-ZSM-5, exchange was 92% complete, and with H-LTA, H-Y, and Cr-ZSM-5, exchange was 96% complete. In all cases, small amounts of iron and magnesium impurities remained.

The sensor devices were studied by SEM (Figure 12.3). The unmodified metal oxide sensors had morphologies typical of screen-printed sensors [22], an arrangement of loosely packed, fused crystallites. The microstructure is quite open, which will allow easy diffusion of gas around the grains. The WO_3 devices were however different from the CTO devices in having more highly fused grains with a substructure of rather fine pores within the grains. Similarly, the zeolite layers have open microstructures. The H-LTA layers are characterized by cube-like zeolite particles up to 1 μm in diameter. The H-ZSM-5 and H-Y layers consist of more irregular particles up to 500 nm in diameter. In both cases, the structures are not densely packed, and this will allow diffusion of gases between zeolite particles. Again, there was some difference between the H-ZSM5 layers deposited onto WO_3 and onto CTO, with the layers on WO_3 appearing to have undergone some grain growth and sintering relative to the layers on CTO. Figure 12.3 shows representative SEM images of the layered sensors.

12.2.3   SENSOR CHARACTERIZATION AND TESTING

Gas sensing experiments on the screen-printed sensors were performed on an in-house test rig (Figure 12.4) [5] designed to maintain up to eight sensors at constant operating temperature via a heater driver circuit connected to each sensor's heater track. The heater circuit was used to set the operating temperature to 350°C, and conductivity measurements were taken via potentiostat circuits. The sensors were tested to mixtures of ethanol, IPA, NO_2, and CO, all in synthetic air and supplied by BOC gases.

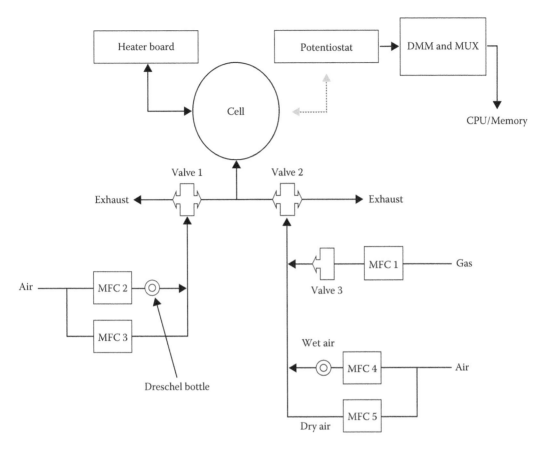

FIGURE 12.4   Schematic of gas analysis rig.

12.2.4 THEORETICAL CONCERNS: DIFFUSION-REACTION MODELING

In an effort to obtain a complete understanding of the processes occurring in the zeolite-modified sensors, diffusion-reaction modeling was performed. The gas sensor response G is generally taken as being related to the gas concentration of gas C_g following [3]

$$G = A_g C_g^\beta \qquad (12.6)$$

G can be either the relative conductivity $G = (\sigma - \sigma_0)/\sigma$ (for materials whose conductivity increases in response to the gas) or the relative resistivity $G = (\rho - \rho_0)/\rho$ (for materials whose resistivity increases in response to the gas) where σ_0 and ρ_0 are the conductivity and resistivity in the absence of the target gas respectively. While the exponent, b, is usually irrational, the assumed chemistry of the response for the materials studied here implies $b = 1$. It has been shown that the observed values reflect the convolution of this basic response with the effects of the microstructure and geometry of the device [23]. We focus here on the effects of gas concentration gradients developed within the porous sensor body: hence, such a simple linear model is a reasonable assumption and serves to show clearly the expected effects. Within the sensing layer, the concentration of a given species will be dependent on the competition between the rate of diffusion through the layer and the rate of reaction at the sensor surface within the pores of the device. By treating the sensor composite as two different macroscopically homogenous layers, characterized by spatially invariant effective diffusion and reaction coefficients, the concentration $C(x,t)$—a function of both time and position within the composite—can be found by solving the diffusion-reaction problem (Equations 12.7 and 12.8,

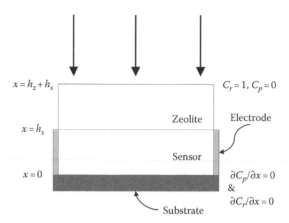

FIGURE 12.5 Schematic summarizing diffusion-reaction model. Subscripts z and s denote the zeolite and sensor layers, respectively.

performed here using the method of finite differences). For illustration, we assume a first-order conversion of the target gas (denoted r) to a product gas (denoted p) where r and p have a different effect upon the conductivity of the sensor materials. The geometry is illustrated in Figure 12.5. The electrode geometry shown is an idealized one that simplifies the calculations and is a good approximation to the case of widely spaced electrodes printed onto the substrate underneath the sensing layer.

$$\frac{\partial C_r(x,t)}{\partial t} = \frac{D_r \cdot \partial^2 C_r(x,t)}{\partial x^2} - k \cdot C_r(x,t) \tag{12.7}$$

$$\frac{\partial C_p(x,t)}{\partial t} = \frac{D_p \cdot \partial^2 C_p(x,t)}{\partial x^2} + k \cdot C_r(x,t) \tag{12.8}$$

where
   D is the diffusivity
   k is the rate constant

These averaged values will themselves depend upon the details of the microstructure, although such effects are not explicitly considered here. The values may also be different in each of the two layers. Values of sensitivity are defined for the reactant and product species such that the reaction may result in an increase or reduction in the net gas sensitivity. The ratio, R, of the sensor material response for the product gas to that for the reactant gas is hence defined.

We also performed atomistic simulations of the structures of IPA and ethanol within the Materials Studio platform with energy minimization being performed using the Smart Minimizer algorithm with the COMPASS force field available as part of the discover module [24]. Bond lengths and bond angles were taken from the optimized structures along with van der Waals corrections [25] and used to calculate critical dimensions of length, width, and height for the two molecules.

12.3   RESULTS

12.3.1   INCREASING DISCRIMINATION

The effect on sensor gas response of two different acidic zeolites (H-LTA and H-ZSM5) layered onto CTO was studied. The zeolites were also chosen for their thermal stability at the temperatures required for both sensor fabrication and operation and to provide a contrast between small- and

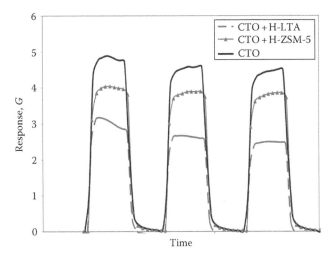

FIGURE 12.6   The gas response of CTO sensors to 30 min exposures of 2000 ppm carbon monoxide in dry air.

larger-pore zeolites. In these results, we present responses to carbon monoxide and ethanol. The test gases were introduced into a stream of synthetic air for a period of 30 min. The sensors were then allowed to recover in a stream of synthetic air for a further 30 min. This process was cycled three times to test reproducibility.

The response of modified and unmodified CTO sensors to carbon monoxide (Figure 12.6) was fast and rose rapidly to a steady state following the gas introduction. The sensors that had zeolite transformation layers showed a diminished response relative to the unmodified sensor, clearly showing discrimination across the sensor array. In all cases, the sensor signal returned to the baseline level during the purge step.

12.3.2   Analyte-Tuned Gas Sensors: Ethanol

The response of modified and unmodified CTO sensors at an operating temperature of 350°C to a series of 30 min long pulses of 28 ppm ethanol gas in a stream of dry air is shown in Figure 12.7. In contrast to the behavior in response to CO, the response was slow. In this case, the effect of the zeolite transformation layer was different: sensors with zeolite transformation layers showed an enhanced response relative to the unmodified sensor. In the case of the sensor modified with H-LTA, the effect was a remarkable 40 times signal enhancement. During the purge step, the sensor response returned to baseline.

To further elucidate these discrimination effects, three different acidic zeolites (H-A, H-ZSM-5, and H-Y) were layered onto CTO, and their sensor performance was examined. An unmodified sensor was also examined for comparative purposes. Specifically, we examine the gas response behavior to 80 ppm ethanol and 80 ppm IPA (Figure 12.8). The two target gases are typically difficult to resolve using traditional metal oxide sensors as they contain the same functional group, and the unmodified sensors are not size or shape selective. The test gases were introduced into a stream of synthetic air for 30 min. The sensors were allowed to recover for 1 h in a stream of synthetic air after exposure to the target gases. The process was cycled and the whole experiment repeated to test reproducibility.

The response of zeolite-modified and zeolite-unmodified CTO sensors to 80 ppm ethanol gas is shown in Figure 12.8a. The sensors gave a *sharkfin* type transient; the response was slow and did not reach a steady state during the time of the gas pulse.

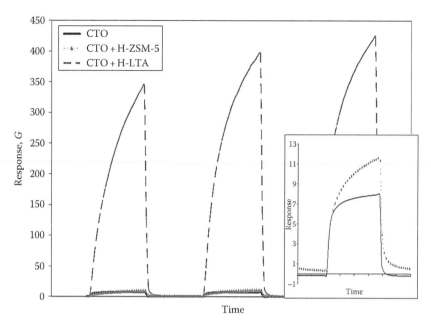

FIGURE 12.7 The gas response of CTO sensors at an operating temperature of 350°C to 30 min exposures of 28 ppm ethanol in dry air. The inset shows magnified data for CTO and CTO + H-ZSM-5 sensors.

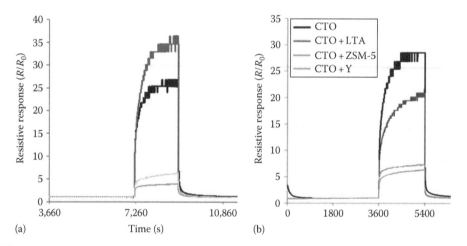

FIGURE 12.8 Response of CTO sensors to 80 ppm ethanol in dry air at 400°C. Response of CTO sensors to 80 ppm IPA in dry air at 400°C.

The zeolite A-modified sensor showed an increased response relative to the unmodified sensor. The other zeolite-modified sensors showed a lower overall response and a corresponding increase in the response speed to the final state. During the gas-purging step, the sensor response returned to its baseline level. The performance of the same sensor set to 80 ppm IPA is shown in Figure 12.8b. The response for the unmodified sensor was similar in magnitude and shape to that found for ethanol. However, the response to IPA was greater than that to ethanol for both the H-ZSM5- and H-Y-modified sensors but smaller for the H-A-modified sensor. Again, an increase in the response rate concomitant with the decrease in response magnitude is found. As for ethanol, the sensor response returns to its baseline level during the gas-purging step.

12.3.3   ANALYTE-TUNED GAS SENSORS: NITROGEN DIOXIDE

Tungsten trioxide sensors overlayered with H-ZSM-5, Cr-ZSM-5, H-A, and Cr-A zeolites were tested to various concentrations of NO_2 ranging from 50 to 400 ppb in dry air. The maximum sensitivity to 400 ppb NO_2 in dry air was found to be at 350°C. The results are shown in Figure 12.9.

The sensors showed a positive response, which is typical of an *n*-type semiconductor to an oxidizing gas. This means that as NO_2 interacts with the sensor surface, conduction band electrons are immobilized creating additional surface acceptor states, and hence the resistance of the material increases [26,27]. It has been suggested that rather than reacting with surface oxygen ions, NO_2 is more likely to directly chemisorb onto the surface and abstract electrons [26]:

$$NO_2 + e^- \rightarrow NO_2^-$$

The various zeolite layers produced very different responses. The H-ZSM-5-modified sensor gave a significantly higher response in comparison to the other sensors, reaching a maximum response of *ca.* 107. This response was nearly 19 times higher than the control sensor. This enhanced response suggests a catalytic effect due to the zeolite, which may have modified the incoming NO_2 gas to form a product to which the underlying WO_3 was more sensitive. Such catalytic effects have been seen and rationalized using diffusion-reaction theory in previous work [4,8]. The response of the Cr-LTA-modified sensor was lower than that of the unmodified sensor. This indicates that the Cr-LTA layer may have modified the NO_2 to form a less-sensitive product. The H-LTA sensor however gave a very similar response to the unmodified sensor. This suggests that the H-LTA layer had no effect on the incoming gas. This could be related to the open microstructure of the

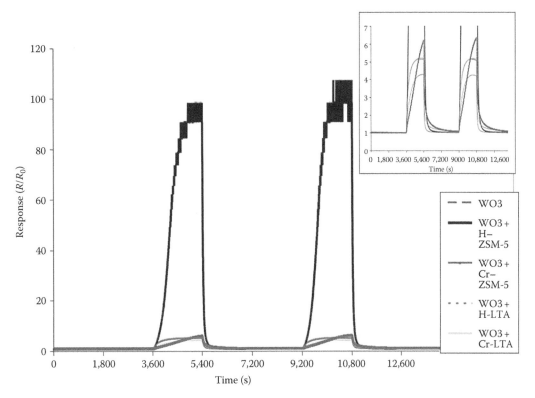

FIGURE 12.9   Resistive response of WO_3 sensors to 400 ppb NO_2 in dry air at an operating temperature of 350°C. Inset shows magnified data of sensor response excluding the WO_3 + H-ZSM-5 sensor.

H-LTA-modified sensor as seen in Figure 12.2. In this case, the test gas could reach the WO_3 sensor by diffusion through the open macroporous zeolite layer without entering the zeolite micropores, and thus not taking part in any catalytic reaction, hence showing response characteristics similar to the control sensor.

It was also found that all sensors apart from the Cr-ZSM-5 sensor reached steady state. The Cr-ZSM-5 sensor showed a slow response with a sharkfin type profile suggesting that the zeolite layer hindered the diffusion of the gas to the sensor element surface. This may have been the case since the SEM images (Figure 12.2) showed densely packed particles. It is also possible that the NO_2 is catalytically converted by the zeolite into a product that could not diffuse easily through the zeolite layer hence giving a slow response. We have previously shown through experiment and modeling that it is possible for a zeolite to catalytically modify the constituents of the gas stream leading to a change in sensor response shape, speed, and magnitude [4,8], as such this behavior is not entirely unexpected.

12.3.4　DIFFUSION-REACTION MODELING RESULTS

To further understand the experimental results, diffusion-reaction modeling was performed. Recall that for the following discussion, the subscripts z and s denote the zeolite and sensing layers respectively. Figure 12.10a and b gives the model gas response as a function of dimensionless time t/T_s where T_s is the characteristic time scale of the sensor sensing layer defined as $T_s = h_s^2/D_s$. Similarly, a characteristic time scale for the zeolite is also defined. Dimensionless rate constants can also be defined for either layer as $K_s = k_s T_s$ and $K_z = k_z T_s$.

From Figure 12.10, one notes that the primary effect of the zeolite is to slow the response dependent on the characteristic time scale of the zeolite layer. Where no reaction occurs within the zeolite layer, the steady-state response will not be altered from the control sensor case although the response reached after the test time may be diminished because of the slowed response time.

If a reaction occurs within the zeolite layer, the response may be enhanced or diminished depending on the sensitivities of the sensor to the reactant and product gases and also on the rate constant K_z (as defined earlier).

Figure 12.10a shows a system where transport through the zeolite layer is slower than through the sensor, and Figure 12.10b the case where the opposite is true. It is observed that in the latter case, the effect of reactions within the zeolite is effectively reduced.

Generally, response time decreases with rate constant and increases with product sensitivity.

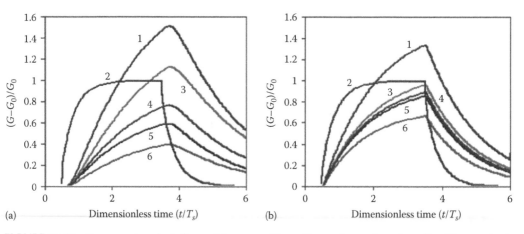

FIGURE 12.10 Response transients for model sensor with zeolite overlayer where the ratio of time scales is (a) $T_z/T_s = 2$ (b) $T_z/T_s = 0.2$. For each, a series of results is present where the rate constant (K_z) and selectivity ratio (R) are varied as given by 1. $K_z = 10$, $R = 2$; 2. No zeolite; 3. $K_z = 1$, $R = 2$; 4. $K_z = 0$; 5. $K_z = 1$, $R = 0.5$; 6. $K_z = 10$, $R = 0.5$. For all these cases, no reaction is assumed within the sensor layer ($K_s = 0$).

12.4 DISCUSSION

12.4.1 EXPERIMENTAL RESULTS

Variables such as gas diffusion to and through the sensor material, the kinetics of surface reaction, and the type of reaction that takes place on the sensor surface can all affect the sensor response. The inclusion of an additional zeolite layer may also have a profound effect on these variables and thus sensor response.

The kinetics of sensor surface reaction is strongly influenced by the surface microstructure. It has been established that a more open and porous microstructure results in a larger surface area and a higher concentration of reactive surface sites that provide a larger contribution to the overall conductivity of the system [5].

Reactions may occur within the entire sensor element that can lead to an increase or decrease in the response of the system. The response will be enhanced if the reaction leads to products to which the sensor material has a higher sensitivity. In contrast, the response of the system will decrease if the reactions form products to which the sensor material is less sensitive.

The addition of a zeolite layer will primarily act to increase the response time, dependent on the diffusion of gas through the layer, which will vary for different gas–zeolite combinations because of the molecular-shaped pores. A zeolite layer can also cause additional reactions to occur within the zeolite framework. Zeolites are well known for their catalytic properties, and so reactions occurring in the zeolite transformation layer can either enhance or diminish the sensor response, in the manner proposed earlier. There is also the possibility of cracking reactions taking place, that is, those where a single molecule is broken into more than one product molecule. As above the occurrence of reactions may lead to increase or decrease in response magnitude due to changes in the sensitivities of the sensing element to constituents of the gas.

The sensor responses of the unmodified CTO sensor toward 80 ppm ethanol and 80 ppm of IPA shown here (Figure 12.8a and b) do not indicate significant discrimination. In both cases, the sensor response is ~27. This is not unexpected as the two gases have the same functional group and are expected to undergo similar reactions on the sensor surface. Indeed the kinetics of each response and magnitude of response are similar: the shape and height of the gas response transient is almost identical in each case.

However, the use of the zeolite overlayer has allowed us to discriminate between the two gases. In the case of the modified CTO sensors (Figure 12.8a and b), the behavior is consistent with the effect of a catalyzed reaction of the target gas within the zeolite layer where the sensing material is less sensitive to the reaction product than it is to ethanol in the case of zeolites Y and ZSM-5 and more sensitive in the case of H-A.

The response rate to IPA on the zeolite-modified devices was observed to be less than that of ethanol (Figure 12.4), and the enhancement of response occurring for ethanol on H-A-modified devices is not observed, suggesting that the IPA is unable to diffuse into the zeolite pores and undergo catalytic reaction. It is likely that this is an effect of size and shape selectivity (IPA is branched, and ethanol is not). As a result, the effect of the H-A overlayer for ethanol is only to slow diffusion to the sensor element surface. Note that the zeolite layer presents a system of two pore sizes: the gas must be able to fit into microscopic zeolite pores to react otherwise it will just pass through the zeolite crystallite macropores.

The zeolite modification of the sensors leads to differential behavior as a result of several competing effects. The most important factors are the diffusivities and reaction rate constants of both the zeolite and the sensor layers. Even in the absence of any gas exchange with the internal molecular channels of the zeolites, all of the zeolites would act as a porous, physical barrier and effectively retard the progress of test gas to the sensor element surface. For IPA, the H-A appears to act in only this manner: the response for the sensor modified with this zeolite is consistently lower than the unmodified material, but there is no significant change to the shape of the transient.

An interpretation is that the IPA gas undergoes a catalyzed total combustion reaction on the surface of the sensor material but that no catalytic reactions are taking place in the H-A layer. For ethanol, this is not the case; the response is enhanced and suggests that a reaction is taking place within the zeolite layer that forms a product to which the sensor material is more sensitive. This hypothesis is consistent with the smaller-pore entries of H-A, and so a simple interpretation of the lack of any specific effect for IPA is that the gas does not permeate into the zeolite to any great extent here whereas the smaller ethanol molecule is able to. The H-Y- and H-ZSM5-modified sensors, as well as having a retardation effect, exhibit a catalytic reaction in the zeolite layer to an insensitive product, indicated by the flattening of the CTO response transient compared to the unmodified and H-A-modified sensor responses.

It is apparent that different reactions occur within the different zeolites—and that shape- and size-selective effects are likely [28,29]. However, the conditions that we use here are not those typically used in catalytic studies of zeolites, nor (due to their unlikely applicability) are the range of zeolites used extensively studied. Although we can state that a variety of different reactions are likely under these conditions and would certainly expect to see a difference between the different gases and zeolites [8], we cannot state with certainty the relative proportions of the likely products. Similarly, we cannot quantify the individual (or composite) response of the sensor to all of the likely products of such reactions. Likely reactions, however, are total oxidation to CO_2, partial oxidation to aldehyde or ketone, and dehydration to propene or ethene.

The rate of any reaction in the zeolite will also influence the response transient. One would therefore expect the response time to decrease with increasing rate constant and increase with increasing product sensitivity. It is apparent that the rate of reaction is not the same for all the zeolites considered: related to their intrinsic properties such as pore size and the density and acidity of the active sites. The CTO H-Y- and H-ZSM5-modified sensors give a fast response to both ethanol and IPA, suggesting either that the reaction in the zeolite layer in both cases is fast or that the sensor is much less sensitive to the products of any reactions (or a combination of both effects). The converse is true for the H-A-modified sensors. Indeed, it is noted that for the majority of cases, the transient speed is faster on the inclusion of the zeolite layer—indicating that the products diffuse more quickly through both layers than the reactant through the sensor layer. This is not the case for IPA in H-A, which correlates to the expected exclusion of IPA from the micropores of H-A due to size selectivity (see the following).

12.4.2 DIFFUSION-REACTION MODELING RESULTS

12.4.2.1 Interpretation of Data

Bearing in mind the earlier discussions and models, there are three types of possible behavior, which can be correlated to the experimental evidence: firstly, a slow diminished response, where the response transient is slow and the level of response is diminished by the zeolite transformation layer (Figure 12.9, though not the H-ZSM-5-modified sensor); secondly, a fast, diminished response, where the response transient is fast but the response is diminished by the zeolite transformation layer (Figure 12.6); And finally, a fast, enhanced response, where the unmodified sensor transient is fast and the addition of a zeolite transformation layer leads to an enhanced response (Figure 12.7 or 12.9 for the H-ZSM-5-modified sensor).

12.4.2.2 Case 1: Slow Diminished Response

In Figure 12.9, the response of WO_3 sensors to 400 ppb NO_2 gas is evaluated. We see that the response has a *sharkfin* type profile, indicating that the sensor element surface did not reach a steady state for the applied gas concentration and the response was therefore slow. This is most likely due to diffusion effects, considering that the microstructure of the WO_3 sensor is one of a highly porous network of rather well-sintered WO_3 particles. Each particle has a number of surface features such as steps and holes on it that the analyte gas must navigate to find a suitable surface site. Since the shape of the unmodified sensor curve is similar to that of the modified sensor, we suggest that the zeolite transformation layer did not significantly reduce diffusion to the sensor element surface, or at least that the rate of diffusion within the zeolite transformation layer is higher than the rate of diffusion in the sensor element. The main effect of the zeolite transformation layer in this instance is to diminish the level of response. This is most likely to be because some of the test gas is diffusing into the zeolite particles and being transformed to a product that the sensor element is less sensitive to.

12.4.2.3 Case 2: Fast Diminished Response

The response of the unmodified CTO sensor to carbon monoxide (Figure 12.6) has a fast response transient, and the signal saturated promptly. The CTO device had an open microstructure, suggesting that diffusion within the sensing layer was sufficiently fast for the observed rapid response. At the sensor operating temperature, catalyzed combustion of CO on the surface of the CTO proceeds at a significant rate. The rapid can be explained by a reduction in the effective thickness of the sensing layer, h_r. The addition of a zeolite transformation layer may have two significant effects. Firstly, the speed of the response was lowered, seen most pronounced in Figure 12.8a. The speed would drop because the response is dependent on the concentration of gas in the sensor element: by putting a transformation layer over the top, which effectively limits diffusion of the test gas to the sensor element, we slow the rate of response. The second important effect of using a zeolite transformation layer was that the overall level of response was decreased. This is most likely to be as a result of catalytic reactions involving the test gas leading to a less-sensitive or insensitive product that diffuses to the sensor element. In both cases, the sensors with H-A transformation layers gave the greatest decrease in response. This may be simply due to the difference in composition with H-A having Si/Al = 1 while H-ZSM-5 is a high-silica system (Si/Al~30). Therefore, it is more likely that a gas molecule will come into contact with a reactive site and undergo transformation to a product, to which, under these specific conditions, the semiconducting oxide is less sensitive to than it is to the original test gas.

12.4.2.4 Case 3: Fast Enhanced Response

The responses of the CTO sensors to 28 ppm ethanol are presented in Figure 12.7. The figure inset shows the response curves for CTO and CTO + H-ZSM-5. The curve for the unmodified CTO sensor shows a fast response that approaches complete saturation very quickly. The response curves for the zeolite-modified sensors are different, being much more *sharkfin* in character indicating a slower response, which is due to diffusion of gas molecules into the zeolite particles, effectively slowing their progress to the sensor element surface. Both zeolite transformation layers lead to enhanced sensor responses. However, the response of the sensor modified with H-LTA is increased by 40 times compared with the unmodified sensor. Clearly, there is a catalytic reaction occurring where the ethanol test gas is converted into a product to which the sensor material is considerably more sensitive to. It is possible that either cracking, dehydration, or partial oxidation could be taking place, and that species such as ethene, acetaldehyde, acetic acid, formaldehyde, and formic acid are formed. Further studies are underway to characterize the reaction products.

Previously, we have discussed reaction-diffusion effects with respect to the response of gas-sensitive resistors [30] and also effects of microstructure on the steady-state response [5,31]. We can think of the composite device as having an effective time scale, h^2/D' (where h denotes the device thickness and D' the effective diffusivity within the device). This time scale can be dominated by the sensor material or by the overlayer, whichever has the lower diffusivity. As noted earlier, within the overlayer, diffusivity may be diminished by gas absorption into the zeolite. If the gas reacts within the overlayer or sensor material, then the resultant local concentration is determined by the ratio kh^2/D' for a first-order reaction with rate constant k. The response is diminished or enhanced depending on the ratio of sensitivities of product and reactant, and

TABLE 12.2

Critical Dimensions of Ethanol and IPA

Calculated from Optimized Structures

	Length (A)	Height (A)	Width (A)
Ethanol	4.99	3.57	3.09
IPA	5.72	4.22	3.35

the time scale of response is modified to (approximately) h_r^2/D^I where h_r is the reaction layer depth, dependent on k and less than the layer thickness. It is noted that a shortcoming of the model here, a limit of the simple reaction scheme used, is that it always predicts the retardation of response.

In Figure 12.8, the response of the unmodified CTO sensors to both ethanol and IPA had a *shark-fin* type profile, indicating that the surface of the sensor element did not reach a steady state for the applied gas concentration; as such, the response is slow. This is most likely attributable to diffusion effects: considering Figure 12.2, the microstructure of the CTO sensor was of a highly porous network of rather well-sintered CTO particles. Each particle has a number of surface features such as steps, which the test gas must navigate to find a suitable surface site. Hence, we compare CTO to Figure 12.10a in terms of the ratios of time scales between the two layers. The model predicts that the effect of reactions in the zeolite will be lessened where diffusion in the zeolite layer is fast compared to the sensor. It is also predicted that the extent of enhancement or diminution depends on the sensitivity. In fact, it is observed that the zeolites have more of an effect on CTO where zeolite diffusion is comparatively faster—this indicates that CTO has greatly more variable sensitivity to the different gases and reactant products tested here.

Figure 12.10 supports the notion that within the zeolite Y layers, the reaction is fast and also that the product is a less-sensitive species, as the response is greatly diminished and the transient speed is fast. From the speed of the transient, it is also suggested that the products are fast diffusing.

The response of CTO to ethanol is greatly enhanced by the addition of an H-A zeolite overlayer, as shown in Figure 12.7, indicating the conversion of ethanol to a species to which the sensor is more sensitive. Comparing these results with Figure 12.10, it is again presumed that the reaction products diffuse quickly through both layers.

IPA does not follow the same behavioral trend as ethanol with H-A as it does for the other two zeolites. Simple measurement of the dimensions of the alcohols (Table 12.2) reveals that IPA will be excluded from the pores of H-A while ethanol would not. Indeed, the dehydration of linear and branched alcohols on zeolite A is the archetypal *shape-selective* reaction in zeolite chemistry [32].

Thus, the signal from IPA is diminished relative to ethanol (as seen in Figure 12.10), and so the effect of the zeolite layer on the sensor response is likely to be diminished, and indeed is observed in Figure 12.8a.

12.5   CONCLUSIONS

Solid-state metal oxide gas sensors have been successfully modified using zeolite overlayers. The sensor microstructure was examined, and their responses to carbon monoxide, NO_2, ethanol, and IPA were tested in dry air. Gas testing revealed that the modified sensors gave varied and different responses. Diffusion-reaction modeling was also performed to gain greater insight into the processes occurring. The modeling results show that these results are from competing reaction and diffusion factors. The rates of catalysis and diffusion are controlled by the intrinsic properties of the zeolites, although the pore dimensions, acidity, and pore size of the zeolites do not control the sensor response in isolation. It has been shown that by using particular combinations of gas-sensitive material and zeolite, it is possible to distinguish between similar target gases. It is clear from the experimental results presented here that using zeolite overlayers leads to selective gas sensor devices that could form a useful array in an electronic nose. However, it is essential to further understand the myriad of processes occurring in these systems in order to be able to successfully design a sensor array for a particular purpose.

ACKNOWLEDGMENTS

I would like to thank the Royal Society for a Dorothy Hodgkin fellowship and the EPSRC for financial support (grant number EP/H005803/1). Mr. Kevin Reeves is thanked for assistance with scanning electron microscopy. I would also like to thank my colleagues Prof. Ivan Parkin and Dr. Dewi Lewis (University College London), Prof. David Williams (University of Auckland), and Prof. Xavier Correig Blanchar (Universitat Rovira i Virgili) for their many helpful discussions. Last but not least, I must thank the students who conducted much of this work. They are Dr. Sheena Dungey, Priya Varsani, Helen Davies, and Ayo Afonja.

REFERENCES

1. R. Martin, *Electronic Structure: Basic Theory and Practical Methods*. Cambridge University Press, Cambridge, U.K., 2004.
2. A. Shriver, *Inorganic Chemistry*, 4th edn. Oxford University Press, New York, 2006.
3. D. E. Williams, Semiconducting oxides as gas-sensitive resistors, *Sensors and Actuators B: Chemical*, 57(1–3), 1–16, 1999.
4. R. Binions, H. Davies, A. Afonja et al., Zeolite-modified discriminating gas sensors, *Journal of the Electrochemical Society*, 156(3), J46–J51, 2009.
5. S. C. Naisbitt, K. F. E. Pratt, D. E. Williams et al., A microstructural model of semiconducting gas sensor response: The effects of sintering temperature on the response of chromium titanate (CTO) to carbon monoxide, *Sensors and Actuators B: Chemical*, 114(2), 969–977, 2006.
6. V. S. Vaishanv, P. D. Patel, and N. G. Patel, Indium tin oxide thin-film sensor for detection of volatile organic compounds (VOCs), *Materials and Manufacturing Processes*, 21(3), 257–261, 2006.
7. A. M. Azad, S. A. Akbar, S. G. Mhaisalkar et al., Solid-state gas sensors. A review, *Journal of the Electrochemical Society*, 139(12), 3690–3704, 1992.
8. R. Binions, A. Afonja, S. Dungey et al., Discrimination effects in zeolite modified metal oxide semiconductor gas sensors, *IEEE Sensors Journal*, 11(5), 1145–1151, 2011.
9. A. Dubbe, G. Hagen, and R. Moos, Impedance spectroscopy of Na^+ conducting zeolite ZSM-5, *Solid State Ionics*, 177(26–32), 2321–2323, 2006.
10. A. Fischerauer, G. Fischerauer, G. Hagen et al., Integrated impedance based hydro-carbon gas sensors with Na-zeolite/Cr_2O_3 thin-film interfaces: From physical modeling to devices, *Physica Status Solidi (A) Applications and Materials*, 208(2), 404–415, 2011.
11. Y. Y. Fong, A. Z. Abdullah, A. L. Ahmad et al., Zeolite membrane based selective gas sensors for monitoring and control of gas emissions, *Sensor Letters*, 5(3–4), 485–499, 2007.
12. D. P. Mann, T. Paraskeva, K. F. E. Pratt et al., Metal oxide semiconductor gas sensors utilizing a Cr-zeolite catalytic layer for improved selectivity, *Measurement Science and Technology*, 16(5), 1193–1200, 2005.
13. D. P. Mann, K. F. E. Pratt, T. Paraskeva et al., Transition metal exchanged zeolite layers for selectivity enhancement of metal-oxide semiconductor gas sensors, *Sensors Journal, IEEE*, 7(4), 551–556, 2007.
14. S. R. Morrison, Selectivity in semiconductor gas sensors, *Sensors and Actuators*, 12(4), 425–440, 1987.
15. M. B. Sahana, C. Sudakar, G. Setzler et al., Bandgap engineering by tuning particle size and crystallinity of SnO_2: Fe_2O_3 nanocrystalline composite thin films, *Applied Physics Letters*, 93(23), 231909, 2008.
16. K. Sahner, G. Hagen, D. Schönauer et al., Zeolites—Versatile materials for gas sensors, *Solid State Ionics*, 179(40), 2416–2423, 2008.
17. A. Satsuma, D. Yang, and K. I. Shimizu, Effect of acidity and pore diameter of zeolites on detection of base molecules by zeolite thick film sensor, *Microporous and Mesoporous Materials*, 141(1–3), 20–25, 2011.
18. P. Varsani, A. Afonja, D. E. Williams et al., Zeolite-modified WO_3 gas sensors, Enhanced detection of NO_2, *Sensors and Actuators B: Chemical*, 160(1), 475–482, 2011.

19. D. Niemeyer, D. E. Williams, P. Smith et al., Experimental and computational study of the gas-sensor behaviour and surface chemistry of the solid-solution $Cr_{2-x}Ti_xO_3$ (x < 0.5), *Journal of Materials Chemistry*, 12(3), 667–675, 2002.

20. N. Magan, A. Pavlou, and I. Chrysanthakis, Milk-sense: A volatile sensing system recognises spoilage bacteria and yeasts in milk, *Sensors and Actuators B: Chemical*, 72(1), 28–34, 2001.

21. G. S. Henshaw, D. H. Dawson, and D. E. Williams, Selectivity and composition dependence of response of gas-sensitive resistors. Part 2.-Hydrogen sulfide response of Cr_2TiO_3, *Journal of Materials Chemistry*, 5(11), 1791–1800, 1995.

22. R. Binions, C. J. Carmalt, and I. P. Parkin, A comparison of the gas sensing properties of solid state metal oxide semiconductor gas sensors produced by atmospheric pressure chemical vapour deposition and screen printing, *Measurement Science and Technology*, 18(1), 190–200, 2007.

23. N. Barsan, M. Schweizer-Berberich, and W. Göpel, Fundamental and practical aspects in the design of nanoscaled SnO_2 gas sensors: A status report, *Fresenius' Journal of Analytical Chemistry*, 365(4), 287–304, 1999.

24. H. Sun, COMPASS: An ab initio force-field optimized for condensed-phase applications overview with details on alkane and benzene compounds, *The Journal of Physical Chemistry B*, 102(38), 7338–7364, 1998.

25. A. Bondi, van der Waals volumes and radii, *The Journal of Physical Chemistry*, 68(3), 441–451, 1964.

26. P. T. Moseley, Solid state gas sensors, *Measurement Science and Technology*, 8(3), 223, 1997.

27. M. Tiemann, Porous metal oxides as gas sensors, *Chemistry—A European Journal*, 13(30), 8376–8388, 2007.

28. R. Binions, A. Afonja, S. Dungey et al., Zeolite modification: Towards discriminating metal oxide gas sensors, *ECS Transactions*, 19(6), 241–250, 2009.

29. R. Binions, H. Davis, A. Afonja et al., Zeolite modified discriminating gas sensors, *ECS Transactions*, 16(11), 275–286, 2008.

30. D. E. Williams, G. S. Henshaw, K. F. E. Pratt et al., Reaction-diffusion effects and systematic design of gas-sensitive resistors based on semiconducting oxides, *Journal of the Chemical Society, Faraday Transactions*, 91(23), 4299–4307, 1995.

31. D. E. Williams and K. F. E. Pratt, Resolving combustible gas mixtures using gas sensitive resistors with arrays of electrodes, *Journal of the Chemical Society, Faraday Transactions*, 92(22), 4497–4504, 1996.

32. P. B. Weisz, Molecular shape selective catalysis, *Pure and Applied Chemistry*, 52(9), 2091–2103, 1980.

Part III

Automotive and Industrial Sensors

13 Micromachined Contactless Suspensions

Kirill V. Poletkin, Christopher Shearwood,
Alexandr I. Chernomorsky, and Ulrike Wallrabe

CONTENTS

13.1 INTRODUCTION

Micromachined contactless suspension (CS) or non-contact bearing technology is a comparatively recent concept. This has already attracted a great deal of attention giving rise to a new generation of micromachined or Micro-Electro-Mechanical Systems (MEMS) inertial sensors [1–6] as well as inertial multisensors [7–9]. Eliminating the mechanical attachment or contact friction in the suspension of a MEMS sensor is the major issue addressed by such technology. This issue leads to, on the one hand, a decreasing value of mechanical–thermal noise [10] and, consequently, an increase in the sensitivity of the sensors and their accuracies [4]. On the other hand, an extension of the lifetime and long-term stability of sensors becomes possible [11].

Although a number of unique solutions have been suggested to provide a CS, most methods are based on either electrostatic or electromagnetic induction suspension. The first prototype of a micromachined inductive CS (ICS) was described in [1] with some experimental results and a suggested application as a gyroscope. The choice over electrostatic suspension is based upon the fact that electromagnetic levitation is intrinsically stable. The device proposed by Shearwood's group is shown in Figure 13.1. A cross section of the device is given in Figure 13.1a. The application of a high-frequency alternating current to the stator coils induces a current in the rotor, which interacts with the excitation current to produce a repulsive force that levitates the rotor and constrains it laterally, essentially acting as a magnetic bearing. The design of a coil to produce levitation only is illustrated in Figure 13.1b, alongside optical micrographs of a micro-fabricated coil without the rotor in place Figure 13.1b. The coil is designed so that it can be micro-fabricated from a single layer of metallization. The rotor was also fabricated by the deposition of 10 μm of aluminum onto a sacrificial layer, followed by micro-fabrication.

The application of micromachined electrostatic CS (ECS) in the inertial multisensor has been studied by the Tokimec and Tohoku universities since 1993. They developed different prototypes of this sensor,

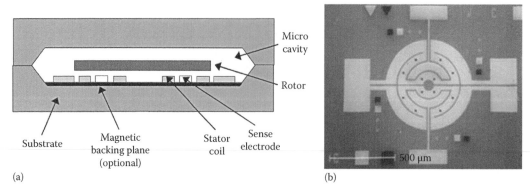

(a)   (b)

FIGURE 13.1 Electromagnetic levitated micro-motor reported by Shearwood: (a) Schematic cross section of a levitated micro-motor; (b) optical micrograph of a four-phase micro-motor stator.

one of which is shown in Figure 13.2a. Here the rotor is ring shaped and capable of detecting three-axis linear acceleration and two-axis angular rate [7]. An electrostatically levitated spherical three-axis accelerometer was reported in [3]. In this accelerometer, a 1 mm diameter spherical PM was completely suspended without any mechanical support by closed-loop controlled electrostatic forcers. Figure 13.2b shows a cross section of a 1 mm diameter, 1.2 mg PM supported electrostatically. The position of the ball is sensed capacitively and closed-loop electrostatic forces maintain its position. Fabrication of the spherical MEMS device was made possible by incorporating ball semiconductor technology and a sacrificial etching process utilizing xenon difluoride gas etching through a gas permeable layer.

As well as allowing passive suspension, these schemes also provide a mechanism for the rotation of a suspended rotor to high angular velocities, often only limited by viscous drag. For instance, it has been reported that for a disk-shaped rotor, of 0.5 mm diameter, the maximum possible speed of rotation achievable in air is 100,000 rpm [2]. Note that in vacuum, the maximum possible speed of a rotor rotation can be significantly greater than in air due to the reduction of viscous drag. Since the mechanism that provides the CS also makes possible rotation, an alternative strategy for improving the accuracy of such gyroscopes is the development of a dynamically tuned gyroscope (DTG), in which the angular stiffness supported by the CS is used. It is possible, given the experimentally achieved speed of rotor rotation in currently developed micromachined gyroscope (MMG) based on CS, and under meeting the rotor tuning condition, that an increased gyroscope gain would lead to several orders of magnitude improvement in the measured angular rate, as was reported in [12].

Moreover, under certain conditions, an electrical spring constant supported by the suspension, for instance, along an input axis of the accelerometer can be minimized or completely eliminated, at the same time keeping the proof mass (PM) at an equilibrium point. It is a new challenge that leads to significantly increasing accuracies of micromachined inertial sensors [13,14].

Indeed, let us consider a micromachined accelerometer consisting of a PM that can be suspended by a mechanical or magnetoelectric CS. The mass of the PM is denoted by m, the effective spring constant of the suspension is denoted by c, and the damping coefficient defined by the friction between the PM and the gas surrounding the PM is μ. Hence, the behavior of the PM can be described by a second-order transfer function, as follows:

$$\frac{y(s)}{a(s)} = \frac{1}{s^2 + (\mu/m)s + (c/m)}, \qquad (13.1)$$

where
   s is the Laplace operator
   a is the external acceleration
   y is the displacement of the PM

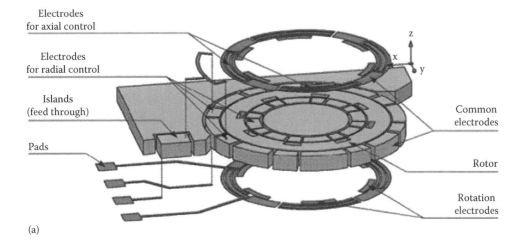

(a)

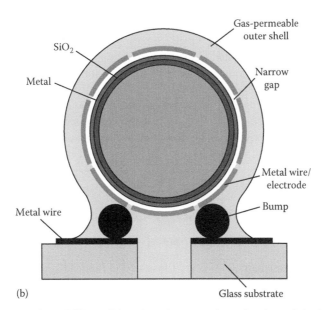

(b)

FIGURE 13.2 Micromachined ECS: (a) Ring-shaped rotor: schematic view of the inertial multisensor; (b) cross-sectional view of accelerometer (1 mm diameter PM).

The static sensitivity of the accelerometer is

$$\frac{y}{a} = \frac{m}{c}. \tag{13.2}$$

As soon as the spring constant trends to zero ($c \to 0$), the static sensitivity (13.2) dramatically increases and, in the limiting case, becomes infinitely large ($m/c \to \infty$). Consequently, the accelerometer transfer function (13.1) can be rewritten as

$$\frac{y(s)}{a(s)} = \frac{1}{s\left(s + \dfrac{\mu}{m}\right)}. \tag{13.3}$$

It is important to note that in Equation 13.3, the integration of the external acceleration a is obtained by eliminating the spring constant c [15].

Obviously, due to the infinitely large sensitivity in such an accelerometer performance, it can operate only under closed-loop control when the feedback forces back the PM to the original equilibrium position. It is known that the static loop gain of a closed-loop accelerometer can be defined as [16,17]

$$K_l = \frac{K_{PO}K_C K_F}{c}, \tag{13.4}$$

where

K_{PO} is the gain of the pick-off circuit
K_C is the controller gain
K_F is the feedback gain

In view of $c \to 0$, it follows from (13.4) that $K_l \to \infty$. As a result, the steady-state errors in such an accelerometer, after eliminating the spring constant of the suspension, will be significantly smaller in comparison with an accelerometer in which the spring constant of the suspension is preserved. Thus, elimination of the spring constant of the suspension provides a dramatic increase in the static sensitivity of the sensor and a significant decrease in the steady-state errors of the closed-loop sensor. Due to these facts, it is expected that applying a suspension with zero spring constant to a micromachined inertial sensor will significantly increase their accuracies.

In this chapter, a micromachined DTG and accelerometer based on a CS and the last one with zero spring constant, respectively, are considered and discussed in details as an alternative strategy for the further improvement of performances of micromachined inertial sensors based on the CS technology.

13.2 MICROMACHINED DTG, BASED ON A CONTACTLESS SUSPENSION

13.2.1 KINEMATICS AND OPERATING PRINCIPLE

The kinematics of a micromachined DTG are illustrated in Figure 13.3. The micromachined DTG consists of two rotors, namely, an inner and an outer one, that are linked together by a pair of torsional springs. The inner rotor is located within an electromagnetic field and suspended there contactlessly at its equilibrium position. Both rotors are considered as rigid bodies in this study. The position of the center of mass of the inner rotor is characterized by the origin O that is also the origin of the coordinate frame (CF) XYZ, fixed to the gyroscope case. Rotation of the rotor system, with respect to the gyroscope case, is provided by the rotating electromagnetic field with a steady angular rate Ω along the Z axis. The Z axis is set as the gyroscope rotational axis. Let us define the rotating CF $x_r y_r z_r$ that is rotated together with the rotating electromagnetic field, and the origin of which coincides with the origin O. The Z and z_r axes are coincident, geometrically. In this study, a steady state is assumed; therefore, the speed of the rotating electromagnetic field and rotors is the same and equal to Ω along the Z axis under the absence of the measuring angular rate (the unperturbed state). In the unperturbed state, the x_r axis of the rotating CF coincides with the axis of torsions as shown in Figure 13.3. It is assumed that the static and dynamic unbalances of the rotors and their linear displacements are negligible, and the centers of the rotors' inertia coincide with the origin O.

Let us assign a CF $x_1 y_1 z_1$ to the inner and a CF xyz to the outer rotors so that the axes of these CF coincide with the principal axes of inertia of the rotors, respectively, and the x_1 and x axes are directed along the axis of the torsional springs, as illustrated in Figure 13.4. The position of the

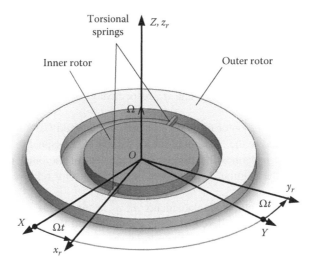

FIGURE 13.3 Kinematics of micromachined DTG: XYZ is the fixed coordinate frame relative to the gyroscope case; $x_r y_r z_r$ is the rotating coordinate frame related to the rotating electromagnetic field; Ω is the steady angular rate of rotation of the electromagnetic field and rotors.

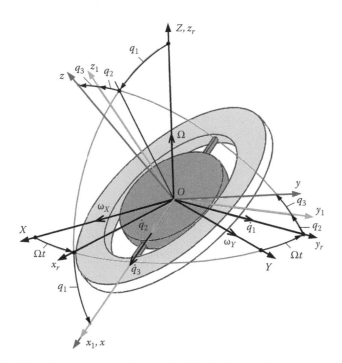

FIGURE 13.4 Coordinate frames $x_1 y_1 z_1$ and xyz are fixed to the inner and outer rotors, respectively.

CF $x_1 y_1 z_1$ with respect to the rotating CF $x_r y_r z_r$ is defined by the angles q_1 and q_2, and the position of the CF xyz by the angles q_1 and q_3, respectively.

Note that the angular coordinate q_2 defines the angular displacement of the inner rotor relative to the x_1 axis and the angle between the surfaces $x_1 z_r$ and $x_1 z_1$ or $x_1 y_r$ and $x_1 y_1$, while the angular coordinate q_3 defines the angular displacement of the outer rotor relative to the x axis and the angle between surfaces $x_1 z_r$ and xz or $x_1 y_r$ and xy. It is seen that angular coordinates

q_2 and q_3 are independent of each other, since the surfaces x_1z_r and x_1y_r do not take part in the angular displacement of the inner as well as the outer rotors.

The rotation of the gyroscope case with the measuring angular rate, which is defined by vector $\bar{\omega}$ lying on the XY surface of the micromachined DTG and having projections ω_X and ω_Y along axes X and Y, respectively, induces motion in the outer rotor relative to the output axes y_r and x (characterized by angular coordinates q_1 and q_3, respectively) that characterizes the value of the measuring input angular rate $\bar{\omega}$.

To define the position of the outer rotor with respect to the fixed CF XYZ, let us introduce the angles α and β that characterize the position of the xy surface with respect to the CF XYZ as shown in Figure 13.5a. The angle α defines the angular displacement of the outer rotor relative to the X axis. The angle β defines the angular displacement of the outer rotor relative to the Y' axis, which lies on the YZ surface.

Since it is assumed that angular displacements q_1, q_3, α, and β are small, the dedicated triangles in Figure 13.5b can be considered as plane triangles, and the angles between the sides of the triangles q_1, q_3, and α, β are 90°; hence, the relationship between angles q_1, q_3, α, and β can be written as

$$\left.\begin{aligned} \alpha &= q_3 \cos \Omega t - q_1 \sin \Omega t; \\ \beta &= q_3 \sin \Omega t + q_1 \cos \Omega t. \end{aligned}\right\} \tag{13.5}$$

Thus, the angular displacements of the outer rotor with respect to the rotating CF (characterized by q_1 and q_3) as well as the fixed CF (characterized by α and β) are a measure of the input rotation of the gyroscope case or the input rates that can be finally transformed into the electrical output signal by means of capacitive angular sensor.

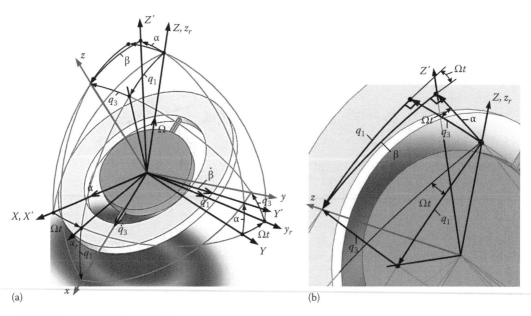

(a)   (b)

FIGURE 13.5 Determination of the position of the outer rotor with respect to the fixed CF, where $X'Y'Z'$ is the ancillary CF, rotating relative to the X axis: (a) The position of the outer rotor with respect to the fixed CF XYZ; (b) The dedicated triangles to define the relationship between angles q_1, q_3, α and β.

13.2.2 MATHEMATICAL MODEL

To develop a mathematical model of the micromachined DTG, the second-kind Lagrange equation is used. In the framework of the problem, the micromachined DTG has four degrees of freedom: the angle of rotors spin Ωt and angles of rotations of rotors q_1, q_2, and q_3 relative to the rotating CF. The angle Ωt is a cyclical angle; with a constant value of Ω, it allows us to construct the following three equations for the description of the micromachined DTG motion, taking q_1, q_2, and q_3 as the generalized coordinates. Hence, we can write

$$\frac{d}{dt}\frac{\partial T}{\partial \dot{q}_i} - \frac{\partial T}{\partial q_i} = -\frac{\partial \Pi}{\partial q_i} - \frac{\partial \Phi}{\partial \dot{q}_i}, \quad (i = 1, 2, 3), \tag{13.6}$$

where
T and Π are the kinetic and potential energies of the system
Φ is the dissipation function
q_i, $\dot{q}_i$ are the generalized coordinates and velocities, respectively

The kinetic energy of the system under consideration is the sum of the kinetic energies of the inner and outer rotors and can be written as

$$T = \frac{1}{2}\left(J_{x1}\omega_{x1}^2 + J_{y1}\omega_{y1}^2 + J_{z1}\omega_{z1}^2\right) + \frac{1}{2}\left(J_x\omega_x^2 + J_y\omega_y^2 + J_z\omega_z^2\right), \tag{13.7}$$

where J_{x1}, J_{y1}, J_{z1} and ω_{x1}, ω_{y1}, ω_{z1} are the three central principal moments of inertia about the x_1, y_1, z_1 axes and projections of angular rate of the inner rotor on the same axes, respectively; J_{x1}, J_{y1}, J_{z1} and ω_{x1}, ω_{y1}, ω_{z1} are the three central principal moments of inertia about the x, y, z axes and projections of angular rate of the outer rotor on the same axes, respectively.

According to the kinematics of the micromachined DTG, for projections of angular rates of rotors on axes of CF $x_r y_r z_r$, $x_1 y_1 z_1$, and xyz, the following equations are true:

$$\left.\begin{aligned}
\omega_{xr} &= \omega_X \cos\Omega t + \omega_Y \sin\Omega t; \\
\omega_{yr} &= -\omega_X \sin\Omega t + \omega_Y \cos\Omega t; \\
\omega_{zr} &= \Omega;
\end{aligned}\right\} \tag{13.8}$$

$$\left.\begin{aligned}
\omega_{x1} &= \omega_{xr}\cos q_1 - \omega_{zr}\sin q_1 + \dot{q}_2; \\
\omega_{y1} &= \left(\omega_{yr} + \dot{q}_1\right)\cos q_2 + \left(\omega_{xr}\sin q_1 + \omega_{zr}\cos q_1\right)\sin q_2; \\
\omega_{z1} &= -\left(\omega_{yr} + \dot{q}_1\right)\sin q_2 + \left(\omega_{xr}\sin q_1 + \omega_{zr}\cos q_1\right)\cos q_2;
\end{aligned}\right\} \tag{13.9}$$

$$\left.\begin{aligned}
\omega_x &= \omega_{xr}\cos q_1 - \omega_{zr}\sin q_1 + \dot{q}_3; \} \\
\omega_y &= \left(\omega_{yr} + \dot{q}_1\right)\cos q_3 + \left(\omega_{xr}\sin q_1 + \omega_{zr}\cos q_1\right)\sin q_3; \\
\omega_z &= -\left(\omega_{yr} + \dot{q}_1\right)\sin q_3 + \left(\omega_{xr}\sin q_1 + \omega_{zr}\cos q_1\right)\cos q_3.
\end{aligned}\right\} \tag{13.10}$$

The CS provides the angular displacement of the inner rotor with the finite angular stiffness that is the same for all equatorial axes of the rotor; the value of this stiffness is denoted by c_s. Hence, the potential energy of the system can be written as follows:

$$\Pi = \frac{1}{2}c_s q_1^2 + \frac{1}{2}c_s q_2^2 + \frac{1}{2}c_t\left(q_3 - q_2\right)^2, \qquad (13.11)$$

where c_t is the stiffness of the torsional springs.

The dissipation function is described as

$$\Phi = \frac{1}{2}k_s \dot{q}_1^2 + \frac{1}{2}k_s \dot{q}_2^2 + \frac{1}{2}k_t\left(\dot{q}_3 - \dot{q}_2\right)^2, \qquad (13.12)$$

where k_s and k_t are the damping coefficients of the CS and the torsional springs, respectively. To write the linear mathematical model of the micromachined DTG, let us substitute (13.7), (13.11), and (13.12), accounting for the kinematical relationships (13.8), (13.9), and (13.10) into Equation 13.6. Since the angles q_1, q_2, and q_3 are small, second-order terms can be neglected, and the mathematical model describing the motion relative to the rotating CF can be written as

$$\left.\begin{aligned}
&\left(J_{y1}+J_y\right)\ddot{q}_1 + k_s\dot{q}_1\left(c_s + \left[\left(J_{z1}+J_z\right)-\left(J_{x1}+J_x\right)\right]\Omega^2\right)q_1 - \left(J_{z1}-J_{y1}-J_{x1}\right)\Omega\dot{q}_2 \\
&\qquad -\left(J_z - J_y - J_x\right)\Omega\dot{q}_3 \qquad\qquad\qquad\qquad\qquad = m_{q1}; \\
&\left(J_{z1}-J_{y1}-J_{x1}\right)\Omega\dot{q}_1 + J_{x1}\ddot{q}_2 + \left(k_s + k_T\right)\dot{q}_2 + \left(c_s + \left[J_{z1}-J_{y1}\right]\Omega^2\right)q_2 - k_t\dot{q}_3 - c_T q_3 = m_{q2}; \\
&\left(J_z - J_y - J_x\right)\Omega\dot{q}_1 - k_t\dot{q}_2 - c_T q_2 + J_x\ddot{q}_3 + k_t\dot{q}_3 + \left(c_t + \left[J_z - J_y\right]\Omega^2\right)q_3 \qquad = m_{q3},
\end{aligned}\right\}$$

$$(13.13)$$

where

$$m_{q1} = \left[\left(J_{z1}+J_z\right)+\left(J_{y1}+J_y\right)-\left(J_{x1}+J_x\right)\right]\Omega\left(\omega_X\cos\Omega t + \omega_Y\sin\Omega t\right);$$

$$m_{q2} = \left[J_{z1}+J_{x1}-J_{y1}\right]\Omega\left(\omega_X\sin\Omega t - \omega_Y\cos\Omega t\right); \qquad (13.14)$$

$$m_{q3} = \left[J_z + J_x - J_y\right]\Omega\left(\omega_X\sin\Omega t - \omega_Y\cos\Omega t\right).$$

Note that if the angular displacements of the inner rotor q_1 and q_2, relative to the y_r and x_1 axes, respectively, are absent ($q_1 = q_2 = 0$, $\dot{q}_1 = \dot{q}_2 = 0$), the last equation of set (13.13) describes the dynamics of a rotor vibratory gyroscope exactly [18–22]. On the other hand, if only the angular displacement of the inner rotor q_2 relative to the x_1 axis is absent ($q_2 = 0$, $\dot{q}_2 = 0$), the first and third equations of set (13.13) are the same as a classic model of a DTG with gimbal [20,23,24].

13.2.3 ANALYSIS OF THE MODEL FOR A PARTICULAR CASE

From a practical viewpoint, let us analyze the motion of the micromachined DTG for the case that follows from the analysis of the experimental study of the MMG prototypes based on CS [1,7,25]. For this case, the following assumptions can be made:

$$J_x = J_y < J_z; \quad J_{x1} = J_{y1} < J_{z1}. \qquad (13.15)$$

Also, the measuring angular rate vector $\bar{\omega}$ is steady and its projections are $\omega_X = \text{const}_1$; $\omega_Y = \text{const}_2$. For further analysis, model (13.13) can be rewritten as follows:

$$\left.\begin{aligned}
J\ddot{q}_1 + k_s\dot{q}_1 + C_{d1}q_1 + h_1\dot{q}_2 + h\dot{q}_3 &= m_{q1}; \\
-h_1\dot{q}_1 + J_{x1}\ddot{q}_2 + 2\left(k_s + k_t\right)\dot{q}_2 + C_{d2}q_2 - k_t\dot{q}_3 - c_t q_3 &= m_{q2}; \\
-h\dot{q}_1 - k_t\dot{q}_2 - c_t q_2 + J_x\ddot{q}_3 + k_t\dot{q}_3 + C_{d3}q_3 &= m_{q3},
\end{aligned}\right\} \tag{13.16}$$

where

$$\left.\begin{aligned}
m_{q1} &= \left(J_{z1} + J_z\right)\Omega\left(\omega_X\cos\Omega t + \omega_Y\sin\Omega t\right); \\
m_{q2} &= J_{z1}\Omega\left(\omega_X\sin\Omega t - \omega_Y\cos\Omega t\right); \\
m_{q3} &= J_z\Omega\left(\omega_X\sin\Omega t - \omega_Y\cos\Omega t\right),
\end{aligned}\right\} \tag{13.17}$$

and $J = J_{y1} + J_y$; $\quad h_1 = (2J_{y1} - J_{z1})\Omega$; $\quad h = (2J_y - J_z)\Omega$; $\quad C_{d1} = c_s + [(J_{z1} + J_z) - (J_{x1} + J_x)]\Omega^2$;

$$C_{d2} = c_s + [J_{z1} - J_{y1}]\Omega^2; \quad C_{d3} = c_t + [J_z - J_y]\Omega^2.$$

It is assumed that the value of angular stiffness of the CS is much greater than the value of the dynamic stiffness, and the following inequalities become true:

$$\left.\begin{aligned}
c_s &\gg \left[\left(J_{z1} + J_z\right) - \left(J_{x1} + J_x\right)\right]\Omega^2; \\
c_s &\gg c_t.
\end{aligned}\right\} \tag{13.18}$$

If conditions (13.18) are satisfied, the CS is best described as "hard." Due to (13.18) and the fact that the value of the moment of inertia of the inner rotor is less than the value of one of the outer rotors, the influence of the motion of the inner rotors relative to the x_1 axis on the motion of the outer rotor is considered negligible. Hence, the mathematical model (13.16) can be simplified as follows:

$$\left.\begin{aligned}
J\ddot{q}_1 + k_s\dot{q}_1 + C_{d1}q_1 + h\dot{q}_3 &= m_{q1}; \\
-h\dot{q}_1 + J_x\ddot{q}_3 + k_t\dot{q}_3 + C_{d3}q_3 &= m_{q3}.
\end{aligned}\right\} \tag{13.19}$$

Note that the model described by set (13.19) is similar to the model of the rotor vibratory gyroscope considered in [26].

New variables are introduced:

$$U = q_1\sqrt{J}; \quad = q_3\sqrt{J_x}. \tag{13.20}$$

Using the variables defined in Equation 13.20, set (13.19) can be rewritten as

$$\left. \begin{aligned} \ddot{U} + \mu_1 \dot{U} + m_1 U + n\dot{V} &= \frac{1}{\sqrt{J}} m_{q1}; \\ -n\dot{U} + \ddot{V} + \mu_3 \dot{V} + m_3 V &= \frac{1}{\sqrt{J_x}} m_{q3}, \end{aligned} \right\}$$
(13.21)

where $\mu_1 = k_s/J$; $\mu_3 = k_t/J_x$; $m_1 = C_{d1}/J$; $m_3 = C_{d3}/J_x$; $n = h/\sqrt{J_x J}$. Taking Laplace transformations of (13.21), under zero initial conditions, set (13.21) can be written in a complex notation as follows:

$$\left. \begin{aligned} \left(s^2 + \mu_1 s + m_1\right) U(s) + nsV(s) &= \frac{1}{\sqrt{J}} m_{q1}(s); \\ -nsU(s) + \left(s^2 + \mu_3 s + m_3\right) V(s) &= \frac{1}{\sqrt{J_x}} m_{q3}(s), \end{aligned} \right\}$$
(13.22)

where s is the Laplace operator. The solution of set (13.22) is

$$U(s) = \frac{\left(s^2 + \mu_3 s + m_3\right) \frac{1}{\sqrt{J}} m_{q1}(s) - ns \frac{1}{\sqrt{J_x}} m_{q3}(s)}{\left(s^2 + \mu_1 s + m_1\right)\left(s^2 + \mu_3 s + m_3\right) + n^2 s^2};$$

$$V(s) = \frac{\left(s^2 + \mu_1 s + m_1\right) \frac{1}{\sqrt{J_x}} m_{q3}(s) + ns \frac{1}{\sqrt{J}} m_{q1}(s)}{\left(s^2 + \mu_1 s + m_1\right)\left(s^2 + \mu_3 s + m_3\right) + n^2 s^2}.$$
(13.23)

Thus, Equation 13.23 describes the motion of the micromachined DTG relative to the rotating CF in the "hard" CS case.

Let us study the behavior of the mathematical model of the micromachined DTG (13.23) under the presence of the measuring angular rate lying on the measuring surface of the gyroscope to define the micromachined DTG gain for the "hard" CS case, relative to both the rotating and the fixed CF.

Equation 13.23 can be rewritten as

$$\left. \begin{aligned} q_1(s) &= W_1(s) \frac{1}{J} m_{q1}(s) - W_2(s) \frac{1}{\sqrt{J_x J}} m_{q3}(s); \\ q_3(s) &= W_2(s) \frac{1}{\sqrt{J_x J}} m_{q1}(s) + W_3(s) \frac{1}{J_x} m_{q3}(s), \end{aligned} \right\}$$
(13.24)

where

$$W_1(s) = \frac{\left(s^2 + \mu_3 s + m_3\right)}{\left(s^2 + \mu_1 s + m_1\right)\left(s^2 + \mu_3 s + m_3\right) + n^2 s^2};$$

$$W_2(s) = \frac{ns}{\left(s^2 + \mu_1 s + m_1\right)\left(s^2 + \mu_3 s + m_3\right) + n^2 s^2};$$
(13.25)

$$W_3(s) = \frac{\left(s^2 + \mu_1 s + m_1\right)}{\left(s^2 + \mu_1 s + m_1\right)\left(s^2 + \mu_3 s + m_3\right) + n^2 s^2}.$$

It is considered that the measuring angular rate creates moments $m_{q1}(s)$ and $m_{q3}(s)$ acting on the inner and outer rotors with a frequency that is equal to the spin speed of the rotor Ω. Hence, let us substitute $j\Omega$ for s into (13.25), to study the magnitude and phase of these transfer functions. We have

$$|W_1(j\Omega)| = \frac{\sqrt{\left(-\Omega^2 + m_3\right)^2 + \mu_3^2 \Omega^2}}{\sqrt{\left[\Omega^4 - \left(m_1 + m_3 + n^2 + \mu_1\mu_3\right)\Omega^2 + m_1 m_3\right]^2 + \left[\left(\mu_1 + \mu_3\right)\Omega^3 + \left(m_3\mu_1 + m_1\mu_3\right)\Omega\right]^2}};$$

$$\arg\left(W_1(j\Omega)\right) = \arctan\left(\frac{\mu_3\Omega}{-\Omega^2 + m_3}\right) - \arctan\left(\frac{\left(\mu_1 + \mu_3\right)\Omega^3 + \left(m_3\mu_1 + m_1\mu_3\right)\Omega}{\Omega^4 - \left(m_1 + m_3 + n^2 + \mu_1\mu_3\right)\Omega^2 + m_1 m_3}\right);$$

(13.26)

$$|W_2(j\Omega)| = \frac{n\Omega}{\sqrt{\left[\Omega^4 - \left(m_1 + m_3 + n^2 + \mu_1\mu_3\right)\Omega^2 + m_1 m_3\right]^2 + \left[\left(\mu_1 + \mu_3\right)\Omega^3 + \left(m_3\mu_1 + m_1\mu_3\right)\Omega\right]^2}};$$

$$\arg\left(W_2(j\Omega)\right) = \frac{\pi}{2} - \arctan\left(\frac{\left(\mu_1 + \mu_3\right)\Omega^3 + \left(m_3\mu_1 + m_1\mu_3\right)\Omega}{\Omega^4 - \left(m_1 + m_3 + n^2 + \mu_1\mu_3\right)\Omega^2 + m_1 m_3}\right);$$

(13.27)

$$|W_3(j\Omega)| = \frac{\sqrt{\left(-\Omega^2 + m_1\right)^2 + \mu_1^2 \Omega^2}}{\sqrt{\left[\Omega^4 - \left(m_1 + m_3 + n^2 + \mu_1\mu_3\right)\Omega^2 + m_1 m_3\right]^2 + \left[\left(\mu_1 + \mu_3\right)\Omega^3 + \left(m_3\mu_1 + m_1\mu_3\right)\Omega\right]^2}};$$

$$\arg\left(W_3(j\Omega)\right) = \arctan\left(\frac{\mu_1\Omega}{-\Omega^2 + m_1}\right) - \arctan\left(\frac{\left(\mu_1 + \mu_3\right)\Omega^3 + \left(m_3\mu_1 + m_1\mu_3\right)\Omega}{\Omega^4 - \left(m_1 + m_3 + n^2 + \mu_1\mu_3\right)\Omega^2 + m_1 m_3}\right).$$

(13.28)

Applying $\omega_Y = 0$ and taking into account (13.26 through 13.28), Equation 13.24 for the steady motion of the outer rotor of the micromachined DTG relative to the rotating (x_r, y_r, z_r) CF can be rewritten as

$$q_1(t) = |W_1(j\Omega)|\frac{J_z + J_{z1}}{J}\Omega\omega_X \cos\left(\Omega t + \arg\left(W_1(j\Omega)\right)\right) - |W_2(j\Omega)|\frac{J_z}{\sqrt{JJ_x}}\Omega\omega_X \sin\left(\Omega t + \arg\left(W_2(j\Omega)\right)\right);$$

$$q_3(t) = |W_3(j\Omega)|\frac{J_z}{J_x}\Omega\omega_X \sin\left(\Omega t + \arg\left(W_3(j\Omega)\right)\right) + |W_2(j\Omega)|\frac{J_z + J_{z1}}{\sqrt{JJ_x}}\Omega\omega_X \cos\left(\Omega t + \arg\left(W_2(j\Omega)\right)\right);$$

(13.29)

Equation 13.29 describes the motion of the untuned gyroscope. To define the rotor tuning condition, let us assume that $\mu_1 = 0$.

$\mu_3 = 0$ and, setting the denominator of transfer functions to zero, we have

$$\Omega^4 - \left(m_1 + m_3 + n^2 \right)\Omega^2 + m_1 m_3 = 0. \tag{13.30}$$

The solutions of (13.30) are the two natural frequencies:

$$\Omega_1 = \sqrt{\frac{m_1 + m_3 + n^2 + \sqrt{\left(m_1 + m_3 + n^2\right)^2 - 4m_1 m_3}}{2}}, \tag{13.31}$$

and

$$\Omega_2 = \sqrt{\frac{m_1 + m_3 + n^2 - \sqrt{\left(m_1 + m_3 + n^2\right)^2 - 4m_1 m_3}}{2}}. \tag{13.32}$$

Furthermore, let us define the parameters m_1, m_3, and n^2 as follows. The parameter m_1 can be written as

$$m_1 = \frac{c_s}{J + \kappa_g \Omega^2}, \tag{13.33}$$

where $\kappa_g = (J_z + J_{z1} - J_x - J_{x1})/(J_x - J_{x1})$ is the gyroscope constructive parameter. Parameters m_3 and n^2 can be expressed as

$$m_3 = \frac{c_t}{J_x + \kappa\Omega^2};$$

$$n^2 = \frac{J_x(1-\kappa)^2\Omega^2}{J}, \tag{13.34}$$

where $\kappa = (J_z - J_x)/J_x$ is the outer rotor constructive parameter. Note that both κ_g and κ are approximately equal to one. For example, the aluminum disk-shaped rotor whose diameter is 500 µm and thickness is 10 µm has a value of κ equal to 0.9998. According to (13.18) and the previous discussion, the following inequality between the parameters of the gyroscope can be written:

$$m_1 \gg m_3 \gg n^2. \tag{13.35}$$

In Equations 13.31 and 13.32, Ω_1 is determined by the parameters of the CS and Ω_2 by the mechanical parameters of the torsional springs, respectively. Hence, to tune the micromachined DTG, the rotors spin speed Ω should be equal to the second natural frequency. The rotor tuning condition becomes

$$\Omega - \Omega_2 = 0. \tag{13.36}$$

The speed of rotor spin that is satisfied by (13.36) is denoted by $\tilde{\Omega}$. Accounting to (13.35), the condition (13.36) can be written as

$$\tilde{\Omega} \approx \sqrt{m_3}, \tag{13.37}$$

or using Equation 13.34,

$$\tilde{\Omega} \approx \sqrt{\frac{c_t}{J_x(1-\kappa)}}. \tag{13.38}$$

The last equation is the classic condition for rotor tuning of a rotor vibratory gyroscope.

If the condition (13.36) is held, then the magnitude and phase of the transfer functions (13.26), (13.27) and (13.28) become

$$\left| W_1(j\tilde{\Omega}) \right| \approx \frac{1}{m_1}; \quad \arg\left(W_1(j\Omega) \right) \approx 0; \tag{13.39}$$

$$\left| W_2(j\tilde{\Omega}) \right| \approx \frac{n}{\mu_3 m_1}; \quad \arg\left(W_2(j\tilde{\Omega}) \right) \approx 0; \tag{13.40}$$

$$\left| W_3(j\tilde{\Omega}) \right| \approx \frac{1}{\mu_3 \tilde{\Omega}};$$

$$\arg\left(W_3(j\tilde{\Omega}) \right) \approx \arctan\left(\frac{\mu_1 \tilde{\Omega}}{m_1} \right) - \frac{\pi}{2}. \tag{13.41}$$

It is assumed that the values of the damping coefficients μ_1 and μ_3 are small. Hence, we can use $\arg(W_3(j\tilde{\Omega})) = -(\pi/2)$. Substituting (13.39), (13.40), and (13.41) into (13.29), we have

$$q_1(t) = \frac{J_z + J_{z1}}{m_1 J} \tilde{\Omega}\omega_X \cos(\Omega t) - \frac{n}{\mu_3 m_1} \frac{J_z}{\sqrt{JJ_x}} \tilde{\Omega}\omega_X \sin(\Omega t);$$

$$q_3(t) = -\frac{J_z}{\mu_3 J_x} \omega_X \cos(\Omega t) + \frac{n}{\mu_3 m_1} \frac{J_z + J_{z1}}{\sqrt{JJ_x}} \tilde{\Omega}\omega_X \cos(\Omega t); \tag{13.42}$$

Equation 13.42 describes the steady motion of the outer rotor of the tuned micromachined DTG relative to the rotating CF. Note that in the second equation of set (13.42), the value of the second term is much less than the first one due to $m_1 \gg n\tilde{\Omega}$ (see (13.33), (13.34), and (13.18)); hence, it can be neglected.

Taking into account (13.5), the steady motion of the outer rotor of the tuned micromachined DTG relative to the fixed (XYZ) CF can be written as follows:

$$\alpha(t) = -K_g\omega_X - K_g\omega_X\cos(2\tilde{\Omega}t) - K_{cc2}\omega_X - K_{cc1}\omega_X\sin(2\tilde{\Omega}t) + K_{cc2}\omega_X\cos(2\tilde{\Omega}t);$$

$$\beta(t) = -K_g\omega_X\sin(2\tilde{\Omega}t) + K_{cc1}\omega_X + K_{cc1}\omega_X\cos(2\tilde{\Omega}t) - K_{cc2}\omega_X\sin(2\tilde{\Omega}t);$$

(13.43)

where $K_g = (J_z/2\mu_3 J_x)$ is the tuned micromachined DTG gyroscope gain for the "hard" CS case and $K_{cc1} = (J_z + J_{z1}/2m_1 J)\tilde{\Omega}$ and $K_{cc2} = (n/\mu_3 m_1)\left(J_z + J_{z1}/\sqrt{JJ_x}\right)\tilde{\Omega}$ are the cross-coupling coefficients.

Thus, the mathematical model of a micromachined DTG, based on a CS, is developed as an alternative strategy for improving the accuracy of MMG. The dynamical analysis of the gyroscope for the case in which the CS provides "hard" electrical spring is conducted, and the rotor tuning condition for this gyroscope is obtained.

13.3 SUSPENSION WITH ZERO SPRING CONSTANT

13.3.1 KINEMATICS AND OPERATING PRINCIPLE OF THE SUSPENSION

Let us consider the following electromechanical, micromachined CS shown in Figure 13.6, which can be used, in particular, to describe an accelerometer. The disk-shaped PM (to be fabricated from a conducting material) is suspended at the equilibrium position characterized by the origin O by means of an ICS. Alternating current i passing through the coil creates in space a variable magnetic flux, which intercepts the surface area of the PM and induces a current, i_2.

In addition, the system of fixed electrodes E_1, E_2, E_3, and E_4, as shown in Figure 13.6, creates an electrical field around the suspended PM. Electrodes E_3 and E_4 are grounded, while potentials u_1 and u_2 are applied to electrodes E_1 and E_2, respectively. The resultant electrostatic forces F_{el} and

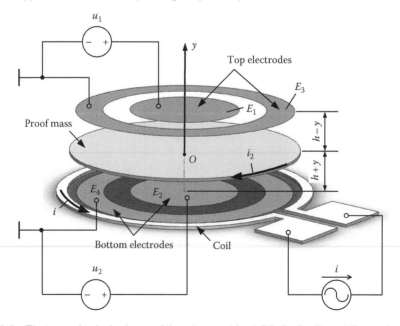

FIGURE 13.6 Electromechanical scheme of the micromachined CS: E_1, E_2, E_3, and E_4 are electrodes.

F_{e2} created by the electrical field and acting on the top and bottom of the PM surfaces help provide a minimization of the spring constant of the inductive suspension along the vertical Oy axis. The conditions required for the complete elimination of the spring constant will be considered later. It is assumed that the PM has only a linear displacement along the Oy axis that is denoted by y from the origin O, which is located in the center characterized by the distance h between the top and bottom electrodes.

13.3.2 MATHEMATICAL MODEL

Lagrange–Maxwell equations are used to compile the mathematical model of the suspension under consideration. At first, let us choose the generalized coordinates of the system.

The suspension can be divided into two parts, namely, the electric part, consisting of a system of electrodes E_1, E_2, E_3, E_4, and PM, and the electromagnetic part, which is the coil and PM. These parts are assumed to be electrically independent of each other; hence, they can be considered separately.

The electrode E_k and the nearest part of the surface of PM to this electrode is considered to be a plane capacitor with capacity C_k, which is dependent on the displacement of PM, where $k = 1..4$ (see Figure 13.7). Assuming that the area of all electrodes is the same and equal to A_e, then from the equation for a plane capacitor, we have

$$C_1 = C_3 = \frac{A}{h-y}; \quad C_2 = C_4 = \frac{A}{h+y}, \tag{13.44}$$

where
$A = \varepsilon_0 \varepsilon A_e$
ε_0 is the relative permittivity
ε is the dielectric constant

According to the design of the electric part of the suspension shown in Figure 13.7, the electric circuit can be represented as shown in Figure 13.8. The capacitor $C(y)$ shown in Figure 13.8 is the sum of capacitors and defined as follows:

$$C(y) = C_3 + C_4 = \frac{2Ah}{h^2 - y^2}. \tag{13.45}$$

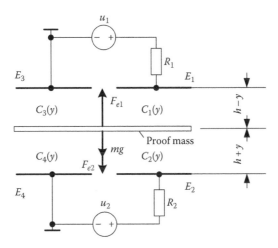

FIGURE 13.7   Scheme of the electric part of suspension: F_{e1} and F_{e2} are the resultant electrostatic forces.

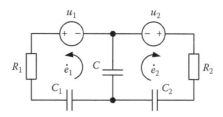

FIGURE 13.8 Electric circuit of the electric part of the suspension: e_1 and e_2 are charges on the electrodes E_1 and E_2, respectively.

The charges e_1 and e_2, on the first and second electrodes, respectively, can be taken as the generalized coordinates of the electric part of the suspension. Note that the current flowing through the capacitor $C(y)$ is equal to $e'_1 + e'_2$; therefore, the charge on the capacitor $C(y)$ is $e_1 + e_2$ [27].

The PM is suspended at the equilibrium position by means of the electromagnetic part of the suspension, the coil of which is fed by an alternating current $i = ie^{j\omega t}$, where ω is a high frequency and $j = \sqrt{-1}$ is an imaginary unit. Due to a mutual inductance denoted by M_{12} between the coil and PM occurs, a current is induced within the PM denoted by i_2, with the same frequency ω (see Figure 13.9). The currents i and i_2 are taken as generalized coordinates of the electromagnetic part of the suspension.

The linear displacement y of the PM is taken as a generalized coordinate of the mechanical part of the suspension. Hence, the following set of the Lagrange–Maxwell equations, describing the system under consideration, can be written as

$$
\begin{cases}
-\dfrac{\partial L}{\partial e_1} + \dfrac{\partial \Psi}{\partial \dot{e}_1} = u_1; \quad -\dfrac{\partial L}{\partial e_2} + \dfrac{\partial \Psi}{\partial \dot{e}_2} = u_2; \\[2mm]
\dfrac{d}{dt}\left(\dfrac{\partial L}{\partial i}\right) + \dfrac{\partial \Psi}{\partial i} = 0; \quad \dfrac{d}{dt}\left(\dfrac{\partial L}{\partial i_2}\right) + \dfrac{\partial \Psi}{\partial i_2} = 0; \\[2mm]
\dfrac{d}{dt}\left(\dfrac{\partial L}{\partial \dot{y}}\right) - \dfrac{\partial L}{\partial y} + \dfrac{\partial \Psi}{\partial \dot{y}} = F_y,
\end{cases}
\tag{13.46}
$$

where
  $L = T(\dot{y}) - \Pi(y) + W_m(y, i, i_2) - W_e(y, e_1, e_2)$ is the Lagrange function of the system
  $T(\dot{y})$ and $\Pi(y)$ are the kinetic and potential energies of the system, respectively
  $W_m(y,i,i_2)$ and $W_e(y,e_1,e_2)$ are energies stored in the magnetic and electric fields, respectively
  $\Psi(\dot{y}, i, i_2, \dot{e}_1, \dot{e}_2)$ is the dissipation function of the system
  $F_y = ma$ is the generalized force acting on the PM

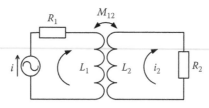

FIGURE 13.9 Electric circuit of the electromagnetic part of the suspension: L_1 and L_2 are self-inductances of the coil and PM, respectively.

The kinetic energy of the PM linear movement is

$$T = \frac{1}{2}m\dot{y}^2. \tag{13.47}$$

The potential energy is

$$\Pi = mgy. \tag{13.48}$$

Taking into account that the mutual inductance M_{12} is dependent on y, the energy stored in the magnetic field of electromagnetic part of the suspension can be written as

$$W_m = \frac{1}{2}L_1 i^2 + M_{12}(y)ii_2 + \frac{1}{2}L_2 i_2^2, \tag{13.49}$$

where L_1 and L_2 are the self-inductances of the coil and PM, respectively. The energy stored in the electric part of the suspension is

$$W_e = \frac{e_1^2}{2C_1} + \frac{e_2^2}{2C_2} + \frac{(e_1 + e_2)^2}{2C} = \frac{e_1^2}{2A}(h-y) + \frac{e_2^2}{2A}(h+y) + \frac{(e_1+e_2)^2}{4Ah}(h^2 - y^2). \tag{13.50}$$

Hence, the Lagrange function becomes

$$L = \frac{1}{2}m\dot{y}^2 - mgy + \frac{1}{2}L_1 i^2 + M_{12}(y)ii_2 + \frac{1}{2}L_2 i_2^2 - \frac{e_1^2}{2A}(h-y) - \frac{e_2^2}{2A}(h+y) - \frac{(e_1+e_2)^2}{4Ah}(h^2 - y^2). \tag{13.51}$$

Neglecting the resistances in the electric part of the suspension, the dissipation function of the system can be written as follows:

$$\Psi = \frac{1}{2}R_1 i^2 + \frac{1}{2}R_2 i_2^2 + \frac{1}{2}\mu\dot{y}^2. \tag{13.52}$$

In practice, μ can be controlled by evacuation of the packaged device.

Substituting (13.51) and (13.52) into (13.46), we have

$$\frac{h-y}{A}e_1 + \frac{h^2 - y^2}{4Ah}(e_1 + e_2) \qquad\qquad = u_1;$$

$$\frac{h+y}{A}e_2 + \frac{h^2 - y^2}{4Ah}(e_1 + e_2) \qquad\qquad = u_2;$$

$$L_1\frac{di}{dt} + \frac{dM_{12}(y)}{dy}\dot{y}i_2 + M_{12}(y)\frac{di_2}{dt} + R_1 i \qquad = 0; \tag{13.53}$$

$$L_2\frac{di_2}{dt} + \frac{dM_{12}(y)}{dy}\dot{y}i + M_{12}(y)\frac{di}{dt} + R_2 i_2 \qquad = 0;$$

$$m\ddot{y} + \mu\dot{y} + mg - \frac{dM_{12}(y)}{dy}ii_2 - \frac{e_1^2}{2A} + \frac{e_2^2}{2A} - \frac{(e_1+e_2)^2}{2Ah}y = F_y.$$

Equation 13.53 is the set of nonlinear equations describing the behavior of the PM suspended by the proposed CS.

13.3.3 CONDITION FOR STABLE LEVITATION

Let us examine the condition under which the PM occupies an equilibrium position at the point O, located at a height, h, above the coil. In other words, the condition required for the stable levitation of the PM in an alternating magnetic field is studied. In this case, the problem is reduced to defining the static position of the PM that is characterized by the height h at which the electromagnetic forces created by the inductive suspension compensate the gravitational force of the PM mass. Hence, the influence of the PM velocity $\dot{y}$, acceleration $\ddot{y}$, and the generalized force F_y acting on the PM is neglected. The mutual inductance M_{12} is considered to have a function dependence on h. Also, it is assumed that the electric part of the suspension is disconnected from the electricity supply. Based on these assumptions, set (13.53) can be rewritten as follows:

$$\begin{cases} L_1 \dfrac{di}{dt} + M_{12}(h)\dfrac{di_2}{dt} + R_1 i = 0; \\[2mm] L_2 \dfrac{di_2}{dt} + M_{12}(h)\dfrac{di}{dt} + R_2 i_2 = 0; \\[2mm] mg - \dfrac{dM_{12}(h)}{dh} i i_2 = 0. \end{cases} \tag{13.54}$$

The current i feeding the coil is supplied by a current generator, and consequently, it is assumed to be constant. Using the second equation of set (13.54), the current i_2 can be expressed in terms of i as follows:

$$i_2 = -i \frac{\sqrt{\omega^4 L_2^2 + \omega^2 R_2^2}}{\omega^2 L_2^2 + R_2^2} M_{12}(h) e^{j\phi}, \tag{13.55}$$

where $\phi = \arctan(R_2/\omega L_2)$. Analysis of Equation 13.55 shows that, firstly, the flow direction of i_2 is in the opposite direction to i, secondly, a phase shift ϕ exists between the currents i_2 and i, caused by the electric resistance of the PM conducting material. For normal operation of the inductive suspension, the phase shift φ must be minimized, made by the following condition:

$$\omega L_2 \gg R_2. \tag{13.56}$$

As a rule, fulfilling (13.56) is provided by adjusting the frequency ω. If condition (13.56) is held, Equation 13.55 can be simplified as

$$i_2 = -i \frac{1}{L_2} M_{12}(h) e^{j\phi}. \tag{13.57}$$

Substituting (13.57) into the third equation of set (13.54), the following equation is obtained:

$$mg + \frac{dM_{12}(h)}{dh} \frac{M_{12}(h)}{L_2} i^2 = 0. \tag{13.58}$$

Hence, Equation 13.58 defines the height h at which the PM occupies the equilibrium position, at the origin O.

On the other hand, at the equilibrium point, the Lagrange–Dirichlet function for this system reaches a maximum:

$$F(h) = \Pi(h) - W_m(h,i) \to \min. \tag{13.59}$$

Hence, the second h-derivative of the function $F(h)$ must have a negative sign and, accounting for (13.58), can be written as

$$\frac{d}{dh}\left(mg + \frac{dM_{12}(h)}{dh} \frac{M_{12}(h)}{L_2} i^2 \right) > 0. \tag{13.60}$$

Differentiating the last equation, Equation 13.60 becomes

$$\frac{i^2}{L_2}\left[\frac{d^2 M_{12}(h)}{dh^2} M_{12}(h) + \left(\frac{dM_{12}(h)}{dh} \right)^2 \right] > 0. \tag{13.61}$$

Since the sign of 13.61 is dependent only on the expression within the brackets, the final condition for stable levitation of the PM can be written:

$$\frac{d^2 M_{12}(h)}{dh^2} M_{12}(h) + \left(\frac{dM_{12}(h)}{dh} \right)^2 > 0. \tag{13.62}$$

Let us study the behavior of the PM near to the equilibrium point O. It is assumed that the linear displacement of the PM y is small in comparison with h; hence, the following inequality can be written:

$$\frac{y}{h} \ll 1. \tag{13.63}$$

Because of (13.63), the function of the mutual inductance $M_{12}(y)$ can be extended by a Taylor series at the point h and, neglecting third-order terms, becomes

$$M_{12}(y) = M_{12}(h) + M_y y + \frac{1}{2} M_{yy} y^2, \tag{13.64}$$

where

$$M_y = \left.\frac{dM_{12}(y)}{dy}\right|_{y=h} \quad \text{and } M_{yy} = \left.\frac{d^2 M_{12}(y)}{dy^2}\right|_{y=h}.$$

Substituting (13.64) into the last equation of set (13.53) and taking into account (13.57) and (13.58), the differential equation of the linear displacement of the PM near to the equilibrium point O (in the absence of the electric part of the suspension) can be written as

$$m\ddot{y} + \mu\dot{y} + \frac{i^2}{L_2}\left[M_{yy}M_{12}(h) + M_y^2\right]y = F_y. \tag{13.65}$$

Equation 13.65 is a linear model of the inductive suspension. The spring constant provided by the inductive suspension is directly proportional to the square M_y and current i feeding the coil, and it is inversely proportional to the self-inductance of the PM L_2.

The self-inductance of the PM is dependent on the induced current circuit within the PM. Due to the high frequency of the supply current i and the disk-shaped PM, the current circuit can be considered to be in the form of a closed ring that has a radius r_{mp}, as shown in Figure 13.10. The mutual induction of two coaxial rings can be approximately described by the following function [28,29]:

$$M_{12}(y) = \mu_0 r_c\left[\ln\frac{8r_c}{\sqrt{y^2 + d^2}} - 2\right], \tag{13.66}$$

where
  μ_0 is the permeability
  r_c is the radius of the coil
  $d = r_c - r_{pm}$

Replacing in (13.66) y by h and substituting into (13.62), the condition for the stable levitation of the disk-shaped PM becomes

$$-\frac{\mu_0^2 r_c^2}{\left(h^2 + d^2\right)^2}\left[\left(d^2 - h^2\right)\left(\ln\frac{8r_c}{\sqrt{h^2 + d^2}} - 2\right) - h^2\right] > 0. \tag{13.67}$$

Once again, the sign of the expression is dependent on the terms within the brackets that must always be positive or

$$\left(d^2 - h^2\right)\left(\ln\frac{8r_c}{\sqrt{h^2 + d^2}} - 2\right) - h^2 < 0. \tag{13.68}$$

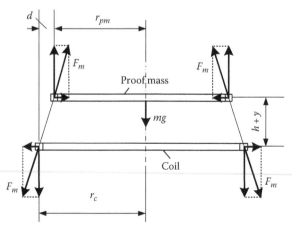

FIGURE 13.10 Scheme of the electromagnetic part of the suspension: F_m—the resultant electromagnetic forces created by the inductive suspension.

Thus, condition (13.68) is necessarily fulfilled for the stable levitation of the disk-shaped PM in an alternating magnetic field induced by the ring-shaped coil.

It follows from the analysis of (13.68) that the value of d must not be equal to zero, or in other words, the radii of the coil and PM must not be equal. In the framework of the feasible system, it can be shown that the function $\ln\left(8r_c/\sqrt{h^2+d^2}\right)-2$ is always positive and much greater than one; hence, Equation 13.68 can be rewritten as

$$\frac{h^2}{d^2} > \frac{\ln\left(8r_c/\sqrt{h^2+d^2}\right)-2}{\ln\left(8r_c/\sqrt{h^2+d^2}\right)-1}. \tag{13.69}$$

Then, Equation 13.69 can be reduced to

$$\frac{h^2}{d^2} > 1 \quad \text{or} \quad h > d. \tag{13.70}$$

Equation 13.70 shows that the levitation height h of the PM is limited by d from the bottom.

13.3.4 Compensation of the Spring Constant

Assuming that the PM is stably levitated (conditions (13.62) and (13.70) are held), the behavior of the PM within the electric field created by the system of the electrodes E_1, E_2, E_3, and E_4 can be described by the following set:

$$\begin{cases} \dfrac{h-y}{A}e_1 + \dfrac{h^2-y^2}{4Ah}(e_1+e_2) = u_1; \\[2mm] \dfrac{h+y}{A}e_2 + \dfrac{h^2-y^2}{4Ah}(e_1+e_2) = u_2; \\[2mm] m\ddot{y} + \mu\dot{y} + \dfrac{i^2}{L_2}\left[M_{yy}M_{12}(h) + M_y^2\right]y - \dfrac{e_1^2}{2A} + \dfrac{e_2^2}{2A} - \dfrac{(e_1+e_2)^2}{2Ah}y = F_y. \end{cases} \tag{13.71}$$

Using the first and second equations of (13.71), the charges e_1 and e_2 can be expressed in terms of the potentials u_1 and u_2 as follows:

$$e_1 = \frac{A}{4h}\left(\frac{3h-y}{h-y}u_1 - u_2\right); \quad e_2 = \frac{A}{4h}\left(\frac{3h-y}{h-y}u_2 - u_1\right). \tag{13.72}$$

Substituting (13.72) into the last equation of set (13.71) and rearranging the equation, the following expression can be written:

$$m\ddot{y} + \mu\dot{y} + \frac{i^2}{L_2}\left[M_{yy}M_{12}(h) + M_y^2\right]y - \frac{A}{4}\left[\frac{u_2^2}{(h+y)^2} - \frac{u_1^2}{(h-y)^2}\right] = F_y. \tag{13.73}$$

In view of (13.63) and the fact that the potentials u_1 and u_2 are assumed to be equal to each other, Equation 13.73 can be linearized and simplified as follows:

$$m\ddot{y} + \mu\dot{y} + \left[\frac{i^2}{L_2} \left[M_{yy}M_{12}(h) + M_y^2 \right] - \frac{Au^2}{h^3} \right] y = F_y. \tag{13.74}$$

Thus, the linear model describing the behavior of the disk-shaped PM suspended by the CS based on combining inductive and electrical suspensions is obtained. Analysis of model (13.74) reveals that the spring constant of the suspension is defined by the difference between the two terms, namely, the spring constants of the inductive (the first term within the brackets) and electric (the second term within the brackets) suspensions. Note that the spring constant of the electric suspension has a negative sign, and its value is inversely proportional to the cubic of the levitation height. To minimize or completely eliminate the spring constant of the suspension, the following condition has to be fulfilled:

$$\frac{i^2}{L_2} \left[M_{yy}M_{12}(h) + M_y^2 \right] - \frac{Au^2}{h^3} \simeq 0. \tag{13.75}$$

From the point of view of the stability of the suspension, condition (13.75) must not be negative.

The developed methodology can be applied to the experimental results of Williams [11]. In this prototype of the inductive suspension, the disk-shaped PM of radius $r_{pm} = 250$ μm and thickness $t_{pm} = 10$ μm was suspended to a height $h = 2$ μm. The measured spring constant along the vertical direction was $4 \cdot 10^{-3}$ N × m^{-1}, with coil current i of 0.35 A.

It is assumed that this inductive suspension is provided by the system of electrodes depicted in Figure 13.6. For further analysis, a dimensionless spring constant is introduced:

$$c_s = \frac{c_m - c_e}{c_m}, \tag{13.76}$$

where $c_m = i^2/L_2 \left[M_{yy}M_{12}(h) + M_y^2 \right]$ and $c_e = Au^2/h^3$. In the case under consideration, the c_m is assumed to be equal to 4×10^{-3} N × m^{-1}, and the area of the electrode is to be 9.82×10^{-8} m^2 that is calculated by the following equation: $A_e = \left(\pi r_{pm}^2 \right)/2$.

Let us plot the dependence of the dimensionless spring constant c_s against the potential u applied to the electrodes E_1 and E_2 as shown in Figure 13.11. Figure 13.11 shows that when the potential u equals $u_0 = 0.1960$ V, the spring constant of the suspension is reduced to zero. The suspension is stable when $u < u_0$ and unstable when $u > u_0$. It is important to note that the value of the applied potential, u, is a tenth of the voltage to eliminate the spring constant of the suspension.

The inverse value of the dimensionless spring constant $1/c_s$ characterizes the static sensitivity to the measuring acceleration, a. Its dependence on the potential u is shown in Figure 13.12. An increase in sensitivity by one order of magnitude occurs after a decrease in the spring constant of the suspension by 90%. At the point $u = u_0$, the static sensitivity becomes infinitely large due to the complete elimination of the spring constant.

Thus, a CS with a zero spring constant is proposed. This leads to, on one hand, a dramatic increase to the static sensitivity of the inertial sensor and, on the other hand, a significant reduction of the steady-state error of the sensor in closed-loop operation. Minimization of the spring constant of the proposed CS is achieved by combining inductive and electrical CSs.

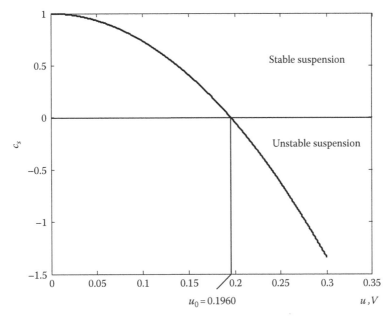

FIGURE 13.11   Dependence of the dimensionless spring constant c_s on the potential u.

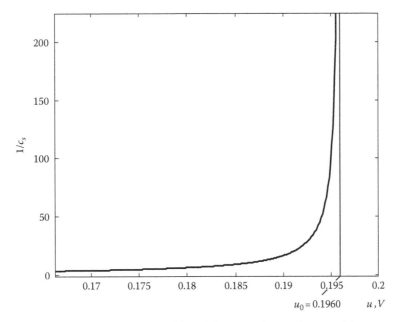

FIGURE 13.12   Dependence of the static sensitivity of the suspension on the potential u.

The mathematical model is developed to study conditions associated with the spring constant elimination as well as the levitational stability. Analysis of the model allows us to define the conditions for stable levitation of an inductive CS in general. In particular case, the condition for the stable levitation of the disk-shaped PM in an alternating magnetic field from a ring-shaped coil is obtained. This condition predicts that for stable levitation of the disk-shaped PM, the radii of the coil and the PM must not be equal to each other and that the height *h* of levitation of the PM is limited form the bottom by the value of the difference between the radii of the coil and the PM.

Based on the data of the experimental study of the ICS prototype developed by Shearwood's group, the performance of the proposed CS is illustrated theoretically. It is shown that the required applied potential for the minimization of the spring constant is a tenth of a volt.

REFERENCES

1. C. Shearwood, C.B. Williams, P.H. Mellor, R.B. Yates, M.R.J. Gibbs, and A.D. Mattingley, Levitation of a micromachined rotor for application in a rotating gyroscope, *Electron. Lett.*, 31(21), 1845–1846, 1995.
2. C. Shearwood, K.Y. Ho, C.B. Williams, and H. Gong, Development of a levitated micromotor for application as a gyroscope, *Sens. Actuators A: Phys.*, 83(1–3), 85–92, 2000.
3. R. Toda, N. Takeda, T. Murakoshi, S. Nakamura, and M. Esashi, Electrostatically levitated spherical 3-axis accelerometer, in *Micro Electro Mechanical Systems, 2002. The Fifteenth IEEE International Conference on*. IEEE, Allentown, PA, January 24, 2002, pp. 710–713.
4. R. Houlihan and M. Kraft, Modelling of an accelerometer based on a levitated proof mass, *J. Micromech. Microeng.*, 12, 495, 2002.
5. B. Damrongsak, M. Kraft, S. Rajgopal, and M. Mehregany, Design and fabrication of a micromachined electrostatically suspended gyroscope, *Proc. Inst. Mech. Eng. C: J. Mech. Eng. Sci.*, 222(1), 53–63, 2008.
6. X. Wu, T. Deng, W. Chen, and W. Zhang, Electromagnetic levitation micromotor with stator embedded (ELMSE): Levitation and lateral stability characteristics analysis, *Microsyst. Technol.*, 17, 5969, 2011.
7. T. Murakoshi, Y. Endo, K. Fukatsu, S. Nakamura, and M. Esashi, Electrostatically levitated ring-shaped rotational gyro/accelerometer, *Jpn. J. Appl. Phys.*, 42(4B), 2468–2472, 2003.
8. F.T. Han, Q.P. Wu, and L. Wang, Experimental study of a variable capacitance micromotor with electrostatic suspension, *J. Micromech. Microeng.*, 20, 115034, 2010.
9. F.T. Han, L. Wang, Q.P. Wu, and Y.F. Liu, Performance of an active electric bearing for rotary micromotors, *J. Micromech. Microeng.*, 21, 085027, 2011.
10. T.B. Gabrielson, Mechanical-thermal noise in micromachined acoustic and vibration sensors, *IEEE Trans. Electron. Dev.*, 40(5), 903–909, 1993.
11. C.B. Williams, C. Shearwood, and P.H. Mellor, Modeling and testing of a frictionless levitated micromotor, *Sens. Actuators A: Phys.*, 61, 469–473, 1997.
12. K.V. Poletkin, A.I. Chernomorsky, and C. Shearwood, A proposal for micromachined dynamically tuned gyroscope, based on contactless suspension, *IEEE Sens. J.*, 12(06), 2164–2171, 2012.
13. K.V. Poletkin, A.I. Chernomorsky, and C. Shearwood, A proposal for micromachined accelerometer, base on a contactless suspension with zero spring constant, *IEEE Sens. J.*, 12(07), 2407–2413, 2012.
14. K. Poletkin, A micromachined contactless suspension with zero spring constant, in *Proceedings of the ASME 2012 International Mechanical Engineering Congress and Exposition (IMECE2012)*. ASME, Houston, TX, November 9–15, pp. 519–527, 2012, accepted with honors.
15. J. Van de Vegte and J.V. Vegte, *Feedback Control Systems*, Prentice Hall, Englewood Cliffs, NJ, 1994.
16. V.I. Mel'nikov, *Electromechanical Transducers Based on Quartz Glass*, Moscow: Mashinostroenie, 1984 (in Russian).
17. M. Kraft, C.P. Lewis, and T.G. Hesketh, Closed-loop silicon accelerometers, in *Circuits, Devices and Systems, IEE Proceedings*. IET, 1998, Vol. 145, pp. 325–331.
18. R. Whalley, M.J. Holgate, and L. Mauder, Oscillogyro, *J. Mech. Eng. Sci.*, 9(1), 55–65, 1967.
19. D. Ormandy and L. Maunder, Dynamics of oscillogyro, *J. Mech. Eng. Sci.*, 15(3), 210–217, 1973.
20. L. Maunder, Dynamically tuned gyroscopes, in *Fifth World Congress on Theory of Machines and Mechanisms*, New York, 1979, pp. 470–473.
21. U.B. Vlasov and O.M. Filonov, *Rotor vibratory gyroscopes applications in navigation systems*, Leningrad: Shipbuilding, 1980 (in Russian).
22. R.J.G. Craig, Theory of operation of a 2-axis-rate gyro, *IEEE Sens. J. Trans. Aerospace Electron. Syst.*, 26(5), 722–731, 1990.
23. R.J.G. Craig, Theory of operation of an elastically supported tuned gyroscope, *Aerospace Electron. Syst. IEEE Trans.*, (3), 280–288, 1972.
24. S. Merhav, *Aerospace Sensor Systems and Applications*, Springer-Verlag, New York, 1996.
25. C.B. Williams, C. Shearwood, P.H. Mellor, A.D. Mattingley, M.R.J. Gibbs, and R.B. Yates, Initial fabrication of a micro-induction gyroscope, *Microelectron. Eng.*, 30(1–4), 531–534, 1996.

26. K.V. Poletkin, A.I. Chernomorsky, and C. Shearwood, Influence of the elastic properties of the spring element on the rotor tuning condition of a rotor vibratory gyroscope, *IEEE Sens. J.*, 11(09), 1856–1860, 2011.

27. Yu.G. Martynenko, *Analytical Dynamics of Electromechanical Systems*, Moscow Power Engineering Institute, Moscow, Russia, 1984 (in Russian).

28. F.W. Grover, *Inductance Calculations: Working Formulas and Tables*, Dover Publications, New York, 2004.

29. C.R. Paul, *Inductance: Loop and Partial*, Wiley-IEEE Press, New York, 2009.

14 Contactless Angle Detection in Automotive, Consumer, and Industrial Applications

Antonio J. López-Martín and Alfonso Carlosena

CONTENTS

14.1 INTRODUCTION

Nowadays, a relevant requirement in many measurement, control, and instrumentation applications is angle measurement, in order to detect and control angular position, displacement, rotation speed, or acceleration. Traditionally, three-terminal potentiometers based on a variable resistive divider whose value is set by a sliding contact have been employed. Their operation is based on the displacement of the sliding contact (wiper) along a resistive element, making electrical contact with it. The resistive element is connected to two electrical terminals at each end of it. The device is completed by a mechanism to move the sliding contact (e.g., a shaft) and a housing containing the resistor and sliding contact. These conventional potentiometers are inexpensive, and the resistive element is usually made of graphite. However, the internal contact between the sliding and resistive elements degrades the device due to wear, making it unsuitable in many applications, specially for automotive and industrial sectors. To remedy this situation, contactless angle detection has been extensively developed in recent years [1–3]. It is based on avoiding the aforementioned internal friction, providing a wear-free operation with increased reliability and lifetime. The avoidance of this physical contact also makes the devices more robust to mechanical degradation and pollution. The resulting contactless potentiometers have become very popular, and there is an ongoing increase in the number of companies producing them and in the number of techniques employed.

A relevant issue in any sensor-based measurement system is the sensor bias and readout electronics that supplies energy to the sensor and processes the sensor signal. Sensors usually require proper DC or AC bias and provide a low-level analog output signal being dependent not only on the parameter to be measured but also on unwanted parameters like pressure or temperature. Other undesirable factors that may affect the sensor output signal are offsets, gain errors, hysteresis, aging, etc. In contactless angle detection, not only electrical offset must be considered. Also the "mechanical" offset, or stated otherwise, the angle error in the zero-reference position, must be compensated. Adequate signal conditioning and calibration are thus required prior to the utilization of the sensor output by, for example, a control system. Typically the final goal is to get an output signal linearly related to the measured parameter and free from dependency on other parameters. Such output is usually required in digital form due to the dominant digital nature of subsequent processing units.

This chapter aims to provide insight into the emerging field of modern contactless angle measurement systems. In Section 14.2, the application of these systems to the automotive, industrial, and consumer fields is outlined. Section 14.3 is devoted to the description of the different sensors employed for contactless angle measurement, highlighting their main features. In Section 14.4, three different fabricated solutions for contactless angle detection based on a giant magnetoresistive (GMR) sensor bridge are presented, and their performance is compared. Finally, some conclusions are drawn in Section 14.5.

14.2 APPLICATIONS OF CONTACTLESS POTENTIOMETERS

Contactless angle detection is an emerging field that is experiencing an ongoing growth in terms of techniques proposed, market sales, and application areas. In the next paragraphs, some of the existing applications in different sectors are outlined.

14.2.1 Automotive Applications

The automotive industry has become one of the main targets for sensor manufacturers. The number of sensors included in vehicles has increased steadily [1] due to the ever-increasing demands on energy efficiency, safety, and comfort.

The design of sensors for automotive applications implies facing stringent requirements that are difficult to balance. The accuracy demanded is typically better than 3% over the entire measurement and temperature ranges. The temperature range is very wide (−40°C to 125°C in the engine compartment), and the vibration experienced may be large (sweeps of up to 10 g during 30 h). The environmental conditions are very adverse in terms of electronic interference, humidity, liquids, dust, and pollution. Moreover, due to the high-volume production and strong competence between companies, cost is a major concern. As a result, automotive sensors must face a complex trade-off between accuracy, robustness, manufacturability, and cost [1].

Among the different sensing requirements in automotive applications, angle measurement is historically the most relevant one [2], specially for measuring angular position and rotation speed. Table 14.1 summarizes some of the main applications in this context. The application area is divided into powertrain, chassis, and body systems, where body system applications include anything not corresponding to the other two categories. Another possible classification often employed is powertrain, body, and safety applications. Powertrain systems include the engine, transmission system, and all the onboard diagnostics elements. Chassis systems include the suspension, braking, lightning, steering, and stability systems. Body systems include safety of occupants, comfort, information services, and in general the rest of systems aimed to fulfill the needs of the vehicle occupants. Although Table 14.1 does not intend to be exhaustive, it can be seen that these sensors are extremely relevant, mainly for powertrain and chassis systems. They are applied for crankshaft and camshaft rotational control of spark and fuel injection timing and in electronically controlled gear shifting to detect transmission input and output shaft speeds. They are also applied to detect wheel speed, playing a major role in electronic braking systems, traction control, and stability systems. They constitute a key element in "drive by wire" systems, active suspension, and automatic headlight leveling, as well as in wiper, mirror, and seat positioning. Another important application is detection of wheel position for automatic navigation systems.

TABLE 14.1
Sensors Used in Automotive Applications

Powertrain

Engine	Crankshaft rotational motion
	Camshaft rotational motion
	Exhaust gas recirculation (EGR)
	Throttle position
Transmission	Gearshift position
	Input/output shaft speeds
	Transmission oil pump

Chassis

Braking	Wheel speed for ABS
	Brake-by-wire pedal angle
Lightning	Automatic headlamp leveling
Steering	Steering wheel angle for electric power Steering/steering by wire
Vehicle	Wheel speed
	Yaw angular rate
	Roll angular rate
	Chassis height/angle
	Wheel-to-wheel variance of rolling speed
	Wiper positioning
	Mirror positioning

Body

Safety	Seat positioning
Navigation	Wheel motion (to zero speed)
	Vehicle yaw rate
Security	Vehicle tilt for anti theft system

14.2.2 INDUSTRIAL APPLICATIONS

Accurate angular position and rotation speed detection are also a key requirement in several industrial applications, mainly for control purposes [3]. As for the automotive applications, the availability of free-of-wear solutions is highly demanded due to the increased reliability and extended lifetime of the device. Requirements in industrial scenarios vary widely depending on the area of activity. For instance, the required temperature range is very small in air-conditioned rooms and extends to temperature between −55°C and +190°C in military applications. However, a common requirement in many of these applications is reducing manufacturing costs. Some of the main application areas are

- Robotic systems
- Process control
- Hydraulic systems
- Energy generation

In process control and robotics, angle and rotation speed are fundamental measurements for incremental and absolute rotary encoders. They are also widely used in several systems to detect valve position and gearwheel position or speed and in electric commutated motors among other applications.

14.2.3 CONSUMER APPLICATIONS

For low-cost domestic applications, conventional potentiometers are widely used since the accuracy and free-of-wear operation requirements are usually not too strict and cost is the key concern. For instance, they are employed in rotary knobs in several electrical appliances, where the reliability and lifetime improvement of contactless devices is often not necessary. However, the increased offer in contactless devices by different companies has reduced their price, making them attractive also for some consumer applications. Some application examples include

- Human–machine interfaces (joysticks, dials, rotary switches)
- Home automation

The inception of the so-named domotic systems is playing a key role in the demands of novel technologies aimed to advanced home automation [4]. These systems are essentially based on the integration of the conventional installations of a house or building aimed to the automated management and control from a single system. Contactless potentiometers are versatile, and accurate devices can find multiple applications in this field, for example, by helping in the control of the angle of the slats of blinds to graduate the amount of natural light that enters a room or office.

14.3 TECHNIQUES FOR CONTACTLESS ANGLE DETECTION

Several techniques have been proposed to achieve angle detection without internal friction. Most of them are based on magnetic sensors, although other devices such as optical and capacitive sensors are also employed. An overview of some of the most popular ones is presented in the following.

14.3.1 OPTICAL SENSORS

A simple way to detect angle by optical means is using an optical-encoded sensor [5,6]. Optical rotary encoders are considered as rotational angle measurement devices with high accuracy, resolution, and reliability. Absolute optical encoders indicate the absolute angular position and can thus be employed for angle measurement, while incremental encoders only provide information about changes in the angle and are more adequate to measure rotation speed or acceleration.

A typical configuration of an optical encoder is a disk with transparent and opaque areas (for instance, a slotted disk). The optical pattern resulting from the position of the disk is read by means of an array of photodiodes placed at one side of the disk, and a light source is applied to the other side.

14.3.2 CAPACITIVE SENSORS

Angle measurements can also be carried out by varying the capacitance of a sensor as a function of the angle to be measured [7]. Capacitive sensors are attractive due to their low power consumption and simple manufacturability, excellent linearity, and noiseless nature of capacitors. A simple approach is employing capacitive sensors of variable area. The angle is measured by detecting the change in capacitance of a movable pole piece relative to a fixed pole piece. As for the optical sensors just mentioned, incremental and absolute capacitive sensors exist. Usually absolute angle measurements are restricted to small angles due to the limited capacitance value, reducing the range of applications. Incremental measurements allow detection of wider angles, by continuously detecting a periodical phase signal of a grid coupling. However, if the power is turned off, the reference position of the movable part must be reset, and this part must be also set to the zero position to calibrate a measurement. An alternative to measure absolute angular positions in a wide range is proposed in [8]. Capacitive sensors usually employ low-level signals, which complicate the electronic readout circuit and make the device more sensitive to electromagnetic interference, humidity, and dust. Thus, they are not very popular for angle measurement in the automotive field (at least for powertrain applications).

14.3.3 INDUCTIVE SENSORS

They are usually based on detecting the time-varying fluctuations of magnetic flux created by the rotating motion of a mechanical element. They are also commonly called variable reluctance sensors [9] and feature advantages such as relatively low size and good temperature insensitivity. However, they have limitations such as small allowed air gaps (typically less than 2 mm), loss of signal at zero speed, and dependence of signal strength and phase with shaft speed. For this reason, they are more popular for rotation speed measurements than for angle measurements [1].

14.3.4 HALL EFFECT SENSORS

They have become one of the most popular solutions in the last years [10]. They are based on exploiting the Hall effect, which occurs when a current flowing through a conductor experiences a magnetic field, generating a small voltage that can provide information about the strength and direction of the magnetic field. In contactless potentiometers, Hall sensors are usually made by semiconductor active devices and generate a voltage signal that reflects the variation of the magnetic flux generated by rotating a mechanical part (typically with a permanent magnet attached). These sensors feature small size, low cost, high linearity, and repeatability and can operate at zero speed. However, the maximum air gap is typically about 2–3 mm and they experience a significant sensitivity to pressure on the sensor package.

14.3.5 MAGNETOTRANSISTORS AND MAGFETS

Another popular alternative, specially when the sensing device requires to be fabricated together with the readout electronics in the same integrated circuit (IC) process, is magnetically sensitive transistors. Initially, bipolar transistors designed to be sensitive to magnetic field were developed, usually called magnetotransistors. They can be divided into vertical magnetotransistors, which depend on the vertical flow of carriers for magnetic detection, and lateral magnetotransistors (LMTs) [11], which depend on the horizontal flow of carriers for such detection. In both cases, they exploit the deflection of the flowing carriers when a magnetic field is applied caused by the Lorentz force.

Today, CMOS technologies are dominant in the IC manufacturing industry due to their low cost, power, and suitability for both analog and digital applications. An advantage of LMTs is that they can be successfully fabricated in standard CMOS technologies. Another advantage is that they are sensitive with the magnetic fields parallel to the chip plane. A particularly useful LMT is the so-named LMTs with suppressed sidewall injection (SSIMT), which features high sensitivity and linearity and can be employed as a magnetic switch.

Typically multiple collector SSIMTs are used, allowing a 2D detection of the magnetic field [11] with a single device. Common disadvantages are relatively large offsets and sensitivity to mechanical misalignments between the sensor and the element generating the magnetic field due to fabrication tolerances or aging.

Another option in CMOS technologies is using MOSFET transistors sensitive to the magnetic field (MAGFETs) [12]. A typical device is the split-drain MAGFET, which uses the same principle as the LMT (deflection of current by an external magnetic field due to the Lorentz force) and senses the amount of current deflected by comparing the current collected in the two drain terminals of the device.

14.3.6 Magnetoresistors

They are devices whose resistance varies with the magnetic flux density. As for the previous devices, operation is based on the deflection of current flow caused by the Lorentz force. A magnetoresistive device can be obtained, for example, if an adequate geometric pattern of uniformly spaced conductive stripes is placed perpendicular to the direction of current flow in semiconductors with high carrier mobility such as InAs or InSb [13]. In the presence of these internal shortings, the deflection of current by the external magnetic field will vary the resistance of the device. Thus, these resistors can be compatible with IC processes and can be fabricated with the readout electronics. Advantages of magnetoresistors include very good repeatability, excellent temperature insensitivity, ability to operate to zero speed and to sense the direction of rotation, and moderate air gaps (up to approximately 3 mm). The main disadvantages are that they require bias current, that nonlinearity often reduces the angular range achievable, and that the cost and size are usually not very low.

14.3.7 Anisotropic Magnetoresistive Sensors

They are magnetoresistive sensors where the variation in resistivity is anisotropic [14]. It makes them useful for detection of the direction of the magnetic field rather than its strength. The most typical material employed is permalloy, a ferromagnetic allow composed of 20% iron and 80% nickel. Typically these sensors are arranged in a four-resistor Wheatstone bridge arrangement that is deposited in a common substrate. Advantages and disadvantages are similar to those mentioned for magnetoresistors.

14.3.8 Giant Magnetoresistive (GMR) Sensors

They are based on the GMR effect [14,15], which consists in the dependence of the resistance of a material on the angle between magnetization directions at different locations in the material. Therefore, to achieve this effect, magnetically inhomogeneous materials are required, which can be achieved by granular structures [16] or, most commonly, by multilayer structures [17]. The value of the resistance of the multilayer structure varies with the angle between the magnetization directions of different ferromagnetic layers in the multilayer. Upon application of an external magnetic field, for example, by a permanent magnet, the direction of at least one of the magnetizations changes, modifying the resistance.

The term "giant" in these sensors is due to the higher sensitivities to variations of the applied magnetic field than in anisotropic magnetoresistive (AMR) sensors (typically about 20 times larger at low temperatures and 3–6 times higher at room temperature). Although the operation principle in GMR and AMR sensors is different, both are more sensitive to the direction of the magnetic field than to its intensity, which is an advantage in angle detection devices.

GMR sensors have less sensitivity to the relative position and distance between the magnet and the sensor as compared to other devices such as Hall sensors or LMTs, thus making mechanical assembly and tolerances due to aging less critical. Besides, the magnet does not require a particular shape. These are important advantages as they reduce production costs. Due to these benefits, focus will be on these devices in the systems presented in the next section.

14.4   CASE STUDY: GMR-BASED CONTACTLESS POTENTIOMETERS

In this section, the practical design of contactless potentiometers based on GMR sensors is presented. Different choices employed by the authors in the last years for the physical arrangement of the potentiometer, the biasing of the sensor, and the readout electronics are discussed and compared.

14.4.1   Physical Arrangement of the Contactless Potentiometer

Figure 14.1 shows a simplified diagram of a possible configuration for a contactless potentiometer based on magnetic detection. It includes a rotary shaft that allows rotating a permanent magnet attached to the axis of the potentiometer. Inside the potentiometer housing, there is also a printed circuit board (PCB) that contains the magnetic sensor (in the top layer). In the bottom layer of the PCB, the readout electronics is implemented. Due to cost and size constraints, an application-specific integrated circuit (ASIC) is usually the optimal choice for integrating all this readout electronics in a single chip, when high-volume applications such as automotive ones are targeted. Minimization of the number of discrete components outside the ASIC reduces system dimensions and cost.

The operation is as follows. When the shaft is rotated, the sensor detects the direction of the magnetic field created by the magnet and hence the angular position of the axis. Then the ASIC processes the sensor signal and delivers it through the connector of the potentiometer in analog and/or digital form.

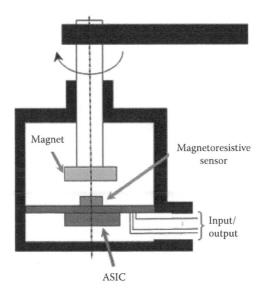

FIGURE 14.1   Diagram of the contactless potentiometer.

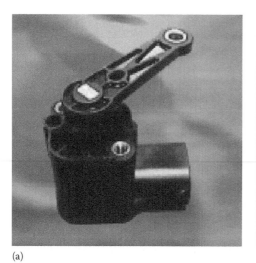

(a)   (b)

FIGURE 14.2 (a) Sealed and shielded housing. (b) Conventional housing.

Figure 14.2a and b shows different housings that can be employed for the contactless potentiometer of Figure 14.1. The housing of Figure 14.2a is aimed to operate in harsh environments, such as powertrain systems in automotive applications or certain industrial applications. It is sealed to avoid the influence of humidity, liquids, dust, or pollutants. It also includes an RFI/EMI metal shielding that reduces the influence of any external electromagnetic interference. The housing of Figure 14.2b is a conventional low-cost metal case housing, which is suitable for most consumer and industrial applications and for some body systems in automotive applications inside the passenger compartment. The figure shows the housing open, where the PCB with the ASIC that contains the readout electronics can be observed.

The contactless potentiometers whose measurements are presented in this chapter have been fabricated according to the arrangement in Figure 14.1 and enclosed in both housings of Figure 14.2. A two-pole rectangular samarium–cobalt magnet with dimensions of $10 \times 5 \times 5$ mm^3 has been employed, and the air gap (approximately 3 mm) is such that the magnetic field at the surface of the magnetic sensor is about 10 kA/m. The magnetic sensor employed is a GMR bridge, which is described in the following.

14.4.2 GMR Sensor Bridge

The GMR sensor employed in these contactless potentiometers is fabricated by Infineon Technologies and is based on a hard–soft multilayer with artificial antiferromagnet (AAF) [18]. A Wheatstone bridge configuration of four GMR devices is implemented by applying a spatially varying magnetic field to the sensor chip after deposition and patterning. Figure 14.3 shows the schematic diagram of the GMR bridge. Resistance values are as follows:

$$R_1 = R_0\left(1 + \alpha_0 T\right) + 0.5\Delta R\left(1 - \alpha_\Delta T\right)$$

$$R_2 = R_0\left(1 + \alpha_0 T\right) - 0.5\Delta R\left(1 - \alpha_\Delta T\right)$$

(14.1)

where
  R_0 is the basic GMR (and bridge) resistance
  ΔR is the variable part, depending on the direction of magnetic field
  α_0 is the absolute temperature coefficient of R_0
  α_Δ is the absolute temperature coefficient of ΔR
  T is the temperature relative to a reference temperature

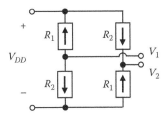

FIGURE 14.3   GMR bridge sensor.

In the GMR sensor employed typically, $\Delta R/R_0 \approx 5\%$ [19]. This large sensitivity is another advantage of GMR sensors. Typical measured values for R_0, α_0, and $-\alpha_\Delta$ are 800 Ω, 0.11%/K, and −0.12%/K, respectively. As shown in Equation 14.1, temperature dependence for R_0 is positive and negative for ΔR, which will be exploited as follows for temperature compensation. The output voltage of the bridge is given by the following expression [20] where V_{DD} is the bias voltage of the bridge:

$$V_{out} = V_2 - V_1 = \frac{\Delta R\left(1 - \alpha_\Delta T\right)}{2R_0\left(1 + \alpha_0 T\right) + \Delta R\left(1 - \alpha_\Delta T\right)} V_{DD} \tag{14.2}$$

Figure 14.4 shows the measured amplified output of four different GMR sensor bridges at a temperature of 25°C. The X axis corresponds to the angular position (°) of the permanent magnet close to the sensor, and the Y axis is the amplified output voltage. Note that the output has a nonlinear (sinusoidal) shape, having sensitivity variations and electrical and mechanical offset errors (noticeable from the vertical and horizontal shift of the curves, respectively). The sensor interface (in ASIC form in this case) is required to compensate for all these errors and for temperature variations as well as to linearize the output signal, thus extracting the information about the angle of the magnet. Different choices for performing these tasks are described in the following.

14.4.3   SENSOR BIASING AND TEMPERATURE COMPENSATION

Depending on the cost and accuracy requirements of the contactless potentiometer, different biasing and temperature compensation techniques can be adequate. Some of them are described in this section.

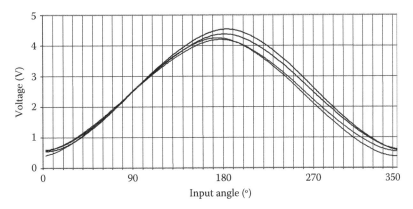

FIGURE 14.4   Amplified response of four GMR bridge sensors at 25°C.

14.4.3.1   Analog Temperature Compensation

Analog temperature compensation techniques are usually employed in analog sensor readout electronics and usually feature simplicity and low to moderate accuracy. For the GMR bridge, note from Equation 14.2 that if $\Delta R \ll R_0$, the two opposite temperature dependencies contribute in the same direction to a sensitivity reduction. The solution to this problem consists basically in biasing the bridge in such a way that the current across it increases with temperature, to compensate for this sensitivity reduction. This can be done either with a temperature-dependent current source [21] as in Figure 14.5a or by voltage biasing the bridge through a negative temperature coefficient (NTC) thermistor, as shown in Figure 14.5b. In both cases, the temperature coefficients of the current source and the NTC need to be precisely adjusted to produce the desired temperature compensation.

A third solution, similar to that of the NTC, consists in biasing the bridge through a negative resistance, implemented, for example, by a negative impedance converter (NIC), in series with the bridge, as shown in Figure 14.5c [20]. If resistance R_{NIC} is negative and its value is larger than R_0, an increase, due to temperature, in the bridge resistance results in the same reduction in the absolute value of the overall resistance $R_0 + R_{NIC}$. Therefore, the bridge current increases, compensating this way the sensitivity reduction. Thus, if R_{NIC} is designed such that

$$R_{NIC} = -R_0 \frac{\alpha_0 + \alpha_\Delta}{\beta + \alpha_\Delta} \tag{14.3}$$

where β is the temperature coefficient of the NIC resistance, then the bridge output voltage becomes independent of the temperature:

$$V_{out} \approx \left(V_{DD} - V_{BIAS}\right) \frac{\Delta R}{\Delta R + 2R_0 \left(1 - \left((\alpha_0 + \alpha_\Delta)/(\beta + \alpha_\Delta)\right)\right)} \tag{14.4}$$

A practical implementation of the NIC is shown in Figure 14.6, where $R_{NIC} = -R_A R_C/R_B$. Adjusting the resistance R_A, the value in Equation 14.3 can be achieved.

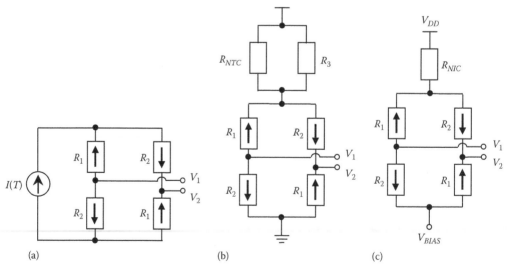

FIGURE 14.5 Temperature compensation techniques (a) using temperature-dependent current biasing, (b) using NTC, and (c) using NIC.

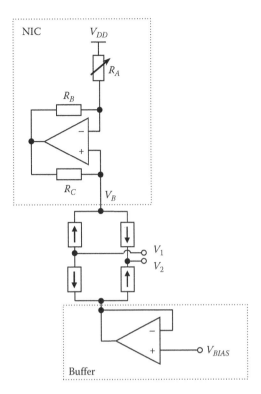

FIGURE 14.6   Detailed diagram of temperature compensation using NIC.

14.4.3.2 Digital Temperature Compensation

Another alternative is using the fact that when the GMR sensor bridge is biased with a temperature-insensitive DC current, the bridge voltage is $R_B = R_0(1+\alpha_0 T)$ so the sensor can also be used for temperature sensing, and the sensed temperature can be used for temperature compensation in the signal conditioning electronics of the sensor readout circuit. This procedure does not require a separate temperature sensor and obviously guarantees that the sensed temperature is that of the GMR sensor. This technique is particularly useful when the compensation is made in the digital domain and the analog to digital (A/D) converter can multiplex at the input the GMR output voltage and the temperature voltage, so that a single A/D converter can be used for both signals.

14.4.4 Amplification and Electrical Offset Compensation

Despite the relatively high sensitivity of the GMR sensor bridge, the differential output voltage is typically of a few mV and requires amplification. The amplifier must have programmable gain to compensate during calibration for sensor-to-sensor sensitivity variations. A possible programmable gain amplifier (PGA) with continuous gain programming is shown in Figure 14.7a. Its output voltage is

$$V_{out} = -\frac{R_{gain}}{R_6}\left(1+\frac{2R_2}{R_1}\right)\frac{R_4}{R_3}\left(V_{in+}-V_{in-}\right)+\left[V_{DC}\left(1+\frac{R_{gain}}{R_6}\right)-\frac{R_{gain}}{R_6}\frac{V_{DD}}{2}\right] \qquad (14.5)$$

Gain can be adjusted by the tunable resistor R_{gain}, and the offset is compensated by adjusting V_{DC}, thus setting the desired DC output. Resistance R_6 should have the same temperature coefficient as R_{gain} to avoid temperature dependence of the PGA gain.

Tunable resistors are usually implemented by discrete off-chip resistors trimmed during calibration, increasing the cost of the system. A better choice to minimize the number of external components is using a PGA with discrete gain programming based on a digital word stored in an internal memory during calibration. Figure 14.7b shows a possible configuration, where gain is set by 4 bits, allowing 16 gain values. One bit allows choosing between two gains in the first stage (8 or 16), and three bits set the gain of the second stage, which can be 52/23, 18/7, 11/4, 56/19, 19/6, 58/17, 59/16, or 4. Switches in the PGA do not contribute any gain error or thermal noise as no current flows through them. The DC output voltage is set by a D/A converter whose input is a digital word also stored in memory during calibration and that allows compensation of the electrical offset.

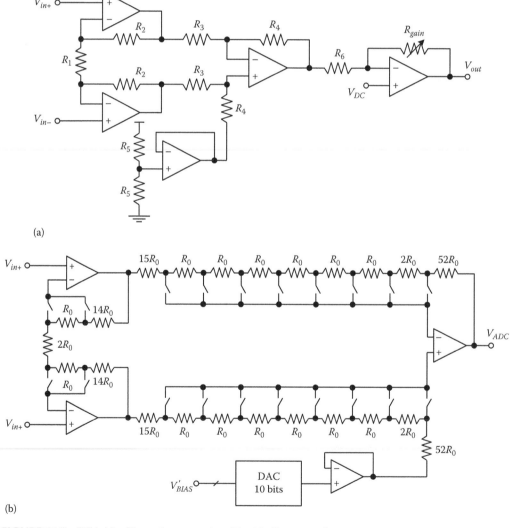

(a)

(b)

FIGURE 14.7 PGA (a) with continuous tuning (b) with discrete tuning.

14.4.5 Sensor Signal Linearization

Note from Figure 14.4 that the GMR sensor bridge output has a sinusoidal dependence with the angle of the rotary shaft. Thus, a linearization technique is required to make the output of the contactless potentiometer linearly dependent on the angle. Such linearization can be made in the analog domain, in the digital domain, or when the signal is converted from A/D domains. These different approaches are addressed in the following.

14.4.5.1 Analog Linearization

A simple approach for linearization of sensors is using analog circuits with active and/or passive components. The main drawbacks of these methods are usually their sensitivity to environmental conditions (mainly temperature) and their lack of flexibility when a different kind of sensor is employed. A simple analog linearization approach readily found in commercial interfaces for bridge sensors is based on the variation of the bridge bias voltage or current according to the output signal using feedback loops [22,23]. For many silicon sensors, this kind of nonlinearity correction may reduce sensor nonlinearity by an order of magnitude. The control of the sensor bias voltage or current is also commonly used for first-order temperature compensation, when an adequate variation of the bias value with temperature is implemented in such a way that thermal variation of the sensor output is compensated [20].

Another alternative is implementing an analog circuit whose input–output transfer characteristic is exactly matched to the inverse of the sensor characteristic. This idealistic situation is very difficult to achieve in practice. A more conventional approach is to implement a piecewise linear (PWL) approximation to the inverse of the sensor characteristic, leading to a much simpler circuit implementation. An adequate choice of the number of segments, the slopes of such segments, and the corner voltages allows achievement of a given accuracy. A trade-off between accuracy and circuit complexity exists as both factors increase with the number of segments. For instance, Figure 14.8 shows a three-segment PWL characteristic. This function is adequate in low-accuracy applications where the sensor characteristic is symmetrical and shows a linear behavior for low signal values, which becomes nonlinear when signal values are higher. This situation is quite common in practice and applies to the sinusoidal nonlinearity of the GMR bridge. In the middle range between corner

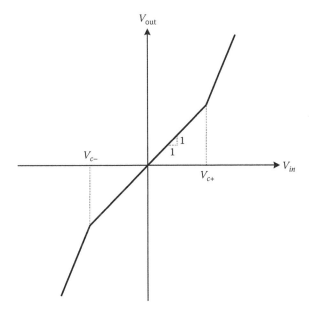

FIGURE 14.8 Three-segment linearization.

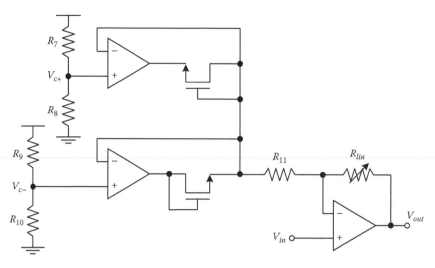

FIGURE 14.9 Implementation of the transfer function of Figure 14.8.

voltages V_{c-} and V_{c+}, the analog linearization block is transparent. Outside this range, the slope of the outer segments is adjusted to minimize the RMS error in the linearized output. In the particular case of the PWL function shown, the loss of sensitivity of the GMR bridge for extreme values is compensated by the slope greater than 1 in the outer segments. A simple circuit that implements the transfer function of Figure 14.8 is the based circuit in Figure 14.9. Each superdiode is composed by a diode-connected MOS transistor in the negative feedback loop of an op-amp. The name is due to the fact that these circuits behave like ideal rectifiers with a 0 V cut-in voltage. Each superdiode is biased by a DC voltage (V_{c+} or V_{c-}) obtained using a resistive divider (R_7–R_8 and R_9–R_{10}). Such DC voltage sets a break voltage in the PWL function implemented. When $V_{c-} < V_{in} < V_{c+}$, the superdiodes are in cutoff and the circuit acts as a simple voltage follower, that is, $V_{out} = V_{in}$. When $V_{in} < V_{c-}$, the common output node of the superdiodes is set to V_{c-} and the circuit corresponds to a noninverting amplifier with

$$V_{out} = -V_{c-}\left(\frac{R_{lin}}{R_{11}}\right) + V_{in}\left(1 + \frac{R_{lin}}{R_{11}}\right) \tag{14.6}$$

Likewise when $V_{in} > V_{c+}$, then

$$V_{out} = -V_{c+}\left(\frac{R_{lin}}{R_{11}}\right) + V_{in}\left(1 + \frac{R_{lin}}{R_{11}}\right) \tag{14.7}$$

Thus, a three-segment PWL transfer characteristic is obtained. Gain in the outer segments is independently set by modifying the slope of the linearization circuit by the external trimmable resistor R_{lin}. Resistor R_{11} is also external and having the same temperature coefficient as R_{lin} to avoid thermal variations of the linearization circuit.

14.4.5.2 Digital Linearization

Digital linearization techniques are currently the most employed ones in almost all sensor signal conditioning approaches, particularly when high performance is demanded. Such techniques achieve as much accuracy as required by the designer, at the expense of more circuit complexity and/or more processing time. Another typical benefit is programmability of the linearization circuit or algorithm, which eases the implementation of general-purpose sensor interfaces able to process signals from different kinds of sensors. These universal interfaces are becoming very common in industry due to their wider market.

The most employed technique is based on the storage of a *lookup table* (LUT) in a read-only memory (ROM) that contains in each entry (row) a digital input value and its corresponding linearized digital output [11,24]. The table is typically acquired by direct measurement of the sensor. This simple approach is completely general, that is, it can linearize any kind of nonlinear sensor dependence, be it monotonic or not. If a programmable memory is employed, then the circuit can be reprogrammed to linearize other responses. However, when high accuracy is required, the LUTs employed used to have many entries, and the digital words stored have many bits. In this case, the silicon area required for the memory is very significant. Alternatively fewer data points can be interpolated (e.g., using PWL, piecewise polynomial, or spline interpolation) [25], but then intensive digital processing is necessary. This method is a trade-off between processing time and silicon area, a situation frequently encountered in digital linearization techniques.

A related approach consists in the implementation of the LUT by means of combinational logic. The input word is applied to the combinational logic, and the linearized word appears at the output. This solution does not require the utilization of a physical memory, so it is very convenient in low-cost ICs, where the need for an internal ROM may significantly increase the price. This approach is particularly effective when the table size is not very large, because in this case, the silicon area occupied by the combinational logic tends to be much lower than the size required for a ROM memory. Again a trade-off between silicon area and processing time appears, as gate count in the combinational logic can be decreased if more logic gate stages are allowed, which increases delay. However, the combinational circuit cannot be reconfigured, so it cannot be rearranged to linearize other types of nonlinear functions. Both in the LUT and in this approach, significant area savings can be achieved with minor extra processing if the table or combinational logic provides the difference between the output word and the corresponding input word. Such procedure corresponds to the storage of the difference between the nonlinear response and a straight line instead of the nonlinear response itself. In the GMR sensor bridge employed, it would correspond to the difference between the arcsin(x) function (in the required angular range) and a linear function. This method is specially useful for weak nonlinearities.

A popular third technique consists in the digital storage of just the coefficients of a polynomial approximation to the nonlinear function [26]. For instance, assuming that a third-order polynomial is a good approximation for the nonlinear function in the range of interest, that is,

$$V_{out}\left(V_{in}\right) \cong a_0 V_{in} + a_1 V_{in}^2 + a_2 V_{in}^3 \tag{14.8}$$

only coefficients a_0, a_1, and a_2 need to be stored. The values of these coefficients are obtained during calibration. When an input word is received, it is processed by arithmetic operations (additions and multiplications) to achieve the desired value. In the typical case that such arithmetic processing is carried out not by dedicated hardware but using the resources of the main processing unit, significant area savings can be achieved. Once more, it represents a trade-off between silicon area consumption and processing time.

14.4.5.3 Mixed-Signal Linearization

This approach is particularly suited to applications where the sensor signal has to be converted to digital form and where the signal processing overhead of linearization in terms of silicon area, processing time, and power consumption needs to be minimized. This is the case, for instance, in low-cost integrated sensor interfaces, where reasonable performances must be obtained at the minimum silicon cost. The common feature of these techniques is that linearization is carried out just at the same time that of A/D conversion and using the same hardware [27–29]. The physical block that performs both these tasks is a nonlinear A/D converter, whose nonlinear A/D conversion characteristic is ideally matched to the inverse of the sensor characteristic. Such a nonlinear A/D conversion gets the best of each bit of resolution or, stated differently, requires the minimum number of bits for achieving a given resolution.

Several nonlinear A/D conversion techniques have been proposed. A well-known method is based on the ratiometric property of most A/D converters [27]. An external ratiometric reference voltage that is made dependent on the input voltage (typically by a simple resistive divider) is employed to achieve the required nonlinearity in the A/D conversion. The digital output voltage corresponds to the ratio of the input voltage to this input-dependent reference voltage. The resulting analog to digital converter (ADC) can be regarded as an analog divider with digital output. Although this technique is simple, the correction of sensor nonlinearity is modest, much lower than that achieved by digital linearization techniques. The reason for the limited accuracy is that it is not possible to exactly implement the ideal A/D conversion characteristic, that is, the inverse of the sensor characteristic. An alternative approach that achieves a better accuracy is the implementation of A/D converters whose conversion characteristic is a PWL approximation to the inverse of the sensor characteristic [30,31]. The circuit can be equivalently seen as an analog linearization circuit with PWL transfer characteristic followed by a linear ADC. If the number and size of the segments in such approximation are properly chosen (i.e., a low RMS error with regard to the ideal A/D characteristic is achieved), high accuracy can be obtained by a relatively simple circuit.

Figure 14.10 shows a PWL characteristic suitable for linearizing the GMR bridge nonlinearity. N break voltages, nonregularly distributed into the input range of the converter, define $N + 1$ conversion intervals of different widths and are chosen so that the resulting characteristic compensates a certain nonlinear dependence of the input signal. In the particular case shown in Figure 14.10, $N = 15$ break voltages have been chosen so that they define the best-fit 16-segment piecewise approximation to the function $\arccos(x)$ in order to linearize the amplified input of the GMR sensor bridge that follows a $\cos(x)$ characteristic. The number of segments and position of the corner voltages was chosen to achieve a maximum fitting error with the ideal $\arccos(x)$ function of ±0.2%.

Figure 14.11 shows the diagram of an ADC that implements the characteristic of Figure 14.10. It is a 4-bit flash stage employed in two conversion steps. In the first conversion step, a PWL ADC conversion takes place, yielding the four most significant bits. Then a second linear ADC conversion step is performed, which provides the four less significant bits.

In the first conversion stage, after the *start-of-conversion* (SOC) signal is received by the controller, the analog input is sampled, and the flash converter stores in Register A the four most significant bits of the output word corresponding to the sampled value. The comparator inputs in this case are

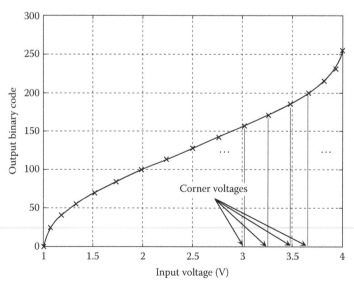

FIGURE 14.10   PWL ADC characteristic.

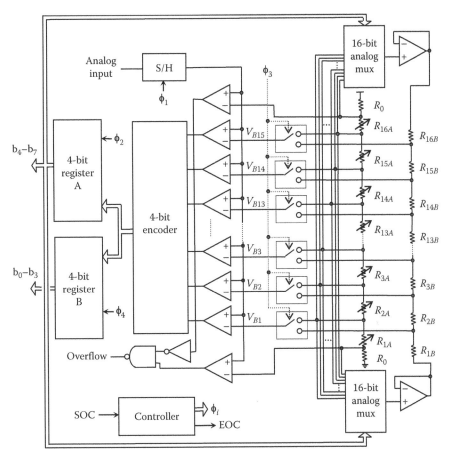

FIGURE 14.11   PWL ADC converter.

the corner voltages V_{Bi} generated by the first resistor string (R_{1A}–R_{16A}), which are nonlinearly distributed into the input range of the converter as shown in Figure 14.10. The resistors of this string are marked as tunable in Figure 14.11. The ith corner voltage is given by

$$V_{Bi} = \frac{R_0 + \sum_{n=1}^{i} R_{nA}}{2R_0 + \sum_{n=1}^{16} R_{nA}} V_{DD} \tag{14.9}$$

where resistances R_0 have been included for setting an input range from V_{SS} + 1 V to V_{DD} − 1 V. Hence, a 4-bit nonlinear conversion is achieved, where the resulting four bits represent the coded value of the nonuniform interval in Figure 14.10 where the sampled input falls into. Using this fact, in the second conversion stage, two 16-to-1 multiplexers controlled by the four bits stored in Register A allow to bias a second string of identical resistances R_{1B}–R_{16B} by selecting the corner voltages that limit the nonuniform interval V_{Bi+1} − V_{Bi} coded by these four bits. Since resistances R_{1B}–R_{16B} are identical, the new corner voltages V'_{Bi} will be given by

$$V'_{Bi} = V_{Bi} + \frac{\sum_{n=1}^{i} R_{nB}}{\sum_{n=1}^{16} R_{nB}} \left(V_{Bi+1} - V_{Bi} \right) = V_{Bi} + \frac{i}{16} \left(V_{Bi+1} - V_{Bi} \right) \tag{14.10}$$

thus being uniformly distributed into the selected conversion interval $V_{Bi+1} - V_{Bi}$. Such voltages are applied to the comparator inputs by switching the 2-to-1 multiplexers, and a linear 4-bit A/D conversion is produced into the selected interval. The resulting four bits are stored in Register B, corresponding to the less significant bits of the output code word. Once data are stable in this register, the controller generates an *end-of-conversion* (EOC) signal. The output is available in Registers A and B at this time. Two additional comparators are included (the upper and lower ones in Figure 14.11) just to ensure that the input signal is in the range $V_{SS} + 1$ V to $V_{DD} - 1$ V. An *overflow* signal is otherwise activated. The values of the resistors R_{1A}–R_{16A} are chosen so that the corner voltages generated by them follow a cosine-like distribution law in order to compensate for the arcos(x) nonlinearity.

14.4.6 MEASUREMENT RESULTS AND PERFORMANCE COMPARISON

Three different contactless potentiometers were developed by the authors using the GMR sensor bridge and the techniques described. They follow the approach in Figure 14.1 and differ in the CMOS ASIC employed for biasing the GMR sensor and processing the sensor signal.

The diagram of the first ASIC is shown in Figure 14.12a. It is a fully analog solution that employs the technique of Figure 14.6 for biasing the GMR bridge and for analog temperature compensation. Programmable amplification and offset compensation are performed by the circuit of Figure 14.7a, and linearization is carried out by the circuit of Figure 14.9. The buffered output is available at the output pin. External trimmable resistors allow adjustment of the temperature compensation circuit, gain, and slope of the linearization circuit. The circuit was designed to measure angles in the range [35°, 145°]. The supply voltage employed was 5 V. Figure 14.13 shows the measured angular error of the potentiometer in the full input range and for temperature ranging −40°C to 120°C. Note that despite the simplicity of the ASIC, worst-case absolute errors are less than 2°, which is an acceptable accuracy for many applications. The main source of error is the reduced number of segments in the PWL function of Figure 14.9.

The diagram of the second ASIC is shown in Figure 14.12b. It is a mixed-signal ASIC that employs the same circuits for GMR bias, temperature and offset compensation, and amplification. However, linearization is carried out by the PWL ADC of Figure 14.11. Mechanical offset due to misalignment of the sensor and the magnet by assembly tolerances is determined in calibration and digitally subtracted after A/D conversion. Finally, a PWM modulator delivers the output to an external pull-up resistor. Figure 14.14 shows the measured PWM output duty cycle versus the input angle. The ideal response is shown in the solid line. The circles correspond to the measured values. The circuit is able to process an input angular range of 160°, leading to maximum errors of about 1% of the output duty cycle. These maximum errors are produced at the extreme values of the angular range.

The diagram of the third ASIC is shown in Figure 14.12c. It is a fully configurable mixed-signal ASIC. Sensor biasing and temperature compensation are carried out as described in Section 14.4.3.2, using a temperature-insensitive DC bias current generated on chip and sensing the temperature-dependent bridge voltage at pin SENS0, which is subsequently digitized and employed for digital temperature compensation. An on-chip voltage regulator allows using nonregulated supply voltages coming, for example, from automotive batteries, which are applied to input VB. Alternatively, a 5 V external voltage can be applied at input VDD in consumer electronics. The circuit also includes an internal EPROM for storing programming and calibration parameters. An external resistance R_{ref} is used to set the internal bias currents and the oscillation frequency of the internally generated clock. The PGA employed is that of Figure 14.7b, set by four EPROM bits. The digital part performs most of the signal conditioning task like offset compensation, fine gain and temperature compensation, linearization, output range setting, PWM modulation, PWM duty cycle limitation, and compensation of mechanical misalignment. Linearization is carried out by a combinational circuit, as described in Section 14.4.5.2. The output is available both in analog and PWM format. Figure 14.15 shows the measured PWM output duty cycle obtained after calibration of the chip, as well as the ideal output. The output PWM duty cycle shows maximum errors within ±0.5° in an angular range of more than 100°. The PWM duty cycle limits were set to 5% and 95%.

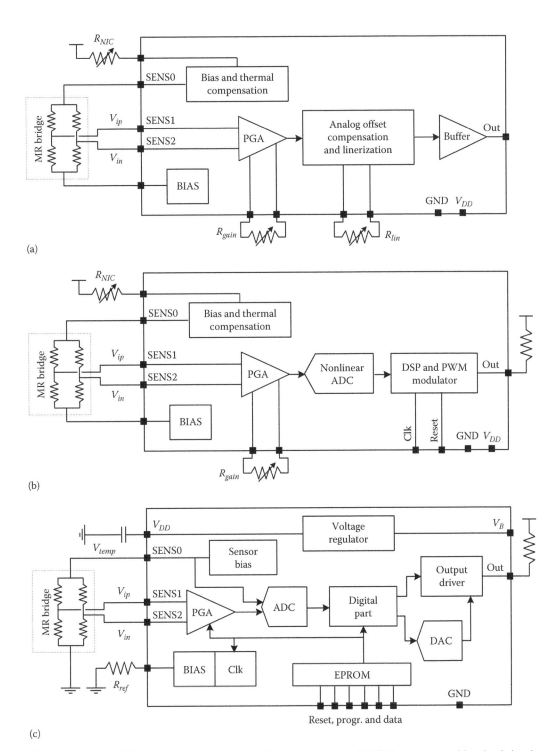

(a)

(b)

(c)

FIGURE 14.12 ASICs implemented (a) analog, (b) mixed-signal, and (c) fully programmable mixed signal.

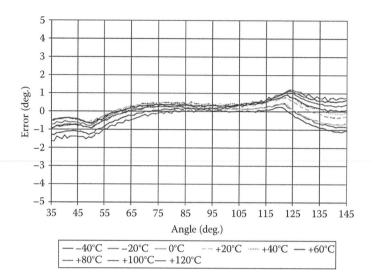

FIGURE 14.13   Measured angular error versus input angle at different temperatures, ASIC of Figure 14.12a.

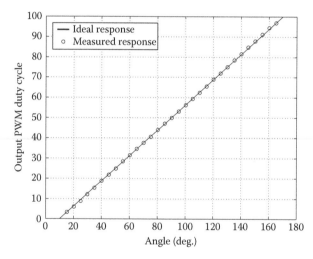

FIGURE 14.14   Measured PWM output duty cycle versus input angle, ASIC of Figure 14.12b.

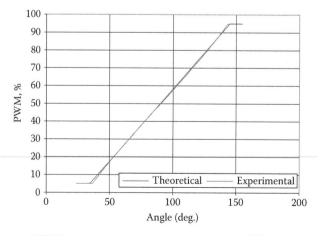

FIGURE 14.15   Measured PWM output duty cycle versus input angle, ASIC of Figure 14.12c.

TABLE 14.2

Performance Comparison of the GMR-Based Contactless Potentiometers Fabricated

	ASIC		
Parameter	Figure 14.12a	Figure 14.12b	Figure 14.12c
Output	Analog	Digital/PWM	Digital/PWM Analog
Sensor bias	Voltage	Voltage	Current
Angular range	35°–145°	10°–170°	10°–170°
Maximum error	±2°	±1.6°	±0.5°, 110° range ±0.9°, 160° range
Silicon area	2 mm^2	6 mm^2	6.9 mm^2

14.5 CONCLUSIONS

An overview of different application areas that benefit from the increased reliability and lifetime of contactless potentiometers has been provided. They are particularly suitable for automotive power-train and chassis applications, as well as in industrial applications such as process control and robot-ics. Their use in consumer applications is increasing, due to the reduction in cost and the inception of new domotic systems.

The most relevant sensing techniques employed for contactless potentiometers have been out-lined. Among them, the GMR sensor bridge has been covered in detail, due to its remarkable advantages for this application: tolerance to mechanical deviations due to assembly errors or aging, relative insensitivity to the strength of the magnetic field generated, and high sensitiv-ity, among others. Different biasing and signal processing techniques for this sensor have been presented, and three contactless potentiometer solutions based on these techniques have been discussed. Table 14.2 shows the main performance features of the three solutions. The first one, using the analog ASIC of Figure 14.12a, is suitable when cost is the primary concern; the output is required in analog form; and moderate errors (about 2°) in a limited range (110°) can be tolerated. The third solution, based on the ASIC of Figure 14.12c, is the best one in terms of performance (errors <1° in a 160° range), but the silicon cost is the highest one among the three options. The second solution using the ASIC of Figure 14.12b is an intermediate solution in terms of cost and performance. Table 14.2 illustrates that the GMR bridge is a good candidate to implement con-tactless potentiometers and that different cost-performance trade-offs are available depending on the application at hand [32].

REFERENCES

1. W.J. Fleming, Overview of automotive sensors, *IEEE Sens. J.*, 1(4), 296–308, December 2001.
2. North American automotive sensor market, Frost & Sullivan, Mountain View, CA, 1999.
3. H. Schewe and W. Schelter, Industrial applications of magnetoresistive sensors, *Sens. Actuators A*, 59, 165–167, 1997.
4. M.A. Zamora-Izquierdo, J. Santa, and A.F. Gomez-Skarmeta, An integral and networked home automa-tion solution for indoor ambient intelligence, *IEEE Pervasive Comput.*, 9(4), 66–77, October 2010.
5. A. Madni and R. Wells, An advanced steering wheel sensor, *Sensors Mag.*, 17, 28–40, February 2000.
6. P.E. Stephens and G.G. Davies, New developments in optical shaft-angle encoder design, *Marconi Rev.*, 46(228), 26–42, 1983.
7. X. Li, G.C.M. Meijer, G.W. de Jong, and J.W. Spronck, An accurate low-cost capacitive absolute angular-position sensor with a full-circle range, *IEEE Trans. Instrum. Meas.*, 45(2), 516–520, April 1996.
8. G. Li and J. Shi, Angle-measuring device with an absolute-type disk capacitive sensor, U.S. Patent 8,093,915 (2007).

9. A. Pawlak, J. Adams, and T. Shirai, Novel variable reluctance sensors, in *SAE Int. Congr. Expo.*, Detroit, MI, February 25, 1991, Paper 910 902.

10. M. Metz, A. Häberli, M. Schneider, R. Steiner, C. Maier, and H. Baltes, Contactless angle measurement using four hall devices on single chip, in *Proc. Transducers '97*, June 16–19, Chicago, IL, 1997, pp. 385–388.

11. A. Häberli, M. Schneider, P. Malcovati, R. Castagnetti, F. Maloberti, and H. Baltes, Two-dimensional magnetic microsensor with on-chip signal processing for contactless angle measurement, *IEEE J. Solid-State Circ.*, 31(12), 1902–1907, 1996.

12. T. Kaulberg and Boganson, G., A silicon potentiometer for hearing aids, *Analog Integr. Circ. Signal Process.*, 9(1), 31–38, January 1996.

13. D. Partin, T. Schroeder, J. Heremans, B. Lequesne, and C. Thrush, Indium antimonide magnetoresistors for automotive applications, in *Proc., Vehicle Displays Microsens. '99*, Ann Arbor, MI, September 22–23, 1999, pp. 183–188.

14. J. Lenz and A.S. Edelstein, Magnetic sensors and their applications, *IEEE Sensors J.*, 6(3), 631–649, June 2006.

15. G. Binasch, P. Grünberg, F. Saurenbach, and W. Zinn, Enhanced magnetoresistance in layered magnetic structures with antiferromagnetic interlayer exchange, *Phys. Rev. B*, 39, 4828–4830, 1989.

16. J.Q. Xiao, J.S. Jiang, and C.L. Chien, Giant magnetoresistance in non-multilayer magnetic systems, *Phys. Rev. Lett.*, 68, 3749–3752, 1992.

17. T.L. Hylton, Limitations of magnetoresistive sensors based on the giant magnetoresistive effect in granular magnetic compounds, *Appl. Phys. Lett.*, 62, 2431–2433, 1993.

18. H.A.M. van den Berg, W. Clemens, G. Gieres, G. Rupp, M. Vieth, J. Wecker, and S. Zoll, GMR angle detector with an artificial antiferro-magnetic subsystem (AAF), *J. Magn. Magn. Mater.*, 165, 524–528, 1997.

19. H.A.M. van den Berg, W. Clemens, G. Gieres, G. Rupp, W. Schelter, and M. Vieth, GMR sensor scheme with artificial antiferromagnetic subsystem, *IEEE Trans. Magn.*, 32, 4624–4626, 1996.

20. A. J. Lopez-Martin, M. Zuza, and A. Carlosena, Analysis of a NIC as a temperature compensator for bridge sensors, *IEEE Trans. Instrum. Meas.*, 52(4), 1068–1072, August 2003.

21. A. Sprotte, R. Buckhorst, W. Brockherde, B. Hostika, and D. Bosch, CMOS magnetic-field sensor system, *IEEE J. Solid-State Circ.*, 29(8), 1002–1005, August 1994.

22. M. Ivanov, Bridge sensor linearization circuit and method, U.S. Patent 6,198,296 (2001).

23. J. Dimeff., Circuit for linearization of transducer, U.S. Patent 4,202,218 (1985).

24. H. J. Ottesen, and G.J. Smith, Method and system for adaptive digital linearization of an output signal from a magnetoresistive head, U.S. Patent 5,283,521 (1994).

25. P. Malcovati, C. Azeredo, P. O'Leary, F. Maloberti, and H. Baltes, Smart sensor interface with A/D conversion and programmable calibration, *IEEE J. Solid-State Circ.*, 29(8), 963–966, August 1994.

26. F. Tarig and T.I. Pattantyus, System and method for sensor response linearization, U.S. Patent 6,449,571 (2002).

27. G.E. Iglesias and E.A. Iglesias, Linearization of transducer signals using an analog-to-digital converter, *IEEE Trans. Instrum. Meas.*, 37(1), 53–57, March 1988.

28. D.H. Sheingold, ed., *Analog-Digital Conversion Handbook*, Analog Devices Inc., Norwood, MA, 1986.

29. L. Breniuc and A. Salceanu, Nonlinear analog-to-digital converters, in *Third Workshop on ADC Modelling and Testing*, Naples, Italy, September 1998, pp. 461–465.

30. G. Bucci, M. Faccio, and C. Landi, The implementation of a smart sensor based on a piece-linear A/D conversion, in *Proc. IEEE Instrum. Meas. Technol. Conf.*, Ottawa, Ontario, Canada, May 1997, pp. 1173–1177.

31. G. Bucci, M. Faccio, and C. Landi, New ADC with piecewise linear characteristic: Case study— Implementation of a smart humidity sensor, *IEEE Trans. Instrum. Meas.*, 49(6), 1154–1166, December 2000.

32. A. J. Lopez-Martin and A. Carlosena, Performance tradeoffs of three novel GMR contactless angle detectors, *IEEE Sensors J.*, 9(3), 191–198, March 2009.

15 Capacitive Sensing for Safety Applications

Thomas Schlegl and Hubert Zangl

CONTENTS

15.1 INTRODUCTION: MOTIVATION, AIMS, AND STATE OF THE ART

A large number of people are injured every year because they or other objects reside in areas where they just should not be. This high number could be reduced by safety devices for detecting objects (i.e., proximity measurement) and classifying these objects (e.g., classify an object as a human). Although there are a lot of sensing technologies that might be applied, only a few can cope with the requirements that exist for most applications. Limitations with respect to spatial dimensions, weight, measurement speed, and power consumption are examples for such requirements.

Research on vision systems targeting safety applications is quite active. However, vision systems (i.e., camera-based systems) do need massive signal processing for detection and classification of objects, which can make them slow and lead to high power consumption. Additionally, it can be challenging to place a camera, for example, the space constraints may only allow very small and thin optical systems (e.g., mountable on a robot grasper). Finally, vision-based systems usually require a free line of sight, which may not always be present.

Optical systems, for example, presented in [1], can be made very small and work within a detection range of 2–40 mm. Such sensor systems can work with a variety of materials, but the performance depends on the color and the surface of such objects. It is shown in [1] that it is difficult to detect transparent objects (e.g., objects made of glass) or reflective objects (e.g., an aluminum can). It is also very hard to distinguish between different approaching objects (i.e., object classification).

259

Magnetic field sensors such as giant magnetic resistor (GMR) sensors are small and fast enough to detect objects within a range of about 30 mm. As shown in [2,3], they can be used for ferromagnetic or conductive objects, respectively.

The resonant frequency of a cavity can also be used as a proximity sensor. This so-called seashell effect presented in [4] uses the changing resonant frequency of the cavity if an object is approaching. This resonant frequency is measured with a microphone, and thus, a distance estimation can be made up to approximately 6 mm. Again, no object classification is possible with this kind of sensor, but in combination with a visual sensor (e.g., Kinect depth sensor), it can improve robot grasping as shown in [5].

A technology that has the potential to do proximity sensing and—to some degree—object classification is capacitive sensing. Thus, the authors believe that capacitive sensing can be a part of the solution to the problem stated in the beginning of this section. In the following section, capacitive sensing is introduced, and some applications are shown where capacitive sensing is already used within a safety system. Also, the difficulties that arise when in this context are emphasized. State-of-the-art measurement circuitry concepts for capacitive sensing are presented and analyzed with respect to safety applications. A new approach founded on the concept of electrical capacitance tomography (ECT) is presented, and an evaluation measurement circuitry is presented.

15.2   CAPACITANCE MEASUREMENT

Capacitance measurement techniques have been known for a long time. One of the first capacitance sensors was the Theremin, which is an electronic music instrument presented in 1920. It is controlled without contact between the player and the instrument [6]. Thus, it can be seen as one of the first capacitive proximity sensors. Although capacitance sensing has been known for such a long time, the breakthrough did not come until the last two decades. With the usage in commercial applications such as touch screens [7,8], in mobile phones in the last decade, capacitance measurement hardware got very cheap and available in integrated circuits [9–11].

The following section aims to give

- An introduction to the physics behind capacitive sensing
- Some example applications where capacitive sensing is used for safety purposes
- An explanation of occurring parasitic effects in open environment measurements

The differences and difficulties when it comes to longer distance sensing (which can be mandatory for safety applications) compared to shorter distance sensing (e.g., used in touch screen applications) will also be explained.

15.2.1   Physics behind Capacitance Measurement

Capacitive sensors consist of at least two conductors (called electrodes) that are separated by a non-conducting material. The distant ground potential can also be seen as one of these two electrodes. An electric field occurs whenever the two electrodes are on different electrical potentials. Capacitive sensing is well described by the Maxwell equations. After transformations and simplifications (e.g., wavelength of sensing signal is much larger than the sensing electrodes), the partial differential equation

$$\nabla \cdot ((\sigma + j\omega\varepsilon)\nabla V) = 0 \tag{15.1}$$

can be obtained, where V denotes the electric scalar potential, σ denotes the conductivity, ω denotes the angular frequency, and ε denotes the dielectric permittivity. This equation possesses a unique solution when boundary conditions (e.g., potentials and perhaps surface current densities on electrodes) are known. More details can be found, for example, in [12,13] and the literature referenced there.

15.2.2 EXAMPLE APPLICATIONS

Capacitive sensing has been used for a long time for a lot of applications as shown in Table 15.1. The table aims to give an overview of possible applications using capacitive sensing. It also shows the properties arising with these applications and gives references for more information. In Table 15.2, the usability of measurement circuitries for each application is evaluated. The weights in Table 15.2 are subjectively chosen factors. Among the numerous applications of capacitive sensing, safety applications are a rather young field. The following examples should give an idea of the variety of applications where capacitive sensors can be used in safety devices. They all have in common that they aim to sense objects when they enter regions where they just should not be. Since other types of objects may be allowed to enter these regions, not only proximity detection but also a classification of objects is of interest. The possibility of having an object classification scheme makes capacitance measurement very interesting for many more applications.

1. *Car bumper*: In [26], a sensor fusion concept is presented that is based on capacitive and ultrasonic (US) techniques for proximity measurement in automotive applications (see Figure 15.1). Although US sensors are a well-accepted technology for distance sensing applications, they reveal drawbacks in the closest vicinity of a vehicle. Capacitive sensors used in this application are suited for proximity measurements of up to 0.3 m and may also provide information about the approaching object itself (e.g., a safety feature in terms of object classification). The measurement range of this fusion concept reaches up to 2 m whereby blind spots are avoided and means for object classification are provided.

2. *Chainsaw*: In [48], a capacitive sensor system is presented that switches off, for example, a chainsaw if an object (e.g., human hand) comes too close. The object of interest has to have a conductive material (e.g., dress with a garment with a wire cloth inserted), which can be connected to a radio-frequency generator. The generator is an 80 kHz Wien bridge oscillator, and the receiver detection unit, mounted on the chainsaw, has to measure this signal level. The 80 kHz signal level directly depends on the distance between the object and the chainsaw. In [49], it could be shown that it is possible to detect humans and animals with a capacitance measurement system mounted only on the chainsaw (no generator on the object of interest is necessary). Thus, the safety feature is not only limited to one object.

3. *Icing*: A different kind of application is shown in Figure 15.2. It shows a capacitive ice sensor for overhead power lines working with a capacitive energy harvesting system [50–52]. Although the sensor is especially used to detect the beginning of icing, it represents a safety device according to our definition because it detects an object (i.e., ice) in a region where it just should not be (e.g., overhead power line). Other sensor systems for icing (e.g., presented in [53,54]) are wired and thus limited to, for example, transformer stations [52].

4. *Cardiac disease detection*: A noncontact proximity sensor that can also be used to detect a cardiographic signal without contact is presented in [55]. It uses an oscillator circuitry and measures the changing displacement current if an object is in front of the sensor surface. It is shown that this object could also be the human heart and be used as a cheap and simple way for cardiac disease detection.

5. *Pretouch for robot grasping*: So-called pretouch sensors are especially useful in robotic applications to close the gap between vision and tactile sensors. Pretouch sensors are not only able to benefit manipulation but also add a safety feature if an object classification is possible (e.g., the robot grasper is not allowed to grasp if a human hand is in the way). In [56,57], a capacitive (also called electric field) pretouch sensor is presented, which is designed to be mounted into the fingers of a robot hand (BarrettHand in the presented case). The sensor is used to align the three fingers of the robot hand around the object to grasp. Thus, when grasping the object, all three fingers make contact with the object without displacing it.

TABLE 15.1

Overview of Capacitive Sensing Applications and Their Properties

	Demanded Resolution	Dynamic Range	Stray Capacitances	Capacitance Range	Susceptibility to EMC	Susceptibility to ESD	Encapsulation Possible	Electrode Topology	Computational Effort	Measurement Rate	Examples
Proximity/distance	Low/high	Medium	High	pF	High	High	No	Planar	Low	Low	[15–18]
Rotary switch	Low	Low	1	fF–pF	High	Medium	Yes	Planar	Low	Low	[17,19,20]
Touch pad	Low	Low	High	pF	High	High	No	Planar	Low	Low	[21–24]
Occupancy detection	Low	High	High	pF	High	High	No	Planar/nonpl.	Low	Low	
Parking aid	Medium	High	High	fF–pF	High	High	No	Planar	Medium	Low	[25,26]
Inclination	High	Low	1	fF	Low	Low	Yes	Nonplanar	High	Medium	[27–29]
Linear position	High	Low	1	fF	2	2	Yes	Planar/nonpl.	High	Medium	[30,31]
Angular position	High	Low	1	fF	2	2	Yes	Planar/nonpl.	High	Medium	[32–35]
Fill level	Medium	Low	High	fF	Low	Low	Yes	Planar	Medium	Low	[36–39]
Thickness	High	Medium	1	fF–pF	High	High	No	Planar/nonpl.	High	Low	[40,41]
Oil quality	Medium	Low	High	Low	Low	Low	Yes	Planar/nonpl.	Medium	Low	[42]
Smart textiles	Low	Low	High	fF–pF	High	High	No	Planar	Low	Low	[17]
Flow measurement	Medium	High	High	fF–pF	Low	Low	Yes	Planar	High	High	[43,44]
Capacitance tomography	High	High	Low	fF–pF	Low	Low	Yes	Planar	High	Medium	[45–47]

Source: Adapted from Zangl, H., Design paradigms for robust capacitive sensors, PhD dissertation, Graz University of Technology, Graz, Austria, May 2005.

Note: 1, High for planar, low for nonplanar topology; 2, low for encapsulated sensor.

TABLE 15.2
Usability of Sensor Circuitry in Selected Target Applications (5 = Best)

	Weight	Oscillators		High Z		Low Z		Bridge
		RC	SC	CF	DC	CA	CF	CF
Proximity switch	1	1	2	1	1	2	4	5
Rotary switch	1	3	5	3	2	5	5	5
Touch pad	1	1	2	1	1	3	4	5
Occupancy detection	1	1	2	1	1	3	4	4
Outdoor/parking aid	0.5	1	2	1	1	3	4	4
Inclination	0.8	5	5	4	2	5	5	5
Position, encapsulated	0.5	5	5	4	2	5	5	4
Position, not encapsulated	1	1	2	1	1	3	4	5
Tank fill level	1	3	4	4	2	4	5	5
Thickness	0.5	1	2	1	1	3	4	5
Oil quality	0.5	3	3	4	3	3	4	5
Smart textiles	0.3	1	2	1	1	3	4	5
Flow measurement	0.5	2	3	4	3	5	5	5
ECT	1	2	2	3	2	4	5	4
Low power	2	5	5	3	5	5	3	2
Total		32.3	41.1	30.5	26.9	48.4	53.2	54

Source: From Zangl, H., Design paradigms for robust capacitive sensors, PhD dissertation, Graz University of Technology, Graz, Austria, May 2005.

Notes: The weights are subjectively chosen factors, which are used to consider the importance of a certain property for the design decision. The total performance figure is the sum of the usability values multiplied by the corresponding weights. Higher values indicate a better average usability.

RC, resistor/capacitor; SC, switched capacitor; CF, carrier frequency; CA, charge amplifier; DC, direct current.

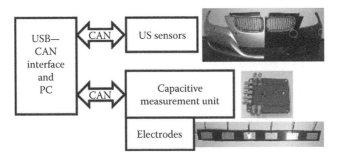

FIGURE 15.1   Measurement setup for a sensor fusion system in automotive applications adopted from [26]. It comprises US and capacitive sensors for proximity sensing and classification.

6. *Object ranging and material type identification*: Not only proximity sensing but also material type identification was done in [58]. It is shown that the complex permittivity ε of an object is a function of the angular frequency of the measurement signal ω, given by the following equation:

$$\varepsilon' = \varepsilon + i\frac{\sigma}{\omega} \tag{15.2}$$

where
   ε denotes the permittivity
   σ denotes the conductivity of the object

FIGURE 15.2 Photo of a capacitive ice sensor mounted on an overhead power line [51]. The energy harvester shell also comprises the measurement circuitry. The sensing electrodes are directly mounted on the power line.

Thus, it is shown that a material identification (i.e., identification of the complex permittivity ε) is possible with a varying measurement signal frequency. This kind of classification is necessary for safety applications since certain objects are allowed to be in areas where others are not.

7. *Protection of power-line contacts*: In [59], a protection system for construction workers to prevent electrocutions is presented. The worker has to wear the proposed electric field sensor, and if he approaches a live power circuit, the system gives an alarm. Since contact with overhead power lines was the most frequently occurring event in the construction industry in the United States from 2003 to 2006 [60], such kind of safety sensor systems is especially needed.

8. *ECT grasper*: A robot grasper is attached with capacitive and GMR sensors in [61] to reconstruct the region of interest (ROI) in an ECT manner (refer to Section 15.3.2). This application uses a sensor fusion approach that is specific to two types of materials. These are dielectric and ferromagnetic materials, which are commonly found in many industrial environments. Electric and magnetic fields are applied in the ROI and the distortion of these fields caused by objects is measured. Thus, a safety feature can be added: the grasper shown in Figure 15.3 only grasps for certain objects (e.g., dielectric objects) and does not grasp for, for example, a human hand in the ROI.

15.2.3 Parasitic Effects in the Open Environment

In order to assess advantages and disadvantages of possible circuitry for capacitive sensing, a model of the sensor front end is necessary. Figure 15.4 shows a model, which is an extension to the equivalent circuit used in [14,62]. It additionally considers an approaching object (if measuring in the open environment) and electromagnetic compatibility (EMC). The three main parasitic effects shown in Figure 15.4 are the following:

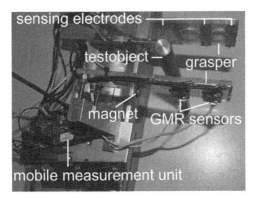

FIGURE 15.3 Robot grasper with GMR and capacitive sensing capability adopted from [61]. The test object (metallic rod) can be detected by the GMR sensors, dielectric objects with the capacitance measurements.

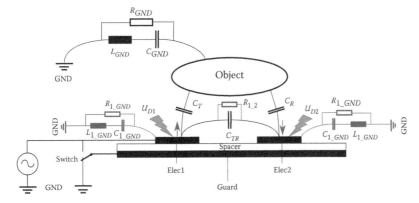

FIGURE 15.4 Sketch of a two-electrode capacitive sensor front end including several parasitic effects [83]. Arrows indicate the displacement current originating electrode 1 (Elec1) and entering electrode 2 (Elec2). U_{D1} and U_{D2} denote capacitive crosstalk from disturbers in the vicinity and ESD to Elec1 and Elec2. The main parasitic effects to ground are shown by the equivalent parallel circuits connected to the electrodes and the object. Depending on the measurement mode (refer to Section III) the guard electrode can be set ground or the excitation signal (i.e., active guarding).

- Parasitic connection to ground through the equivalent parallel circuits (R_{GND}, L_{GND}, C_{GND}, R_{1_GND}, L_{1_ND}, C_{1_GND} and R_{2_GND}, L_{2_GND}, C_{2_GND}) connected to the sensing electrodes 1 and 2 and the approaching object. Thus, only a part of the displacement current (indicated by red arrows) originating from electrode 1 is entering electrode 2 and is measured in the mutual capacitance mode (see Section 15.3).
- Capacitive crosstalk from disturbers and electrostatic discharge (ESD) to the sensing electrodes indicated by U_{D1} and U_{D2}. This is especially a problem in open environment measurements, and its influence can be reduced by, for example, methods shown in [63].
- Resistive path R_{1_2} parallel to the capacitance of interest C_{TR}.

The used measurement circuitry has to deal with these parasitic effects [64]. Table 15.3 tries to give an overview of how these parasitic effects influence the different measurement circuitries.

15.2.4 SHIELDING AND COUPLING

Two other effects that can be observed with capacitive sensing are the so-called coupling and shielding effects [14]. They occur for certain objects and depend on the properties of these approaching objects. Besides others, the capacitive connection of the objects to the distance ground is an

TABLE 15.3

Comparison of Different Capacitive Sensor Front-End Circuitry

Class	Oscillators		High Z		Low Z		Bridge
Circuit	RC	SC	CF	DC	CA	CF	CF
Guarding	Active	Passive	Active	Active	Passive	Passive	Passive
ADC required?	No	No	Yes	Yes	Yes	Yes	Yes
BP filtering	No	No	Difficult	Diff.	Possible	Possible	Possible
Complexity	Low	Low	High w. guarding	Low	Medium	High	High
Sens. to res. shunt	Yes	Minor	Yes	Yes	Medium	Minor	Minor
Extended wire length?	Minor	Minor	No	No	Minor	Yes	Yes
Long-time stability	Moderate	Moderate	Low to medium	Low to medium	Moderate	Low to medium	Medium/low
Short-time stability	Good	Good	Good	Good	Good	Good	Good
EMC emission	Low	SR limit.	SR limit.	Very low	SR limit.	SR limit.	SR limit.
EMC sensitivity	High	Medium	Low (Freq. shifting)	High	Medium	Low (Freq. shifting)	Low (Freq. shifting)
Spark discharge sens.	Low	Low	High	High	Low	Low	Low
Measurement rate	Low	Low	Medium	High	High	Medium	Medium
Matching	Medium	Good	Medium	Medium	Good	Good	Good
Power consumption	Low	Low	Moderate	Low	Low	Moderate	Medium
Suppression of:							
R_{1_GND}	+	+	+	+	+	+	+
C_{1_GND}	+	+	+	+	+	+	+
R_{1_2}	−	+	−	−	+	+	+
R_{2_GND}	+	+	−	−	+	+	+
C_{2_GND}	+	+	−	−	+	+	+
U_{D1}	−	+	+	+	+	+	+
U_{D2}	+	−	−	−	−	+	+
ESD 1	−	+	+	+	+	+	+
ESD 2	+	+	−	−	+	+	+

Source: Adapted from Zangl, H., Design paradigms for robust capacitive sensors, PhD dissertation, Graz University of Technology, Graz, Austria, May 2005.

Notes: The lower part of the table indicates whether a circuit is sensitive (−) or insensitive (+) to one of the parasitic effects shown in Figure 15.4.

RC, resistor/capacitor; SC, switched capacitor; CF, carrier frequency; CA, charge amplifier; DC, direct current; SR limit, slew rate limitation required.

important property. As shown in Figure 15.4, it mostly depends on the capacitances C_{GND}, C_T, C_{TR}, and C_R. If a capacitive sensor system is measuring in mutual capacitance mode (see Section 15.3) and an object approaches, the displacement current originating from electrode 1 can go to electrode 2 and to the distance ground *GND*. How much displacement current is entering electrode 2 depends on the relation between C_{GND} (which stays nearly constant for an approaching object) and the capacitance of the parallel circuit of C_T, C_{TR}, and C_R (which increases for an approaching object). A bigger portion of the displacement current goes from electrode 1 to C_T and C_{GND} to the distance ground for an object further away (since the capacitance of the parallel circuit C_{GND}, C_T, C_{TR} is rather small). Thus, at first, the measured capacitance decreases with an approaching object. This is called shielding mode. At a certain distance to the sensor surface, the capacitance of the parallel circuit C_{GND}, C_T,

C_{TR} gets a higher influence than C_T and C_{GND}, and more displacement current goes from electrode 1 to electrode 2 than to the distance ground. The measured capacitance increases and this is called the coupling mode. Because these effects strongly depend on the approaching object, it can also be used for classification of the approaching object as shown in [26]. Both effects can be observed in the presented measurements in Sections 15.4.2 and 15.4.3.

15.3   MEASUREMENT CIRCUITRIES AND MODES

There exist a huge variety of measurement circuitries for measuring electrical capacitances. A coarse classification presented in [12] can be as follows:

- Direct DC
- Oscillators
- Single ended
- High Z
- Low Z
- Bridge

Table 15.3 gives an overview of the circuitries most commonly used for proximity sensing (i.e., direct DC and single-ended measurement systems are not taken into account). This work focuses on the effects arising when capacitive sensing gets into the open environment (which most often is the case in safety applications) rather than on the different properties of the circuitries. The interested reader can refer to [12,47,65] and to the literature referenced there for more information on capacitance measurement circuitries.

It is also possible to distinguish the sensing system by the used measurement mode. The two different modes are often denoted as

- Mutual capacitance mode
- Self-capacitance mode

The first mode utilizes measurements of the capacitance between two electrodes by applying a voltage on one electrode and measuring, for example, the displacement current on the other electrode (i.e., low-Z circuitry). The second mode utilizes measurements of the displacement current originating from one electrode to the distance ground.

A difficulty associated with the self-capacitance mode is the fact that the sensitivity is quite high at the edges of the electrodes in particular when conductive objects reside in the vicinity, for example, as the carrier of the electrode. In this case, moisture and contamination may significantly affect the measurement, and no reliable proximity determination may be possible. A commonly used method to cope with this problem is active guarding where a guard is placed between the actual electrode and a metallic carrier. Thus, the sensitivity moves away from the edges of the electrodes. However, this also leads to a reduced sensitivity with respect to small objects in the vicinity of the electrodes. On the other side, the self-capacitance mode usually offers a higher signal-to-noise ratio (SNR) compared to the mutual capacitance mode and, in conjunction with active guarding, a high robustness. The mutual capacitance mode usually has a worse SNR but has the capability to detect objects in situations where the self-capacitance mode is blind. Thus, a measurement circuitry that combines both measurement modes is preferable for applications where objects of different sizes and permittivities in different distances to the sensor surface have to be measured (e.g., in safety applications).

In the following section, the different types of measurement hardware used in the example applications in Section 15.2.2 are presented. Measurement results are shown where applicable, and the different approaches for safety applications are compared.

15.3.1 MEASUREMENT SYSTEMS IN EXAMPLE APPLICATIONS

1. *Car bumper*: The measurement hardware used in this application was a commercially available capacitance-to-digital converter integrated in the Analog Devices IC AD7143 [66]. Figure 15.5 shows the estimated approaching line of the sensor fusion system (capacitance measurement system in combination with a US system) for an approaching human. Several objects were measured when approaching and the capacitance measurement traces were stored. To simulate real-world situations, the capacitance measurement trace of the approaching human was deleted. As can be seen from Figure 15.5, the estimated approaching line nearly matches the true approaching line. The algorithm is based on a Kalman filter with a maximum likelihood (ML) criterion that estimates the objects that are most similar to the one of the human (in the presented case, it was the object fence). Thus, the proximity sensor works as desired and also features a classification scheme that can be used to differ between different approaching object classes.

2. *Chainsaw*: This safety application uses a transmitter circuit mounted on the object of interest and a receiver circuitry mounted on the chainsaw as presented in Section 15.2.2. It could be shown in [48] that with a rectified mean value detector, the signal value can be measured through the capacitive connection between the transmitter and the receiver. Thus, a proximity switch could be realized. It switches off the chainsaw at a distance of about 100 mm (equals a threshold of 300 mV) that is early enough expecting a maximum blade speed of 2 m/s and a response time of 10 ms of the whole system.

3. *Icing*: The ice detection system presented in [51] uses an integrated capacitance-to-digital converter operating at a nominal frequency of 240 kHz. The measurement system works in the mutual capacitance mode with one transmitter and two receiver electrodes. It was shown that occurring icing on an overhead power line can be detected in laboratory measurements (e.g., climate room) as well as in a field test. There, the sensor system was mounted on a power line at a hilltop location in Austria. In both cases, early icing could be detected and distinguished from melting. This is especially important for the de-icing process. The safety system comprises a capacitance measurement system for object detection (i.e., ice detection) and object classification (i.e., distinguish between ice and water).

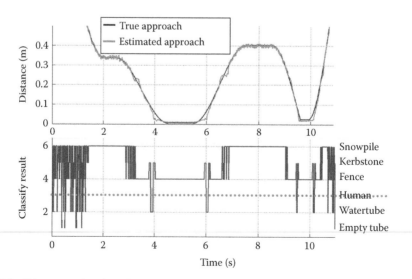

FIGURE 15.5 Distance estimation of an approaching and leaving human and selected object class based on a Kalman filter with an ML criterion adopted from [26]. The object class human was deleted from the stored measurements to simulate real-world situations and demonstrate the robustness of the used algorithm.

Further investigations will focus on a measurement system with more than three electrodes to reliably reconstruct the thickness of the ice on the power line.

4. *Cardiac disease detection*: The presented noncontact proximity sensor that can also be used to detect the motion of the human heart uses an oscillator circuit and one electrode [55]. The presented measurements show that not only proximity measurement is possible but also the measurement of the motion of the human heart. The sensor utilizes a measurement frequency of 21 MHz and a sampling time of 2 ms. The proposed minimum capacitance change is 7.5 fF. Because only one electrode is used, the sensor can be used to detect objects but not to classify them. If the sensor is used as a cardiac disease detection system, the measurement signals correspond to the basic wave of the cardiograph. However, the validity of the cardiac signal has to be proven in the clinic [55].

5. *Pretouch for robot grasping*: The so-called pretouch system presented in [56,57] is mounted in the fingertips of a three-finger robot hand (BarrettHand). Each fingertip consists of four electrodes (two transmitter and two receiver electrodes) that are used for short-range (<2 cm) and mid-range (<5 cm) sensing, respectively. Another electrode is positioned in the palm and is used as transmitter electrode. Using this palm transmitter and the fingertip receiver, long-range sensing (10–15 cm) is possible. The robot hand was able to pick up an object it was tuned for. Additionally, it was able to grasp for an object that was brought into the vicinity by a human. As soon as the human disengages the object, the robot hand moves to a certain position with the object. Little information about the measurement circuitry and speed is given in [57]. Although the sensor system showed promising performances for objects it was tuned for, it failed for objects that differed in size. However, the experiment with the human interaction showed the possibilities of such a capacitive sensing system for safety applications in robot applications.

6. *Object ranging and material type identification*: The proximity and classification sensor presented in [58] uses two electrodes in a mutual capacitance system. The used measurement hardware was presented in [16,67,68]. It uses a sinusoidal signal as transmitter (first electrode) and a charge amplifier as receiver (second electrode). The measurements are done at three frequencies to obtain a material classification. Four types of material were tested: human, concrete, wood, and metal. The promising results are shown in Table 15.4. The material classification result is used in the proximity determination algorithm resulting in a better distance estimation compared to the distance sensing without a classification.

TABLE 15.4
Material Classification Results (%)

Material under Test	Concrete	Metal	Wood	Human
Concrete	**100**	0	0	0
Painted mild steel	0	**99.7**	0	0.3
Aluminum	0	**99.3**	0	0.7
Thick mild steel	0	**100**	0	0
Wood	0	0	**88.3**	11.7
Thick wood	0	0	**97.3**	2.7
Human	0	0	0	**100**

Source: Kirchner, N. et al., *Sens. Actuators A: Phys.*, 148(1), 96, 2008.
Note: Bold values highlight the correct classification.

7. *Protection of power-line contacts*: A protection system for construction workers uses capacitance measurement hardware comprising a variable high-gain preamplifier, a narrow 60 Hz bandpass filter, a fixed gain post amplifier, an analog-to-digital converter (ADC) and a communication unit for the connection with a host computer [59]. The measurement results presented in [59] show that an approach toward a power line with both 120 and 9000 V can be detected starting at a distance of approximately 1 m. Thus, a proximity sensor for an energized power circuit for the safety of construction workers was presented in [59].

15.3.2 Different Approach: Electrical Capacitance Tomography

A related technology is ECT. It is used in industrial processes to obtain 2D cross-sectional image of the material (i.e., permittivity) distribution within pipes [69]. ECT is essentially an array of capacitive sensors with heavy signal processing to calculate an image of the ROI [46,70,71]. The calculation has to deal with a nonlinear and ill-posed inverse problem [72] with a higher number of unknowns (i.e., number of pixels) than independent measurements (i.e., number of capacitance measurements). Thus, the reconstruction method typically needs some kind of regularization or prior knowledge (e.g., Tikhonov regularization, total variation). The calculation or reconstruction methods for online reconstruction can typically be divided into two types [73]: noniterative algorithms (such as offline iteration/online reconstruction [74], optimal approximation [75], and singular value decomposition) and iterative algorithms (such as Gauss–Newton methods [76] also in combination with statistical methods like particle filter [77] or Kalman filter [78,79]). Other approaches presented in [80,81] use neural networks for solving this inverse problem.

Figure 15.6 shows the idea of using an ECT approach for capacitive safety applications. The enclosed structure of the capacitive array is opened and attached to the surface of interest. The measurements obtained by the measurement circuitry are processed in an ECT manner. Compared to ECT, the environment for such a safety device is very uncertain in most cases. Additional parasitic effects (shown in Figure 15.4) can have a huge influence on the measurement results and also the measurement circuitry has to have the ability to deal with these effects (compare Table 15.3).

Recently in [61], it could be shown that such an ECT approach can be transferred to the open environment for, for example, safety applications. Although it showed promising results, limitations due to the measurement hardware and open environment effects (described in Section 15.2.3) were found and are presented in the following.

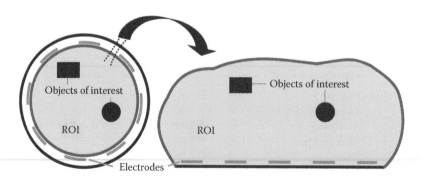

FIGURE 15.6   Sketch of the proposed idea to transfer the ECT approach to capacitive sensing for the open environment. The enclosed structure of an ECT system is opened and attached to the surface of interest. The ROI changes from the well-known inside of, for example, a pipe to the uncertain open environment.

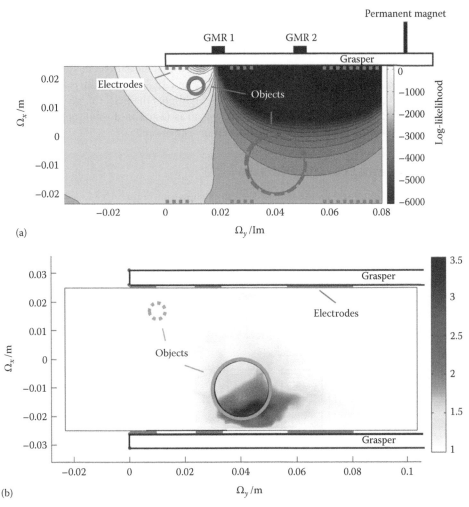

FIGURE 15.7 Reconstruction results of the ECT robot grasper for two different objects (ferromagnetic and dielectric) adopted from [61]. (a) Likelihood for the position of an iron rod. The small circle with continous outline indicates the true position of the iron rod and the dashed circle the true position of a PVC rod which is not recognized by the GMR sensors. (b) Reconstruction of the region of interest in an ECT manner. The spatial permittivity distribution is reconstructed. The true position of the PVC rod (indicated by the dashed circle) matches the reconstruction result. However, the shape cannot be reconstructed.

ECT robot grasper: As mentioned previously, the ECT approach could be transferred to a robot grasping application in [61]. The measurement system uses capacitive and GMR sensors to reconstruct the ROI in an ECT manner. Although the reconstruction algorithms were taken from an ECT application without adaptation for the open environment, the reconstruction showed promising results (shown in Figure 15.7). Dielectric objects (e.g., rod made of polyvinyl chloride, PVC) and ferromagnetic objects (e.g., iron rod) can be detected by this measurement system. However, because of the open environment parasitic effects (explained in Section 15.2.3), the iron rod could not be reconstructed using only the capacitive sensors. The capacitance measurement system uses the mutual capacitance mode in a low-Z scheme. With an additional self-capacitance measurement (i.e., measuring the displacement current originating the sensing electrodes), it could be possible to overcome the parasitic effects, which originally resulted in blind spots for certain objects in Figure 15.7.

15.4 PROPOSED MEASUREMENT SYSTEM

Taking into account the parasitic effects occurring in open environment measurements (presented in Section 15.2.3) and following the proposed ECT approach in Section 15.3.2, a new capacitance measurement hardware is presented in [82]. It is shown how this measurement system is beneficial compared to the example applications in Sections 15.2.2 and 15.3.1. A comparison with two commercially available capacitance measurement systems highlights the performance of the presented system. Furthermore, the presented measurement system is tested in a robot application, and its feasibility is demonstrated by means of experimental investigations.

15.4.1 DESIGN OF THE EVALUATION CIRCUITRY

An overview of the presented measurement system is shown in Figure 15.8a. A sinusoidal signal, generated by a direct digital synthesizer (DDS), is applied to one or more electrodes through a switch circuitry. The displacement current originating the electrodes used as transmitters is measured by a transmitter circuitry [82]. Each electrode is also connected to a receiver circuitry. If an electrode is not used as a transmitter, the receiver circuitry measures the displacement current entering this electrode. Since each electrode can be used as a transmitter or a receiver, a total of $N_{elec}(N_{elec} - 1)/2$ independent measurements, where N_{elec} is the number of electrodes, can be

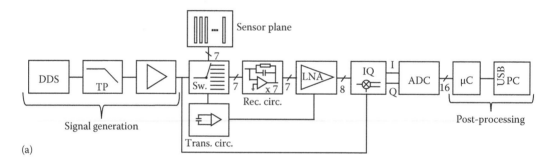

(a)

(b)

FIGURE 15.8 Proposed capacitance evaluation circuitry. (a) Overview of the different parts of the measurement system. (b) Picture of the three stacked PCBs of the evaluation circuitry.

obtained with the mutual capacitance mode. With the additional self-capacitance mode, a total of $(N_{elec}(N_{elec} - 1)/2) + N_{elec}$ independent measurements can be obtained. Additionally, the backside of the sensor can be connected to ground (mutual capacitance mode) or to the excitation signal (i.e., active guarding in self-capacitance mode). This function is also possible with each electrode if the electrode is not used as transmitter or receiver. After amplification, an IQ-demodulator is used to get phase and amplitude information of the measured signals with respect to the excitation signal. The postprocessing consists of an ADC and a microcontroller (μC). The μC is used to control the measurement hardware (e.g., ADC, IQ-demodulator, DDS), store the measurement signals, and communicate with a host computer to do further postprocessing (e.g., reconstruction algorithms).

The proposed measurement system (shown in Figure 15.8b) is able to work in the mutual capacitance mode and the self-capacitance mode. It provides a high measurement rate (>1 kHz). The measurement frequency can be changed between 10 kHz and 1 MHz to any frequency value of interest. Thus, it is possible to obtain additional information about parasitic effects, due to their frequency dependency as shown in Figure 15.4. This also gives additional information for material classification [58]. Furthermore, a change in the measurement frequency can be used to deal with EMC problems as shown in [63].

As shown in Figure 15.8b, the measurement hardware consists of three stacked printed circuit boards (PCB). The top PCB is a commercially available microcontroller evaluation board. The PCB positioned central comprises all digital parts (e.g., clock generator, DDS, ADC). The bottom PCB comprises the analog circuitries such as transmitter and receiver circuitries and IQ-demodulator.

Since each electrode can be used as transmitter and receiver, the proposed measurement system can also be used for ECT applications. Thus, it is appropriate for the stated approach for capacitive safety applications in Section 15.3.2.

15.4.2 COMPARISON WITH STATE-OF-THE-ART CAPACITIVE SENSORS

Table 15.5 gives an overview of the proposed measurement system [82] and two commercially available systems [66]. One of the commercially available measurement systems is working in the self-capacitance mode (AD7148), and the other one is working in the mutual capacitance mode (AD7746). Although there exist a huge variety of capacitance measurement systems (compare Section 15.2.2), these systems are appropriate as state-of-the-art systems by means of resolution and speed (i.e., measurement update rate).

Several experiments (Figure 15.9a through c) were carried out in [82] with the proposed measurement system and compared to the two commercially available ones (Analog Devices AD7148 and AD7746 [66]). In the first experiments shown in Figure 15.9a, a human hand approaches the sensor surface and leaves again. The human hand can be detected in self-capacitance mode as well as in mutual capacitance mode with all three measurement systems

TABLE 15.5
Properties of the Proposed Measurement System Compared to State-of-the-Art Sensors

Excitation Signal	Proposed Sensor Sinusoidal Signal	AD7746 Square Wave	AD7148 Square Wave
Frequency	Tunable from 10 kHz to 1 MHz	32 kHz	250 kHz
Measurement rate	1.25 kHz (max. 6.25 kHz @ 1 MHz)	10–90 Hz	40 Hz
Measurement method	Self-cap. and mutual cap. mode	Mutual cap. mode	Self-cap. mode
Shielding	Active guarding and grounded shielding	Grounded shielding	Active guarding
Number of electrodes	$N_{elec} = 7$	$N_{elec} = 3$	$N_{elec} = 8$
Number of independent measurements	$28 \left(= \dfrac{N_{elec}(N_{elec} - 1)}{2} + N_{elec} \right)$ for each frequency	2	8

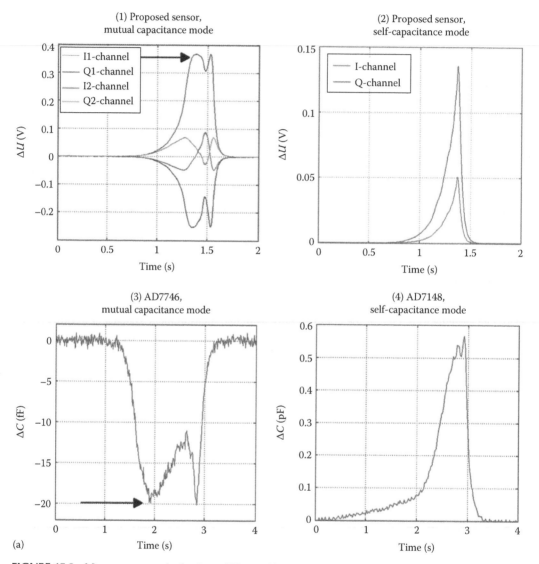

FIGURE 15.9 Measurement results for three different objects obtained with the proposed capacitance measurement system and two commercially available systems [82]. (a) A human hand approaching and leaving the sensor surface. Black arrows indicate the transition from shielding to coupling mode (refer to Section 15.2.4).

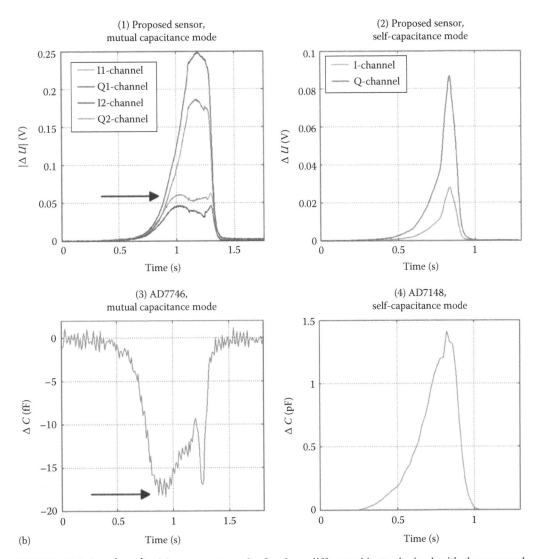

FIGURE 15.9 (continued) Measurement results for three different objects obtained with the proposed capacitance measurement system and two commercially available systems [82]. (b) A metal rod approaching and leaving the sensor surface.

(continued)

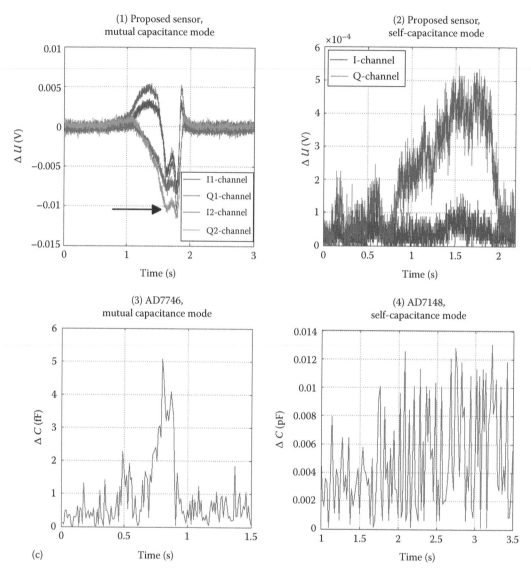

FIGURE 15.9 (continued) Measurement results for three different objects obtained with the proposed capacitance measurement system and two commercially available systems [82]. (c) An empty plastic box approaches and leaves the sensor surface. With the self-capacitance mode, it is difficult to detect the approaching box. However, the proposed measurement system working in mutual capacitance mode is able to detect even objects that have such a low permittivity and small volume.

(proposed sensor and commercially available ones). Although due to shielding and coupling effects (described in Section 15.2.4) at a certain distance to the sensor surface, the measured capacitance increases (marked with arrows in Figure 15.9a). This effect can yield to ambiguities in proximity determination. An approaching metal rod shows similar SNRs as a human hand. It can be detected by all three measurement systems. Objects with low permittivity ε_r (i.e., close to 1) are difficult to detect with a self-capacitance mode measurement system. As can be seen in Figure 15.9c, the plastic box can be detected by the proposed measurement system in the mutual capacitance mode and for close distances in the self-capacitance mode. With the two commercially available measurement systems, it is very difficult to detect this kind of objects (i.e., low permittivity and low volume).

15.4.3 Measurements for a Highly Reactive Proximity Sensor on a Robot Arm

As stated in [61], special precautions are required to avoid injuries when robots and humans share the same environment. In the future, we can expect that more and more autonomous systems and robots will become a part of our lives. This also means that robots will operate in fairly undefined environments, where little prior knowledge is available. It is therefore important that these systems also gather information about the environment in a similar fashion as humans explore an unknown environment. Vision will be quite important for this task. However, it will also require other senses. It is also quite attractive to add sensing capabilities that are beyond the abilities of humans [61]. Thus, in the presented application, a robot arm (Kuka LWR 4) is attached with the presented capacitance measurement system to avoid a human–robot collision [83]. Figure 15.10 shows a picture of the setup comprising the robot arm and the sensing electrodes.

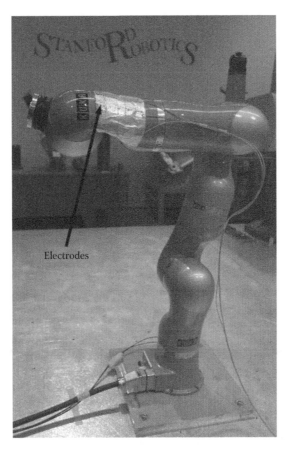

FIGURE 15.10 Proposed capacitance measurement system mounted on the 7 doF robot arm (Kuka LWR 4).

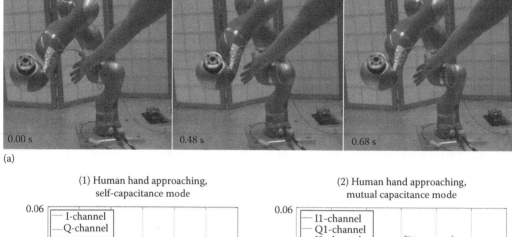

(a)

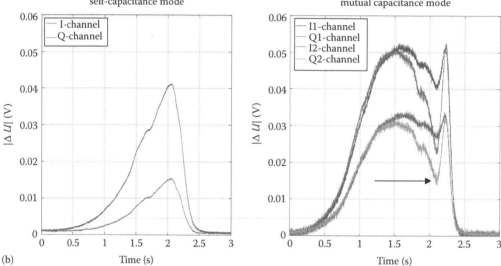

(b)

FIGURE 15.11 Experiments with the proposed measurement system on a robot arm [83]. (a) The robot arm reacts on the capacitance measurements instantaneously and avoids a collision with a human hand. (b) Capacitance measurement results for a human hand approaching and leaving the sensor surface.

An experiment and measurement results are shown in Figure 15.11. The robot arm is moving in its workspace. As soon as an object (e.g., human hand) is detected by the capacitance measurement system, the robot arm reacts instantaneously to the measurements and tries to avoid contact with the object [84] (shown in Figure 15.11a). Figure 15.11b shows the measurements of the proposed measurement system for an approaching human hand.

The comparison with other capacitance measurement systems and the example application on the robot arm shows promising performances of the presented measurement system. With this kind of capacitance measurement system, it should further be possible to realize a complete ECT sensor system for the open environment. Additional information due to both self-capacitance and mutual capacitance measurements will allow an object classification. Furthermore, possible calibration techniques and the available frequency hopping of the excitation signal will allow to handle parasitic effects and improve capacitive sensing for safety applications.

15.5 CONCLUSION

In the last years, research in the field of capacitive sensing has more and more addressed collision avoidance in order to protect humans, animals, and objects. The technology offers complementary features compared to other technologies that make it particularly interesting for fusion concepts.

The strength of capacitive sensing is the capability to perform well for short distances without requiring a free line of sight. Furthermore, it is also capable to provide classifications of objects.

The chapter started in Section 15.1 with an introduction to the state-of-the-art technology for safety applications and an introduction to capacitance measurement in Section 15.2. The physics behind capacitance measurement were shown, example applications were presented, and the problems when it comes to open environment measurements were shown. Section 15.3 presented state-of-the-art measurement circuitries that were also used in the example applications. Furthermore, we proposed a new approach using capacitive sensing for safety applications in an ECT manner.

Finally, a measurement system was presented in Section 15.4, and it has been shown that evaluation circuitry is now capable to comply with the short response times as are needed in many safety applications. In particular, the technology addresses the need for such sensors in human–robot interaction. Current research focuses on the application of the devices in open environments, that is, environments that have almost no constraints with respect to object types or environmental conditions.

ACKNOWLEDGMENT

This research was partially funded by the Austrian Science Fund (FWF): P21855-N22. Parts of the presented work were performed while the first author was visiting the Stanford Robotics Laboratory at Stanford University, whose hosting is gratefully acknowledged.

REFERENCES

1. Hsiao, K., P. Nangeroni, M. Huber, A. Saxena, and A. Y. Ng, Reactive grasping using optical proximity sensors, in *IEEE International Conference on Robotics and Automation, 2009, ICRA '09*, Kobe, Japan, May 12–17, 2009, pp. 2098–2105.
2. Schlegl, T., M. Moser, and H. Zangl, Directional human approach and touch detection for nets based on capacitive measurement, in *Proceedings of the IEEE Instrumentation and Measurement Technology Conference (I2MTC)*, Graz, Austria, May 13–16, pp. 81–85, 2012.
3. Renhart, W., M. Bellina, C. Magele, and A. Köstinger, Hidden metallic object localization by using giant magnetic resistor sensors, in *14th IGTE Symposium*, Graz, Austria, September 19–22, pp. 465–469, 2010.
4. Jiang, L.-T. and J. Smith, Seashell effect pretouch sensing for robotic grasping, in *IEEE International Conference on Robotics and Automation (ICRA), 2012*, St. Paul, Minnesota, May 14–18, 2012, pp. 2851–2858.
5. Jiang, L.-T. and J. R. Smith, A unified framework for grasping and shape acquisition via pretouch sensing, in *IEEE International Conference on Robotics and Automation (ICRA)*, May 2013.
6. Salter, C., *Entangled: Technology and the Transformation of Performance*, The MIT Press, Cambridge, MA, 2010.
7. Sears, A. and B. Shneiderman, High precision touchscreens: Design strategies and comparisons with a mouse, *International Journal of Man-Machine Studies*, 34(4), 593–613, 1991 [Online]. Available: http://www.sciencedirect.com/science/article/pii/0020737391900378.
8. Gould, J. D., S. L. Greene, S. J. Boies, A. Meluson, and M. Rasamny, Using a touchscreen for simple tasks, *Interacting with Computers*, 2(1), 59–74, 1990 [Online]. Available: http://www.sciencedirect.com/science/article/pii/0953543890900149.
9. Puers, R., Capacitive sensors: When and how to use them, *Sensors and Actuators A: Physical*, 37–38, 93–105, 1993 [Online]. Available: http://www.sciencedirect.com/science/article/pii/092442479380019D.
10. Kim, H.-K., S. Lee, and K.-S. Yun, Capacitive tactile sensor array for touch screen application, *Sensors and Actuators A: Physical*, 165(1), 2–7, 2011 [Online]. Available: http://www.sciencedirect.com/science/article/pii/S0924424709005573.
11. Tiwana, M. I., S. J. Redmond, and N. H. Lovell, A review of tactile sensing technologies with applications in biomedical engineering, *Sensors and Actuators A: Physical*, 179, 17–31, 2012 [Online]. Available: http://www.sciencedirect.com/science/article/pii/S0924424712001641.
12. Baxter, L., *Capacitive Sensors, Design and Applications*, IEEE Press, Washington, DC, 1997.
13. Dyer, S. A., *Wiley Survey of Instrumentation and Measurement*, Wiley, New York, 2004.
14. Zangl, H., Design paradigms for robust capacitive sensors, PhD dissertation, Graz University of Technology, Graz, Austria, May 2005.
15. Volpe, R. and R. Ivlev, A survey and experimental evaluation of proximity sensors for space robotics, in *Proceedings of IEEE International Conference on Robotics and Automation*, vol. 4, San Diego, CA, May 8–13, 1994, pp. 3466–3473.

16. Novak, J. and J. Wiczer, A high-resolution capacitative imaging sensor for manufacturing applications, in *Proceedings of IEEE International Conference on Robotics and Automation*, vol. 3, Sacramento, CA, April 9–11, 1991, pp. 2071–2078.

17. Wijesiriwardana, R., K. Mitcham, W. Hurley, and T. Dias, Capacitive fiber-meshed transducers for touch and proximity-sensing applications, *IEEE Sensors Journal*, 5(5), 989–994, 2005.

18. Neumayer, M., B. George, T. Bretterklieber, H. Zangl, and G. Brasseur, Robust sensing of human proximity for safety applications, in *IEEE Instrumentation and Measurement Technology Conference (I2MTC), 2010*, Austin, TX, May 3–6, 2010, pp. 458–463.

19. Krein, P. and R. Meadows, The electroquasistatics of the capacitive touch panel, *IEEE Transactions on Industry Applications*, 26, 529–534, May–June 1990.

20. Castelli, F., An integrated tactile-thermal robot sensor with capacitive tactile array, *IEEE Transactions on Industry Applications*, 38, 85–90, January–February 2002.

21. Karlsson, N. and J.-O. Jarrhed, A capacitive sensor for the detection of humans in a robot cell, in *IEEE Instrumentation and Measurement Technology Conference*, Irvine, CA, May 18–20, 1993, pp. 164–166.

22. Lucas, J., C. Bâtis, S. Holé, T. Ditchi, C. Launay, J. Da Silva, H. Dirand, L. Chabert, and M. Pajon, Morphological capacitive sensors for air bag applications, *Sensor Review*, 23, 345–351, 2003.

23. George, B., H. Zangl, T. Bretterklieber, and G. Brasseur, A combined inductive–capacitive proximity sensor for seat occupancy detection, *IEEE Transactions on Instrumentation and Measurement*, 59(5), 1463–1470, 2010.

24. Zeeman, A., M. Booysen, G. Ruggeri, and B. Lagana, Capacitive seat sensors for multiple occupancy detection using a low-cost setup, in *IEEE International Conference on Industrial Technology (ICIT), 2013*, Cape Town, South Africa, February 25–27, 2013, pp. 1228–1233.

25. Snell, D., A. Moon, C. Clatworthy, and L. Jones, Capacitive proximity sensor, U.K. Patent Application GB 2400666, March 27, 2003.

26. Schlegl, T., T. Bretterklieber, M. Neumayer, and H. Zangl, Combined capacitive and ultrasonic distance measurement for automotive applications, *IEEE Sensors Journal*, 11, 2636–2642, 2011.

27. Bretterklieber, T. and H. Zangl, Impacts on the accuracy of a capacitive inclination sensor, in *Proceedings of the IEEE Conference on Instrumentation and Measurement Technology (IMTC/2004)*, Como, Italy, May 18–20, pp. 2315–2319, 2004.

28. Zangl, H. and T. Bretterklieber, Dynamic inclination estimation with liquid based sensors, in *International Workshop on Robot Sensing, 2004, ROSE 2004*, Graz, Austria, May 24–25, 2004, pp. 61–64.

29. Bretterklieber, T., On the design of a robust liquid based capacitive inclination sensor, PhD dissertation, Graz University of Technology, Graz, Austria, July 2008.

30. Castelli, F., The thin dielectric film capacitive displacement transducer to nanometer, *IEEE Transactions on Instrumentation and Measurement*, 50, 106–110, February 2001.

31. Jeon, S., H.-J. Ahn, D.-C. Han, and I.-B. Chang, New design of cylindrical capacitive sensor for on-line precision control of AMB spindle, *IEEE Transactions on Instrumentation and Measurement*, 50(3), 757–763, June 2001.

32. Brasseur, G., A capacitive 4-turn angular-position sensor, *IEEE Transactions on Instrumentation and Measurement*, 47, 275–279, February 1998.

33. Gasulla, M., X. Li, G. Meijer, L. van der Ham, and J. Spronck, A contactless capacitive angular-position sensor, *IEEE Sensors Journal*, 3, 607–614, October 2003.

34. Falkner, A., The use of capacitance in the measurement of angular and linear displacement, *Transactions on Instrumentation and Measurement*, 43, 939–942, 1994.

35. Zangl, H., S. Cermak, B. Brandstätter, G. Brasseur, and P. Fulmek, Simulation and robustness analysis for a novel capacitive/magnetic full turn absolute angular position sensor, *IEEE Transactions on Instrumentation and Measurement*, 54, 436–441, February 2005.

36. Toth, F., G. Meijer, and M. van der Lee, A new capacitive precision liquid-level sensor, in *Conference on Precision Electromagnetic Measurements*, Braunschweig, Germany, June 17–21, 1996, pp. 356–357.

37. Sawada, R., J. Kikuchi, E. Shibamura, M. Yamashita, and T. Yoshimura, Capacitive level meter for liquid rare gases, *Cryogenics*, 43, 449–450, August 2003.

38. Holler, G., A. Fuchs, and G. Brasseur, Fill level measurement in a closed vessel by monitoring pressure variations due to thermodynamic equilibrium perturbation, in *Instrumentation and Measurement Technology Conference Proceedings, 2008, IMTC 2008, IEEE*, Victoria, BC, Canada, May 12–15, 2008, pp. 641–646.

39. Neumayer, M. and H. Zangl, Electrical capacitance tomography for level measurements of separated liquid stacks, in *Sensor+Test Conference*, Nuremberg, Germany, 2011, pp. 433–438.

40. Irani, K., M. Pekkari, and H.-E. Ångström, Oil film thickness measurement in the middle main bearing of a six-cylinder supercharged 9 litre diesel engine using capacitive transducers, *Wear*, 207, 29–33, June 1997.

41. Pinto, J., J. Monteiro, R. Vasconcelos, and F. Soares, A new system for direct measurement of yarn mass with 1 mm accuracy, in *Proceedings of IEEE International Conference on Industrial Technology*, vol. 2, Bangkok, Thailand, December 11–14, 2002, pp. 1158–1163.

42. Schröder, J., S. Doerner, T. Schneider, and P. Hauptmann, Analogue and digital sensor interfaces for impedance spectroscopy, *Measurement Science and Technology*, 15(7), 1271–1278, July 2004.

43. Zangl, H., A. Fuchs, and T. Bretterklieber, A novel approach for spatially resolving capacitive sensors, in *4th World Congress on Industrial Process Tomography*, Aizu, Japan, September 5–8, 2005, pp. 36–41.

44. Fuchs, A., H. Zangl, and G. Brasseur, A sensor fusion conception for precise mass flow measurement, in *Proceedings of the 8th International Conference on Bulk Materials Storage, Handling and Transportation (ICBMH'04)*, New South Wales, Australia, July 5–8, 2004, pp. 366–370.

45. Byars, M., Developments in electrical capacitance tomography, in *Proceedings of Second World Congress on Industrial Process Tomography*, Hannover, Germany, August 2001, pp. 542–549.

46. Yang, W. and L. Peng, Image reconstruction algorithms for electrical capacitance tomography, *Measurement Science and Technology*, 14(1), R1–R13, January 2003.

47. Cui, Z., H. Wang, Z. Chen, Y. Xu, and W. Yang, A high-performance digital system for electrical capacitance tomography, *Measurement Science and Technology*, 22, 2011.

48. Norgia, M. and C. Svelto, RF-capacitive proximity sensor for safety applications, in *Instrumentation and Measurement Technology Conference Proceedings, 2007, IMTC 2007*, Warsaw, Poland, *IEEE*, Warsaw, Poland, May 1–3, 2007, pp. 1–4.

49. George, B., H. Zangl, and T. Bretterklieber, A warning system for chainsaw personal safety based on capacitive sensing, in *IEEE International Conference on Sensors*, Lecce, Italy, October 26–30, 2008, pp. 419–422.

50. Moser, M., H. Zangl, T. Bretterklieber, and G. Brasseur, An autonomous sensor system for monitoring of high voltage overhead power supply lines, *e&i Elektrotechnik und Informationstechnik*, 126, 214–219, 2009.

51. Moser, M., T. Bretterklieber, H. Zangl, and G. Brassuer, Capacitive icing measurement in a 220 kV overhead power line environment, *IEEE Sensors*, 1–4 Nov, pp.1754–1758, Kona District, Hawaii, 2010, doi: 10.1109/ICSENS.2010.5689885.

52. Moser, M., T. Bretterklieber, H. Zangl, and G. Brasseur, Strong and weak electric fields interfering: Capacitive icing detection and capacitive energy harvesting on a 220 kV high-voltage overhead power line, *IEEE Transactions on Industrial Electronics*, 58, 2597–2604, 2011.

53. Di Santo, M., A. Vaccaro, D. Villacci, and E. Zimeo, A distributed architecture for online power systems security analysis, *IEEE Transactions on Industrial Electronics*, 51(6), 1238–1248, 2004.

54. Blais, A., M. Lacroix, and L. Brouillette, Hydro Quebecs de-icing system: Automated overhead line monitoring and de-icing system, in *Cigre Session*, vol. B2–211, Paris, France, August 2008, pp. 1–7.

55. Benniu, Z., Z. Junqian, Z. Kaihong, and Z. Zhixiang, A non-contact proximity sensor with low frequency electromagnetic field, *Sensors and Actuators A: Physical*, 135(1), 162–168, 2007 [Online]. Available: http://www.sciencedirect.com/science/article/pii/S0924424706004663.

56. Mayton, B., L. LeGrand, and J. Smith, An electric field pretouch system for grasping and co-manipulation, in *IEEE International Conference on Robotics and Automation (ICRA), 2010*, Anchorage, AK, May 3–8, 2010, pp. 831–838.

57. Liang-Ting Jiang, J. R. S., Pretouch sensing for pretouch sensing for manipulation, in *Robotics: Science and Systems (RSS) Workshop: Alternative Sensing Techniques for Robotic Perception*, Sydney, Australia, July 11–12, 2012.

58. Kirchner, N., D. Hordern, D. Liu, and G. Dissanayake, Capacitive sensor for object ranging and material type identification, *Sensors and Actuators A: Physical*, 148(1), 96–104, 2008 [Online]. Available: http://www.sciencedirect.com/science/article/pii/S0924424708004184.

59. Zeng, S., J. R. Powers, and B. H. Newbraugh, Effectiveness of a worker-worn electric-field sensor to detect power-line proximity and electrical-contact, *Journal of Safety Research*, 41(3), 229–239, 2010 [Online]. Available: http://www.sciencedirect.com/science/article/pii/S0022437510000368.

60. Janicak, C. A., Occupational fatalities due to electrocutions in the construction industry, *Journal of Safety Research*, 39(6), 617–621, 2008 [Online]. Available: http://www.sciencedirect.com/science/article/pii/S0022437508001448.

61. Schlegl, T., M. Neumayer, S. Muhlbacher-Karrer, and H. Zangl, A pretouch sensing system for a robot grasper using magnetic and capacitive sensors, *IEEE Transactions on Instrumentation and Measurement*, 62(5), 1299–1307, May 2013, doi: 10.1109/TIM.2013.2238034.

62. Bracke, W., P. Merken, R. Puers, and C. van Hoof, On the optimization of ultra low power front-end interfaces for capacitive sensors, *Sensors and Actuators A: Physical*, 117, 273–285, January 14, 2005.

63. Brasseur, G., Design rules for robust capacitive sensors, *IEEE Transactions on Instrumentation and Measurement*, 52, 1261–1265, August 2003.
64. Zangl, H., Capacitive sensors uncovered: Measurement, detection and classification in open environments. *Procedia Engineering*, 5, 393–399, 2010.
65. Wegleiter, H., A. Fuchs, G. Holler, and B. Kortschak, Analysis of hardware concepts for electrical capacitance tomography applications, in *IEEE Sensors Conference*, Irvine, CA, October 30–November 5, pp. 688–691, 2005.
66. Analog Devices. Capacitance to digital converters. http://www.analog.com/en/analog-to-digital-converters/capacitance-to-digital-converters/products/index.html, May 2013.
67. Novak, J. L. and J. Feddema, A capacitance-based proximity sensor for whole arm obstacle avoidance, in *Proceedings, 1992 IEEE International Conference on Robotics and Automation*, vol. 2, Nice, France, May 1992, pp. 1307–1314.
68. Feddema, J. and J. Novak, Whole arm obstacle avoidance for teleoperated robots, in *Proceedings, 1994 IEEE International Conference on Robotics and Automation*, vol. 4, May 8–13, San Diego, California, 1994, pp. 3303–3309.
69. Neumayer, M., G. Steiner, and D. Watzenig, Electrical capacitance tomography: Current sensors/algorithms and future advances, in *2012 IEEE International Instrumentation and Measurement Technology Conference (I2MTC)*, Graz, Austria, May 13–16, 2012, pp. 929–934.
70. Neumayer, M., H. Zangl, D. Watzenig, and A. Fuchs, Current reconstruction algorithms in electrical capacitance tomography, in *New Developments and Applications in Sensing Technology*, Springer, Berlin, Germany, 2011.
71. Watzenig, D. and C. Fox, A review of statistical modelling and inference for electrical capacitance tomography, *Measurement Science and Technology*, 20(5), 052 002+, 2009 [Online]. Available: http://dx.doi.org/10.1088/0957-0233/20/5/052002.
72. Soleimani, M. and W. Lionheart, Nonlinear image reconstruction for electrical capacitance tomography using experimental data, *Measurement Science and Technology*, 16, 1987–1996, October 2005.
73. Isaksen, O., A review of reconstruction techniques for capacitance tomography, *Measurement Science and Technology*, 7, 325–337, 1996.
74. Liu, S., L. Fu, W. Yang, H. Wang, and F. Jiang, Prior-online iteration for image reconstruction with electrical capacitance tomography, *IEE Proceedings-Science Measurement and Technology*, 151, 195–200, May 2004.
75. Zangl, H., D. Watzenig, G. Steiner, A. Fuchs, and H. Wegleiter, Non-iterative reconstruction for electrical tomography using optimal first and second order approximations, in *Proceedings of the 5th World Congress on Industrial Process Tomography*, Bergen, Norway, September 3–6, 2007, pp. 216–223.
76. Brandsttter, B., G. Holler, and D. Watzenig, Reconstruction of inhomogeneities in fluids by means of capacitance tomography, *COMPEL: The International Journal for Computation and Mathematics in Electrical and Electronic Engineering*, 22(3), 508–519, 2003.
77. Watzenig, D., G. Steiner, and M. Brandner, A particle filter approach for tomographic imaging based on different state-space representations, *Measurement Science and Technology*, 18, 30–40, 2007.
78. Trigo, F., R. Gonzalez-Lima, and M. Amato, Electrical impedance tomography using the extended Kalman filter, *IEEE Transactions on Biomedical Imaging*, 51, 72–81, January 2004.
79. Soleimani, M., M. Vauhkonen, W. Yang, A. Peyton, B. Kim, and X. Ma, Dynamic imaging in electrical capacitance tomography and electromagnetic induction tomography using a Kalman filter, *Measurement Science and Technology*, 18, 3287–3294, November 2007.
80. Nooralahiyan, A., B. Boyle, and J. Bailey, Performance of neural network in image reconstruction and interpretation for electrical capacitance tomography, in *IEE Colloquium on Innovations in Instrumentation for Electrical Tomography*, London, May 11, 1995, pp. 5/1–5/3.
81. Zang, L., H. Wang, M. Ma, and X. Jin, Image reconstruction algorithm for electrical capacitance tomography based on radial basis functions neural network, in *Proceedings of the Fourth International Conference on Machine Learning and Cybernetics*, Guangzhou, China, August 18–21, 2006, pp. 4149–4152.
82. Schlegl, T. and H. Zangl, Sensor interface for multimodal evaluation of capacitive sensors, in *Journal of Physics: Conference Series*, 450 012018, 1–5, 2013, doi: 10.1088/1742-6596/450/1/012018.
83. Schlegl, T., T. Kröger, A. Gaschler, O. Khatib, and H. Zangl, Virtual Whiskers—Highly Responsive Robot Collision Avoidance. In: *Intelligent Robots and Systems, 2013. IROS 2013. IEEE/RSJ International Conference*. Tokyo, Japan, November 3–8, 2013 (in press).
84. Kröger, T., *On-Line Trajectory Generation in Robotic Systems*, 1st ed., vol. 58, Springer Tracts in Advanced Robotics Series, Springer, Berlin, Germany, January 2010.

16 Conformal Microelectromechanical Sensors

Donald P. Butler and Zeynep Çelik-Butler

CONTENTS

16.1 INTRODUCTION

Flexible electronic systems have been an interest of scientists and engineers since the birth of the electronic age in 1906 [1]. Flexible electronics has been a subject of science fiction but has been steadily evolving into science fact [2]. Flexible circuit boards are extensively used in the consumer electronics industry with conventional, rigid integrated circuits resulting in a savings in size, weight, and cost. A variety of flexible electronic devices and systems have been developed to date with flexible displays emerging into the consumer marketplace and a strong interest in flexible solar panels. Thinned silicon electronics has been used to produce *Smart Cards* that have memory, communications, data processing, display, and data entry functions. Although thinned silicon integrated circuits are fabricated in crystalline silicon wafers, when the thin layer is separated away by fracturing, wafer polishing, or wafer etching, the remaining crystalline film is sufficiently thin such that it is flexible and can be bent [3–6]. Amorphous and polysilicon thin-film transistors [7,8], polymer diodes [9], and polymer transistors [10] have also been developed. All of these devices are being applied to produce flexible integrated circuits with varying degrees of performance for different applications.

A large variety of flexible sensors has been developed over the years as well. The earliest flexible sensors of the modern age were developed with circuit board technology. More recently, flexible microsensors that take advantage of modern microfabrication technology have emerged. Micromachining has simultaneously been utilized on rigid, primarily silicon substrates to produce a variety of physical, chemical, and biological sensors that are members of the broad family of microelectromechanical devices and systems (MEMS) [11]. The development of surface micromachining techniques has to some degree made the fabrication of MEMS devices independent of the substrate used. So instead of using rigid silicon wafers to fabricate a *rigid* MEMS integrated circuit, flexible polymer substrates can be employed to produce flexible MEMS sensor integrated circuits that can conform to nonplanar surfaces; hence, conformal microsensors have been developed. The future merger of flexible electronics for signal processing, data storage, and communications with flexible microsensors will result in a technology that can be referred to as a *Smart Skin* where a *dense* collection of sensors can exist over large areas. Batch, parallel, simultaneous fabrication of large numbers of conformal, flexible sensors will result in comparatively low production costs just as it has resulted in state-of-the-art sub-15-microcent 32-mn complementary metal oxide semiconductor (CMOS) transistors in rigid silicon integrated circuits [12]. In addition, many MEMS sensors can be produced without the need for comparatively expensive deep-UV or e-beam lithography, though they usually consume larger areas on the substrate.

As flexible transistors continue to evolve, a wider range of monolithic to heterogeneous integration options with flexible MEMS sensors will emerge. Simultaneously, the materials employed in flexible, compliant MEMS sensors will continue to evolve to increase performance while allowing for greater flexibility of the substrate and perhaps even significant stretch ability. At present, monolithic or heterogeneous integration [13–16] of flexible MEMS sensors with thinned silicon CMOS electronics seems to be the most likely near-term solution. For example, monolithic thinned silicon CMOS imagers have already been developed where silicon photodiodes are used to sense the optical radiation [17]. The choice between monolithic or heterogeneous integration of the components of a particular electronic system will remain an economic one, as always. This chapter will discuss different methodologies for conformal MEMS sensor fabrication, strategies for conformal sensor packaging, and preliminary results on different types of conformal MEMS sensors and suggest future approaches toward advancing the technology.

16.2 CONFORMAL MICROELECTROMECHANICAL SENSING

There is an advantage to being able to have the sensor devices and integrated circuits have the ability to conform to nonplanar surfaces since most surfaces we encounter in everyday life are not flat. One can easily see this by looking at the surface of a car, an aircraft, a robot, and many tools. Curved surfaces are more common. If one has the desire to *sensorize* the surface or interior parts of a machine with electronic sensors, then there is an advantage to having the sensor circuits conform to nonplanar surfaces. With this goal in mind, a number of researchers have been working on the development of flexible sensors. In our research group, we have focused on the development of flexible, surfaced micromachined MEMS sensors that can sense pressure [19], force [20,21], touch [22], acceleration [23], strain, acoustic emissions, IR radiation [24,25], and sound. Flexible MEMS energy harvesting devices are currently under development to serve at a source of power for isolated flexible sensor circuits. Figure 16.1 shows the contact side of a typical flexible sensor fabricated as a surface-mount *chip* for easy integration on to a flexible circuit board. The total thickness of the integrated circuit is approximately 100 µm. At this time, any surface micromachined MEMS device can be fabricated on flexible substrates with the constraint that the materials used to produce the device need to be deposited at low temperatures (<400°C). This temperature limit is similar to the temperature constraint for monolithic integration of MEMS sensors on traditional rigid CMOS substrates. Flexible chemical and biological sensors are also possible and currently being investigated by other research groups. Many MEMS sensors that are fabricated on rigid silicon substrates can be migrated to flexible substrates. The result or benefit would be a reduction in size, perhaps cost with volume production, and the ability to bend or conform to nonplanar surfaces.

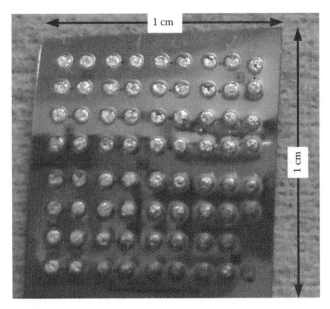

FIGURE 16.1 A flexible, conformal surface-mount die containing an array of temperature sensors showing the electrical contact side of the device. (© 2012 IEEE. Reprinted, with permission, from Temperature sensor in a flexible substrate by Ahmed, M., Chitteboyina, M.M., Butler, D.P., and Celik-Butler, Z., in *IEEE Sens. J.*, **12**, 864, 2012.)

16.3 FABRICATION APPROACHES

To produce conformal MEMS sensors, two types of microfabrication equipment could be utilized. The first type of equipment has been developed especially for processing large-area flexible devices and circuits and is called reel-to-reel or roll-to-roll processing [26]. Although it is capable of fabricating large-area devices and circuits, feature size and the number of lithographic level that can be aligned for device fabrication are typically much less than what can be achieved with conventional CMOS processing equipment. The second approach utilizes conventional CMOS processing equipment by either attaching or spin casting the flexible substrate onto a silicon carrier wafer to fabricate the device, then remove the flexible sensor circuits at the end of device fabrication [27,28]. Our research group has employed spin casting the flexible substrate onto the silicon carrier wafer because substrates produced in this manner seem to encounter less geometry distortion with thermal cycling in the processing equipment than polymer films attached to the wafer carrier with an adhesive [16]. A release layer between the carrier wafer and the spin cast substrate aids in the removal of the flexible film at the end of device fabrication [15,17]. Figure 16.2 shows a piece of the flexible substrate at the stage where it has been removed from the 4 in. carrier wafer used so conventional process equipment could be employed in the fabrication of the test structures, in this case.

MEMS sensors fabricated by surface micromachining with flexible substrates enjoy an equivalent performance to devices fabricated on rigid silicon wafers. The use of flexible polymer substrates does necessitate the utilization of low-temperature processing such that the wafer is kept below the glass transition temperature of the polymer used as the flexible substrate. In some cases, it may be possible to exceed the glass transition temperature for a short period of time (Table 16.1). However, the maximum temperature of all known polymers is going to be less than that of crystalline silicon so the choice of materials and processes that can be employed is more limited. Nevertheless, high-performance, conformal MEMS sensors are possible. Polyimide, with a glass transition temperature up to 400°C, is one choice that is attractive due to its relatively high glass transition temperature.

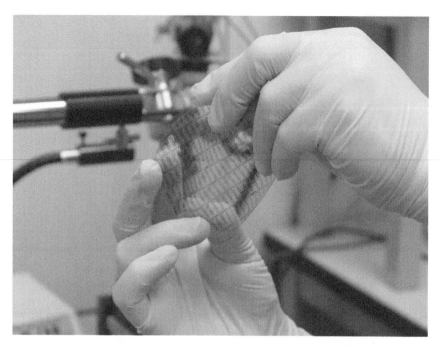

FIGURE 16.2 A flexible, conformal substrate/superstrate combination containing test structures after removal from a silicon carrier wafer. The film is ready to be cut into individual die. (Reprinted with permission from Ahmed, M., and Butler, D. P., *Journal of Vacuum Science and Technology B*. 31(5), 050602, Sep/Oct 2013. Copyright 2013, American Vacuum Society.)

A third approach would be to fabricate the MEMS sensors on conventional glass or silicon wafers or substrates with conventional fabrication equipment, then thin the wafer by backside etching/polishing to achieve a thickness less than 50 μm where the wafer becomes flexible and conformal [29]. In this case, high-temperature fabrication can be used reducing the restrictions on the materials employed. This approach would also permit the monolithic integration with silicon electronics for signal processing and communications but would require the relatively expensive wafer-thinning process.

16.4 PACKAGING

Since conformal sensors are flexible by definition, one does not take advantage of the full potential of the technology by placing them in a conventional, rigid integrated circuit package. A new packaging approach is required. Despite the widespread effort to develop flexible, conformal electronics, there has not been a strong effort in packaging. For the most part, the packaging can be described as device-level where a fully packaged flexible device is fabricated across the entire substrate [24]. In this case, the package is fabricated along with the device. Another approach can be described as wafer level where wafer bonding is employed across the substrate to form the packaged devices [41]. Simultaneously, a packaging technique for thinned silicon die has been developed where the thinned silicon integrated circuit is surrounded with a polymer substrate and superstrate that has been referred to as an *ultrathin chip package* or *ultrathin-film package* [42–45]. In all these cases, flexible, very-low-profile integrated circuits containing the sensors can be fabricated with thicknesses in the range of 70–150 μm thick. The techniques also utilize the basic principle developed by Suo et al. [46] that strain or stress on the devices with bending can be minimized if the device is surrounded by a substrate and superstrate to place the device at a low-stress plane near the center of the flexible film (Figure 16.3).

TABLE 16.1

Typical 300 K Properties of Selected Commercially Available Flexible Substrate Materials

Material Commercial Name (Manufacturer)	Young's Modulus (GPa)	Tensile Strength (GPa)	Glass Transition Temp. (°C)	Coefficient of Linear Thermal Expansion α, (m/m/°C) (−18°C–150°C)	Volume Resistivity (Ω-cm)	Chemical Resistance	Reference
PEI—polyetherimide Ultem (GE)	3.3–8.9	0.097–0.193	190–215	5.5×10^{-5}	1.0×10^{17}	Fair to good	[30]
PEEK—polyetheretherketone Victrex (ICI)	3.0–16.5	0.097–0.283	250–315	$1.8–4.7 \times 10^{-5}$	5×10^{16}	Good to excellent	[31]
PPS— Polyphenylene sulfide Supec (GE) Ryton (Phillips)	3.8–20.0	0.097–0.193	260–350			Good to excellent	[32]
Polyimide For example, Kapton (DuPont)	2.6	0.231	~400	2×10^{-5}	1.5×10^{17}	Excellent	[33]
PET—polyethylene terephthalate Mylar (DuPont)	5.1	0.2	~200	3×10^{-5}	10^{18}	Good to excellent	[34]
Polyamic acid			107				[35]
PDMS			125				[36]
PMMA			<124				[37]
Polyurethane			<65				[38]
Parylene	2.76	68.9	<90				[39,40]

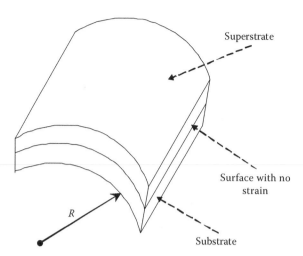

FIGURE 16.3 The structure containing the substrate, electronic layer, and the superstrate is bent to a radius R. In the middle, there is a plane with no strain determined by the thickness and Young's modulus of the substrate and superstrate.

MEMS devices often require some form of vacuum or hermetic packaging, which can be accomplished by using the device-/wafer-level packaging approach in a manner similar to the way wafer-level packaging has resulted in a reduction in cost for rigid MEMS devices. Hermetic packaging can be achieved across a wafer by wafer bonding for devices fabricated both on rigid and flexible substrates. Forehand and Goldsmith showed that a polymer could be used to hermetically seal a MEMS device in a cavity [47]. Vacuum packaging is more difficult since specialized equipment is required to perform wafer-level packaging to seal the cavity under vacuum with wafer bonding. Device-level packaging can also be employed where the vacuum cavity is made and sealed during device fabrication where the vacuum deposition of materials is commonplace [24,48]. The packaging of sensors is generally more difficult than conventional CMOS electronics, for example, because the sensors require some form of environmental access to sense the parameter they are designed to measure. This environmental access needs to be designed into the package with the ability to reject or filter material from the *real-world* environment that may be harmful to the sensor. This is a common problem encountered by both flexible and rigid sensing devices that means the package needs to be appropriately engineered for the sensor.

Once the flexible conformal MEMS sensors are fabricated with wafer-level or device-level packaging, the resulting integrated circuits, which most likely would be surface-mount devices, can be placed and mounted onto flexible circuit boards in a manner similar to what is done today with conventional, rigid integrated circuits on flexible circuit boards to form the desired electronic sensing system. There will always be the economic trade-off in determining what functionality to integrate into a single *chip* versus combining different integrated circuits to produce the electronic system. One interesting aspect of flexible, conformal integrated circuits is that the flexible die or chips can be cut up with a tool as simple as a X-Acto knife.

16.5 DESIGN AND SPECIAL CONSIDERATIONS

The design of flexible, conformal microsensors not only involves the ability to achieve high performance with materials that are deposited at low temperatures but also requires the ability to manage the stress within the device with bending. The properties of many devices vary with applied stress including thinned silicon circuits [49] and MEMS sensors are not an exception. Extensive modeling and simulation using finite element analysis is required to balance and minimize the stress in the flexible film containing the devices with bending. Flexible electronics, like all things man-made, can be broken if it is bent too sharply. The limits of bending depend upon both the design of the device and the materials employed. A basic principle in engineering is the utilization of flexible, perhaps even stretchable, regions to connect rigid regions. This idea was utilized by Jiang et al. [50] to create a flexible shear stress sensor array by joining silicon islands containing the devices with a polyimide film. The silicon islands were formed by bulk micromachining the silicon substrate. The approach was also applied by Sterken et al. [45] to create a stretchable ECG system using thinned silicon integrated circuits. The principle has also been applied by the Rogers Group at the University of Illinois to produce a variety of flexible and even stretchable electronic systems on a variety of scales [51]. The stress and strain management in flexible electronics introduces the idea of conformal substrate engineering that must be considered simultaneously with the electronic or electromechanical functions to produce viable electronic systems.

16.6 EXAMPLES OF CONFORMAL MICROELECTROMECHANICAL SENSORS

At this point, any physical, chemical, and biological MEMS sensor that can be produced with low-temperature surface micromachining could be realized as a flexible, conformal sensor. Temperature [25], force/tactile/pressure [18–21], infrared [23,24], and acceleration [22] sensors have been produced as examples of physical MEMS sensors [35]. Although, the temperature sensors are not strictly MEMS sensors, MEMS processing steps were employed to produce the flexible, conformal temperature sensors. As examples of the methods employed and performance of the respective sensors, the following discussion is presented as examples of what can be accomplished with conformal sensors.

16.6.1 Temperature Sensors

The temperature sensors produced in Ref. [25] were thermistors that utilized an amorphous silicon transducer (Figure 16.1). The resistance of the amorphous silicon was observed to obey the Arrhenius relation near room temperature with a temperature coefficient of resistance of –2.7%/K at 30°C. The temperature sensors were fabricated on a low-stress plane near the center of a polyimide substrate and polyimide superstrate with the aid of a silicon carrier wafer. The sensors were thin flexible surface-mount devices intended for integration on flexible circuit boards. The total thickness of the sensors was approximately 70 μm, representing a thin, low-profile integrated circuit with an array of 35 temperature sensors per die. The contact pads openings were fabricated on the bottom of the sensor while the top of the sensor was considered to be the sensing surface. Since polyimide does not possess a particularly high thermal conductivity, the top surface of the sensors was coated with aluminum, and a via in the polyimide superstrate was used with aluminum to conduct heat to the amorphous silicon thermistor at the center of the membrane. This arrangement provided an engineered differential thermal conductivity to the sensing thermistor. The thermistor resistance was approximately a mega Ohm at room temperature. The 1/f-noise of the temperature sensors and an average normalized Hooge coefficient of 1.2×10^{-11} was measured. In the Johnson noise regime, the noise equivalent temperature, where the signal-to-noise ratio (SNR) is unity, was approximately 70 $\mu K/\sqrt{Hz}$.

16.6.2 Pressure/Force/Tactile Sensors

Two types of pressure/force/tactile sensors have been produced on flexible substrates. The first is force/tactile sensors that were designed to respond to the normal force or pressure if considered per unit area. These sensors utilized an open cavity to increase their sensitivity and simultaneously remove their sensitivity with respect to changes in the atmospheric pressure of the environment. The open cavity design allows for the sensing membrane to be easily deflected into the open space the cavity represents with a small force creating the high sensitivity to small forces. The cavity is open, as opposed to being sealed, so the constantly changing pressure of the environment backs the membrane. One disadvantage of the design is that the open cavity could become contaminated with small, microscopic particles from the environment over time. The size of the access hole can be made quite small to keep out larger particles, but as long as the cavity is not sealed so air has access, other particles also have access. The deflection of the membrane into the cavity with an applied force from touch, for example, causes the simultaneous creation of stress and strain of the membrane, which was sensed with the use of piezoresistors that were located by design at positions of maximum strain. The piezoresistors were placed in a half-Wheatstone bridge configuration with two identical resistors placed on the unstrained substrate surface to serve as reference resistors. The use of identical reference resistors provides a built-in compensation for changes in resistance with temperature. One disadvantage of using piezoresistors is that the noise floor of the sensor is limited by the Johnson noise of the resistors. The 1/f-noise of the piezoresistors can also be a factor. The dynamic range of the sensors investigated has varied from 10^4 to 10^6, depending upon the design. Force/tactile sensors have been fabricated with membranes ranging in size from 70 to 400 μm across and cavity depths ranging from 2 to 7 μm. The cavity depth determines the maximum force that can be sensed since the membrane will be deflected until it reaches the bottom of the cavity. Simultaneously, the cavity depth with the membrane thickness determines the maximum stress and strain the membrane experiences with full deflection, and these parameters need to be engineered to prevent the membrane from fracturing (Figure 16.4).

In our investigations, polysilicon, polymer nanocomposites, and Nichrome piezoresistors have been investigated. There are relative advantages and disadvantages to the different materials employed. Other piezoresistive materials can also be used as long as low-temperature (<400°C) processing can be employed. Some other desirable properties of the piezoresistive material would be the ability to form low-resistance contacts, a resistivity that yields a reasonable value of resistance, low 1/f-noise, a low-temperature coefficient of resistance, uniformity and repeatability in fabrication, and, lastly, a reasonable gauge factor. The gauge factor provides a measure of the relative change in resistance with strain. Using the piezoresistance of metal films is attractive due to their ease of fabrication by sputtering or evaporation at room temperature in most cases. Although the gauge factor of metals is generally lower than semiconductors, so is their 1/f-noise, mostly attributed to their higher carrier density [53–56]. Very thin metals, on the order on nanometers, can have higher gauge factors [57]. However, the ability to achieve reproducible behavior from device to device as well as the linearity in the change in resistance becomes difficult to achieve in very thin, nanometer thick metal films. Most piezoresistive semiconductors have higher gauge factors but also require considerably higher deposition temperatures that may exclude their easy application to flexible sensors. Combined with generally

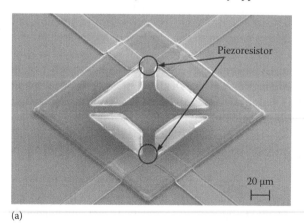

(a)   (b)

FIGURE 16.4 (a) An SEM micrograph of the force/tactile sensor before the application of the polyimide superstrate showing the active piezoresistors on the suspended membrane. (b) A 2 × 2 cm² section of flexible substrate containing an array of 48 force/tactile sensors. (© 2011 IEEE. Reprinted, with permission, from *MEMS relative pressure sensor on flexible substrate* by Ahmed, M., Butler, D.P., and Celik-Butler, Z., in *Proceedings of the 2011 IEEE Sensors Conference*, Limerick, Ireland, October 28–31, 2011, pp. 460–463.)

higher 1/f-noise and a higher temperature coefficient of resistance, a better piezoresistive sensor does not always result from higher gauge factor alone. Thin metal film piezoresistors can make effective piezoresistors as a result of their comparatively low 1/f-noise, reasonable gauge factor, linearity of their response, and low-temperature coefficient of resistance producing a reasonable response and large dynamic range resulting from the low noise floor.

It would also be relatively easy to produce capacitive sensors where the deflection of the membrane with applied force or pressure is measured by the change in capacitance of a parallel plate capacitor. A capacitive sensor has the advantage of eliminating the 1/f-noise and Johnson noise from the sensor so the electronic noise of the sensor would be determined by the thermomechanical noise associated with the motion of the suspended membrane [58]. This usually represents a substantially lower noise floor than produced by the Johnson noise of a piezoresistor allowing for the measurement of very small forces and pressures. However, to achieve the ability to measure forces and pressures at the thermomechanical noise limit, the readout electronics for the capacitive sensor needs to be carefully designed [59]. Some of the disadvantages of capacitive sensing are the inherent nonlinearity of the capacitance change with the displacement of the membrane for a parallel plate capacitor and their susceptibility to electromagnetic interference. Piezoelectric sensors are also possible though the material employed would need to be able to be reliably and reproducibly fabricated at low temperatures. These constraints would limit the selection of the material.

For the Nichrome piezoresistive force/tactile/pressure sensors that have been investigated, force sensitivities or response on the order of 1 V/N have been demonstrated with the ability to sense a maximum force of 2.5–3 mN and a minimum force of approximately 10 µN, as determined by the noise equivalent force in the 1/f-noise regime with a bandwidth of 1–8 Hz. The maximum force being in the millinewton range appears to be a limitation. However, one needs to consider the size of the sensor. In this case, it may be better to express the quantities as a pressure where the maximum pressure that could be measured by this particular design would be 30 kPa, which is in the range of many pressures encountered in everyday life such as touch, the pressure on a standing person's feet, or human blood pressure. The specific performance characteristics of this type of sensor can be varied by engineering the geometry of the sensors. The small size of the sensors allows for the possibility of obtaining a relatively high degree of spatial resolution in the measurements by arrays of the sensors.

Force/tactile sensors embedded into the flexible membrane have also been investigated [22]. In this case, the cavity no longer remained open to the ambient but became sealed at atmospheric pressure through the application of a thin polyimide membrane on top of the sensor. In this case, the sensor would have some sensitivity to variations in atmospheric pressure, but since the cavity is essentially sealed at atmospheric pressure, the sensitivity would not be large. The advantages of this approach are that the cavity would not become contaminated with debris from the environment while the presence of the cavity still allows for a high sensitivity to small applied forces by its lack of resistance permitting the membrane to be deflected more easily. The capping superstrate layer was kept thin so it would not absorb the applied force, and the applied force would then be effectively transferred to the sensing membrane.

16.6.3 Absolute Pressure Sensors

Flexible, conformal MEMS absolute pressure sensors have also been developed. In this case, the absolute pressure sensors utilized a sealed vacuum cavity backing the moveable sensing membrane. The vacuum cavity was sealed at a pressure of 5 mTorr of argon, so the sensors are capable of measuring pressures greater than this value. The fixed cavity height established the maximum pressure that can be measured by the sensor. For the designs in Ref. [19], the maximum pressure was approximately 8–10 MPa. The sensors were designed to be approximately 400 µm across and utilized Nichrome piezoresistors located at a position of maximum strain in the moving diaphragm. As with the force/tactile/pressure sensors described earlier, the piezoresistors were configured into a half-Wheatstone bridge circuit to provide compensation for variations in temperature. The sensors had a sensitivity or response of about 1 nV/Pa and a noise equivalent pressure of about 10 Pa in the Johnson noise regime. The response of the sensors was linear up to the maximum pressure where the membrane is deflected to the bottom of the cavity by the applied pressure (Figure 16.5).

As with the force/tactile/pressure sensors described earlier, other means of measuring the deflection of the diaphragm with applied pressure could be employed such as by using piezoelectric materials or through changes in capacitance. The absolute pressure sensors are very similar to the force/tactile/pressure sensors described earlier except that the absolute pressure sensors employed a fixed pressure, vacuum cavity as a reference pressure for the sensor while the force/tactile/pressure sensors had an open cavity backing the diaphragm to make that sensor insensitive to changes in atmospheric pressure.

16.6.4 ACCELEROMETERS

MEMS accelerometers have been developed on flexible substrates so that they can be conformal to nonplanar surfaces. In this work, *x*-, *y*-, and *z*-axis accelerometers were simultaneously fabricated of the same die using surface micromachining [23]. The accelerometers were hermetically packaged

(a)

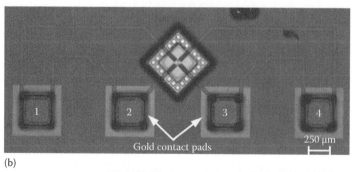

(b)

FIGURE 16.5 (a) A 2 × 2 cm² section of flexible substrate containing an array of 75 absolute pressure sensors. (b) An confocal micrograph of a completed absolute pressure sensor. (© 2012 IEEE. Reprinted, with permission, from MEMS absolute pressure sensor on a flexible substrate, by Ahmed, M., Butler, D.P., and Celik-Butler, Z., in *Proceedings of the 2012 IEEE MEMS Conference*, Paris, France, January 29–February 2, 2012, pp. 575–578.)

by bonding a polyimide superstrate with cavities to surround the accelerometers using a wafer-level packaging approach [41]. The accelerometers consisted of a proof mass suspended by springs with air providing the damping. Capacitive sensing was employed to measure the change in the position of the proof mass by the acceleration it experienced. The accelerometers were designed to be able to move more easily in the direction of the acceleration they were to measure while being stiff in the other two perpendicular directions so they would not respond to any acceleration in those directions. The size and geometry of the proof mass were designed to achieve the desired damping factor (~0.6–0.7) as well as the acceleration sensitivity and range. In-plane (x and y) accelerometers were identical but oriented perpendicular to each other on the die. The x and y accelerometers utilized differential comb capacitors. The acceleration perpendicular to the substrate was measured with a z-axis accelerometer that utilized the parallel plate capacitance between the proof mass and an electrode on the substrate. The accelerometers were designed to measure up to $20g$ (Figure 16.6).

The accelerometers were fabricated by surface micromachining that utilized a UV-LIGA process to fabricate the nickel proof mass and spring in a photoresist mold by electroplating the nickel. After electroplating, the mold and sacrificial layer that suspended the proof mass above the substrate were removed. The accelerometers were characterized by using a capacitance readout amplifier and a shaker to provide the acceleration. Accelerations up to $7g$ were provided by the shaker at frequencies up to 800 Hz. A commercially available reference accelerometer was used to provide a calibrated measurement of the acceleration provided by the shaker. The sensitivity was measured to be 10 fF/g for the x- and y-axis accelerometers with the comb capacitors and 20 fF/g for the z-axis accelerometers with the parallel plate capacitor between the substrate and the proof mass. The minimum acceleration that can be measured was limited by the noise from the capacitance readout circuit to be 4.8 mg/$\sqrt{\text{Hz}}$ while the thermomechanical noise floor for the accelerometers was 63–145 µg/$\sqrt{\text{Hz}}$ for the x- and y-axis accelerometers and 189–271 µg/$\sqrt{\text{Hz}}$ for the z-axis accelerometers. To reach the thermomechanical noise limit of the accelerometers, a lower noise readout circuit would be required as done by Tsai and Fedder [11]. In this investigation, it was observed that the accelerometers fabricated on flexible, conformal substrates behave equally well to accelerometers with identical design fabricated on rigid silicon substrates.

To package the accelerometers in a hermetic air cavity, a wafer-level packaging was employed with a polyimide superstrate that contained cavities to surround the devices [34]. The superstrate was fabricated on a separate carrier wafer patterning the device cavities with a photo-definable polyimide. The superstrate was then thermally bonded to the substrate using a wafer bonder. The superstrate sealed the accelerometers into a hermetic cavity without affecting their performance and provided a stable atmosphere for them to operate.

FIGURE 16.6 Electroplated z-axis (right image) and integrated x-, y-axes (left image) accelerometers. (© 2011 IEEE. Reprinted, with permission, from Surface Micromachined MEMS Accelerometers on Flexible Polyimide Substrate by Gönenli, I.E., Celik-Butler, Z., and Butler, D.P., in *IEEE Sens. J.*, 11, 2318, 2011.)

16.7 CONCLUSIONS AND FUTURE DIRECTIONS

These examples demonstrate that flexible, conformal MEMS sensors with a variety of functions can be fabricated using flexible films cast onto a rigid carrier wafer. MEMS sensors on conformal substrates can be realized with an equivalent performance to those fabricated on rigid silicon substrates. This approach allows conventional fabrication facilities to be used for the fabrication of the flexible sensors. The flexible films containing the sensors can be cut into die or flexible integrated circuits for integration with other devices and circuit boards. The use of flexible substrates allows for the fabrication of ultrathin sensor circuits that can lead to a reduction in system size. In addition, a reduction in fabrication costs may also be possible since the carrier wafers can be reused after the flexible film containing the MEMS sensors is removed.

At present, conventional metals, insulators, and semiconductor materials have been employed in device fabrication. The continuous development of new materials such as conducting rubbers, nanostructured materials, and nanocomposites may enable greater bending ability and even a high level of stretching ability in the future. The investigation of 2D materials such as graphene and silicene may make the incorporation of high-performance active circuits for signal processing and communication in flexible films easier in the future.

ACKNOWLEDGMENTS

This material is based in part upon work supported by the National Science Foundation and Air Force Office of Scientific Research.

REFERENCES

1. Using the invention of the vacuum triode in 1906 by Robert von Lieben and independently Lee De Forest as the event triggering the electronic age. The first flexible circuit board patent is reported to be issued to Albert Hanson in 1903.
2. S. M. Venugopal, D. R. Allee, M. Quevedo-Lopez, B. Gnade, E. Forsythe, and D. Morton, Flexible electronics: What can it do? What should it do? *Proceedings of 2010 IEEE International Reliability Physics Symposium (IRPS)*, May 2–6, 2010, Anaheim, CA, pp. 644–649.
3. K. J. Lee, M. J. Motala, M. A. Metil, W. R. Childs, E. Menard, A. K. Shim, J. A. Rogers, and R. Z. Nuzzo, Large-area, selective transfer of microstructured silicon: A printing-based approach to high-performance thin-film transistors supported on flexible substrate, *Adv. Mater.*, 17, 2332–2336, 2005.
4. D. H. Kim, J. H. Ahn, W. M. Choi, H. S. Kim, T. H. Kim, J. Song, Y. Y. Huang, Z. Liu, C. Lu, and J. A. Rogers, Stretchable and foldable silicon integrated circuits, *Science*, 320, 507–511, 2008.
5. R. L. Chaney, and D. R. Hackler, High performance single crystal CMOS on flexible polymer substrate, *36th Annual GOMACTech Conference*, March 21–24, 2011, Orlando, FL.
6. D. Shahrjerdi, S. W. Bedell, A. Khakifirooz, K. Fogel, P. Lauro, K. Cheng, J. A. Ott, M. Gaynes, and D. K. Sadana, Advanced flexible CMOS integrated circuits on plastic enabled by controlled spalling technology, *Proceedings of the 2012 IEEE International Electron Devices Meeting (IEDM)*, December 10–12, 2012, San Francisco, CA, pp. 5.1.1–5.1.4.
7. P. G. LeComber, W. E. Spear, and A. Ghaith, Amorphous silicon field-effect device and possible application, *Electron. Lett.*, 15, 179–181, 1979.
8. T. Afentakis, M. Hatalis, A. T. Voutsas, and J. Hartzell, Design and fabrication of high-performance polycrystalline silicon thin-film transistor circuits on flexible steel foils, *IEEE Trans. Electron Dev.*, 53, 815–822, 2006.
9. W. Helfrich, and W. Schneider, Recombination radiation in anthracene crystals, *Phys. Rev. Lett.*, 14, 229–231, 1965.
10. H. Koezuka, A. Tsumura, and T. Ando, Field-effect transistor with polythiophene thin film, *Synth. Met.*, 18, 699–704, 1987.
11. V. K. Varadan and V. V. Varadan, Microsensors, microelectromechanical systems (MEMS), and electronics for smart structures and systems, *Smart Mater. Struct.*, 9, 953–972, 2000.
12. Based on a simple calculation that an Intel six-core i7 microprocessor retails for ~$300 in 2013 and has ~2.3 billion 32 nm transistors.

13. B. Dang, P. Andry, C. Tsang, J. Maria, R. Polastre, R. Trzcinski, A. Prabhakar, and J. Knickerbocker, CMOS compatible thin wafer processing using temporary mechanical wafer, adhesive and laser release of thin chips/wafers for 3D integration, *Proceedings of the 2010 60th Electronic Components and Technology Conference (ECTC)*, June 1–4, 2010, Las Vegas, NV, pp. 1393–1398.
14. C. Banda, R. W. Johnson, T. Zhang, Z. Hou, and H. K. Charles, Flip chip assembly of thinned silicon die on flex substrates, *IEEE Trans. Electron. Packaging Manuf.*, **31**, 1–8, 2008.
15. T. Loher, D. Schutze, W. Christiaens, K. Dhaenens, S. Priyabadini, A. Ostmann, and J. Vanfleteren, Module miniaturization by ultra thin package stacking, *Proceedings of the 2010 3rd Electronic System-Integration Technology Conference (ESTC)*, September 17–20, 2010, Amsterdam, the Netherlands, pp. 1–5.
16. S. Priyabadini, T. Sterken, L. Van Hoorebeke, and J. Vanfleteren, 3-D stacking of ultrathin chip packages: An innovative packaging and interconnection technology, *IEEE Trans. Compon. Packaging Manuf. Technol.*, **3**, 1114–1122, 2013.
17. G. C. Dogiamis, B. J. Hosticka, and A. Grabmaier, Investigations on an ultra-thin bendable monolithic Si CMOS image sensor, *IEEE Sens. J.*, **13**, 3892–3900, 2013.
18. M. Ahmed, M. M. Chitteboyina, D. P. Butler, and Z. Celik-Butler, Temperature sensor in a flexible substrate, *IEEE Sens. J.*, **12**, 864–869, 2012.
19. M. Ahmed, D. P. Butler, and Z. Celik-Butler, MEMS absolute pressure sensor on a flexible substrate, *Proceedings of the 2012 IEEE MEMS Conference*, Paris, France, January 29–February 2, 2012, pp. 575–578.
20. S. K. Patil, Z. Celik-Butler, and D. P. Butler, Characterization of MEMS piezoresistive pressure sensors using AFM, *Ultramicroscopy*, **110**, 1154–1160, 2010.
21. M. Ahmed, M. Chitteboyina, D. P. Butler, and Z. Celik-Butler, Micromachined force sensor in a flexible substrate, *IEEE Sensors J*, **13**, 4081–4089, 2013.
22. R. Kilaru, Z. Celik-Butler, D. P. Butler, and İ. E. Gönenli, NiCr MEMS tactile sensors embedded in polyimide, towards a smart skin, *IEEE/ASME J. Microelectromech. Syst.*, **22**, 349–355, 2013.
23. İ. E. Gönenli, Z. Çelik-Butler, and D. P. Butler, Surface micromachined MEMS accelerometers on flexible polyimide substrate, *IEEE Sensors J.*, **11**, 2318–2326, 2011.
24. S. A. Dayeh, D. P. Butler, and Z. Celik-Butler, Micromachined infrared bolometers on flexible polyimide substrates, *Sensors Actuat. A*, **118**, 49–56, 2005. Erratum: *Sensors Actuat. A*, **125**, 597–598, 2006.
25. A. Mahmood, D. P. Butler, and Z. Celik-Butler, Device-level vacuum-packaging scheme for microbolometers on rigid and flexible substrates, *IEEE Sens. J.,* **7**, 1012–1019, 2007.
26. A. Drost, G. Klink, M. Feil, and K. Bock, Studies of fine pitch patterning by reel-to-reel processes for flexible electronic systems, *Proceedings of the International Symposium on Advanced Packaging Materials: Processes, Properties and Interfaces*, Irvine, CA, March 16–18, 2005, pp. 130–135.
27. M. Ahmed, and D. P. Butler, Flexible substrate and release layer for flexible MEMS devices, *Journal of Vacuum Science and Technology B.*, **31**(5), 050602, Sep/Oct 2013.
28. A. Yildiz, Z. Çelik-Butler, and D. P. Butler, Microbolometers on a flexible substrate for infrared detection, *IEEE Sens. J.*, **4**, 112–117, 2004.
29. L. Wang, K. M. B. Jansen, M. Bartek, A. Polyakov, and L. J. Ernst, Bending and stretching studies on ultra-thin silicon substrates, *Proceedings of the 2005 6th International Conference on Electronic Packaging Technology*, Shenzhen, China, September 2, 2005, pp. 1–5.
30. For example, Aetna Plastics, Valley View, Ohio, PEI polyetherimide data sheet.
31. For example, Victrex plc, Lancashire, United Kingdom, Victrex PEEK polymer datasheet.
32. For example, Chevron Phillips Chemical Co., The Woodlands, Texas, Ryton PPS datasheet.
33. For example, DuPont Inc., USA, Kapton polyimide film datasheet.
34. For example, DuPont Inc., USA, Mylar polyester film datasheet.
35. M. Kotera, T. Nishino, and K. Nakamae, Imidization processes of aromatic polyimide by temperature modulated DSC, *Polymer*, 41, 3615–3619, 2000.
36. W. W. Y. Chow, K. F. Lei, G. Shi, W. J. Li, and Q. Huang, Microfluidic channel fabrication by PDMS-interface bonding, *Smart Mater. Struct.*, 15, S112–S116, 2006.
37. J. Biros, T. Larina, J. Trekoval, and J. Pouchly, Dependence of glass transition temperature of poly on their tacticity, *Colloid Polym. Sci.*, 260, 27–30, 1982.
38. Estane Thermoplastic Polyurethane, general brochure, Lubrizol Advance Material Inc., Cleveland, OH.
39. H.-S. Noh, Y. Huang, and P. J. Hesketh, Parylene micromolding, a rapid and low-cost fabrication method for parylene microchannel, *Sensors Actuat. B Chem.*, 102(1), 78–85, September 2004.
40. P.-Y. Li, T. K. Givrad, D. P. Holschneider, J.-M. I. Maarek, and E. Meng, A parylene MEMS electrothermal valve, *J. Microelectromech. Syst.*, 18, 1184–1197, 2009.

41. M. Ahmed, I. E. Gonenli, G. S. Nadvi, R. Kilaru, D. P. Butler, and Z. Celik-Butler, MEMS sensors on flexible substrate towards a smart skin, *Proceedings of the IEEE Sensors Conference*, Taipei, Taiwan, pp. 881–884, October 28–31 2012.

42. W. Christiaens, E. Bosman, and J. Vanfleteren, UTCP: A novel polyimide-based ultra-thin chip packaging technology, *IEEE Trans. Compon. Packaging Technol.*, **33**, 754–760, 2010.

43. T.-Y. Kuo, Z.-C. Hsiao, Y.-P. Hung, W. Li, K.-C. Chen, C.-K. Hsu, C.-T. Ko, and Y.-H. Chen, Process and characterization of ultra-thin film packages, *Proceedings of the 2010 5th International Microsystems Packaging Assembly and Circuits Technology Conference (IMPACT)*, Taipei, Taiwan, October 20–22, 2010, pp. 1–4.

44. M. Op de Beeck, A. La Manna, T. Buisson, E. Dy, D. Velenis, F. Axisa, P. Soussan, and C. Van Hoof, An IC-centric biocompatible chip encapsulation fabrication process, *Proceedings of the 2010 3rd Electronic System-Integration Technology Conference (ESTC)*, September 17–20, 2010, Amsterdam, the Netherlands, pp. 1–6.

45. T. Sterken, J. Vanfleteren, T. Torfs, M. Op de Beeck, F. Bossuyt, and C. Van Hoof, Ultra-thin chip package (UTCP) and stretchable circuit technologies for wearable ECG system, *Proceedings of the 2011 Annual International Conference of the IEEE Engineering in Medicine and Biology Society (EMBC)*, August 30–September 3, 2011, Boston, MA, pp. 6886–6889.

46. Z. Suo, E. Y. Ma, H. Gleskova, and S. Wagner, Mechanics of rollable and foldable film-on-foil electronics, *Appl. Phys. Lett.*, **74**, 1177–1179, 1999.

47. D. I. Forehand and C. L. Goldsmith, Wafer level micropackaging for RF MEMS switches, *Proceedings of IPACK2005, ASME InterPACK '05*, San Francisco, CA, July 17–22, 2005, pp. 1–5.

48. M. S. Rahman, M. Chitteboyina, D. P. Butler, Z. Celik-Butler, S. Pacheco, and R. McBean, Device-level vacuum packaged MEMS resonator, *IEEE/ASME J. Microelectromech. Syst.*, **19**, 911–917, 2010.

49. C.-Y. Hsieh, J.-S. Chen, W.-A. Tsou, Y.-T. Yeh, K.-A. Wen, and L.-S. Fan, A biocompatible and flexible RF CMOS technology and the characterization of the flexible MOS transistors under bending stresses, *IEEE 22nd International Conference on Micro Electro Mechanical Systems (MEMS 2009)*, Sorrento, Italy, January 25–29, 2009, pp. 627–629.

50. F. Jiang, Y.-C. Tai, K. Walsh, T. Tsao, G.-B. Lee, and C.-M. Ho, A flexible MEMS technology and its first application to shear stress sensor skin, *Proceedings of the IEEE Tenth Annual International Workshop on Micro Electro Mechanical Systems*, Nagoya, Japan, January 26–30, 1997, pp. 465–470.

51. J.A. Rogers, Semiconductor devices inspired by and integrated with biology, *Proceedings of the 2012 IEEE 25th International Conference on Micro Electro Mechanical Systems (MEMS)*, Paris, France, January 29–February 2, 2012, pp. 51–55.

52. M. Ahmed, D. P. Butler, and Z. Celik-Butler, MEMS relative pressure sensor on flexible substrate, *Proceedings of the 2011 IEEE Sensors Conference*, Limerick, Ireland, October 28–31, 2011, pp. 460–463.

53. A. van der Ziel, *Fluctuation Phenomena in Semiconductors*, London, U.K.: Butterworth, 1959.

54. F. N. Hooge, 1/f noise is no surface effect, *Phys. Lett. A*, 29, 139–140, 1969.

55. F. N. Hooge, 1/f noise, *Phys. B+C*, 83, 14–23, 1976.

56. F. N. Hooge, The relation between 1/f noise and number of electrons, *Physica B Condens. Matter*, **162**, 344–352, 1990.

57. S. U. Jen, C. C. Yu, C. H. Liu, and G. Y. Lee, Piezoresistance and electrical resistivity of Pd, Au, and Cu films, *Thin Solid Films*, **434**, 316–322, 2003.

58. T. B. Gabrielson, Mechanical-thermal noise in micromachined acoustic and vibration sensors, *IEEE Trans. Electron Dev.*, **40**, 903–909 (1993).

59. J.M. Tsai and G. K. Fedder, Mechanical noise-limited CMOS-MEMS accelerometers, *Proceedings of the 18th IEEE International Conference on MEMS*, Miami Beach, FL, January 30–February 3, 2005, pp. 630–633.

17 Embedded Temperature Sensors to Characterize RF and mmW Analog Circuits

Josep Altet, Diego Mateo, and Jose Silva-Martinez

CONTENTS

17.1 INTRODUCTION

Technology scaling has involved advances in the integration of radio frequency (RF) circuits that have allowed the development of high-performance communication systems. As the complexity of these systems-on-chip (SoCs) grows, their complexity and performances increase, and their debugging and testing becomes more difficult, mainly due to the limited observability of their RF nodes. Being able to observe the performances of individual blocks that constitute the transceiver chain is beneficial for the identification of faulty devices that would result in improved efficiency during production testing. At the same time, technology scaling has provoked the increase of technological parameter variation and more severe aging effects; furthermore, as the technology scales down, both time-zero variability (process variation) and time-dependent variability (aging) gain in importance as their effect on both digital and analog/RF circuitry becomes greater [Yid11,Garg13]. The availability of reliable internal RF power measurements promises to ease the pass–fail testing of transceivers and also may enable the use of active knobs embedded in self-healing schemes to enhance system performance and to correct electrical performance modifications due to process variation and aging [Ona12]. In conventional built-in self test (BIST) characterization strategies, power detectors are strategically placed at nodes of the circuit under characterization in order to measure the power of a test signal along the signal path. The paradigm is that electric power detectors have to be reliable and must operate at RF frequencies to be useful. Although small, the finite input impedance of RF detectors degrades the performance of the transceiver chain, especially if the auxiliary devices are placed at critical nodes such as the low-noise amplifier (LNA) input, mixer, and frequency synthesizer where additional parasitic capacitances cannot be tolerated [Bow13]. Evidently, the degradation becomes more critical as system performance and operating frequency increase. In this direction, a novel noninvasive test and characterization strategies suitable for RF SoCs employing DC and low-frequency noninvasive temperature measurements are described in this chapter.

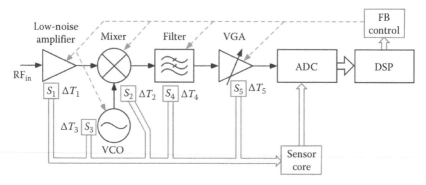

FIGURE 17.1 Example of a receiver diagram with thermal sensing strategy.

Thermal coupling through the semiconductor substrate generates an increase in the temperature around the circuit/device under test/characterization that is proportional to its electric power dissipation [Ant80]. Thermal power has two components: the first one is due to the DC biasing of the system, and the second one is a low-frequency signature that indicates the behavior of the RF signals. A remarkable property of these signals is that they predict both the biasing conditions of the RF system and the strength of the RF signal through DC and low-frequency temperature measurements [Mat06,Alt13].

Figure 17.1 shows a common structure for an RF receiver chain. The observability of critical blocks such as the LNA, voltage-controlled oscillator (VCO), or mixer outputs is not straightforward, so the characterization of these blocks is difficult. By adding thermal sensors to each one of these circuits, it is possible to estimate the power it is dissipating or the voltages/currents associated with the device under test; hence, it becomes possible to characterize the electrical performance of the circuits by analyzing the noninvasive thermal measurements. In the example of Figure 17.1, this information is used to correct the circuits' performance variations by means of a feedback control loop that drives tuning knobs [Ona12].

This chapter is organized as follows: Section 17.2 presents the physical principles of the thermal monitoring to test/characterize analog/RF circuits. Aspects of the thermal coupling mechanism affecting the temperature test are detailed as well as a discussion on how the Joule effect behaves as a frequency down-conversion mixer that allows DC and low-frequency measurements to characterize the electrical high-frequency performances of the circuit under test (CUT). Section 17.3 further elucidates the relationship between electrical high-frequency performances and DC and low-frequency temperature variations. By using a common-source tuned-load amplifier, the relations between electrical performances (gain, frequency response, linearity, and power efficiency) and temperature variations are analytically derived. Section 17.4 presents different methods to measure temperature, which focus on the use of built-in differential temperature sensors to measure electrical performances at low cost. Section 17.5 shows experimental results of the electrical performances obtained through thermal measurements by means of built-in thermal sensors, validating the analysis presented in previous sections. Finally, Section 17.6 summarizes the chapter's key points in a conclusion.

17.2 PHYSICAL PRINCIPLES OF TEMPERATURE MONITORING TO TEST RF CIRCUITS

The goal of this section is twofold: (1) to review the physical mechanisms that link CUT operation and the temperature increase that this circuit generates in its nearby silicon surface and (2) to develop suitable stimuli, which allow reliable temperature measurements for monitoring CUT high-frequency characteristics.

Let's consider a silicon die such as the one depicted in Figure 17.2a. The high-frequency CUT and the temperature sensor are placed close to each other; T_S is the sensed temperature. For simplicity, let's assume that the CUT is linear and that it is biased with a DC voltage source and driven with a high-frequency sinusoidal AC voltage. Before the CUT is powered (i.e., the voltage given by

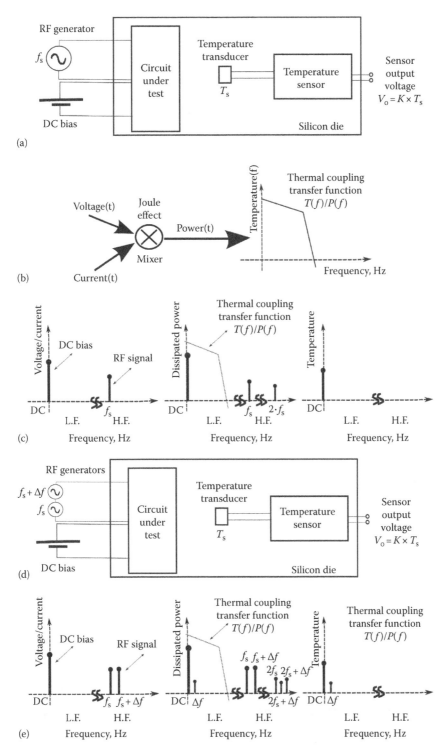

FIGURE 17.2 Description of the physical mechanisms that take place in built-in temperature measurements: (a) a linear CUT and a temperature sensor are placed in the same silicon die; (b) circuital model of the thermal coupling mechanism; (c) spectral components of the CUT electrical signals, power dissipated by the CUT, and temperature at the transducer location with homodyne excitation; (d) heterodyne CUT excitation; and (e) spectral components under heterodyne CUT excitation.

all the sources applied to this circuit is zero) and assuming that the temperature sensor has negligible self-heating, the temperature at the silicon surface is homogeneous and equal to the ambient temperature. When the CUT is biased, there is a temperature increase in its nearby silicon surface. Moreover, if now the RF stimulus is applied, the CUT surrounding temperature changes.

Two physical mechanisms are involved in this process: the Joule effect and thermal coupling through the silicon substrate. When a voltage difference is applied across a conducting material (i.e., a device), an electric field is generated within this conductor. The electric field generates a force that acts upon free carriers (electrons and holes), which experience acceleration in the direction of the electric field favoring their polarity. Since the carriers travel across the device, they experience collisions with impurity atoms, phonons, and each other. Then, all the kinetic energy of the carrier is transferred into the conductive material, which results in an increase of the internal vibration of the atoms (so named internal heat for historical reasons), resulting in an increase of its temperature [Lun90]. If there is no electrical energy storage due to the creation of electric fields (in the case of capacitors) or magnetic fields (in the case of inductances), the energy rate (J/s or W) at which the device transfers the electrical energy into internal heat (or thermal energy) is given by the Joule effect:

$$P(t) = V(t) \cdot I(t) \tag{17.1}$$

where
   $P(t)$ is the instantaneous power dissipated by the device [J/s or W]
   $V(t)$ is the voltage difference applied across the device
   $I(t)$ is the current flowing through it (C/s or A)

Let's now review the principles of thermal coupling. Let us divide the integrated circuit (IC) substrate into small cubes. The cubes that form the device receive thermal energy (heat) due to the Joule effect. Part of this transferred thermal energy is stored in each cube (by means of an increase of its internal atom vibration) causing an increase in its temperature. The amount of thermal energy delivered to the unity mass of a certain material that increases its temperature by 1° (Celsius or Kelvin) defines the specific heat capacity of the material (J/(K · kg)) [Tho92]. Then,

$$Cth \cdot M \cdot \Delta T = \Delta E \tag{17.2}$$

where
   ΔT is the temperature increase of the cube due to the increase of energy ΔE stored in the cube
   Cth is the specific heat of the material
   M is the mass of the cube

It is interesting to note the parallelism between the mathematical expression that relates temperature–specific heat capacitance–thermal energy with the one that relates voltage–capacitance–charge in an electrical capacitor [Tho92].

Part of the heat that the cubes are receiving due to the Joule effect is stored in the cube itself, while the other thermal energy portion is transferred to the neighboring cubes. Fourier's law of heat conduction states that there is a flow of heat (J/s or W) from the hot cube to the cold cube that is proportional to the difference of temperatures between the cubes and inversely proportional to the thermal resistance between them both. The thermal resistance (its reciprocal is the thermal conductance) is a property of each material and depends on the cube dimensions [Tho92]. There is again a parallelism between the mathematical expression that relates energy flow (or power), thermal resistance, and temperature difference with Ohm's law that relates charge flow (or current), electrical resistance, and potential difference (or voltage).

The mathematical expression of the temperature variation as a function of the power dissipated by the CUT, the IC dimensions, and time can be obtained considering the energy conservation principle. In each cube, the net balance of thermal energy must be zero. Each cube receives thermal energy from two sources: from neighboring hotter cubes (due to heat conduction) and due to the Joule effect (electrical energy dissipated by the CUT). In addition, each cube delivers thermal energy to neighboring cooler cubes. Finally, part of the thermal energy received is not transferred, but absorbed causing a cube temperature increase. This is mathematically expressed in the following equation:

$$\frac{\text{Thermal energy stored}}{\text{Time}} = \text{Power Joule effect} + \text{Net heat flow conduction} \qquad (17.3)$$

where
   the term at the left of the equation is the thermal energy that absorbs the cube to increase its
      temperature
   the second term is the thermal energy originated by the Joule effect
   the last term is the difference between the thermal energy received from neighboring hotter
      cubes and the one delivered to neighboring cooler ones

This equation eventually can be written as an ordinary differential equation. Nevertheless, due to the parallelism between the thermal and electrical equations, the thermal coupling can be electrically modeled with a resistor–capacitor network where each node represents a small cube of the silicon die. The voltage at each node represents the mean temperature of the cube. The thermal energy absorbed by the cube is modeled by the charge stored in a capacitor connected from this node to a voltage source representing the ambient temperature. The thermal energy generated within the cube per unit of time is equivalent to the charge per unit of time delivered by a current source connected to this node, while the balance of energy through the cubes edges due to heat conduction is modeled by the electrical resistances that connect the nodes corresponding to neighboring cubes [Tho92,Lee93,Ona11]. The energy conservation principle in this model becomes current conservation represented by the Kirchhoff current law. Boundary conditions have to be specified in this model to obtain the solution; for that purpose, usually, the bottom surface of the silicon die is assumed to be at a constant temperature [Tho92,Lee93,Ona11]. Two of the properties of the thermal coupling become apparent: (1) the relationship between power dissipation and temperature gradients is linear and holds for low values of temperature increase [Bon95] since both the thermal resistance and the specific heat are temperature dependent, and (2) this function has a low-pass filter behavior attenuating the high-frequency components. The work reported in [Nen04] shows temperature measurements performed up to 100 kHz, whereas the work reported in [Alt12] reaches 1 MHz.

Figure 17.2b shows an electrical model of a complete thermal coupling mechanism. A key property of the Joule effect is that the electrical power dissipation behaves like a conventional signal mixer wherein the low-frequency components of the dissipated power (i.e., the ones that generate temperature increase) carry on a replica of the high-frequency electrical signals flowing through the CUT.

To achieve an efficient temperature signature on the silicon surface that can be effectively used to monitor the high-frequency behavior of the CUT, two possible measurement setups are possible:

   1. Homodyne temperature measurements (Figure 17.2a–c): The CUT (for simplicity, let's assume
      that is linear and purely resistive) is DC biased and driven with a single RF sinusoidal tone of
      frequency f_s. The voltage and current flowing through the CUT device can be written as

$$v(t) = V_{DC} + A \cdot \cos\left(2\pi f_s t\right)$$

$$i(t) = I_{DC} + B \cdot \cos\left(2\pi f_s t\right) \qquad (17.4)$$

Then, the power dissipated by this device generates low-frequency and high-frequency signatures given by

$$P(t) = \left[V_{DC}I_{DC} + \frac{AB}{2} \right] + \left[BV_{DC} + AI_{DC} \right] \cos\left(2\pi f_s t\right) + \frac{AB}{2}\cos\left(4\pi f_s t\right) \qquad (17.5)$$

For the case $f_s \gg 100$ kHz, then only the DC component of the dissipated power will generate a temperature increase over the silicon surface around the CUT, and that can be detected by the temperature transducer placed very close to the CUT:

$$T(t) = T_{DC} = R_{TH} \cdot \left[V_{DC}I_{DC} + \frac{AB}{2} \right] = R_{TH} \cdot P_{DC} \qquad (17.6)$$

where

R_{TH} is the thermal coupling resistance between the CUT and the temperature transducer
P_{DC} is the DC component of the power in Equation 17.5

According to the aforementioned simplified RC electrical model, when the power dissipated has only DC components, then all the capacitors behave just as open circuits and the relationship between the DC current and the DC voltage is only due to thermal resistances. It is interesting to notice that the DC temperature increase depends on both the DC bias of the CUT and the amplitude of the RF signals, but insensitive to the RF signal frequency f_s. Figure 17.2c shows the frequency components.

2. Heterodyne temperature measurements: The CUT is DC biased and driven with a two-tone RF sinusoidal signal of equal amplitude at f_s and $f_s + \Delta f$, respectively. This scenario is depicted in Figure 17.2d. The CUT (assumed linear and resistive) voltage and current can be written as

$$v(t) = V_{DC} + A \cdot \left[\cos\left(2\pi f_s t\right) + \cos\left(2\pi\left(f_s + \Delta f\right)t\right) \right]$$
$$i(t) = I_{DC} + B \cdot \left[\cos\left(2\pi f_s t\right) + \cos\left(2\pi\left(f_s + \Delta f\right)t\right) \right] \qquad (17.7)$$

The power dissipated by the CUT generates several spectral components (see Figure 17.2e). Among them, the two most interesting are

$$P(t) = \left[V_{DC}I_{DC} + AB \right] + AB\cos\left(2\pi\Delta f t\right) + \text{Higher frequency terms} \qquad (17.8)$$

The low-frequency components generate measurable temperature increase:

$$T(t) = R_{TH} \cdot \left[V_{DC}I_{DC} + AB \right] + Z_{TH}\left(\Delta f\right) \cdot AB \cdot \cos\left(2\pi\Delta f t\right)$$
$$= R_{TH} \cdot P_{DC} + Z_{TH}\left(\Delta f\right) \cdot P_{\Delta f}\left(t\right) = T_{DC} + T_{\Delta f}\left(t\right) \qquad (17.9)$$

where

$P_{\Delta f}(t)$ is the spectral component at Δf of the power in Equation 17.8
$Z_{TH}(\Delta f)$ is the thermal coupling impedance between the CUT and the temperature transducer at the frequency Δf, which in turn is usually complex since the nodal equivalent impedance of the RC model is in general complex

The DC temperature increase (T_{DC}) depends on both the DC bias and the power of the RF stimuli. According to Equation 17.9, there is an AC component of the temperature increase generated at Δf, $T_{\Delta f}(t)$, whose amplitude depends only on the power of the test RF signals present in the CUT, and this AC component is independent of f_S. The measurement of this spectral component of the temperature is known as lock-in temperature measurements [Bre03].

Although lock-in temperature measurements are more complex in terms of the amount of instrumentation needed, they provide some advantages over homodyne temperature measurements. First, the heterodyne component of the temperature increase only depends on the RF CUT characteristics; after converting this information into digital format, precise measurements allow better characterization of the CUT. Second, by controlling the value of Δf, temperature measurements become independent of the particular IC mounting configuration (package, socket, etc.), that is, the silicon die can be considered as a semi-infinite medium (considering the RC model where for a given high frequency, big capacitances behave as short circuits). For example, this has been achieved in [Alt01] when the spectral component of the temperature has a frequency higher than 100 Hz. This fact makes the calibration setup easier when a close relationship between the temperature measurement and the dissipated power is desired. Moreover, since thermal coupling is a diffusion-like physical mechanism, the value of Δf chosen has a direct impact on the spatial range of the heterodyne temperature increase generated by the power dissipated by the CUT. Suitable high-frequency values of the heterodyne frequency ensure that the temperature increase generated by the CUT only affects the temperature sensor placed in close proximity to the CUT, and other temperature sensors placed far away from the CUT will experience only the background temperature [Alt08]. This provides an intrinsic partitioning of the silicon substrate, that is, a natural isolation when testing multiple devices at the same time.

17.3 THE RELATIONSHIP BETWEEN ANALOG CIRCUITS' ELECTRICAL PERFORMANCES AND THE TEMPERATURE INCREASE GENERATED

In this section, we analyze some figures of merit of a simple amplifier whose performance can be tracked through temperature monitoring. The CUT, shown in Figure 17.3, is a peak resonant common-source amplifier, that is, a common-source amplifier with an RLC load that resonates at the frequency ω_0.

Assuming a first-order linear model for the transistor, the small signal amplifier's voltage gain is [Raz01]

$$A_v = -g_m Z_L \rightarrow |A_v| = g_m |Z_L|$$ (17.10)

where
   Z_L is the impedance of the RLC amplifier's load
   g_m is the transistor small signal transconductance

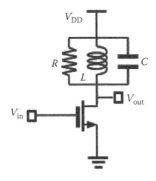

FIGURE 17.3 Simplified schematic of a tuned-load common-source amplifier.

The instantaneous drain–source power can then be obtained as

$$Pot_M = V_{out}I = \left(V_{outDC} + \Delta V_{out}\right)\left(I_{bias} + \Delta I\right)$$

$$= V_{outDC}I_{bias} + \Delta V_{out}I_{bias} + V_{outDC}\Delta I + \Delta V_{out}\Delta I \tag{17.11}$$

where ΔV_{out} and ΔI are, respectively, the variation of the amplifier output voltage and the transistor drain current due to the RF signal applied to the amplifier's input. Assuming a sinusoidal input voltage $V_{in} = A\cos(\omega t)$ and a first-order linear model for the transistor, then,

$$\Delta I = g_m\Delta V_{in}$$

$$\Delta V_{out} = -\Delta I Z_L \tag{17.12}$$

According to Equation 17.12 and for the input voltage applied, the first and last terms of the Pot_M expression (17.11) will be the only terms that contribute to the DC power. After some straightforward algebra, we can obtain that the dissipated power's DC component can be computed as

$$Pot_M\big|_{DC} = V_{outDC}I_{bias} - \frac{1}{2}g_m A|Z_L|g_m A\cos\varphi = Pot_{Mbias} - \frac{1}{2}|A_v|g_m A^2\cos\varphi \tag{17.13}$$

where φ is the phase associated with the complex load impedance Z_L, which is 0 around the center frequency f_0 where Z_L peaks. The variation of the DC power when we apply a signal at the CUT input is proportional to the voltage gain of the circuit:

$$\Delta Pot_M\big|_{DC} = \frac{1}{2}|A_v|g_m A^2\cos\varphi \tag{17.14}$$

By sweeping the frequency around the center frequency and tracking the temperature variation at the same time, we will be able to track the variation of the power dissipated at DC and then to obtain CUT's shape, that is, the center frequency and the bandwidth of the resonant amplifier. To measure the amplifier bandwidth, the correction factor $\cos\varphi$ must be considered. Since the phase of the load at the frequency corresponding to the limits of the bandwidth defined as −3 dB power gain (i.e., when the voltage gain is reduced by a factor of $1/\sqrt{2}$) is ±45°, then the corrector factor is $\cos\varphi = 1/\sqrt{2}$. When measuring the temperature, just by using a $10\cdot\log(\text{Temp})$ scale and measuring the 3 dB bandwidth, we will get the 3 dB power gain bandwidth of the amplifier.

If two tones are used, then looking at the component at Δf (heterodyne method), we obtain

$$Pot_M\big|_{\Delta\omega} = \frac{1}{2}g_m A g_m A\left(\left(|Z_L|_{\omega_1}\cos\varphi_1 + |Z_L|_{\omega_2}\cos\varphi_2\right)\cos\left(\omega_2 - \omega_1\right)t\right.$$

$$\left. + \left(|Z_L|_{\omega_1}\sin\varphi_1 - |Z_L|_{\omega_2}\sin\varphi_2\right)\sin\left(\omega_2 - \omega_1\right)t\right) \tag{17.15}$$

where
$\Delta\omega = 2\pi\Delta f = (\omega_2 - \omega_1)$
φ_i is the phase of the load at ω_i
$|Z_L|_{\omega_i}$ is the module of the load at ω_i

Assuming that $\Delta\omega$ is much lower than the bandwidth of the load to be within the sensor's thermal bandwidth, then we can approximate:

$$\varphi_1 \approx \varphi_2$$
$$\left.|Z_L|\right._{\omega_1} \approx \left.|Z_L|\right._{\omega_2} \tag{17.16}$$

and the previous expression simplifies to

$$Pot_M\big|_{\Delta w} \cong g_m A\, g_m A \left.|Z_L|\right._{\omega_{1,2}} \cos\varphi_{1,2} \cos\left(\omega_2 - \omega_1\right)t \cong |A_v|\, g_m A^2 \cos\varphi_{1,2} \cos\left(\omega_2 - \omega_1\right)t \tag{17.17}$$

Therefore, the baseband dissipated power is proportional to the voltage gain:

$$Pot_M\big|_{\Delta\omega} \propto |A_v| \tag{17.18}$$

By sweeping the frequency of the two applied tones around the working frequency (keeping $\Delta\omega$ constant) and tracking the amplitude of the power variations of the frequency component at $\Delta\omega$, we should be able to monitor amplifier's frequency response, that is, the center frequency as well the bandwidth of the resonant amplifier.

Let us now consider the measurement of the power efficiency of a class A high-frequency power amplifier (PA). The circuit is depicted in Figure 17.4. In this case, the inductor acts as a choke inductor, and we assume that the inductance is high enough to be considered as an open circuit for RF frequencies. The capacitor is also a choke element, that is, its value is large enough to be considered a short circuit at RF frequencies.

For an input voltage, $V_{in} = A\cos(\omega_0 t)$, and assuming that the current flowing through the output is $I_{R_L} = B\cos(\omega_0 t)$, then we have that the power P_L dissipated in the load as

$$P_L = \frac{B^2 R_L}{2} \tag{17.19}$$

Notice that in this analysis, we are ignoring the effect of the parasitic capacitors; this assumption is questionable in RF circuits, but it serves the purpose of illustrating the proposing techniques. If the current flowing through the transistor is considered to be constant $I_L \approx I_{DC}$, then the power delivered by the power supply is constant $P_{V_{DD}} \approx I_{DC}V_{DD}$, while the power dissipated in the transistor P_M is

$$P_M \approx P_{V_{DD}} - P_L = I_{DC}V_{DD} - \frac{B^2 R_L}{2} \tag{17.20}$$

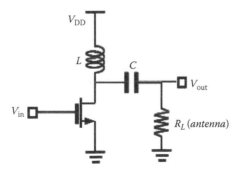

FIGURE 17.4 Simplified schematic of a class A PA.

When the signal is applied at the amplifier's input, the RF power is transferred to the load and both drain current and drain–source voltage are modulated by the signal. By tracking the temperature gradients around the channel of the transistor, we are able to track the variation in its power dissipated and then monitor the amplifier's power efficiency.

This technique does not require system calibration, but simply requires a couple of thermal measurements. Let us consider in more detail the temperature increase by defining T_S as the temperature at the monitoring spot, which is close to the transistor. Similarly to (17.6), the temperature increase at the monitoring spot can be expressed as KP_M, where K is the coupling thermal resistance (in °C/W) and P_M stands for the transistor's power dissipation.

If we perform a first thermal measurement when neither biasing nor signal is applied, the background temperature is obtained:

$$T_S = T_0 \tag{17.21}$$

If no other circuitry is turned on within the die, T_0 will be the ambient temperature. If now we perform a second thermal measurement when only the biasing circuitry is turned on, the transistor's temperature yields

$$T_S = T_1 = T_0 + K I_{DC} V_{DD} \tag{17.22}$$

If we perform a third thermal measurement when the biasing and the input signal are on, then the PA efficiency can be evaluated; in this case, the temperature is obtained as

$$T_S = T_2 = T_0 + K \left(I_{DC} V_{DD} - P_L \right) \tag{17.23}$$

By defining the power efficiency η as

$$\eta = \frac{P_L}{P_{V_{DD}}} = \frac{P_L}{I_{DC} V_{DD}} \tag{17.24}$$

it can be shown that the amplifier's power efficiency is obtained from the thermal measurements by computing

$$\eta = -\frac{T_2 - T_1}{T_1 - T_0} \tag{17.25}$$

A more in-depth analysis has been done in [Alt13], where a cascode structure is considered with similar results. Thermal measurements can indeed provide the required information to compute both power dissipated and power delivered to the load in all types of amplifiers such as the class A PA considered in this section.

The last electrical performance to be considered is the linearity of an amplifier. More specifically, the challenge is to measure the 1 dB compression point. In [Ona11], an in-depth analysis of the relation between the electrical 1 dB compression point and the behavior of the DC temperature measurement when the amplifier enters into the compression regime is considered. In such analysis, it is found that the DC component of the dissipated power when the input amplitude is swept presents a minimum (that the temperature will have as well) and that the measure of such a minimum can be used to infer the 1 dB compression point by means of the previous quantification of the shift from one to the other by simulation.

17.4   TEMPERATURE SENSING STRATEGY: DIFFERENTIAL TEMPERATURE SENSORS

The temperature increase generated by the CUT can be efficiently used to extract its high-frequency figures of merit. Two general monitoring strategies can be used: off-chip and on-chip temperature sensors.

Several off-chip temperature sensing techniques that have the sensitivity, spatial resolution (i.e., diameter of the area where the temperature is measured), and bandwidth to perform the required temperature measurements are available and include the infrared cameras [Bre03], laser reflectometer [Cla93], Internal infra red laser deflection (IIRLD) [Per09], and laser interferometer [Alt06]. A common limitation in these techniques, when used for testing purposes, is that they require direct optical access to the silicon die and an entire laboratory environment to perform the measurements. Moreover, metal layers and passivation layers placed over the silicon die can affect their measurement accuracy (except for the IIRLD technique). Nevertheless, these techniques are successfully used to monitor the characteristics of high-frequency CUT during failure analysis and during the debugging product stage.

On-chip (or built-in) temperature sensors, albeit the cost of an area overhead, provide enough flexibility for the test procedure, reduce the cost of the equipment needed, and allow in-field testing. Many of these devices can be deployed in SoCs since no direct visual access to the silicon die is required. Also, direct thermal coupling measurements can be taken without being affected by any layer placed over the silicon. The output signal of the typical built-in temperature sensor is proportional to the absolute temperature (implying only one temperature transducer); then,

$$\text{Signal}_{\text{OUT}} = G \cdot T_{\text{S}} \tag{17.26}$$

where

G is the sensor sensitivity
T_{S} is the temperature sensed by the transducer

On the other hand, the differential sensor outputs a signal that is proportional to the difference of temperature at two points of the silicon surface, whose output is described by the following equation:

$$\text{Signal}_{\text{OUT}} = S_{\text{Td}} \cdot \left(T_2 - T_1 \right) \tag{17.27}$$

where

T_2 and T_1 are the temperatures at two points of the silicon surface
S_{Td} is the sensor's differential sensitivity

Differential temperature measurements present two major advantages:

1. They are low sensitive to temperature increase that may offset the thermal map of the silicon surface, for example, overall surface temperature changes or different IC mounting configurations with different packaging thermal resistance.
2. This technique ensures faster measurements, as the thermal steady state has to be reached only in the silicon volume affected by the sensors.

Figure 17.5 shows the schematic of a differential temperature sensor [Ald10]. The sensor's circuit is basically an operational transconductance amplifier (OTA) composed by the emitter-coupled *npn* bipolar transistors Q_{S1} and Q_{S2} as the core, which are the temperature transducer devices. One of these devices is placed very close to the CUT to record temperature variations due to its power dissipation.

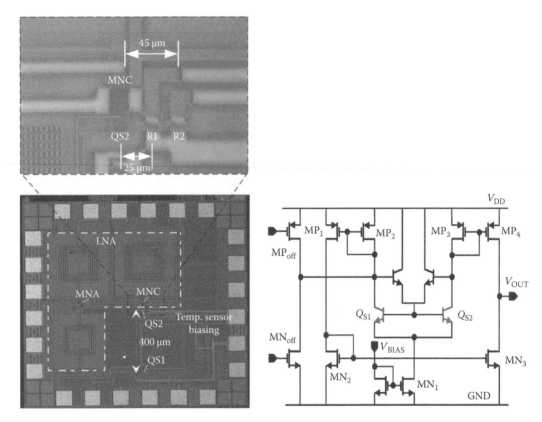

FIGURE 17.5 The images on the left show the layout of an LNA with the sensor integrated. The schematic on the right shows a differential temperature sensor. The detail shows the placement of some devices belonging to the LNA (the cascade transistor) and the sensor (the bipolar transistor Q_{S2}).

The second sensing device is placed 400 μm away from CUT. The temperature difference detected by Q_{S1} and Q_{S2} imbalances the current flowing through the collectors of the differential pair. The imbalance generated by the temperature difference is then detected and analyzed by using the variation of the collector current as a function of the temperature, g_T defined as

$$g_T = \frac{\partial I_C}{\partial T} \qquad (17.28)$$

where
I_C is the collector current of the bipolar transistor used as transducer
T is the absolute temperature

The current mirror pairs, MP_2–MP_1, MN_2–MN_3, and MP_3–MP_4, and the high output impedance of the node V_{OUT} r_o (given by the parallel of the MN_3 and MP_4 output impedances) transform the temperature difference detected by Q_{S1} and Q_{S2} into changes of the output voltage V_{OUT}. Assuming a 1:1 ratio in all current mirrors, variations at the output voltage ΔV_{OUT} due to temperature imbalances are computed using the following expression:

$$\Delta V_{OUT} = g_T \cdot r_o \cdot \Delta T \qquad (17.29)$$

This expression assumes that the metal oxide semiconductor (MOS) transistor used as current source MN_1 has infinite output impedance. Cascode current source topology can be used if better performance is required. The finite output impedance of the DC current source MN1 usually reduces the sensor differential sensitivity. Moreover, the differential sensor output voltage becomes a function of the differential temperature as well as dependent on the common-mode temperature (average) variations:

$$\Delta V_{OUT} = S_{Td} \cdot (T_2 - T_1) + S_{Tc} \cdot \frac{(T_2 + T_1)}{2} \tag{17.30}$$

where S_{Tc} is the sensitivity of the common temperature. If we assume that the silicon surface is at the ambient temperature T_a prior the CUT activation, then

$$\begin{aligned} T_1 &= T_a + \Delta T_1 \\ T_2 &= T_a + \Delta T_2 \end{aligned} \tag{17.31}$$

where ΔT_1 and ΔT_2 is the temperature increase at the temperature transducers' locations due to CUT activity. The differential sensor output becomes

$$\Delta V_{OUT} = S_{Td} \cdot (\Delta T_2 - \Delta T_1) + S_{Tc} \cdot \left(T_a + \frac{(\Delta T_2 + \Delta T_1)}{2} \right) \tag{17.32}$$

If a differential temperature sensor is used for built-in testing purposes, its common-mode sensitivity must be as small as possible, whereas the differential sensitivity must be as high as possible to improve its common-mode rejection ratio; this is a natural advantage when using differential pair–based sensing elements. In such a case, it will be only sensitive to the temperature increase generated by the CUT operation. The placement of the temperature transducers inside the silicon die will provide an output voltage proportional to $(\Delta T_2 - \Delta T_1)$ when the CUT is activated, according to Equation 17.6:

$$\Delta T_1 = R_{TH1} \cdot P_{CUT}$$

$$\Delta T_2 = R_{TH2} \cdot P_{CUT} \tag{17.33}$$

$$\Delta V_{OUT} = S_{Td} \cdot (\Delta T_2 - \Delta T_1) = S_{Td} (R_{TH2} - R_{TH1}) \cdot P_{CUT} = S_{Td} \Delta R_{TH} \cdot P_{CUT}$$

where
   P_{CUT} is the power dissipated by the CUT
   R_{THi} is the thermal coupling resistance between the CUT and the temperature transducer i
      (i being either 1 or 2)

For the sake of illustrative simplicity, we have assumed DC thermal coupling, but the analysis can be easily generalized to the case of baseband signals. Assuming that the common sensitivity is negligible, the sensor output voltage can be rewritten as

$$\Delta V_{OUT} = S_{Td} \cdot \Delta R_{TH} \cdot P_{CUT} = S_{Pd} \cdot P_{CUT} \tag{17.34}$$

where S_{Pd} is the sensor's differential sensitivity to the power dissipated by the CUT. It is interesting to highlight that although S_{Td} only depends on the temperature sensor topology, S_{Pd} depends as well on the placement of the temperature transducers with respect to the CUT location in the layout.

   To compensate for any unwanted transistor mismatch due to process voltage temperature (PVT) variations and temperature gradients from surrounding circuits, an offset calibration mechanism composed by transistors MP_{Off} and MN_{Off} (see schematic of Figure 17.5) is used to balance the differential pair emitter current, as well as to calibrate V_{OUT} within its linear operation range (e.g., $V_{DD}/2$) prior to performing the measurement. This procedure can be made automatic by adding an offset control loop. Finally, the other two bipolar devices (in the middle of Figure 17.5) are used to bias the base of the temperature transducers. A detailed analysis of this sensor topology can be found in [Alt97,Alt01].

Figure 17.5 (bottom left) shows the microphotograph of the 1.25×1.25 mm^2 test chip, which has been fabricated in TSMC 0.25 μm MS/RF complementary MOS (CMOS) technology. The placement of the differential temperature sensor is divided into three parts: the sensing device Q_{S1} (used as temperature reference device), the temperature sensing device Q_{S2}, and the biasing circuitry that includes all other sensor transistors apart from Q_{S1} and Q_{S2}. As annotated in the figure, the distance between Q_{S1} and Q_{S2} is 400 μm. This distance ensures that any temperature increase generated by the power dissipation in a device placed close to Q_{S2} will not heat up Q_{S1}, which maximizes the resolution of the differential temperature sensing element. The sensing devices are deep n-well vertical bipolar transistors available in this CMOS technology, and their layout area is only 15 μm × 15 μm.

The inset of Figure 17.5 shows the sensing device Q_{S2} and two n-well resistors, R_1 and R_2, each with an aspect ratio of [8 μm/4.9 μm] and a nominal resistance value of 300 Ω. R_1 and R_2 are located at 25 and 45 μm from Q_{S2}, respectively. These resistors are used to characterize the sensor performances for both DC and AC conditions. Figure 17.6a shows how the value of the sensor's DC output voltage V_{OUT} changes as a function of the static power dissipated by these resistors (with sensor biasing: $V_{DD} = 3.3$ V, $V_{Bias} = 0.68$ V, $V_{OUT} = 1.65$ V). Focusing on its linear range, the sensor's differential sensitivities to the power dissipated by R_1 and R_2 are computed from Figure 17.6b as 117 and 64 V/W, respectively. It is expected that the CUT that is farther away from the sensor device presents the lower thermal coupling resistance with the temperature transducer Q_{S2} and, therefore, the lower differential sensitivity to its dissipated power. Notice that in both cases, the temperature sensor presents linear characteristics up to an output voltage of 800 mV. In terms of power dissipated by the resistors, the sensor's linear range is 10 and 15 mW, respectively. This linear range can be extended to cover more power range, at the expense of reducing the sensitivity of the temperature sensor.

The measurements shown in Figure 17.6b were obtained by stimulating resistors R_1 and R_2 with a sinusoidal signal given by $v(t) = (1 + \cos(2\pi f_x t))$ V and measuring the amplitude of the fundamental frequency component at the temperature sensor output while sweeping the signal frequency f_x. This plot represents the magnitude response of the cascade of two transfer functions. The first one is due to the thermal coupling between the CUT (either R_1 or R_2) and the temperature transducers that represents the frequency variation of ΔR_{TH}. The second transfer function is due to the frequency response of the sensor's differential sensitivity to temperature S_{Td}. To obtain this data, the sensor

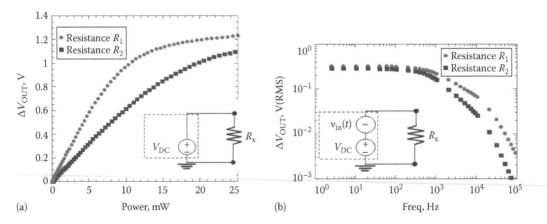

(a)   (b)

FIGURE 17.6 (a) Sensor output voltage variation (DC) as a function of the DC power dissipated by the resistors R1 and R2 (Figure 17.5); (b) Bode (amplitude) that relates the small signal sensor output voltage variation as a function of the frequency of the power dissipated by the resistors R1 and R2.

was loaded by the 1 MΩ input impedance of the lock-in amplifier. The results in Figure 17.6b reveal a sensing setup bandwidth of approximately 1 kHz. It is worth mentioning that this bandwidth and the sensitivity functions are a strong function of the distance between the CUT and the temperature sensing device (Q_{S2}). The results suggest that in this particular case, the bandwidth limitation is imposed by the transfer function of the thermal coupling rather than by the sensor's transfer function.

Before concluding this section, it is important to note that the aforementioned example uses vertical bipolar transistors as temperature transducers because the technology used provides a deep n-well layer that allows us to stack the *npn* bipolar junction transistor (BJT) over the substrate. However, it is not mandatory to use vertical devices to built-up differential temperature sensors; even conventional PN diodes should be enough [Sen12]. Vertical *pnp* bipolar transistors [Ona11] or lateral parasitic bipolar transistors [Ald07] have been successfully used for the realization of efficient thermal sensing devices. As a matter of a fact, bipolar devices have been preferred because their design kits provide better modeling of the dependency of the devices' electrical characteristics as a function of the temperature. However, ordinary MOS transistors have also been used as temperature transducers [Sya02] or thermocouples [Ald07].

Finally, the sensors reported in the previous example present a 0–800 mV linear range for power variations around 10–15 mW. This range can be easily expanded, taking into account that BJTs allow collector current to change by several decades, and then current-based temperature sensors usually provide more linear range [Ona11]; another option is using voltage mode processing with a smaller sensor's sensitivity [Gon11].

17.5 EXPERIMENTAL EXAMPLES

In this section, experimental results for two case studies are discussed. The first one illustrates the use of heterodyne temperature measurements to characterize the center frequency and the 1 dB compression point of a tuned 1 GHz LNA. The second case shows how homodyne temperature measurements can be used for the characterization of a 2.4 GHz class A PA.

Figure 17.7a shows the schematic of the CUT used in the first example. It is a conventional narrow-band common-source LNA with inductive source degeneration and tuned load [Ald10] for operation at 1 GHz. MOS transistors MNC and MNA are composed by 20 interdigitated fingers to yield an effective transistor size of 200 µm/0.35 µm. To perform temperature measurements and extract the characteristics of the CUT, a critical issue is the placement of the temperature sensors. The placement of the different devices is shown in Figure 17.5. Notice that the sensing device Q_{S2} of the differential temperature sensor is placed close to cascode transistor MNC. This transistor has stronger RF power dissipation than the driver MNA. Both transistors share the same current, but the drain–source voltage swing across MNC is significantly larger because the LNA output presents a larger signal swing due to the higher load impedance needed to provide large gain. To experimentally demonstrate this fact, MNC and MNA were separated from each other by 350 µm in the layout.

To perform heterodyne measurements, two tones of equal amplitude at frequencies f_1 and $f_2 = f_1 + \Delta f$ were applied at LNA's input. Two different experiments are reported in Figure 17.7: the measurements of the LNA center frequency and the measurement of the 1 dB compression point.

To measure the LNA center frequency, the two frequencies f_1 and f_2 are simultaneously swept, and the temperature increase at frequency Δf as a function of f_1 was monitored. For this characterization, it is important to maintain Δf constant such that the measurements are consistent and are not affected by the frequency response of the thermal measurement setup. Thermal coupling is sensitive to the frequency of operation as shown in Figure 17.6.

The experimental results shown in Figure 17.7b compare the frequency response of the LNA obtained using spectrum analyzer measurements (E44443A PSA) with the results obtained using the embedded temperature sensor. In this case, the input power was set at −10 dBm, $V_{DD} = V_{BIAS} = 3.3$ V, and $\Delta f = 1012$ Hz. Comparing both plots, a remarkable correlation on the frequency response characterizations is observed in the frequency band from 500 MHz up to 1.5 GHz. The peak

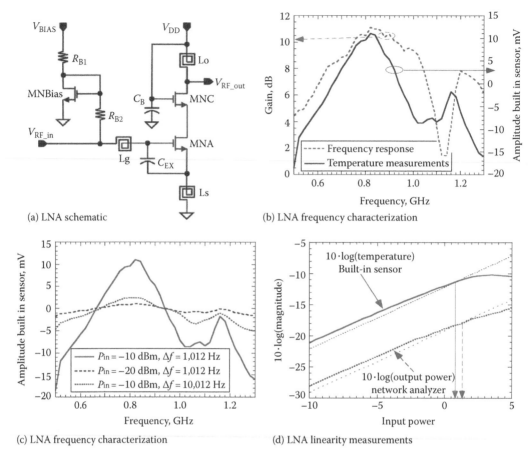

(a) LNA schematic

(b) LNA frequency characterization

(c) LNA frequency characterization

(d) LNA linearity measurements

FIGURE 17.7 (a) Schematic of the LNA. (b) Comparison of the LNA frequency response with the temperature measurements at 1 kHz. (c) Effect of the heterodyne frequency and the RF input power level on the temperature measurements. (d) Comparison of the 1 dB compression point measurement employing conventional methodologies with temperature measurements performed at 1 kHz.

temperature increase corresponds to the estimated LNA center frequency of 830 MHz, whereas a value of 880 MHz was obtained based on the conventional electrical frequency response measurement. Even the notch observed at 1.18 GHz using the spectrum analyzer is visible around 1.04 GHz when using the proposed thermal methodology.

The sensor is biased in such a way that it is working in its linear region along all the measurements. For frequency characterization where obtaining the shape of the transfer function is the goal, we do not have to calibrate the sensing elements since the measurements are relative and the absolute value of the sensor's sensitivity is not required. However in other cases, a calibration phase might be required.

As indicated in Figure 17.6, the magnitude of the thermal coupling is a strong function of the heterodyne frequency Δf. In addition, as indicated in Equation 17.17, the heterodyne component of the dissipated power detected by the temperature sensor is a direct function of the RF signal amplitude applied to the CUT. Figure 17.7c compares the built-in sensor response presented in Figure 17.7b for three different cases: (1) two-tone test signal with $P_{in} = -10$ dBm, $\Delta f = 1,012$ Hz; (2) two-tone test signal with $P_{in} = -20$ dBm, $\Delta f = 1,012$ Hz; and (3) two-tone test signal with $P_{in} = -10$ dBm, $\Delta f = 10,012$ Hz. As expected, the sensor's sensitivity decreases for higher values of the heterodyne frequency Δf. Evidently, lower sensor's output is read for reduced input power, which is also visible in the figure. Notice that the LNA's shape is still visible, and further signal processing should allow us to retrieve the information if low-noise test circuitry is used. Appropriate values of Δf and P_{in} can be used to ensure that the built-in sensor operates in its linear regime.

Sweeping the input power while keeping the test frequency constant is of practical relevance when the objective is to estimate the CUT linearity and its 1 dB compression point [Lee04]. Figure 17.7d shows the temperature increase of the magnitude of the tone ($10 \cdot \log$ of the built-in sensor's spectral component) at Δf as a function of the input power; for this test, $f_1 = 800$ MHz and $\Delta f = 1012$ Hz. The output power delivered by the LNA (in dBm) was also measured using a commercially available spectrum analyzer. At 800 MHz, the measured 1 dB compression point is 1.3 dBm. The on-chip temperature sensor predicts a 1 dB compression point of around 0.8 dBm; a remarkable precision was obtained with this very simple methodology. These results demonstrate that temperature measurements can indeed be used to track the power delivered by the CUT and to estimate CUT large signal linearity.

Figure 17.8 shows the schematic of a 2 GHz PA used as a second test bench. The class A PA employs a pseudo differential structure where each branch is a common-source configuration with cascode device; the architecture was implemented in a CMOS 65 nm technology. The inductors and coupling capacitors C_{DC} are off-chip components and used to center the PA to one of four possible subbands in the 2–2.5 GHz wideband spectrum.

Figure 17.8 shows the location of the differential temperature sensors embedded in the PA. The two temperature transducers are implemented using bipolar transistors, Q_1 and Q_2. Q_2 was placed 25 μm away from the PA cascode transistor M_3. Electrothermal simulations of this PA-sensor placement [Gom12] ensure that this distance is enough for the sensor output voltage to track the changes in the power dissipated by the PA. Q_1 was placed together with the other devices that compose the sensor core 240 μm away from M_3. The experimental results are depicted in Figure 17.8 [Alt12b]. The DC readings obtained from the differential sensor are plotted as a function of the RF electrical power applied to the PA input. The larger the V_{BIAS}, the larger the transistor's bias current is, leading to larger transconductance values. Then, higher power is dissipated in M_3 when increasing the V_{BIAS}, which leads to higher voltage readings at the output of the sensing device when the input power is swept from −10 to 0 dBm.

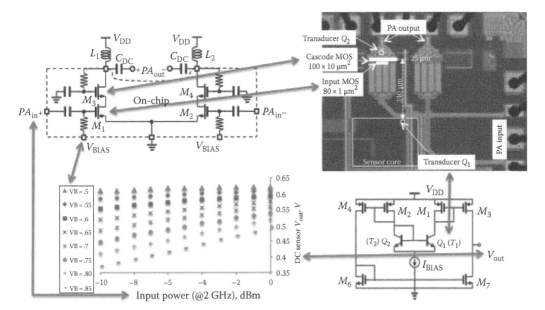

FIGURE 17.8 PA and temperature sensor schematic and layout. Measurements show the evolution of the sensor output voltage (DC) as a function of the RF power (2.4 GHz) applied to the PA input for several PA bias.

Notice that the temperature measurements can be used to track the AC power delivered to the PA load and then evaluate PA's power efficiency [Alt13]. Figure 17.8 also shows that the higher the gain of the PA, the higher the sensitivity of the sensing device, making evident that the PA gain becomes measurable employing thermal methodologies [Alt12]. Indeed, the temperature evolution as a function of the RF input power is a precise indication of the PA gain.

17.6   CONCLUSIONS

Compact yet efficient sensors can be embedded within the CUT to monitor its relevant figures of merit. We have discussed some of them throughout this chapter, for example, center frequency in tuned amplifiers, efficiency and output power in PAs, and linearity in LNAs. A major goal was to use economical test equipment and to reduce the testing time required to perform system tests for user-friendly maintenance after it is manufactured. On the other hand, it is highly desirable to develop monitoring circuits for in-field characterization that could eventually drive a feedback strategy altogether with tuning strategies of key block performances to correct for environmental, process variations, aging, or any cause of degradation in system performance [Gom12b], [Ona11b]. The target was to increase yield and system reliability.

Conventional monitoring circuits (voltage, current, or power sensors) electrically load the system's testing nodes. This loading, even if minimal, introduces performance degradation, especially in RF front ends. The sensitivity of SoC performance to the presence of these monitors increases with the operating frequency of the system being monitored. Moreover, as the SoC frequency increases, a redesign of the conventional monitoring systems may be needed.

These drawbacks can be overcome if temperature sensors are used to monitor system performances. The inherent thermal coupling through the silicon surface grants that temperature sensing devices and the CUT are electrically disconnected. In addition, the information carried by high-frequency electrical signals is down-converted into low-frequency temperature increases; that information is used for testing and healing purposes.

The use of differential temperature sensors as monitoring circuits presents two major advantages: First, with suitable device placement, the sensor is highly sensitive to variations in the power dissipated by the CUT, obtaining a signature independent of the ambient temperature or the particular IC mounting setup (package, socket, etc.). Second, the thermal steady state has to be reached only in the silicon volume around the two temperature transducers. For instance, in [Abd13], a settling time is reported in homodyne temperature measurements of 80 µs when the temperature transducers are at 14 and 150 µm from the CUT.

As a proof of concept, we have revised two design examples: the measurement of the center frequency of a tuned LNA (50 MHz error—6%—with respect to classical electrical measurements) and measurement of the 1 dB compression (0.5 dB error). These measurements do not need calibration of the differential temperature sensor.

There are still some areas of research in this topic. To improve the temperature sensor, a new calibration scheme is needed if other figures of merit (such as gain and output RF power) are the primary goal in an industrial environment. Some recommendations for using the thermal monitoring approach are provided in [Alt03,Abd12,Abd13]. Future research includes the study of challenges that would appear when several temperature sensors and CUT are placed in a complete SoC.

ACKNOWLEDGMENTS

This work has been partially founded by funded by EU-Feder and Spanish MINECO under the TEC2008-01856 project and by AGAUR SGR 1497 funds.

The authors would like to acknowledge the work performed in this field by Dr. Eduardo Aldrete, Dr. Marvin Onabajo, and Dr. Dídac Gómez.

REFERENCES

[Abd12] Abdallah, L., H. G. Stratigopoulos, S. Mir, J. Altet, Testing RF circuits with true non-intrusive built-in sensors, *Proceedings of the 2012 Design, Automation & Test in Europe Conference & Exhibition (DATE)*, Dresden, Germany, 2012, March 12–16, 2012, pp. 1090–1095.

[Abd13] Abdallah, L., H.-G. Stratigopoulos, S. Mir, J. Altet, Defect-oriented non-intrusive RF test using on-chip temperature sensors, *Proceedings of the IEEE VLSI Test Symposium*, Berkeley, California, April 29–May 2, 2013, pp. 57–62.

[Ald07] Aldrete-Vidrio, E., D. Mateo, J. Altet, Differential temperature sensors fully compatible with a 0.35 μm CMOS process, *IEEE Transactions on Components and Packaging Technologies*, 30(4), 2007, 618–626.

[Ald10] Aldrete-Vidrio, E., D. Mateo, J. Altet, M. Amine Salhi, S. Grauby, S. Dilhaire, M. Onabajo, J. Silva-Martinez, Strategies for built-in characterization testing and performance monitoring of analog RF circuits with temperature measurements, *Measurements Science and Technology*, 21, 2010, 075104.

[Alt97] Altet, J., A. Rubio, Differential sensing strategy for dynamic thermal testing of ICs, *15th IEEE VLSI Test Symposium*, Monterrey, CA, 1997, pp. 434–439.

[Alt01] Altet, J., A. Rubio, E. Schaub, S. Dilhaire, W. Claeys, Thermal coupling in integrated circuits: Application to thermal testing, *IEEE Journal of Solid-State Circuits*, 36(1), January 2001, 81–91.

[Alt03] Altet, J., A. Rubio, J. L. Rosselló, J. Segura, Structural RFIC device testing through built-in thermal monitoring, *IEEE Communications Magazine*, 41(9), September 2003, 98–104.

[Alt06] Altet, J., W. Claeys, S. Dilhaire, A. Rubio, Dynamic surface temperature measurements in ICs, *Proceedings of the IEEE*, 93(8), 2006, 1519–1533.

[Alt08] Altet, J., E. Aldrete-Vidrio, D. Mateo, X. Perpiñà, X. Jordà, M. Vellvehi, J. Millan, A. Salhi, S. Grauby, W. Claeys, S. Dilhaire, A heterodyne method for the thermal observation of the electrical behavior of high-frequency integrated circuits, *Measurement Science and Technology*, 19, 2008, 115704 (8 pp.).

[Alt12] Altet, J., J. L. González, D. Gómez, X. Perpinyà, S. Grauby, C. Dufis, M. Vellvehi, D. Mateo, S. Dilhaire, X. Jordà, Electro-thermal characterization of a differential temperature sensor and the thermal coupling in a 65 nm CMOS IC, *Proceedings of the 18th International Workshop on Thermal Investigations of ICs and Systems (THERMINIC)*, Budapest, Hungary, September 25–27, 2012, pp. 61–65.

[Alt12b] Altet, J., D. Mateo, D. Gomez, X. Perpiñà, M. Vellvehi, X. Jordà, DC temperature measurements for power gain monitoring in RF power amplifiers, *2012 International Test Conference (ITC)*, Anaheim, CA, 2012.

[Alt13] Altet, J., D. Gomez, X. Perpinyà, D. Mateo, J. L. González, M. Vellvehi, X. Jordà, Efficiency determination of RF linear power amplifiers by steady-state temperature monitoring using built-in sensors, *Sensors and Actuators A: Physical*, 192, April 2013, 49–57.

[Ant80] Antognetti, P., G. R. Bisio, F. Curatelli, S. Palarar, Three-dimensional transient thermal simulation: Application to delayed short circuit protection in power ICs, *IEEE Journal of Solid State Circuits*, SC-15, June 1980, 277–281.

[Bon95] Bonani, F., G. Ghione, On the application of the Kirchhoff transformation to the steady-state thermal analysis of semiconductor devices with temperature-dependent and piecewise inhomogeneous thermal conductivity, *Solid State Electronics*, 38(7), July 1995, 1409–1412.

[Bow13] Bowers, S. M., K. Sengupta, K. Dasgupta, B.D. Parker, A. Hajimiri, Integrated self-healing for mm-wave power amplifiers, *IEEE Transactions on Microwave Theory and Techniques*, 61(3), March 2013, 1301, 1315.

[Bre03] Breitenstein, O., M. Langenkamp, *Lock-In Thermography: Basics and Use for Evaluating Electronic Devices and Materials*, Advanced Microelectronics Series, Springer, Berlin, Germany, 2003.

[Cla93] Claeys, W., S. Dilhaire, V. Quintard, J. P. Dom, Y. Danto, Thermoreflectance optical test probe for the measurement of current-induced temperature changes in microelectronic components, *Quality and Reliability Engineering International*, 9(4), July/August 1993, 303, 308.

[Garg13] Garg, S., D. Marculescu, Mitigating the impact of process variation on the performance of 3-D integrated circuits, *IEEE Transactions on Very Large Scale Integration (VLSI) Systems*, 2013, Vol. 21, No. 10, 2013, pp. 1903–1914, doi 10.1109/TVLSI.2012.2226762.

[Gom12] Gómez, D., C. Dufis, J. Altet, D. Mateo, J. L. González, Electro-thermal coupling analysis methodology for RF circuits, *Microelectronics Journal*, 43(9), September 2012, 633–641.

[Gom12b] Gómez, D., J. Altet, D. Mateo, On the use of static temperature measurements as process variation observable, *Journal of Electronic Testing*, 28(5), October 2012, 686–695.

[Gon11] Gonzalez, J. L., B. Martineau, D. Mateo, J. Altet, Non-invasive monitoring of CMOS power amplifiers operating at RF and mmW frequencies using an on-chip thermal sensor, *Proceedings of the 2011 IEEE Radio Frequency Integrated Circuits Symposium*, Baltimore, MD, 2011.

[Lee93] Lee, S.-S., D. J. Allstot, Electrothermal simulations of integrated circuits, *IEEE Journal of Solid-State Circuits*, 28(12), December 1993, 1283–1293.

[Lee04] Lee, T.H., *The Design of CMOS Radio-Frequency Integrated Circuits*, 2nd Edition, Cambridge University Press, New York, 2004.

[Lun90] Lundstrom, M., *Fundamentals of Carrier Transport*, Volume X, Modular Series on Solid State Devices, Addison-Wesley Pub. Co., Boston, MA, 1990.

[Mat06] Mateo, D., J. Altet, E. Aldrete-Vidrio, J. L. Gonzalez, Frequency characterization of a 2.4 GHz CMOS LNA by thermal measurements, *IEEE Radio Frequency Integrated Circuits (RFIC) Symposium*, San Francisco, CA, June 2006, pp. 517.

[Nen04] Nenadovic, N., S. Mijalkovic, L. K. Nanver, L. K. J. Vandamme, V. d'Alessandro, H. Schellevis, J. W. Slotboom, Extraction and modeling of self-heating and mutual thermal coupling impedance of bipolar transistors, *IEEE Journal of Solid-State Circuits*, 39(10), 2004, 1764–1772.

[Ona11] Onabajo, M., J. Altet, E. Aldrete-Vidrio, D. Mateo, J. Silva-Martínez, Electrothermal design procedure to observe RF circuit power and linearity characteristics with homodyne differential temperature sensor, *IEEE Transactions on Circuits and Systems I: Regular Papers*, 58(3), 2011, 458–469.

[Ona11b] Onabajo, M., D. Gómez, E. Aldrete-Vidrio, J. Altet, D. Mateo, J. Silva-Martinez, Survey of robustness enhancement techniques for wireless systems-on-a-chip and study of temperature as observable for process variations, *Journal of Electronic Testing*, 27(3), June 2011, 225–240.

[Ona12] Onabajo, M., J. Silva-Martinez, *Analog Circuit Design for Process Variation-Resilient Systems-on-a-Chip*, Springer, New York, 2012.

[Per09] Perpiñà, X., X. Jordà, J. Altet, M. Vellvehi, N. Mestres, Laser beam deflection-based perimeter scanning of integrated circuits for local overheating location, *Journal of Physics D: Applied Physics*, 42, 2009, 012002.

[Raz01] Razavi, B., *Design of Analog CMOS Integrated Circuits*, International Edition, MacGrawHill, New York, 2001.

[Sen12] Sengupta, K., K. Dasgupta, M. S. Bowers, A. Hajimiri, On-chip sensing and actuation methods for integrated self-healing mm-wave CMOS power amplifier, *2012 IEEE MTT-S International Microwave Symposium Digest*, June 17–22, 2012, Montreal, CA.

[Sya02] Syal, A., V. Lee, A. Ivanov, J. Altet, CMOS differential and absolute thermal sensors, *Journal of Electronic Testing: Theory and Applications*, 18(3), 2002, 295–304.

[Tho92] Thomas, L. C., *Heat Transfer*, Prentice Hall, Upper Saddle River, NJ, 1992.

[Yid11] Liu, Y., J.-S. Yuan, CMOS RF power amplifier variability and reliability resilient biasing design and analysis, *IEEE Transactions on Electron Devices*, 58(2), February 2011, 540, 546.

Part IV

Software and Sensor Systems

18 Multisensory System Integration Dependability

*Omid Sarbishei, Majid Janidarmian, Atena Roshan Fekr,
Benjamin Nahill, Zeljko Zilic, and Katarzyna Radecka*

CONTENTS

In the past few years, we have observed increasing interest in sensing and monitoring devices motivated by their dropping cost, size, and power consumption paired with improved reliability. Consequently, sensing devices can be noticed in increasing number of industrial and biomedical applications. The demand for even better solutions has magnified the importance of designing sensing systems with high accuracy and tolerance to external noise and potential faults [8]. Numerous researches have aimed to improve such parameters in sensing systems [4,5].

The accuracy of a sensor system is mostly evaluated in terms of the error measures mean square error (MSE) or maximum absolute error (maximum mismatch (MM)), where the error is defined as the difference between the actual reference value x_{ref} and the final sensor readout $x_{readout}$:

$$error = x_{ref} - x_{readout}.$$

The error metrics MSE and MM are defined as

$$MM = \max\left(|error|\right), \quad MSE = E(error^2),$$

where $E(x)$ and $|x|$ return the expected and absolute values of x, respectively. Note that the error measure MSE captures the overall quality of sensor readouts, while MM indicates the worst case of sensor measurements.

The error occurring at a single sensor's readout can be distinguished as a systematic offset and gain as well as a random noise [4]. Calibration, which is defined as the process of mapping raw sensor readings into corrected values [4], can be used to compensate the systematic offset and gain. It can be done using online or offline methods. Offline methods are mostly based on curve fitting, for example, the least-square method [6]. On the other hand, online methods are based on time series and prediction in real time, such as Kalman filters [22].

Multisensor data fusion [1,2] is also a common approach, which combines data from multiple sensors to achieve more accurate readouts compared to the case where a single sensor is used [10–14]. Data fusion methods can also be used to detect faulty sensors [14] and deliver fault-tolerant measurements. This is crucial, since the technology trends indicate that sensing has by far the highest fault rates [17]. Furthermore, high-risk applications such as medical and control necessitate delivering fault-tolerant sensor readouts in real time.

In this chapter, after reviewing some of the previous work on calibration and sensor fusion, we present an efficient data fusion algorithm with tolerance to multiple faults, which not only minimizes MSE but also keeps the precision high by bounding MM. Our solution is applicable to a central multisensor architecture, for which the postcalibration statistical characteristics of sensors can be measured experimentally. Note that in a central multisensor architecture, the sensors communicate through a central processor, and they capture the same reference data. Certain applications such as wireless systems often make use of a distributed multisensor architecture [27], where each sensor is capable of communicating with all (some) of the other sensors. Such applications are not studied in this chapter.

Since industrial and biomedical applications have high demands for the accuracy of temperature measurement and control [7], as well as accelerometers [22], in our experiments, we illustrate the performance of our algorithms on the example of two multisensor systems:

- A multisensor system consisting of eight STTS751 temperature sensors
- A multisensor system consisting of five MMA8451Q 3-axis accelerometers

The rest of this chapter is organized as follows. Section 18.1 presents the background and related work on calibration, sensor fusion, and fault detection in multisensor systems. Section 18.2 presents our screening process to detect the potentially faulty sensors in a multisensor system. Section 18.3 introduces the proposed data fusion algorithm, which minimizes MSE while bounding MM. Finally, experimental results are presented and discussed in Section 18.4.

18.1 BACKGROUND AND RELATED WORK

Calibration is a crucial step in improving the accuracy of individual sensors in a multisensor system. Certain systems might require dedicated calibration procedures, which in general can be categorized under online or offline methods. Offline methods are mostly based on curve fitting, for example, the least-square method [6], to map raw sensor readings into corrected values [4] and compensate the systematic offset and gain. On the other hand, online methods are based on time series and prediction in real time, such as Kalman filters [22]. Here, we go through a few examples.

In [3], Bychkovskiy et al. suggest a methodology for localized calibration of the light sensors. It first considers the physically close sensors and then tries to obtain the most consistent way to provide all pair-wise relationships. Feng et al. [5] have focused on a time-variant actuator-based method in order to calibrate the sensors. The results are shown on a set of light intensity measurements recorded by deployed sensors. There are two scenarios in this study: one where only two neighboring sensors have to communicate to achieve the calibrated readouts and the other where a provably minimum number of communications is achieved. In [1], authors consider four types of temperature sensors. They present an analysis on the biases and errors of the specified temperature sensors, signal conditioning circuitry, and data achievement system [7]. The authors in [22] have used an extended Kalman filtering approach to calibrate three-axis accelerometers in real time. The online calibration approach presented in [22] can be further improved when combined with offline methods such as curve fitting and the least-square method.

While calibration aims to improve the accuracy of sensors individually [9], sensor fusion [1,2,23,24] combines data from multiple sensors to improve the accuracy of the whole system [10–14]. An example is to use accelerometers and gyroscopes together for position tracking. A straightforward approach to increase the accuracy of sensor readouts in terms of the error measure MSE is to perform an average computation over the current results [15,6,10]. This technique can improve the MSE by the factor of n, if all sensors are identical in terms of the statistical characteristics, where n is the total number of sensors. Otherwise, a weighted average computation can be performed to minimize MSE. The solution in [16] uses a neural network–based training heuristic to optimize the weights for the purpose of average computations.

Data fusion methods are useful for detecting potentially faulty sensors [14] and provide fault-tolerant measurements as well. To enhance the method of fault detection, the sensor fault models have been derived experimentally [18–20]. In the process to determine the faulty sensors, the sensor readouts are captured for a reasonably long period of time, and then the results are compared with the given statistical characteristics of the sensors in their normal operation mode. Next, if the results of a sensor deviate from the expected characteristics by a particular threshold, then the sensor is assumed to be faulty [18]. The approach in [18] proposes a technique to find the optimal threshold value for such a purpose. Albeit useful, these techniques are not appropriate for real-time applications like health systems [10]. Here, sensor failures may occur within relatively short intervals [10]. The approaches in Refs. [10,6] are more suitable for such purposes. However, they can only handle single faults and are not efficient when multiple faults occur.

Multiple faults are usually hard to detect and tolerate. Some approaches such as [21] focus on diagnosing multiple faults in a multisensor system using a conventional fuzzy soft clustering. This method requires no prior knowledge or information about the used sensors.

Through the rest of this chapter, we first present a screening process to exclude the potentially faulty sensors from the fusion algorithm. This method makes it possible to perform the data fusion algorithm with tolerance to multiple faults. Furthermore, the potentially faulty sensors are detected in real time. The screening process is then followed by an efficient linear data fusion algorithm.

The proposed data fusion in this chapter minimizes MSE while bounding MM. This indicates that our solution maximizes accuracy in the average case while keeping the precision high in the worst case of sensor measurements. This is not the case with the previous work on sensor fusion, for example, [1,15], which aim to minimize MSE only. The statistical characteristics of the calibrated sensor errors required by our screening and data fusion processes are found through experimental measurements. The proposed data fusion in this chapter is a linear algorithm (weighted average computation), where the optimal coefficients (weights) are found using a convex optimization scheme and a simple formula. The approach is useful for heterogonous multisensor systems, where the sensor errors are not necessarily identical in terms of statistical characteristics. This assumption is due to the inevitable physical and fabrication issues, as well as calibration [3], which could result in sensor errors having different statistical characteristics.

18.2 FAULT-TOLERANT SCREENING PROCESS

This section presents the proposed screening process to quickly detect the potentially faulty sensors online. The approach is an extension to the single-fault detection technique in [6]. In [6], the authors find the sensor readout x_j, which is furthest from the average of the other sensors' readouts, and if the distance is higher than a threshold, it is assumed that the sensor is faulty. Although the approach in [6] can only be used to detect a single faulty sensor, we can extend it to an iterative process, to detect multiple faults along with single ones.

Through the rest of this chapter, we use the following notation:

x_{ref}: The reference data to be measured.

x_j: The jth sensor readout ($j = 1,...,n$) after calibration, where n is the total number of sensors.

Note that the solutions presented in this chapter are only applicable to the multisensor systems, where different sensors aim to capture the same reference data x_{ref}:

e_j: The error of jth sensor readout after calibration, that is,

$$e_j = x_{ref} - x_j, \tag{18.1}$$

M_j: The maximum absolute error of the jth sensor readout after calibration, that is, $M_j = \max(|e_j|)$. Note that M_j is obtained through experimental measurements.

E_j: The expected value of the error e_j, that is, $E_j = E(e_j)$. Note that we can set $E_j = E(e_j) \approx 0$, by tuning the offset coefficient in the calibration process.

S_j: The average square error, that is, $S_j = E\left(e_j^2\right)$, which is obtained through experimental measurements. Due to the condition $E_j = E(e_j) \approx 0$, the value of $S_j = E(e_j^2)$ also represents the variance of error e_j.

The fault-tolerant screening process is addressed in Algorithm 18.1. The *for* loop in step 1 is executed k times, where $k < n$ is the number of potentially faulty sensors, which deviate from the average of other sensor readouts by a distance higher than M_i (step 8).

Algorithm 18.1: Fault Screening ($M_{1:n}$, $x_{1:n}$)
//Inputs: $x_{1:n}$, $M_{1:n}$, Output: returns the nonfaulty sensors

1. for ($m = 1$; $m < n$; m++) {

2. $sum = \sum_{j=1}^{n} x_j$;

3. for ($i = 1$; $i \leq n$; i++)

4. $\{a_i = \dfrac{sum - x_i}{n-1}$; *//Average computation excluding x_i*

5. $d_i = |x_i - a_i|$;} *//Deviation from the average of others*

//Find the furthest from average of others

6. for ($i = 1$; $i \leq n$; i++)

7. {if $d_i = \max(d_{1:n})$ break;}

8. if $d_i > M_i$ {throw away x_i; $n = n - 1$; continue;}

9. else return $x_{1:n}$;} *//Return the nonfaulty sensors*

18.3 OPTIMAL LINEAR DATA FUSION

In this section, we propose an optimal homogenous linear data fusion technique, which is performed over nonfaulty sensors to minimize MSE while keeping the precision high. We assume that the inputs to the data fusion algorithm have passed the screening step in Algorithm 18.1. The problem is formulated as a convex optimization scheme, for which we propose the deterministic solution.

18.3.1 PROBLEM FORMULATION

The linear data fusion is defined as follows:

$$x_{est} = \sum_{j=1}^{n} c_j x_j, \tag{18.2}$$

where the coefficients c_j should be found, such that the MSE between the reference data x_{ref} and the estimated one x_{est} is minimized, and at the same time, the MM between x_{ref} and x_{est} is bounded as well (high precision).

Let us compute the error $x_{\text{ref}} - x_{\text{est}}$:

$$error = x_{\text{ref}} - x_{\text{est}} = x_{\text{ref}} - \sum_{j=1}^{n} c_j x_j \quad \overset{\text{Equation 18.2}}{\Rightarrow}$$

$$= x_{\text{ref}} - \sum_{j=1}^{n} c_j \left(x_{\text{ref}} - e_j \right) = x_{\text{ref}} \left(1 - \sum_{j=1}^{n} c_j \right) + \sum_{j=1}^{n} c_j e_j. \tag{18.3}$$

The error in Equation 18.3 is not only a function of the individual sensor errors e_j but also depends on the reference data x_{ref}. For large values of $|x_{\text{ref}}|$ in particular corner cases, the error term $x_{\text{ref}} \left(1 - \sum_{j=1}^{n} c_j \right)$ in Equation 18.3 results in a high MM (low precision). Hence, it is preferable to make the error function in Equation 18.3 independent from x_{ref}. This can be achieved by the following condition:

$$x_{\text{ref}} \left(1 - \sum_{j=1}^{n} c_j \right) = 0 \Rightarrow \sum_{j=1}^{n} c_j = 1. \tag{18.4}$$

We consider the condition given by Equation 18.4 as the basis of our fusion process to ultimately achieve high precision.

Next, we aim to find the coefficients c_j, such that MSE is minimized, and at the same time, the condition in Equation 18.4 is satisfied keeping the precision high. We later in Section 18.3.3 discuss the overall precision of the system using the proposed fusion. Based on Equation 18.3, we can compute MSE as follows:

$$\text{MSE} = E\left(\left(x_{\text{ref}} - x_{\text{est}} \right)^2 \right) = E\left(\left(\sum_{j=1}^{n} c_j e_j \right)^2 \right) = E\left(\sum_{j=1}^{n} c_j^2 e_j^2 + 2 \sum_{j=1}^{n} \sum_{k=j+1}^{n} c_j c_k e_j e_k \right).$$

Next, assuming that the calibrated sensor errors are independent from each other, we find MSE as

$$\text{MSE} = \sum_{j=1}^{n} c_j^2 S_j. \tag{18.5}$$

The goal is to set the coefficients c_j, such that the MSE in Equation 18.5 is minimized, subject to the constraint given by Equation 18.4. Note that S_j in Equation 18.5 are given values found by experimental measurements. In summary, the problem is formulated as follows:

$$\begin{array}{|l|}
\hline
\text{Objective function} \left(S_{1:n} \right): \\[2mm]
\text{Find min of } f_1 = MSE = \sum_{j=1}^{n} c_j^2 S_j, \\[2mm]
\text{Constraint}: f_2 = \sum_{j=1}^{n} c_j = 1 \\
\hline
\end{array} \tag{18.6}$$

18.3.2 DETERMINISTIC SOLUTION

A real-valued function $f(x)$ on an interval is called *convex*, if and only if (iff) for any two points within the interval, for example, x_1 and x_2, and any $t \in [0, 1]$, the following property holds [32]:

$$f\left(tx_1 + (1-t)x_2\right) \leq tf\left(x_1\right) + (1-t)f(x_2).$$

Note that for a convex function, any local minimum is also the global minimum [32].

Both functions f_1 and f_2 in Equation 18.6 are convex functions of the variables c_j, since we have

$$\frac{\partial^2 f_1}{\partial c_j^2} = 2S_j \geq 0 \quad \text{and} \quad \frac{\partial^2 f_2}{\partial c_j^2} = 0.$$

We apply the Lagrange multiplication to define a single objective function f, which is a linear function of f_1 and f_2 as follows:

$$f = f_1 + \lambda f_2 = \sum_{j=1}^{n} S_j \times c_j^2 + \lambda \sum_{j=1}^{n} c_j, \tag{18.7}$$

where $\lambda > 0$. The function f in Equation 18.7 is also a convex function, since it is a positive linear combination of the convex functions f_1 and f_2. The global minimum of f can be found using derivative computations as

$$\frac{\partial f}{\partial c_j} = 0 \Rightarrow 2c_j S_j + \lambda = 0 \Rightarrow c_j = \frac{-\lambda}{2S_j} \quad j = \{1, \ldots, n\}. \tag{18.8}$$

Next, using Equation 18.8, we set λ, such that the condition $\sum_{k=1}^{n} c_k = 1$ (the constraint in Equation 18.6) is satisfied:

$$\sum_{j=1}^{n} c_j = 1 \Rightarrow \sum_{k=1}^{n} \frac{-\lambda}{2S_k} = \frac{-\lambda}{2} \sum_{k=1}^{n} \frac{1}{S_k} = 1 \Rightarrow \lambda = \frac{-2}{\sum_{k=1}^{n} (1/S_k)}. \tag{18.9}$$

By substituting the value of λ from Equation 18.9 into Equation 18.8, we find the optimal coefficients as follows:

$$c_j = \frac{1}{S_j \sum_{k=1}^{n} \frac{1}{S_k}} \quad (j = 1, \ldots, n). \tag{18.10}$$

Lemma 18.1: The Solution in Equation 18.10 is the Global Minimum of the Problem in Equation 18.6

Proof: The c_j values in Equation 18.10 give the deterministic optimal solution of $f = f_1 + \lambda f_2$, when $f_2 = 1$. Now, we aim to prove that Equation 18.10 minimizes $f_1 = \text{MSE}$ in Equation 18.6 as well. We make use of contradiction reasoning to tackle this problem. Let us assume that C_j is the solution vector $C_j = [c_1, \ldots, c_n]$, where each c_j value is found using Equation 18.10. We know from the convexity of $f = f_1 + \lambda f_2$ that C_j is the global minimum of f, when $\lambda = -2/\sum_{k=1}^{n} (1/S_k)$, that is, $f_2 = 1$. Further, assume that C_j is *not* the global minimum of $f_1 = \text{MSE}$ in Equation 18.6. This means that there exists another solution vector $\hat{C}_j \neq C_j$, which minimizes $f_1 = \text{MSE}$ at $f_2 = 1$. Since $\hat{C}_j$ is the global minimum of $f_1 = \text{MSE}$ at $f_2 = 1$, it is also the global minimum of $f = f_1 + \lambda f_2$, when $f_2 = 1$. Hence, it is true that $\hat{C}_j = C_j$. ∎

Based on Equation 18.10, we present the optimal linear data fusion process in Algorithm 18.2. We assume that the sensors have initially passed the screening process in Algorithm 18.1, and the potentially faulty sensors are excluded from the fusion.

Note that if all sensor errors have identical statistical characteristics, then we obtain $c_1 = c_2 = \cdots = c_n = 1/n$. In this boundary case, the data fusion algorithm degrades to a simple average computation process [15]. Similar approaches in Refs [10,6] utilize the average computation for the purpose of data fusion. This gives the optimal solution, only if all sensors have identical implementation statistical characteristics. However, due to numerous fabrication and physical flows, as well as calibration procedures performed on individual sensors at different time using data poised with the imprecision errors, the sensors are not expected to have identical statistical characteristics for the same input stimuli. Under such conditions, our data fusion in Algorithm 18.2, which assumes that the sensor errors are not necessarily identical in terms of the statistical characteristics, becomes more useful by delivering a much lower value of MSE. The experimental results in Section 18.4 illustrate this issue in more details.

Algorithm 18.2: Data Fusion $(S_{1:n}, x_{1:n})$
//Inputs: $x_{1:n}$, $S_{1:n}$, Output: x_{est} in Equation 18.2

1. for $(j = 1; j \leq n; j++)$

2. $\left\{ c_j = \dfrac{1}{S_j \sum_{k=1}^{n} (1/S_k)}; \right\}$ *//Finding the optimal coefficients, Equation 18.10*

3. return $x_{est} = \sum_{j=1}^{n} c_j x_j$; *//Return the estimated readout*

Both the screening process (Algorithm 18.1) and the data fusion approach (Algorithm 18.2) are executed periodically and online corresponding to each set of sensor readouts x_j ($j = 1, \ldots, n$). Algorithm 18.2 returns the minimum MSE for any arbitrary n. However, for the screening process in Algorithm 18.1 to be successful in detecting potentially faulty sensors, it is required that at least two sensors are not faulty, and they return readouts that are close to each other. Under such conditions, using Algorithm 18.1, we can assume that the rest of the sensors, which deviate from the two nonfaulty sensors, are in fact faulty.

18.3.3 PRECISION ANALYSIS

Although the quality of sensor measurements and a data fusion algorithm is mostly evaluated in terms of the error measure MSE, it is also crucial to look into the precision and MM of the fusion process, that is, $\max(|x_{ref} - x_{est}|)$, where x_{est} is given by Equation 18.2. Note that the error measure MM corresponds to the worst case of sensor measurements.

The proposed data fusion (Algorithm 18.2) in addition to minimizing MSE bounds MM and does not compromise the overall precision of the system due to the condition given by Equation 18.4.

Lemma 18.2: Using the Data Fusion Coefficients in Equation 18.10, the Condition $\max(|x_{ref} - x_{est}|)$ $\leq \max(M_j)$ ($j = 1, \ldots, n$) Always Holds

Proof: An upper bound on the MM of x_{est} is found by applying the triangle inequality:

$$\max\left(|x_{ref} - x_{est}|\right) = \max\left(\left|\sum_{j=1}^{n} c_j e_j\right|\right) \leq \sum_{j=1}^{n} \max\left(|c_j e_j|\right).$$

Since $S_j > 0$, and hence, $c_j > 0$ (see Equation 18.10), we get

$$\sum_{j=1}^{n} \max\left(\left|c_j e_j\right|\right) = \sum_{j=1}^{n} c_j \max\left(\left|e_j\right|\right) = \sum_{j=1}^{n} c_j M_j.$$

Next, we apply the triangle inequality again to

$$\sum_{j=1}^{n} c_j M_j \leq \sum_{j=1}^{n} c_j \max\left(M_j\right) = \max\left(M_j\right) \sum_{j=1}^{n} c_j \overset{\text{Equation 18.4}}{\Rightarrow} = \max\left(M_j\right).$$

This means that the MM of x_{est} in our data fusion never exceeds the MM of the individual sensor with lowest precision. Note that the condition $\max(|x_{ref} - x_{est}|) \leq \max(M_j)$ could be violated, if the condition in Equation 18.4 is not satisfied.   ∎

18.4   EXPERIMENTAL SETUP AND RESULTS

In this section, we present the experimental results on evaluating the proposed sensor fusion approach. Since accelerometers and temperature sensors are vital parts of many biomedical and industrial applications [7,22], we evaluate our experiments on two multisensor systems consisting of temperature sensors and accelerometers.

First, in Sections 18.4.1 and 18.4.2, we discuss the hardware configuration of the two multisensor systems used in our experiments. Sections 18.4.3 and 18.4.4 discuss the calibration process for the two systems. Finally, in Section 18.4.5, we compare the results obtained by our data fusion (Algorithm 18.1) with previous work. Furthermore, online detection of multiple faulty sensors using Algorithm 18.2 is addressed.

18.4.1   SYSTEM CONFIGURATION WITH TEMPERATURE SENSORS

The STTS751 is a six-pin digital temperature sensor that supports different slave addresses [28]. The STTS751 communicates over a two-wire serial interface compatible with the SMBus 2.0 standard. Temperature data, alarm limits, and configuration information are communicated over the bus. The STTS751 is available in two versions. Each version has four slave addresses determined by the pull-up resistor value connected to the Addr/Therm pin. In our experiments, the configurable temperature reading precision is set to 12 bits, or 0.0625°C per LSB.

The data of eight temperature sensors is collected by an STM32F407 microcontroller [25], which uses an ARM Cortex-M4 32-bit [7] core, as the I^2C master. To accommodate eight sensors with only four distinct addresses, a second I^2C bus is used. Figure 18.1 depicts the multisensor system architecture with temperature sensors.

18.4.2   SYSTEM CONFIGURATION WITH ACCELEROMETERS

A system of five MMA8451Q three-axis accelerometers on FRDM-KL25Z development boards [29] has been chosen for the experiments. Accelerometers are networked together and synchronized by the STM32F4 board [25] as shown in Figure 18.2a. The accelerometers sample 14-bit readings at 800 Hz. They deliver 32-sample packets using a Serial Peripheral Interface to the STM32F4 board, which buffers them in its SRAM for later delivery to a PC over USB interface.

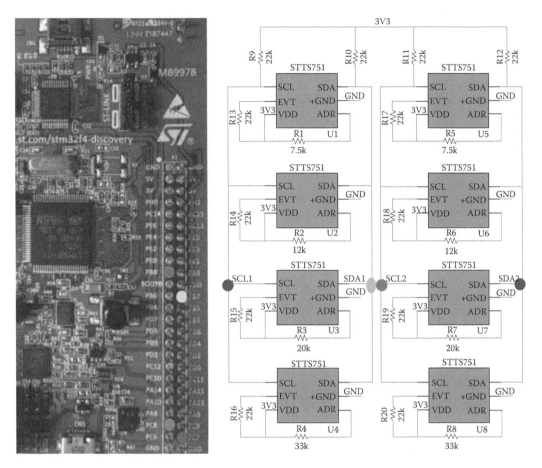

FIGURE 18.1 Schematic of the multisensor system with temperature sensors.

18.4.3 Calibration of Temperature Sensors

Each STTS751 temperature sensor is calibrated using the least-square method and a linear curve-fitting process [30] using the Temptronic TP4500 environmental thermal chamber as the reference model [31]. The Temptronic TP4500 temperature environmental thermal chamber (shown in Figure 18.3) is ideal for lab testing and failure analysis of microsystems due to its fast temperature transitions and high air-flow over its wide operating range between −45°C and +225°C. It is also capable of traversing the full range within 12 s [31]. The TP4500 works by placing a thermal shroud around a device under test and directing a flow of air over the sample at a controlled temperature. After collecting raw data from eight sensors, the linear curve-fitting step is used to map the raw sensor readouts to the reference temperatures provided by TP4500.

In order to make sure that the sensors' readings are stabilized for the purpose of calibration, the measurements for each set point were taken in 12 min intervals. In generating a dual-slope temperature ramp with Temptronic TP4500, the temperatures were changing in 4° steps between 10°C and 30°C with starting points 10°C (or alternatively 12°C). Note that after calibration, most of the data converges to the reference temperature. This is depicted in Figure 18.4 for a small portion of the measurements. The calibration process, which can be similarly performed for other temperature ranges, delivers a significant improvement in terms of accuracy compared to the raw sensor readouts.

We have also evaluated the confidence levels of the calibrated results with respect to different confidence intervals as addressed in Table 18.1. Note that the confidence level shows the probability

 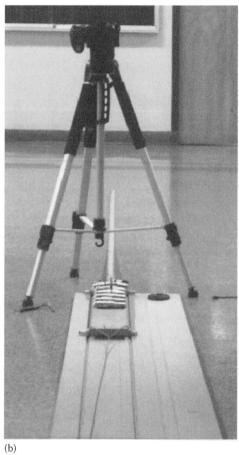

(a)   (b)

FIGURE 18.2 System configuration with accelerometers: (a) STM32F4 board and sensor network on the test platform (train). (b) The test platform being monitored by CASIO EX-F1 high-speed camera.

FIGURE 18.3 The Temptronic TP4500 with STTS751 temperature sensors under test.

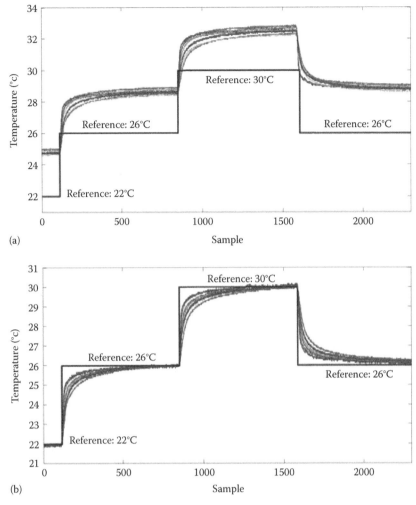

FIGURE 18.4 Temperature readouts (a) and calibrated values (b) for eight sensors using a dual-slope ramp for the reference chamber temperatures.

TABLE 18.1

Confidence Level of the Error of Calibrated Sensors with Respect to the Reference Temptronic Chamber over Different Intervals

Confidence Interval (°C)	Confidence Level (%)
[−0.1,0.1]	45.32
[−0.2,0.2]	79.15
[−0.3,0.3]	97.58
[−0.4,0.4]	99.4
[−0.5,0.5]	99.4
[−0.6,0.6]	99.7
[−0.6535,0.6535]	100

of the sensor errors lying within the range given by the confidence interval. As can be seen, the confidence interval of [−0.3°C,0.3°C] covers most of the measurements, that is, 97.58% of the data.

To get the distribution of sensor errors e_j, we have evaluated the difference between reference and calibrated temperatures for more than 100,000 reference temperature samples for all the eight temperature sensors. The results indicate that the error values e_j can be mapped to Gaussian distributions with a zero mean, but with different variances (see Figure 18.5). This means that the sensors do not have identical statistical characteristics, and hence, the proposed data fusion in Algorithm 18.2 becomes more useful compared to the conventional solutions such as the normal average computation approach [10], which assumes that the sensors have identical statistical characteristics.

18.4.4 CALIBRATION OF ACCELEROMETERS

The calibration of MMA8451Q accelerometers has been performed in a two-step process. First, the offline linear least-square method in [26] is used to calibrate the three-axis raw sensor readouts based on six stationary positions. We have considered more than 30,000 raw sensor readouts for this purpose.

Next, we perform another linear least-square-method calibration on the rail setup shown in Figure 18.2. All five FRDM-KL25Z development boards are placed and fixed with identical positions on the train model. The STM32F4 board as well as a lithium-polymer battery has also been placed on the train. The track and platform are made of machined medium-density fiberboard. Ball bearings are used as wheels to ensure smooth travel on the track.

Several accelerations have been generated manually to collect test data. Next, the initially calibrated sensor readouts are passed through a ninth-order infinite impulse response (IIR) low-pass filter with the cutoff frequency of 40 Hz.

The filtered acceleration measurements are then compared with reference values obtained by CASIO EX-F1 high-speed camera reading 1200 frames per second. We have filtered the reference accelerations found by the camera with a similar low-pass filter with the cutoff frequency of 40 Hz. We have decimated both the reference and accelerometer readings to the common rate of 240 Hz.

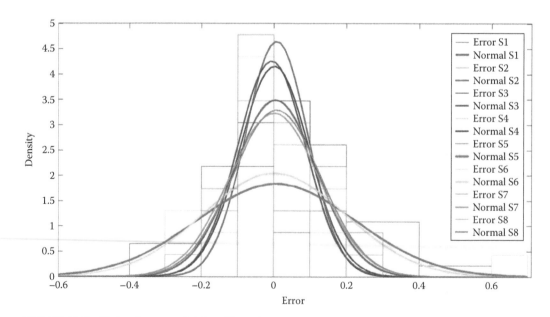

FIGURE 18.5 Fitting the error of eight calibrated temperature sensors to separate normal distributions.

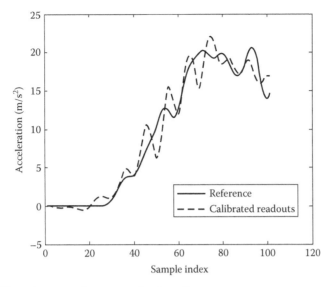

FIGURE 18.6 Calibrated versus reference accelerations for one sensor.

Finally, another linear least-square curve-fitting approach is used to match the five sensor readouts to the reference values obtained by the camera. Figure 18.6 depicts some of the calibrated and filtered sensor readouts for one of the sensors compared to the reference values. As can be seen, the calibrated sensor readouts almost match the reference accelerations.

Using the aforementioned procedure, we have computed the postcalibration parameters S_j and M_j. This is used later for data fusion in real time (see Algorithm 18.2). It has been interesting to observe that all five sensors return identical statistical characteristics with $M_j = 4.6081$ m/s^2 and $S_j = 2.73$ m^2/s^4. This not only shows the consistency of the MMA8451Q accelerometers but also emphasizes on the robustness of our test platform.

18.4.5 ACCURACY COMPARISON

In this section, we compare the accuracy of the proposed data fusion and screening process with previous work. We compare four different fault-tolerant data fusion methods for the experiments in this subsection.

The first approach (M1), which is introduced in [10], uses an average computation for the data fusion process while ignoring the sensor readout, which is furthest from the average of other sensors (potentially faulty sensor). The second method (M2) presented in [6] also uses the average computation for the data fusion process, ignoring a sensor readout, which is not only furthest from the average of others but also the deviation is higher than a threshold. Both methods in [6] can detect only a single faulty sensor and cannot handle multiple faults.

The third approach (M3) uses the proposed screening process described by Algorithm 18.1 to detect multiple faults. Data fusion coefficients c_j in Equation 18.6 are found using the following genetic algorithm in MATLAB®. A chromosome in the genetic algorithm represents a coefficient vector $C_j = [c_1, \ldots, c_n]$, which is a possible solution to the fusion problem. Each coefficient c_j is considered as a gene, which should satisfy the constraint given by Equation 18.6. We have used the default options in MATLAB on population size, selection algorithm, and genetic operators including crossover and mutation. The default crossover operator randomly chooses a gene at the same position from one of the two parents and assigns it to the child. The crossover is a useful operator to avoid getting stuck at local minimums. Mutation, on the other hand, adds a random vector from a Gaussian distribution to the parent. The data fusion coefficients are optimized offline using genetic algorithm considering all the possible potentially faulty sensors.

Finally, the fourth method (M4) makes use of the screening process in Algorithm 18.1 to detect multiple faulty sensors, as well as Algorithm 18.2 to perform the data fusion based on the formula in Equation 18.10 to find the optimal data fusion coefficients c_j.

The results of comparing these methods in terms of the error measure MSE are summarized for temperature sensors and accelerometers in Tables 18.2 and 18.3, respectively.

The first column in both Tables 18.2 and 18.3 shows the number of faulty sensors, which are randomly chosen among the available ones for each case. We make use of the fault model in [11] to estimate the sensor readout in its failure mode. MSE is also obtained by Monte Carlo simulations using 100,000 samples. As can be seen, regarding temperature sensors in all cases, the proposed approach (M4) significantly improves MSE compared to the other methods regardless of the number of faulty sensors. However, the results are different for accelerometers, since the postcalibration statistical characteristics of sensor errors have been measured as almost identical. As a consequence, the proposed solution M4 does not improve MSE compared to the other methods, when no faults exist. However, when multiple faults occur, the solutions M3 and M4, which both make use of the proposed screening process (Algorithm 18.1), become the superior solution in terms of MSE.

It is also worth noting that the screening process proposed by Algorithm 18.1 was able to online diagnose 95.66% of faulty sensors in its single execution. This success rate is improved to 100% when two consecutive executions of the screening process for two separate sets of sensor readouts are taken into account. This shows the robustness of the screening process in detecting the potentially faulty sensors, even when multiple faults occur.

TABLE 18.2

MSE Comparison of Different Data Fusion Methods on Eight Temperature Sensors

	MSE			
No. of Faulty Sensors	M1	M2	M3	M4
0	0.0026	0.0024	0.0021	**0.0016**
1	0.0029	0.0029	0.0029	**0.0019**
2	1.3706	1.3583	0.0031	**0.0026**
3	4.3513	4.3701	0.0044	**0.0038**
4	8.9762	8.9197	0.0075	**0.0064**
Average MSE improvement w.r.t. M1	1.76%	63.76%	**74.52%**
Average MSE improvement w.r.t. M2		62.42%	**73.49%**
Average MSE improvement w.r.t. M3			**20.54%**

TABLE 18.3

MSE Comparison of Different Data Fusion Methods on Five Accelerometers

	MSE		
No. of Faulty Sensors	M1	M2	M3 and M4
0		**0.6199**	
1	0.8414		**0.802**
2	5.112	5.1449	**1.8778**
Average MSE improvement w.r.t. M1		1.35%	**22.65%**
Average MSE improvement w.r.t. M2			**21.17%**

TABLE 18.4
MM Comparison of Different Data Fusion
Methods on Eight Temperature Sensors

		MM		
No. of Faulty Sensors	max(M_j)	M1	M2	M4
0	0.9451	0.226	0.2166	**0.1641**
1	19.998	0.2589	0.2507	**0.1926**
2			2.9362	**0.4258**
3			5.7483	**0.3759**
4			8.3847	**0.5172**
Average MM improvement w.r.t. M1			1.47%	**65.16%**
Average MM improvement w.r.t. M2				**64.04%**

TABLE 18.5
MM Comparison of Different Data Fusion
Methods on Five Accelerometers

		MM	
No. of Faulty Sensors	max(M_j)	M1 and M2	M3 and M4
0	8.6323	**4.3277**	
1	19.5998	**5.8347**	
2		18.8952	11.6238
Average MM improvement w.r.t. M1 (M2)			**12.83%**

The complexity of executing both Algorithm 18.1 and the fusion method (Algorithm 18.2) is simple enough to be implemented in real time in a low-power system.

We have also evaluated the precision of the aforementioned data fusion methods $M1$, $M2$, and $M4$ based on the error measure MM over 100,000 samples. Table 18.4 summarizes the results for temperature sensors, while Table 18.5 addresses the results for accelerometers. The proposed data fusion $M4$ delivers the superior precision compared to the other methods. Furthermore, $M4$ is capable of significantly improving the overall precision (MM) compared to the individual sensor with maximum error, that is, max(M_j) (see Column 2 in Tables 18.4 and 18.5). This confirms the reasoning provided by Lemma 18.2, which claims that our data fusion algorithm not only gives the optimal solution in terms of MSE but also it results in a high precision (low MM).

18.5   SUMMARY AND DISCUSSION

In this chapter, we introduced a minimum MSE linear sensor fusion algorithm, which delivers a bounded MM (high precision) as well. The approach is applicable to any multisensor system, for which the post-calibration statistical characteristics of sensor errors can be measured experimentally. The coefficients to the data fusion process are found using convex optimization. The approach is enhanced by a preprocessing screening step to detect the potentially faulty sensors and exclude them from the fusion. The screening technique makes it possible to detect multiple sensor faults in real-time. The proposed generic data fusion approach is evaluated on a system of eight temperature sensors as well as a system of five three-axis accelerometers. Experiments focus on the efficiency of the proposed data fusion algorithm in terms of MSE (accuracy), MM (precision), and tolerance to multiple faults compared to previous work.

A promising future work avenue would be to extend the proposed data fusion and screening methods (Algorithms 18.1 and 18.2) to an adaptive fusion for applications dealing with high risks and major uncertainties. For instance, regarding glucose sensors, the readouts and distributions of errors could be different depending on the patient's activities, medical history, etc. Another important factor, which could affect the distribution of sensor errors, is aging. This should be addressed in the future work as well.

REFERENCES

1. Hall, D., *Mathematical Techniques in Multisensor Data Fusion*. Boston, MA: Artech House, 1992.
2. Klein, L.A., *Sensor and Data Fusion Concepts and Applications*, vol. 14. Washington, DC: SPIE Optical Engineering Press, 1993.
3. Bychkovskiy, V., Megerian, S., Estrin, D., and Potkonjak, M., Calibration: A collaborative approach to in-place sensor calibration, *2nd International Workshop on Information Processing in Sensor Networks (IPSN'03)*, April 22–23, 2003, Palo Alto, CA, pp. 301–316.
4. Feng, J., Qu, G., and Potkonjak, M., Sensor calibration using nonparametric statistical characterization of error models, *Proceedings of IEEE Sensors*, vol. 3, October 24–27, 2004, Vienna, Austria, pp. 1456–1459.
5. Feng, J., Megerian, S., and Potkonjak, M., Model-based calibration for sensor networks, *Proceedings of IEEE Sensors*, vol. 2, Toronto, Canada, October 22–24, 2003, pp. 737–742.
6. Roshan fekr, A., Janidarmian, M., Sarbishei, O., Nahill, B., Radecka, K., and Zilic, Z., MSE minimization and fault-tolerant data fusion for multi-sensor systems, *IEEE ICCD*, Montreal, Canada, September 30–October 3, 2012, pp. 445–452.
7. Xianjun, Y. and Cuimei, L., Development of high-precision temperature measurement system based on ARM, *9th International Conference on Electronic Measurement & Instruments, ICEMI '09*, August 16–19, 2009, Beijing, China, pp.1-795–1-799.
8. Yang, G.Z., *Body Sensor Networks*. London, U.K.: Springer-Verlag, 2006.
9. Feng, J.; Megerian, S.; Potkonjak, M., "Model-based calibration for sensor networks," Sensors, 2003. Proceedings of IEEE, vol.2, Toronto, Canada, no., pp. 737- 742 Vol.2, 22-24 Oct. 2003.
10. Zilic, Z. and Radecka, K., Fault tolerant glucose sensor readout and recalibration, *Proceedings of Wireless Health, WH 2011*, October 10–13, 2011, San Diego, CA.
11. Balzano, L., Addressing fault and calibration in wireless sensor networks, Master's thesis, University of California, Los Angeles, CA, 2007.
12. Feng, J.; Qu, G.; Potkonjak, M., "Sensor calibration using nonparametric statistical characterization of error models," Sensors, 2004. Proceedings of IEEE, vol., no., pp. 1456-1459 vol.3, 24-27 Oct. 2004.
13. Waltz, E., Data fusion for C3I: A tutorial, in *Command, Control, Communications Intelligence (C3I) Handbook*. Palo Alto, CA: EW Communications, 1986, pp. 217–226.
14. Llinas, J. and Waltz, E., *Multisensor Data Fusion*. Boston, MA: Artech House, 1990.
15. Xiao, L., Boyd, S., and Lall, S., A scheme for robust distributed sensor fusion based on average consensus, *International Conference on Information Processing in Sensor Networks*, April 25–27, 2005, Los Angeles, CA, pp. 63–70.
16. Fan, C., Jin, Z., Zhang, J., and Tian, W., Application of multisensor data fusion based on RBF neural networks for fault diagnosis of SAMS, *International Conference on Control, Automation, Robotics and Vision, ICARCV 2002*, December 2–5, 2002, Singapore, pp. 1557–1562.
17. F. Koushanfar, S. Slijepcevic, M. Potkonjak, A. Sangiovanni-Vincentelli, "Error-tolerant multi-modal sensor fusion", *IEEE CAS Workshop on Wireless Communication and Networking*, 2002, Pasadena, CA.
18. Mehranbod, N. and Soroush, M., Probabilistic model for sensor fault detection and identification, *AIChE J.*, 49(7), 1787–1802, July 2003.
19. Aradhye, H.B., Sensor fault detection, isolation, and accommodation using neural networks, fuzzy logic, and Bayesian belief networks, MS dissertation, University of New Mexico, Albuquerque, NM, 1997.
20. Rojas-Guzman, C. and Kramer, M.A., Comparison of belief-networks and rule-based expert systems for fault-diagnosis of chemical, *Eng. Appl. Artif. Intell.*, 6, 191, 1993.
21. Kareem, M.A., Langari, J., and Langari, R., A hybrid real-time system for fault detection and sensor fusion based on conventional fuzzy clustering approach, *IEEE International Conference on Fuzzy Systems, FUZZ*, May 22–25, 2005, Reno, NV, pp. 189–194.
22. T. Beravs, J. Podobnik, M. Munih, "Three-axial accelerometer calibration using Kalman filter covariance matrix for online estimation of optimal sensor orientation", *IEEE Trans. On Instrumentation and Measurements*, Vol. 61, No. 9, Sept. 2012, pp. 2501–2511.

23. Sarbishei, O., Roshan Fekr, A., Janidarmian, M., Nahill, B., and Radecka, K., A minimum MSE sensor fusion algorithm with tolerance to multiple faults, *IEEE European Test Symposium (ETS)*, May 27–31, 2013, Avignon, France, Accepted.

24. Sarbishei, O., Nahill, B., Roshan Fekr, A., Janidarmian, M., Radecka, K., Zilic, Z., and Karajica, B., An efficient fault tolerant sensor fusion algorithm for accelerometers, *IEEE Body Sensor Network Conference*, May 6–9, 2013, Cambridge, MA, pp. 1–6.

25. STMicroelectronics User Manual UM1472 (2012, January). STM32F4DISCOVERY, STM32F4 high performance discovery board [Online]. Available: http://www.st.com/internet/com/TECHNICAL_RESOURCES/TECHNICAL_LITERATURE/USER_MANUAL/DM00039084.pdf.

26. STMicroelectronics (2010, April). Tilt measurement using a low-g 3-axis accelerometer, application note AN3182 [Online]. Available: http://www.st.com/internet/com/TECHNICAL_RESOURCES/TECHNICAL_LITERATURE/APPLICATION_NOTE/CD00268887.pdf.

27. Clouqueur, T., Saluja, K.K., and Ramanathan, P., Fault tolerance in collaborative sensor networks for target detection, *IEEE Trans. Comput.*, 53(3), 320–333, March 2004.

28. STMicroelectronics Datasheet (Doc ID: 16483 Rev 5, July 2010). STTS751, 2.25 V low-voltage local digital temperature sensor [Online]. Available: http://www.st.com/st-web-ui/static/active/en/resource/technical/document/datasheet/CD00252523.pdf.

29. Freescale Semiconductor FRDM-KL25Z User's Manual, Revision 1.0, September 2012 [Online].

30. Xuezhen, C., Fen, Y., and Maoyong, C., Study of mine dust density sensor output characteristic based on normal linear regression method of Excel, *2010 International Conference on Computer Application and System Modeling (ICCASM)*, vol. 8, October 22–24, 2010, Taiyuan, China, pp. V8-144–V8-148.

31. inTEST Thermal Solutions, Temptronic TP4500, 2011 [Online]. Available: http://www.temptronic.com/Products/TP4500.htm.

32. Boyd, S. and Vandenberghe, L., *Convex Optimization*, 7th edn. Cambridge, England: Cambridge University Press, 2009.

19 Usable Signal Processing
Constructing Models of Signals Corrupted by Nonstationary Interference for Its Detection and Removal

Brett Y. Smolenski and Catherine M. Vannicola

CONTENTS

LIST OF ACRONYMS

EM	Expectation maximization
FFT	Fast Fourier transform
HMM	Hidden Markov model
k-NN	k-Nearest neighbors
KL	Korhunen–Loeve
KS	Kolmogorov–Smirnov
LDA	Linear discriminant analysis
LPC	Linear prediction coefficients
LSE	Least square error
MAD	Maximization absolute deviation
MAP	Maximum *a posteriori*
MFCC	Mel-frequency cepstral coefficients
ML	Maximum likelihood
MMSE	Minimum mean square error
MRAM	Modified rank autocorrelation measure
pdf	Probability density function
PCA	Principal component analysis
QDA	Quadratic discriminant analysis
SAPVR	Spectral autocorrelation peak-to-valley ratio
SNR	Signal-to-noise ratio
TEO	Teager energy operator
TIR	Target-to-interferer ratio
VQ	Vector quantization
ZCR	Zero-crossing rate

19.1 GENERAL STATISTICAL ANALYSIS OF USABLE SIGNAL SEGMENTS

19.1.1 INTRODUCTION

In order to identify usable speech segments as accurately as possible, it is necessary to fully characterize the statistics of the random processes generating the segment sequence. In a cochannel environment, the usable speech segments are produced when one speaker's voiced speech overlaps with the other speakers' silence or unvoiced speech segments. Hence, if one has a statistical model of voiced, unvoiced, and silence segments, one can use this information to obtain a model of the usable speech segments. To accomplish this, statistical models that account for the observed voiced, unvoiced, and silence segment lengths are developed in Section 19.1.2. There are a number of ways a segment's length can be measured. For example, one could use the number of samples or the number of frames, but using these units would result in a measure that depended on the sampling frequency and the size of the frame. To avoid any confusion, all segment lengths are measured in units of time in this chapter.

In Section 19.1.3, Markov models are used to account for dependencies between voiced, unvoiced, and silence segments. Only contiguous segments are considered, not frames of segments. In order to estimate the segmental target-to-interferer ratio (TIR) for the detection of usable speech segments, the sampling distribution of the segmental TIR is developed in Section 19.1.4. The short- and long-term correlation present in the segmental TIR signal is also explored in this section.

The speech data used in all of the examples that follow were taken from the TIMIT database (Appendix 19.A). Twelve male and 12 female utterances, 2 from each of the 6 dialect regions, were chosen. The labeling was performed by the Air Force Research Laboratory in Rome, New York. The data were downsampled from 44.1 to 16 kHz, and each sample of the data was labeled as either voiced, weak-voiced, transition (between voiced and unvoiced), unvoiced, and silence. In this chapter, weak-voiced is lumped into the unvoiced class, since these segments have lower energy and contain little observable phonation. Also, the voiced portions of the transitions are lumped into the voiced class and the unvoiced portions are lumped into the unvoiced class.

19.1.2   MODELING SEGMENT LENGTHS

In this section, the randomness of speech segments is first explored using the nonparametric *runs* test. If the speech segments were completely random, the statistical analysis would not be very helpful. Some necessary results from the combinatory analysis of the speech class segments are also discussed. Lastly, a model of the three speech class segment durations based on the gamma probability density function (pdf) is developed and tested.

19.1.2.1   Runs Test for Randomness

The randomness of speech and silence segments can be studied using the runs test. A nonparametric test for randomness is provided by the theory of runs. To understand what a run is, consider a sequence made up of two symbols, **s** and **n**, such as

$$s\,s \mid n\,n\,n \mid s \mid n\,n \mid s\,s\,s\,s\,s \mid n\,n\,n \mid s\,s\,s\,s$$

where
  s represents a speech frame
  n represents a silent frame

A run is defined as a set of identical symbols contained between two different symbols or no symbol. Proceeding from left to right in the earlier sequence, the first run, indicated by a vertical bar, consists of two **s**'s. Similarly, the second run consists of three **n**'s, the third run consists one **s**, etc.

It seems clear that some relationship would exist between randomness and runs. Thus, for the sequence

$$s \mid n \mid s \mid n \mid s \mid n \mid s \mid n$$

there is a cyclic pattern, in which the sequence goes from **s** to **n**, back to **s** again, etc., which one could hardly believe to be random. In such cases, there are too many runs. On the other hand, for the sequence

$$s\,s\,s\,s\,s\,s \mid n\,n\,n\,n \mid s\,s\,s\,s\,s \mid n\,n\,n$$

there is a pattern, in which the **s**'s and **n**'s are grouped together. In such case, there are too few runs, and one would not consider the sequence to be random.

Thus, a sequence would be considered nonrandom if there are either too many or too few runs, and random otherwise. To quantify this idea, suppose that all possible sequences consisting of N_1 **s**'s and N_2 **n**'s are formed, for a total of N symbols in all. The collection of all possible sequences provides one with a sampling distribution. Each sequence has an associated number of runs, denoted by V. In this way, one is led to the sampling distribution of the statistic V. It can be shown that this sampling distribution has a mean and variance, respectively, given by the formulas

$$\mu_V = \frac{2N_1 N_2}{N_1 + N_2} \tag{19.1}$$

$$\sigma_V^2 = \frac{2N_1N_2(2N_1N_2 - N_1 - N_2)}{(N_1 + N_2)^2(N_1 + N_2 - 1)} \tag{19.2}$$

Using the earlier formulas, one can test the hypothesis of randomness at appropriate levels of significance. Fortunately, if both N_1 and N_2 are at least equal to 8, then the sampling distribution of V is very nearly normal. Thus,

$$z = \frac{V - \mu_V}{\sigma_V} \tag{19.3}$$

is normally distributed with mean 0 and variance 1. For a two-tailed test at the $\alpha = 0.05$ significance level, one would accept the null hypothesis H_0 of randomness if $-1.96 \le z \le 1.96$ and would reject if otherwise. For the data set referred to in Section 19.1.1, a z value of $-2.61 < -1.96$ is obtained, indicating that there are too few runs resulting in a clustering of speech and silence segments. One could potentially exploit this result to help improve the classification of usable speech segments when one considers two overlapping sequences. In addition, when using the multinomial distribution, the runs test can be formulated for more than two classes, such as for voiced, unvoiced, and silence, which will be performed in the following section with the help of a Markov model.

19.1.2.2 Combinatory Analyses

Combinatory techniques form the basis on which many of the statistical analysis methods used throughout this chapter are founded. For example, if one wants to know the theoretical probability of obtaining a particular pair of contiguous segments, combinatory techniques can easily be used to answer the question. Contiguous here refers to one complete voiced, unvoiced, silence, usable, or unusable segment, that is, no speech data of the same speech class can exist at the boundaries of the segment.

Considering all pairs of contiguous segments of a particular speech class, one can have **VU**, **VS**, **UV**, **SV**, **US**, and **SU**, where **V**, **U**, and **S** represent voiced, unvoiced, and silence segments, respectively. How many ways can any one segment pair go to another pair of segments? The answer is four, since only matchings where the last segment in the pair is different than the first segment in the pair are possible, assuming only contiguous segments are being considered.

How many combinations of three or more contiguous segments are allowed? Using enumeration, it can be shown that for combinations of size n, there are $3 \cdot 2^{n-1}$ possible combinations. Further, there would be 2^n ways of pairing one of these combinations with the other. If there were more than three classes, say j classes, then the number of possible contiguous combinations would be $j(j-1)^{n-1}$ possible combinations with $(j-1)^n$ ways of pairing them. In addition, if one is considering combinations of frames of speech with the constriction of contiguity, then each frame can be from any of the classes, that is, there would be j^n possible combinations with no restriction on pairing them.

19.1.2.3 Gamma Distribution

The gamma distribution has often been used as a model of product lifetime in reliability theory. The gamma distribution is a generalization of the exponential distribution, which is often used to model the lifetime of objects that do not depend on their age. This is why an exponentially distributed random variable is referred to as memoryless. However, the effect of aging can be accounted for with the gamma distribution. The gamma pdf is defined as

$$f(t \mid a,b) = \begin{cases} \dfrac{1}{b^a \Gamma(a)} t^{a-1} e^{-t/b} & x > 0 \\[2mm] 0 & x \le 0 \end{cases} \tag{19.4}$$

where
  t is the independent variable, which for these examples represents time
  a and b are shape parameters
  $\Gamma()$ is the complete gamma function

The gamma function is a generalization of the factorial function for noninteger arguments and it is defined as

$$\Gamma(a) = \int_0^\infty e^{-\tau} \tau^{a-1} d\tau \tag{19.5}$$

To see if the gamma distribution can accurately model the observed histograms of segment durations of voiced, unvoiced, and silence segments, the maximum likelihood (ML) estimates of the gamma density parameters (a and b) are computed for all the durations observed for each segment class. Next, using the ML estimated parameters, one can compare an observed histogram of segment durations to a theoretical gamma density having the estimated parameters. To quantitatively compare the observed and theoretical distributions, a chi-squared goodness of fit test can be used.

Figure 19.1 shows the histogram obtained for all the voiced speech segments. To be able to compare the histogram to the estimated theoretical gamma density, the histogram is normalized so that the sum of all the bins is unity. The black curve represents the theoretical gamma density evaluated at the bin centers using the ML estimated parameters. As is customary, the number of bins used is equal to the rounded square root of the number of observations, which includes 301 voiced segments. The grey curves represent the gamma densities with the 90% upper and lower confidence

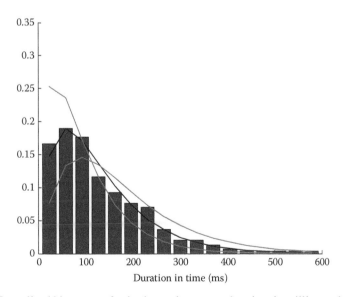

FIGURE 19.1   Normalized histogram of voiced speech segment durations in milliseconds.

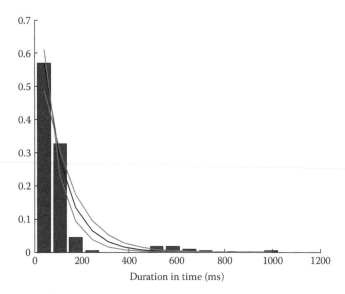

FIGURE 19.2 Normalized histogram of unvoiced speech segment durations in milliseconds.

bounds on the ML estimates of the parameters. Hence, assuming the gamma distribution is a good model of the duration of voiced segments, one would expect that 90% of the observations would fall between these two grey lines.

Figure 19.2 shows the histogram obtained for the unvoiced speech segments. With a median of 61 ms, the unvoiced segments are substantially shorter than voiced speech segments, which had a median duration of 98 ms. Since the gamma distribution is highly skewed, median values are used, as opposed to mean values, to indicate central tendency. The gamma does not appear to be as good of a fit in this case, especially in the tails of the distribution. However, the exponential distribution does turn out to be a much better fit. Perhaps another distribution, such as the Weibull distribution, which is also commonly used to predict product lifetimes, would produce an even better fit.

Figure 19.3 shows the histogram obtained for the silence segments. The median duration is 50 ms, the lowest of the three segment classes. The estimated gamma distribution appears to be a good fit to the observed relative frequency distribution. A more quantitative evaluation of goodness of fit will be carried out next using the chi-squared goodness of fit test.

Suppose that in a particular experiment a set of possible events $E_1, E_2, E_3, \ldots, E_k$, are observed to occur with frequencies $o_1, o_2, o_3, \ldots, o_k$, called the observed frequencies, and that according to probability rules, the outcomes are expected to occur with frequencies $e_1, e_2, e_3, \ldots, e_k$, called the theoretical frequencies (Table 19.1).

A measure of the discrepancy existing between the observed and expected frequencies is supplied by the statistic χ^2 given by

$$\chi^2 = \frac{(o_1 - e_1)^2}{e_1} + \frac{(o_2 - e_2)^2}{e_2} + \cdots + \frac{(o_k - e_k)^2}{e_k} = \sum_{j=1}^{k} \frac{(o_j - e_j)^2}{e_j} \qquad (19.6)$$

If $\chi^2 = 0$, the observed and theoretical frequencies agree exactly, while if $\chi^2 > 0$, they do not agree exactly. The larger the value of χ^2 implies there exists a greater discrepancy between the observed and the expected frequencies.

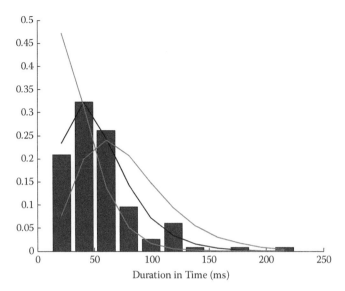

FIGURE 19.3 Normalized histogram of silence segment durations in milliseconds.

If the number of expected frequencies is at least equal to 5, the sampling distribution of χ^2 is approximated very closely by the chi-squared distribution:

$$
f(x) = \begin{cases} \dfrac{1}{2^{v/2}\Gamma(v/2)} x^{v/2-1} e^{-x/2} & x > 0 \\ 0 & x \le 0 \end{cases}
\tag{19.7}
$$

The approximation improves for larger values. This criterion can be used to determine the goodness of fit of the earlier histograms to the proposed theoretical distributions. Note that the chi-squared distribution is a special case of the gamma distribution with the parameters $a = v/2$ and $b = 2$. The number of degrees of freedom, v, is given by $v = k - l - 1$, if the expected frequencies can be computed by estimating l population parameters from the sample statistics. Since the a and b parameters of the fitted gamma distribution have to be estimated from the data, $l = 2$ in these examples. The number of bins used to generate the histograms corresponds to k. The observed frequencies correspond to relative counts in each bin of the histogram, and the evaluation of the estimated gamma distribution at the bin centers corresponds to the expected frequency in the table earlier.

From the test data, χ^2 values of 24.7, 24.3, and 15.8 are obtained for the voiced, unvoiced, and silence histograms, respectively. Substituting the numbers of bins (17, 15, and 11) for the corresponding degrees of freedom and using 0.95 percentile values result in critical values of 27.6, 25.0, and 19.7, respectively. Since the calculated values of the statistic χ^2 are less than the corresponding

TABLE 19.1

Formularization of Chi-Squared Goodness of Fit Test

Event	E_1	E_2	E_3	...	E_k
Observed frequency	o_1	o_2	o_3	...	o_k
Expected frequency	e_1	e_2	e_3	...	e_k

Technologies for Smart Sensors and Sensor Fusion

critical values, it can be concluded at the 0.05 significance level that the gamma distribution is indeed a good fit to the distribution of voiced and silence segment durations and the exponential is a good fit to the unvoiced segment durations.

The χ^2 distribution can also be used to derive a confidence interval for the estimate of the distribution. There exists other goodness of fit tests that do not require binning of the data, such as the Anderson–Darling and Kolmogorov–Smirnov tests, but these require modification when applied to the gamma distribution or when estimates of the distribution parameters have been made.

19.1.3 MODELING SEGMENT DEPENDENCIES

In this section, a Markov chain is first used to model the statistical dependence between adjacent segments. Next, an attempt to generalize the Markov model to incorporate any dependence between more than two consecutive segments is considered. However, a chi-squared test of independence shows that little dependence exists when more than two adjacent segments are considered. Lastly, the hidden Markov model (HMM) is briefly introduced, since in reality, the state sequence is not observable and only the features from each state would be observable.

19.1.3.1 Segment Sequence and Markov Chains

A Markov chain is comprised of a set of possible states and a set of probabilities of transitioning from one state to the next (Figure 19.4). For the current situation, one requires only three states: one for the voiced speech state, another for the unvoiced speech state, and lastly a state for the silence/background noise state. Since only contiguous segments are being considered, as opposed to frames, the transition probabilities of staying in the same state are always zero.

The state transition probabilities, p_{vu}, p_{vs}, etc., are used to populate a state transition matrix, where the rows of the matrix represent the current state (first subscript) and the columns represent the next state (second subscript). The state transition matrix for English speech is estimated using the relative frequency definition of probability (number of observed occurrences divided by number of possible occurrences) to be

$$\begin{bmatrix} 0 & p_{vu} & p_{vs} \\ p_{uv} & 0 & p_{us} \\ p_{sv} & p_{su} & 0 \end{bmatrix} = \begin{bmatrix} 0 & 0.71 & 0.29 \\ 0.62 & 0 & 0.38 \\ 0.67 & 0.33 & 0 \end{bmatrix} \tag{19.8}$$

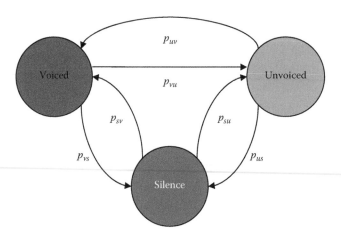

FIGURE 19.4 State transition diagram of voiced, unvoiced, and silence segments.

Since each row of a state transition matrix must sum to one, the earlier state transition matrix can be written in the more compact form:

$$\begin{bmatrix} 0 & p_{vu} & 1-p_{vu} \\ p_{uv} & 0 & 1-p_{uv} \\ p_{sv} & 1-p_{sv} & 0 \end{bmatrix} \tag{19.9}$$

A square matrix $\mathbf{M}$ is called *regular* if all of its elements are greater than zero for some finite power of $\mathbf{M}$. One can easily verify, by taking the second power, that the earlier state transition matrix is regular if all of the allowable state transitions are nonzero. If the state transition matrix of a Markov process is regular, then the Markov process is said to be regular.

An interesting property of regular Markov processes is that the state distributions become stationary. That is, as the process unfolds, the probability of being in any particular state becomes constant. One can use this property of regular Markov processes to see if the state probabilities predicted by the Markov model are significantly different than the observed state probabilities. A regular Markov process will produce a state probability vector $[p_v \; p_u \; (1-p_v-p_u)]$ that satisfies the following linear system of equations:

$$\begin{bmatrix} p_v & p_u & (1-p_v-p_u) \end{bmatrix} \begin{bmatrix} 0 & p_{vu} & 1-p_{vu} \\ p_{uv} & 0 & 1-p_{uv} \\ p_{sv} & 1-p_{sv} & 0 \end{bmatrix} = \begin{bmatrix} p_v \\ p_u \\ 1-p_v-p_u \end{bmatrix} \tag{19.10}$$

Using the earlier system, the state transition probabilities can be expressed in terms of the state transition probabilities as

$$p_v = \frac{p_{uv}}{1+p_{uv}} \tag{19.11a}$$

$$p_u = \frac{1+p_{vu}p_{uv}-p_{uv}}{(1+p_{uv})(2-p_{uv})} \tag{19.11b}$$

$$p_s = \frac{1-p_{uv}p_{vu}}{(1+p_{uv})(2-p_{uv})} \tag{19.11c}$$

Alternatively, one can use the system of equations (19.10) to express the state transition probabilities in terms of the state probabilities. In the case of frames, the main diagonal is no longer zero.

The state transition matrix for Spanish speech is estimated to be

$$\begin{bmatrix} 0 & p_{vu} & p_{vs} \\ p_{uv} & 0 & p_{us} \\ p_{sv} & p_{su} & 0 \end{bmatrix} = \begin{bmatrix} 0 & 0.77 & 0.23 \\ 0.66 & 0 & 0.34 \\ 0.69 & 0.31 & 0 \end{bmatrix} \tag{19.12}$$

As one can see that the state transition matrix depends on language. If there was no dependence between adjacent segments, the nonzero entries would be more uniformly distributed along each row. Given the size of the data set and the fact that multiple male and female speakers were used, it is unlikely that these differences occurred by chance. One would also expect that there could exist some contextual dependence in the state transition matrix as well. For example, the pattern of silences should be different for a telephone conversion versus a lecture.

Although the statistics of the lengths of the segments depend on the state, the probabilities of being in a particular state are nearly the same for both English and Spanish:

$$\begin{bmatrix} p_v & p_u & p_s \end{bmatrix} = \begin{bmatrix} 0.44 & 0.37 & 0.19 \end{bmatrix}_{\text{English}} \tag{19.13a}$$

$$\begin{bmatrix} p_v & p_u & p_s \end{bmatrix} = \begin{bmatrix} 0.49 & 0.33 & 0.18 \end{bmatrix}_{\text{Spanish}} \tag{19.14a}$$

It is expected that these probabilities have some dependence on context as well. The predicted probabilities of the regular Markov process are closely matched to the earlier observed probabilities:

$$\begin{bmatrix} p_v & p_u & p_s \end{bmatrix} = \begin{bmatrix} 0.47 & 0.35 & 0.18 \end{bmatrix}_{\text{English}} \tag{19.13b}$$

$$\begin{bmatrix} p_v & p_u & p_s \end{bmatrix} = \begin{bmatrix} 0.49 & 0.32 & 0.119 \end{bmatrix}_{\text{Spanish}} \tag{19.14b}$$

Such a close match between the predicted and observed proportions gives strong evidence to the accuracy of the Markov model.

This information can now be used to obtain a model for usable cochannel speech segments by overlapping the voiced, unvoiced, and silence model with itself and considering the segments of voiced over unvoiced or silence as usable and the rest as unusable. One should note that the case of unvoiced over silence is considered unusable in this model, since the usable speech features are not applicable on unvoiced speech.

19.1.3.2 Modeling Long-Term Dependence

One of the limitations of the earlier model is that it assumes that the current segment depends solely on the previous segment. It is not difficult to incorporate dependence on more than one segment into the Markov model. For example, one could simply assign each pair of possible segments to a state, which for contiguous segments would yield six states with only four ways of leaving any particular state (see Table 19.2). Each cell in the table represents the number of occurrences of the segment pair in the corresponding row to the segment pair in the corresponding column. A total of 2530 segments were used.

One can observe that the transitional probabilities become nearly uniform when more than one state is incorporated into the model, which implies there is little statistical dependence between more than two consecutive segments. Statistical evidence for this is obtained using the following chi-squared test of independence.

TABLE 19.2

Contingency Table between All Possible Pairs of Current State Segments (Rows) to All Possible Pairs of Next State Segments (Columns)

Current/Next	VU	VS	US	UV	SV	SU	Total
VU	113	107	0	0	98	103	421
VS	102	96	111	106	0	0	415
US	104	97	117	103	0	0	421
UV	0	0	112	95	108	107	422
SV	0	0	109	118	107	99	433
SU	103	107	0	0	110	98	418
Total	422	407	449	422	423	407	2530

The chi-squared test of independence tests the null hypothesis H_0 that there is no dependence between current pairs of segments and the next pair of observed segments. The chi-squared dependence statistic is calculated as

$$\chi^2_{df} = \sum \frac{(E-O)^2}{E} \qquad (19.15)$$

where
the E are the expected cell frequencies
the O are the observed cell frequencies
the summation is the overall of the cells in the contingency table

Note that this test is structurally equivalent to the chi-squared goodness of fit test. The df stands for the number of degrees of freedom, which is equal to $(R - 1)(C - 1)$, where R is the number of rows in the contingency table and C is the number of columns. To compute the expected cell frequencies, the following formula is used:

$$E_{ij} = \frac{T_i T_j}{N} \qquad (19.16)$$

where
E_{ij} is the expected frequency for the cell in the ith row and the jth column
T are the totals for the ith row and the jth column
N is the grand total of all state transitions (total in the last row and column of the contingency table)

One should note that the expected frequencies should be set to zero for the cells where it would not be possible to have the state transition and maintain contiguous segments.

The computed χ^2 value is found to be 22.4, which at 25 degrees of freedom corresponds to a probability value of only 0.0023 of there being a relationship between the current and next states. Although it is known that long-term dependence with respect to what words are being spoken and the context of the dialog, these dependences are not known and hence are not observable in the model. It may be possible that language models and speech recognition could be used to improve performance.

In addition, a quantitative measure of the dependence between the two groups can also be obtained using the correlation of attributes:

$$r = \sqrt{\frac{\chi^2}{N(k-1)}} \qquad (19.17)$$

where k is the number of rows or columns. This measure is simply a categorical version of the linear correlation coefficient. Substituting the corresponding values from the earlier contingency table yielded a correlation of 0.00023, a small amount of correlation indeed.

19.1.3.3 Hidden Markov Model

In an operational environment, one cannot directly observe the state sequence of voiced, unvoiced, silence, usable, or unusable segments. As the speech signal occurs, one has access only to the speech signal. The phoneme classes producing the observed characteristics of the signal are not directly observable. In this situation, one has what is referred to as a HMM.

Having observed the state sequences of training data and assuming the models formed to characterize these state sequences are consistent for a given language, the model's parameters in conjunction with the Viterbi algorithm can be used to improve the performance of the segment classification system.

19.1.4 Modeling Segmental Target-to-Interferer Ratio

Since the aim of this chapter is to describe a method used to classify usable speech segments using an estimate of the TIR, it would be prudent to have a statistical model of the segmental TIR signal. The goal of this section is to develop a statistical model of the segmental TIR for cochannel speech. In the first subsection, the F distribution is used to model the pdf of the segmental TIR values. In the second subsection, the correlation in the segmental TIR is analyzed and modeled.

19.1.4.1 Distribution of Segmental Interference Level

If one assumes that the two speech signals forming the cochannel speech are approximately normally distributed and are independent processes, which is usually not a bad approximation, then the segmental TIR sequence can be easily modeled using the F distribution. It can be shown that the ratio F of the estimates of the variances of two normally distributed random variables with corresponding variances σ_A^2 and σ_B^2 follows an F distribution:

$$F = \frac{\hat{S}_A^2 / \sigma_A^2}{\hat{S}_B^2 / \sigma_B^2} \tag{19.18}$$

where the subscripts refer to the two speakers A and B that produced the cochannel speech. The unbiased variance estimates are obtained using

$$\hat{S}^2 = \frac{(X_1 - \bar{X})^2 + (X_2 - \bar{X})^2 + \cdots + (X_m - \bar{X})^2}{m - 1} \tag{19.19}$$

with sample mean estimate calculated using

$$\bar{X} = \frac{X_1 + X_2 + \cdots + X_m}{m} \tag{19.20}$$

Since speech is a zero mean process, the sample mean can be set to zero in most applications. The frame size m represents the number of samples used to calculate the corresponding quantities.

Expressing Equation 19.18 in decibel values, one obtains

$$TIR_0 + 10\log_{10} F = TIR_s \tag{19.21}$$

where
TIR_0 represents the overall TIR
TIR_s represents the segmental TIR for a given frame of speech
F is the F distribution

The F pdf is defined as

$$f(x) = \begin{cases} \dfrac{\Gamma(v)}{\Gamma^2(v/2)} x^{(v/2)-1}(1+x)^{-v} & x > 0 \\ 0 & x \le 0 \end{cases} \tag{19.22}$$

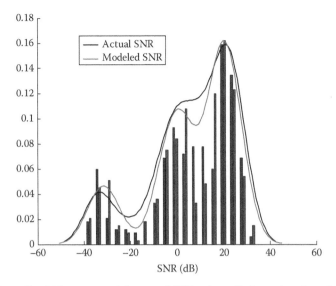

FIGURE 19.5 Normalized histograms of the actual TIR values (dark grey) and modeled segmental TIR values (light grey) for one utterance.

where v is the number of degrees of freedom, which is equal to $m - 1$. Those familiar with the F distribution may recall that in general, it has two parameters for the number of degrees of freedom: one for the number of samples used in the numerator variance estimate and another for the number of samples used in the denominator variance estimate. Since typically one does not use different sample sizes to calculate the segmental TIR_s, both parameters are set equal to v in the earlier definition.

From Equation 19.18, it is observed that as a random variable, the segmental TIR_s in dB is distributed as a scaled log F distribution with mean shifted by the value in dB of the overall TIR_0. This information could prove valuable in estimating the overall and segmental TIR values.

Figure 19.5 shows the normalized histograms of the actual and modeled segmental TIR values for one cochannel utterance. The three modes in the histograms are the result of the speech signal energy being approximately in one of three states: voiced, unvoiced, or silence. To better compare the two histograms, nonparametric probability density estimation is performed on both the simulated and actual TIR values using unit variance Gaussian kernel functions with the range of the data divided up equally into 100 blocks. The dark grey curve corresponds to the actual TIR density estimate and the light grey curve corresponds to the simulated TIR density estimate.

The simulated data were obtained by using log F distributed random number generators with the degrees of freedom set to match the frame size of the actual data, 64 sample frames. Three mixtures were generated with proportions set to the proportions of voiced, unvoiced, and silence in the original utterance with corresponding means of 20 dB, 0 dB, and −35 dB, respectively. Since the actual segmental TIR is highly correlated, the variances of the simulated distributions had to be appropriately scaled. The correlation in the segmental TIR will be analyzed next.

19.1.4.2 Correlation in Segmental Interference Level

Figure 19.6 shows an 8 kHz sampled cochannel speech utterance with an overall TIR of 20 dB (top panel) and the corresponding segmental TIR signal computed using 20 ms rectangular window frames with no overlap (bottom panel). The utterance is *we'll serve rhubarb pie after Rachel's talk.* Notice that the segmental TIR signal is highly correlated and takes on approximately three states, which correspond to when the speech was voiced, unvoiced, or silent. When the speech was voiced, the segmental TIR hovers around 25 dB, while for unvoiced and silence, the TIR hovers around 0 dB and −30 dB, respectively. Changing the overall TIR simply shifts the segmental TIR curve up or down depending on whether the interferer was decreased or increased.

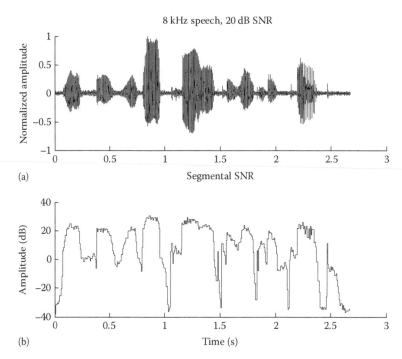

(a)

(b)

FIGURE 19.6 Cochannel speech utterance with interferer added at 20 dB (a) and the segmental TIR signal computed using 20 ms rectangular window frames with no overlap (b).

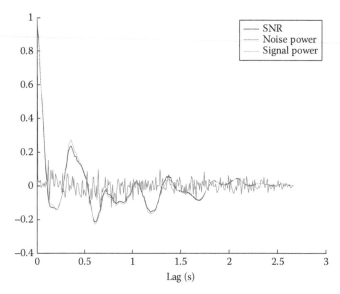

FIGURE 19.7 Normalized autocovariance functions of the segmental TIR, segmental speech power, and segmental signal power for a typical cochannel utterance.

Figure 19.7 shows the normalized autocovariance functions of the segmental TIR, segmental speech power, and segmental signal power for a typical utterance. The autocovariance should be used, since these signals are not zero mean. The overall TIR is 0 dB, but this has no effect on the correlations observed. From this figure, one can see that since the added interferer is nonstationary, the segmental TIR and segmental signal power are almost identical. Another important observation is that the short-term positive correlation lasts for about 100 ms, which from Section 19.1.2.3 is the median duration of the voiced segments. There also exists long-term correlation that peaks at around 350 ms, which is likely due to the pattern of voiced and unvoiced segments in the utterance.

19.1.5 Concluding Remarks

In this section, the statistical properties of the voiced, unvoiced, and silence segment sequence of speech were analyzed and modeled. Based on this information, a model of the usable segments was developed. The goal being that with a better understanding of the statistical characteristics of the segment sequence, one can design a better system for classifying the segments. In the next section, features that indicate the voiced, unvoiced, and silence classes will be developed along with techniques of classifying these feature vectors.

19.2 EXAMPLE PHYSICAL MODELING OF INTERFERENCE ROBUST SIGNAL SEGMENTS

19.2.1 Introduction

Segmentation of the speech signal into voiced, unvoiced, and silence/background noise segments is an important first step in usable speech extraction, since speaker identification technology performs best on voiced speech segments. Several reliable voiced/unvoiced/background classifiers have been reported in the speech processing literature. Most of the research performed in the 1970s and early 1980s centered on supervised linear discriminant analysis (LDA) classifiers using no more than five features. The most common features that were used were average zero-crossing rate (ZCR), normalized first autocorrelation coefficient, and a low-to-high-frequency power ratio. Often, features that were commonly used for pitch detection were included, such as long-term autocorrelation and cepstral peak picking. Other features were often just variations of the earlier basic features, for example, peak picking the autocorrelation of the linear prediction coefficient (LPC) residual or autocorrelation of center-clipped signal. In the 1990s, wavelets and higher-order spectral analysis techniques, such as those derived from the bispectrum, were used as features.

Typical percentages of correct classifications for the early LDA classifiers ranged from 90% to 95% for clean speech and 80%–90% for moderately noisy speech (~20 dB SNR). In the 1980s, research into more sophisticated classifiers began, such as those using neural networks, which yielded modest improvement. However, one should note that when the speech was noisy, training and testing of the classifier was done using the same kind of noise. When the training and testing environments were different, the supervised classifiers produced equal error rates (EER) in the range of 50%–60%.

To improve the robustness of voiced/unvoiced/background classifiers on mismatched training and testing data sets, research into unsupervised clustering-based classifiers began in the late 1990s and remains an active area of research. One approach based on self-organizing maps (SOM) boasted that it could accurately classify voiced segments in 10 dB signal-to-noise ratio (SNR). However, discriminating between unvoiced and background noise in highly degraded and cochannel environments still remains an unsolved problem.

In this section, a novel unsupervised cochannel voiced, unvoiced, and silence/background segmentation preprocessing system is developed. The qualitative properties of both voiced and unvoiced speech are first studied in Section 19.2.2. Section 19.2.3 discusses several traditional segmentation features, as well as two novel features derived from a nonlinear state-space representation. Principal component analysis (PCA) is also performed to observe the amount of complimentary information available in the feature set and the number of features that should be used. These features are then used as input to both supervised and unsupervised classifiers, which include k-means clustering, quadratic discriminant analysis (QDA)-, Gaussian mixture model (GMM)-, and k-nearest neighbor (*k*-NN)-type classifiers. Further, an HMM is also used to incorporate contextual information into the classification process. The supervised classifiers are able to obtain 92% accurate speech segment classification in 10 dB SNR, but perform poorly when the noise conditions during training are different than those used during testing. The unsupervised approaches are able to obtain a 15% EER for SNR values as low as 10 dB.

One should note that although the cochannel speech is assumed to be recorded in a low-noise environment, noise is added to the data for the examples in this section to demonstrate the noise robustness of the segmentation algorithm. If the cochannel data also include additive stationary noise, then the background segments detected with the segmentation algorithm could be used with traditional speech enhancement algorithms to reduce its effects.

19.2.2 Classification of Speech Sounds

Speech can be modeled as a sequence of quasi-stationary sounds called phonemes. All phonemes can be classified as being either *voiced*, such as vowel sounds like $\bar{a}$, $\bar{e}$, and $\bar{o}$, or *unvoiced*, such as consonant sounds like *sh*, *f*, and *s*. In addition to voiced and unvoiced speech, about 20% of conversational speech is silence or background noise (in the case of a noisy recording environment).

19.2.2.1 Voiced Speech

Voiced speech occurs when the glottis rapidly opens and closes as air is forced through it from the lungs. This process is called phonation by linguists. All vowel sounds are voiced speech sounds. Vowels are referred to as continuant, since their characteristics change relatively slowly with time. There are also voiced phonemes called semivowels, which are noncontinuant, since their characteristics change rapidly with time. Further, there are nasalized voiced speech sounds (*n*, *m*, and *ng*) that are created when some portion of the oral cavity is closed off and the sound is forced to resonate in the nasal cavity.

19.2.2.2 Unvoiced Speech

Unvoiced speech sounds occur when the vocal tract is constricted at some point and the air that is forced through the constriction becomes turbulent, thus producing a noise-like sound. These continuant sounds are called consonants or fricatives by linguists. There also exist fricative sounds that have a voicing component in addition to the turbulent flow, for example, the sounds produced when speaking *v* and *z*. These sounds are called voiced fricatives. In addition, there are noncontinuant unvoiced speech sounds called stops or plosives, which occur when the vocal tract is closed off and pressure is allowed to build up behind the constriction. The constriction is then suddenly released resulting in sounds like *t* and *k*. Plosive sounds can include a voiced component as well, for example, *b* and *d*. Linguists also distinguish classes of both voiced and unvoiced noncontinuant speech sounds known as diphthongs and affricatives, which are formed approximately by the rapid concatenation of two speech sounds, for example, the sound *oy* as in *boy*.

From the earlier discussion, one observes that there are several examples of speech sounds that contain both voiced and unvoiced components. In fact, even in uncorrupted voiced speech, there exists some noise-like (unvoiced-like) components, especially above 2 kHz. Given this information, it may be more prudent to measure voicing as opposed to classifying it; however, for the extraction of usable speech segments, it is desirable to consider just the continuant vowel sounds as the voiced speech segments.

19.2.3 Speech Segmentation Features

In this section, several features for the classification of voiced, unvoiced, and background noise segments are developed. Three classes of features are examined: traditional speech segmentation features, modern features, and two novel features. The traditional speech segmentation features studied were average ZCR, autocorrelation coefficients, and a low-to-high-frequency energy ratio. For the modern features, higher-order spectral analysis techniques, such as the bispectrum, become necessary. The novel features that were developed include measures of stationarity and state-space embedding features.

19.2.3.1 Voiced Speech Features

The average ZCR has been used as a feature for voiced/unvoiced/silence classification for over 30 years. It is computed by counting the number of times the signal crosses the time axis and dividing by the number of samples in the frame. The average ZCR indicates the amount of high-frequency information in the signal. For voiced speech, most of the signal energy is concentrated in the low-frequency range (100–2000 Hz), while for unvoiced speech, the opposite holds. Hence, for voiced speech, one should observe a small ZCR, while for unvoiced speech, the ZCR should be high.

Figure 19.8 shows conditional pdf estimates for the ZCR feature with 20 dB additive white noise. A Parzen technique using Gaussian kernels was used to estimate the densities. To distinguish voiced vowels from other voiced sounds that contain strong unvoiced components, weak-voiced and transition densities have been displayed. For the usable speech detection system, only the voiced class should be processed. Figure 19.9 shows the same set of density estimates, but with 10 dB additive white noise. One can clearly see that noise has a significant effect on the conditional density functions.

In addition to the ZCR, the autocorrelation coefficients have also been traditionally used as segmentation features. Most of the parametric features used in speech processing, such as LPCs, log-area ratios, line spectral pairs, and reflection coefficients, can be obtained from the autocorrelation coefficients of the speech signal. While speaking a particular phoneme, the speech signal can be well modeled by a linear autoregressive moving average (ARMA) process. In general, the parameters associated with an ARMA(p, z) model with p poles and z zeros can be obtained from the first $(1 + p + q)$ autocorrelation values starting from zero lag. The inclusion of the zero lag autocorrelation value $r(0)$ is needed to match the power of the model to that of the signal. When one divides the autocorrelation values by the signal variance $r(0)$, the autocorrelation coefficients are obtained.

Reliable estimates of the first autocorrelation coefficients can be obtained provided the number of signal samples used to estimate them is greater than four times the number of coefficients required.

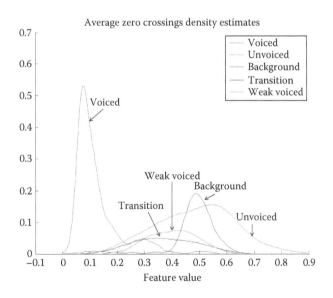

FIGURE 19.8 Conditional pdf estimates for the ZCR feature with 20 dB additive white noise.

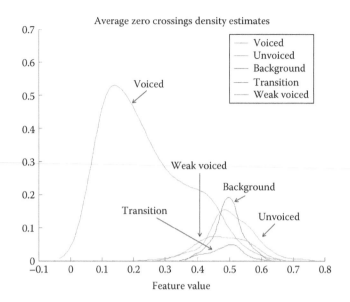

FIGURE 19.9 Conditional pdf estimates for the ZCR feature with 10 dB additive white noise.

In addition, since the speech signal is only quasi-stationary, the number of signal samples used to calculate the autocorrelation coefficients cannot be arbitrarily large. Since it has been shown that a 20 ms. segment from a speech signal sampled at up to 10 kHz can be accurately modeled with 5 zeros and 10 poles, 16 autocorrelation coefficients should contain most of the explainable linear variation in the signal. Hence, a frame size of at least 64 samples should suffice to obtain reliable estimates of these 16 autocorrelation coefficients. It should be noted that a different frame length will be used for the usable speech detection algorithm.

An important advantage with using autocorrelation coefficients as features is that they would more flexibly allow for the varying model order inherent in any speech signal. For example, during unvoiced speech, the signal is best modeled by an ARMA(5,3) system, while for a nasalized vowels (*m*, *n*, and **ng**) the signal is best modeled by an ARMA(10,5) system. In addition, using Fisher's *Z* transformation, the

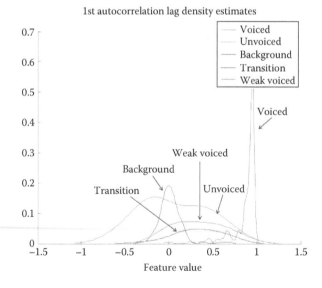

FIGURE 19.10 Conditional pdf estimates for the first autocorrelation lag $r(1)$ feature with 20 dB additive white noise.

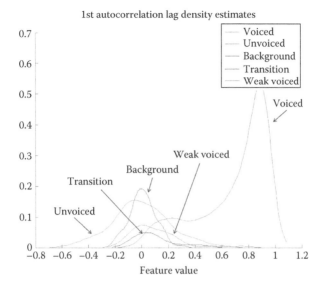

FIGURE 19.11 Conditional pdf estimates for the first autocorrelation lag $r(1)$ feature with 10 dB additive white noise.

autocorrelation coefficient estimates can be transformed into features that are asymptotically Gaussian distributed. Hence, these transformed autocorrelation coefficient features would be a good match for a parametric classification system. Figures 19.10 and 19.11 show the conditional pdf estimates for the first autocorrelation lag $r(1)$ feature with 20 dB and 10 dB additive white noise, respectively. The autocorrelation coefficients show similar class separation for larger lags as well.

Another traditional speech segmentation feature is the nonparametric low-to-high-frequency energy ratio. Voiced speech can have as much as 30 dB more energy than unvoiced speech. Using this fact, one may want to use the average energy in a frame as a feature. However, it is difficult to use the raw energy as a feature because the signal's overall energy can vary wildly depending on recording conditions and the nature of the speaker. Hence, if one wants to use energy, energy ratios should be used. Since voiced speech is known to be more of a low-pass signal while unvoiced speech is more of a high-pass signal, the ratio of the energy in the band from 200 Hz to 1800 kHz to energy in the band 2000 kHz $- f_s/2$, where f_s is the sampling frequency of the signal, should be a good indicator of voicing state. Figure 19.12 shows conditional pdf estimates for the low-to-high-frequency energy ratio feature with 20 dB additive white noise.

The previous traditional segmentation features are related to the second-order statistics of the signal. Modern segmentation also includes information from higher-order statistics. Higher-order statistics are simply generalizations of first- and second-order statistics. For example, the first four moments of a strict-sense stationary random process $x(t)$ are defined as

$$m_1 = E[x(t)] \tag{19.23a}$$

$$m_2(\tau_1) = E[x(t)x(t - \tau_1)] \tag{19.23b}$$

$$m_3(\tau_1, \tau_2) = E[x(t)x(t - \tau_1)x(t - \tau_2)] \tag{19.23c}$$

$$m_4(\tau_1, \tau_2, \tau_3) = E[x(t)x(t - \tau_1)x(t - \tau_2)x(t - \tau_3)] \tag{19.23d}$$

where E is the expectation operator. The first two moments are the familiar mean and autocorrelation function. By the Wiener–Khintchine theorem, taking the Fourier transform of

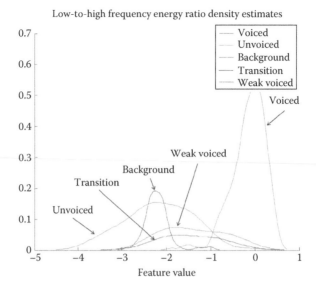

FIGURE 19.12 Conditional pdf estimates for the low-to-high-frequency energy ratio feature with 20 dB additive white noise.

the autocorrelation function produces the power spectral density of the random process $x(t)$. Similarly, if one takes a 2D or 3D Fourier transform of the third or fourth moments, then one obtains the bispectrum and trispectrum, respectively, of the random process. Usually, only the first four moments are used, since large quantities of data are required to accurately estimate higher-order moments. If a random process is Gaussian, then the process is completely characterized by its first two moments. Since speech is known to be a non-Gaussian process, higher-order moments can yield additional information that could be used to improve speech segmentation.

The higher-order moments are generally regarded to contain nonlinear (non-Gaussian) information about the signal; however, one would not want to use only nonlinear features any more than they would want to use only linear features. The impetus for using nonlinear features is due to the complementary information they would provide when used in conjunction with linear features. In the following paragraphs, two novel segmentation features will be developed based on nonlinear processing approaches that are deterministic in nature, that is, are not based on probabilistic models.

State-space embedding is a nonparametric approach to analyzing nonlinear chaotic systems. It should be noted that state-space embedding is not a feature, but features will be derived from the embedding process. The set of state-space trajectories of a system can completely describe the system. A state-space embedding of a signal is typically used to qualitatively study any nonlinearities of the system generating a particular signal. However, while it may be easy for one to observe patterns in the state-space trajectories of a system, it is often difficult to quantify what is observed. To address this issue, two novel features are extracted from a state-space embedded signal using concepts from differential geometry. These features are computed iteratively on the 1D speech signal and they completely characterize the state-space trajectories formed by the signal.

It is well known that the speech production mechanism, in general, is a nonlinear system, which can only be approximated using linear models. The Navier–Stokes equation is a nonlinear partial differential equation that represents one of the most general models of the human speech production mechanism. For example, unvoiced speech corresponds to turbulent flow that is described by chaotic solutions of the Navier–Stokes equation. A chaotic signal is a signal that appears random, but is actually the result of a deterministic nonlinear system. As a note, one also observes chaotic signal properties when the speech is cochannel.

The evolution of a nonlinear dynamical system can be described by a point moving along a trajectory in its state-space, where the coordinates of the point are independent degrees of freedom of the system (memory elements). The signal state-space embedding method was developed for analyzing chaotic signals generated by nonlinear systems. When a signal is embedded in state-space, it is transformed into a trajectory in an m-dimensional space. The number of necessary dimensions corresponds to the number of state variables necessary to describe the system. Unfortunately, these state variables are not directly observable. However, according to Takens' embedding theorem, it is possible to reconstruct a state-space representation topologically equivalent to the original state-space of a system using the 1D observable signal (Takens, 1981). Topologically equivalent means that there exists a one-to-one transformation between the embedded signal and the actual state-space trajectory.

Using Takens' method of delays, points $x(i)$ in an m-dimensional space are formed from time-delayed values of a signal $s(i)$:

$$x(i) = [s(i), s(i-d), s(i-2d), \ldots, s(i-(m-1)d)] \tag{19.24}$$

where

    m is the embedding dimension
    d is the chosen delay value in samples

In this chapter, a constant embedding dimension of 3 is used, since it has been shown that voiced speech can be adequately embedded in three dimensions. The choice of an optimal delay parameter d depends on the sampling rate and the mutual information between samples in the signal. The delay should be large enough so that adjacent points $x(i)$ have a minimum of mutual information between them. However, one cannot make the delay arbitrarily large, since one would sacrifice time resolution. A constant d value of 12 samples has been found to produce good embedding results and thus is the delay value used in the examples found in this chapter.

Figure 19.13 shows a segment of a hypothetical state-space trajectory in three dimensions, which could describe the dynamics of a nonlinear system having three memory elements. As the state-space trajectory of the system unfolds, one can consider the rectangular **TNB** frame of reference moving along the curve characterized by the three perpendicular *osculating, normal*, and *rectifying* planes.

The Serret–Frenet theorem from differential geometry states that any 3D space curve can be completely characterized by the following matrix equation relating the **T**, **N**, and **B** vectors:

$$\begin{bmatrix} \dot{T} \\ \dot{N} \\ \dot{B} \end{bmatrix} = \begin{bmatrix} 0 & \tau & 0 \\ -\tau & 0 & \kappa \\ 0 & -\kappa & 0 \end{bmatrix} \begin{bmatrix} T \\ N \\ B \end{bmatrix} \tag{19.25}$$

where

    κ is the *curvature*
    τ is the *torsion* (defined in the following)

The derivatives of the vectors on the left side of Equation 19.25 are with respect to s, the arc length of the curve.

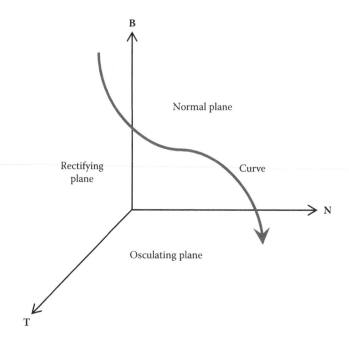

FIGURE 19.13 Illustration of the rectangular TNB frame of reference along a segment of a state-space trajectory.

The curvature and torsion are defined as

$$\kappa = \lim_{\Delta s \to 0} \frac{\Delta \theta}{\Delta s} \tag{19.26}$$

$$\tau = \lim_{\Delta s \to 0} \frac{\Delta \Phi}{\Delta s} \tag{19.27}$$

where
$\Delta \theta$ is the angle between tangents **T** to the curve
$\Delta \Phi$ is the angle between the binormals **B** to the curve

Thus, κ is the rate at which the tangent at a point P rotates as it moves along the curve. The reciprocal of κ is the radius of curvature. The torsion τ is the rate at which the unit binormal **B** at the point P rotates as it moves along the curve. Since the torsion and curvature parameters completely describe the curve, it is likely that they would serve as useful nonlinear features that would complement the linear features.

It should be noted that the space curve formed by the state-space embedding procedure is really an estimate and sampled version of the actual state-space trajectory. Hence, the curvature, torsion, and necessary derivatives in Equation 19.25 must be estimated from the discrete embedded state-space curve. This was accomplished by using the following formulas:

$$S_n = |A_n| \tag{19.28}$$

where
$|\cdot|$ represents the Euclidean norm
the vector A_n is defined as

$$A_n = \left\langle x_n - x_{n-1}, y_n - y_{n-1}, z_n - z_{n-1} \right\rangle \tag{19.29}$$

which is the elemental arc length of the discrete embedded state-space curve. To approximate the curvature K_n and torsion T_n, the formulas in the following were derived:

$$K_n = \cos^{-1}\left(\frac{-A_n \cdot A_{n+1}}{|A_n||A_{n+1}|} \right) \tag{19.30}$$

and

$$T_n = \cos^{-1}\left(\frac{\left\langle -A_n \times A_{n+1} \right\rangle \cdot \left\langle -A_{n+1} \times A_{n+2} \right\rangle}{|A_n \times A_{n+1}||A_{n+1} \times A_{n+2}|} \right) \tag{19.31}$$

where $\times$ and $\cdot$ represent the vector cross and dot products, respectively.

Figures 19.14 and 19.15 show the state-space embedding for unvoiced and voiced speech frames, respectively. The unvoiced embedding was obtained using 500 samples of the phoneme/s/, and the voiced embedding was obtained using 500 samples from the word *we'll*. One can observe that the embedded signal for unvoiced speech is highly chaotic and random, while the voiced speech generates a very structured embedded signal. This is because voiced speech can be described by a system having a lower number of state variables, whereas the turbulence in unvoiced speech requires a large number of state variables.

For voiced speech, it has been shown that the embedded trajectories should be in the general form of an ellipse with small loops attached. The ellipse corresponds to the pitch period and the smaller loops correspond to the resonances in the vocal tract. The torsion for any space curve that can be represented in a plane, such as an ellipse, is always zero. Hence, for a particular vowel, one would expect the curvature signal to be consistently large (near the value of pi) and the torsion signal to be consistently

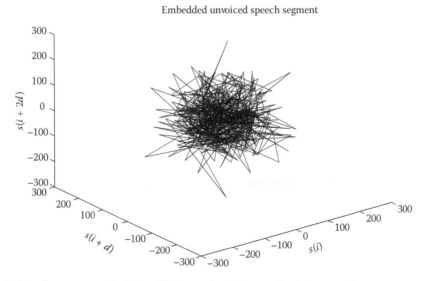

FIGURE 19.14   State-space embedding of 500 samples of the unvoiced phoneme/s/.

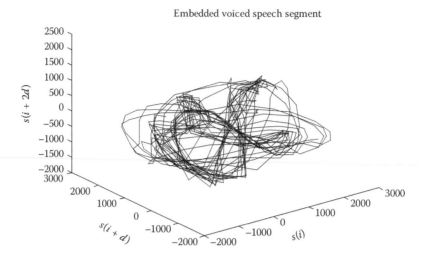

FIGURE 19.15   State-space embedding of 500 samples from the word *we'll*.

small (near the value of zero). Since the 500 samples used for Figure 19.15 were taken across the three voiced phonemes comprising the word *we'll*, several elliptical structures can be observed.

Figure 19.16 shows the raw curvature as well as the median filtered curvature histogram for the entire data set prior to adding noise. This resulted in a bimodal histogram—one mode for unvoiced speech and the other for voiced speech. Since curvature values are higher for voiced segments than unvoiced, the mode representing voiced speech is on the right, while the mode representing unvoiced speech is on the left. The curvature signal was a 79-point median filtered in order to smooth it and reduce the overlap between the two classes. Since it is well known that median filters preserve jump discontinuities, a median filter was necessary in order to preserve the temporal resolution of the feature. In addition, the individual modes are approximately exponentially distributed. In this situation, it can be shown that the best predetection filter would be a median filter.

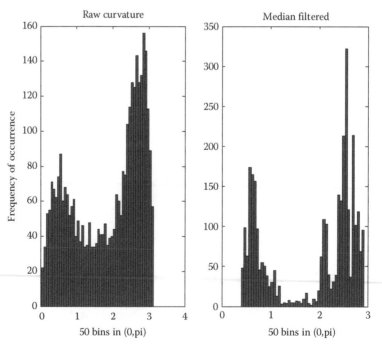

FIGURE 19.16   Histogram of raw curvature values for entire data set with no added noise.

It is interesting to note the level of bimodality seen in the raw curvature signal. The energy and ZCRs have traditionally been used to classify voicing state; however, even their averaged values do not have such a high level of separation. Figure 19.17 shows the 79-point median filtered curvature signal plotted with the corresponding speech signal uttering the words *we'll serve*, where the voiced fricative /v/ has been truncated. For ease of viewing, both signals have been normalized between (−1, 1) and the curvature signal had a constant of 1 added to it. One can clearly see how the curvature signal decreases significantly during the unvoiced /s/ phoneme.

Figure 19.18 shows the 79-point mean filtered torsion signal plotted with the same speech as in Figure 19.17. As expected, the torsion signal increases during unvoiced speech; however, the transition is not as pronounced as with the curvature signal and there exists more variability. Since median filters preserve impulses as well, mean filtering was necessary for the torsion signal due to its impulsive structure.

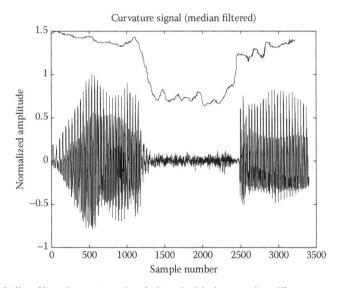

FIGURE 19.17   Median filtered curvature signal plotted with the speech *we'll ser....*

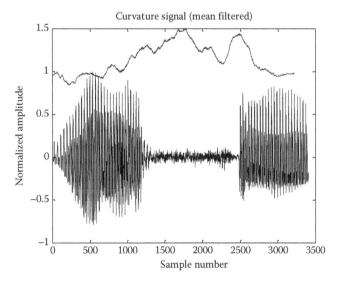

FIGURE 19.18   Mean filtered torsion signal plotted with the speech *we'll ser....*

One should note that alternative approaches of performing the state-space embedding do exist. For example, one alternative approach to embedding a signal can be accomplished using singular value decomposition (SVD) of a Hadamard matrix formed from the lagged versions of the signal. This approach has been shown to be more robust in the presence of additive noise (Kubin, 1995). Further improvement may be obtainable by using state-space with the LPC residual or, equivalently, using generalized SVD (GSVD), which prewhitens the signal. Since the effects of the vocal track resonances would be removed, prewhitening should *straighten out* the embedded signal for voiced speech so that only quasi-elliptical orbits would remain as the state-space trajectories. This should yield more consistent curvature and torsion values during usable voiced speech.

In addition, the average parameters of the ellipses could be linearly transformed into circles having an average unit radius, hence, producing lower variance curvature and torsion values. This is possible, since it can be shown that for a circle, the curvature is the radius of the circle and the torsion is zero. Further, these transformations and the whitening process would have the effect of making the curvature and torsion features for unvoiced speech more uniformly distributed on the interval $(0, \pi)$. A model for the nonlinearities of speech production can be found by applying the Teager energy operator to the resonances of the speech signal.

19.2.3.2 Unvoiced Speech Features

One may have noticed from the figures earlier that the features that discriminate voiced speech from unvoiced speech and background do not have much separation between the unvoiced and background classes. Hence, it is necessary to have additional features that can separate the unvoiced speech and background noise segments. Although most background noise processes are statistically similar to the unvoiced speech, the characteristics of unvoiced speech usually change much more rapidly than most background noise processes. Thus, if one could measure the degree to which a random process was nonstationary, then this should be a good feature for discrimination between background noise and unvoiced speech. The following measure of stationarity, defined for the pth frame, was used as an unvoiced/background feature:

$$\Delta Y(pL) = \left[\frac{1}{\pi} \int_0^\pi |Y(pL,w) - Y((p-1)L, w)|^2 \, dw \right]^{1/2} \tag{19.32}$$

19.2.4 Principal Component Analysis

The objective of PCA is to reduce the number of features necessary for a classifier. Principal components are obtained by diagonalizing the sample covariance matrix Σ of all the training data feature vectors. The square matrix Σ can be diagonalized by

$$D = U\Sigma U^{-1} \tag{19.33}$$

where
    U is the matrix having the eigenvectors of Σ as column vectors
    D is a diagonal matrix containing the corresponding eigenvalues of Σ

Note that changing the order of the columns of Σ changes the order of the eigenvalues along the diagonal of D. Since the trace of D equals the total variance of the training data, by arranging the columns of U such that the eigenvalues of D are in descending order from the first row to the last, one can observe which features contribute the most to the total data set's variability. Furthermore, by transforming each observed feature vectors x by U^{-1}, one obtains the principal components of the data.

Principal components are decorrelated features, but they are not necessarily statistically independent features. However, for the Gaussian case, decorrelation is all that is required to satisfy independence of the transformed variables, since the multivariate Gaussian distribution is completely determined by its mean vector and covariance matrix. Further, it is well known that if a set of Gaussian variables undergoes a nonsingular linear transformation, then the resultant variables are also Gaussian. If the original variables were transformed in such a way as to make the resulting covariance matrix diagonal, then the resulting multivariate normal distribution would factor, which is the definition of independence.

PCA was performed on the entire data set to see if the dimensionality of the feature vector could be reduced and to yield decorrelated features for the classification algorithms to follow. Figure 19.19 displays a stem plot of the eigenvalues of the sample covariance matrix Σ. From this figure, one can observe that almost all of the variance in the original variables would be accounted for if one used only the first three principal components. The list on the right of Figure 19.19 displays the cumulative percentage of the total variance accounted for by these eigenvalues, which indicates that over 97% of the total variance is accounted by the first three largest eigenvalues. Given this information, it was decided to use only the first three principal components in the classification algorithms.

Figure 19.20 displays a histogram of the training data projected in the direction of the first principal component for a 3 s utterance (197 128-sample frames). To help one see the contributions of the different classes to the histogram, a Gaussian curve (not normalized) was fit to each portion of the data corresponding to each class. The curve on the left corresponds to background noise frames, the curve in the middle corresponds to unvoiced speech frames, and the curve on the right corresponds to voiced speech frames. The square root of the number of data values in each class was used to determine the number of bins for each class.

Figure 19.21 displays a scatter plot of the z values (the data transformed in the direction of the principal components) for the same 3 s utterance used in Figure 19.20. Each point has been mapped to a color depending on the corresponding class label for the point. The medium grey points correspond to background noise, the green points correspond to unvoiced speech frames, and the dark grey points correspond to voiced speech frames.

Given the results in Figure 19.21, it would appear that if one had labeled data available, a reliable classifier could be constructed to separate future observed frames. However, if the data from

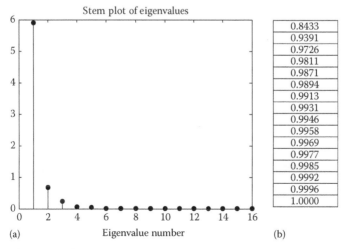

FIGURE 19.19 (a) Stem plot of eigenvalues of Σ_{xx} in descending order. (b) Cumulative percentage of total variance accounted for by these eigenvalues.

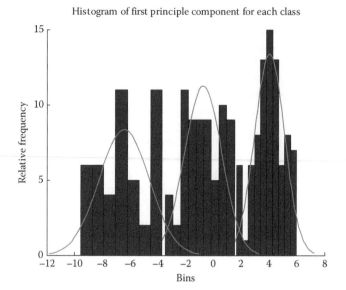

FIGURE 19.20 Histogram of the values of the first principal component from a 3 s utterance. The three curves correspond to background noise (left), unvoiced speech (middle), and voiced speech (right).

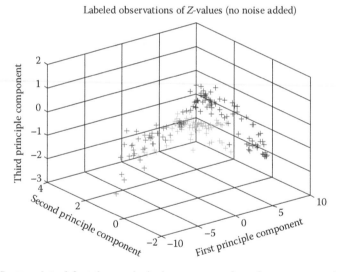

FIGURE 19.21 Scatter plot of first three principal components for a 3 s utterance with no noise added. Medium grey points correspond to background noise frames, green points correspond to unvoiced speech frames, and dark grey points correspond to voiced speech.

a different speaker was plotted or the same speaker was in a different environment, the clusters could be in different locations. For example, Figure 19.22 uses the same data as in Figure 19.21, but with the addition of white noise at a 20 dB SNR. Although the three classes still appear to be separable, the class boundaries that would be determined for this data would clearly be different than the class boundaries that would be determined for the data in Figure 19.21.

Since the usable speech extraction system will have to work for different speakers and in different environments, unsupervised classification should be a more robust approach to voiced, unvoiced, and background noise discrimination than supervised classification. The experimental results in the following section should prove this point.

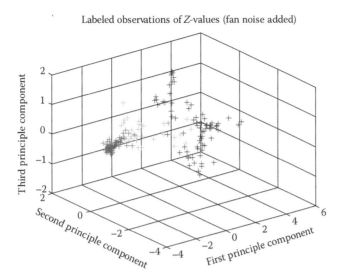

Labeled observations of Z-values (fan noise added)

FIGURE 19.22 Scatter plot of first three principal components for a 3 s utterance with the addition white noise at a 20 dB SNR. Medium grey points correspond to background noise, green points correspond to unvoiced speech, and dark grey points correspond to voiced speech.

19.2.5 Speech Segment Classification

All classifiers roughly fall into one of two groups, those that are supervised and those that are unsupervised. Supervised classifiers require at least some initial labeled training data in order to classify any unlabeled test data. On the other hand, unsupervised classifiers search the feature space of unlabelled data points for any clumps or clusters in the data that would represent different classes. Unsupervised classifiers cannot run in real time, since some data must first be collected for clusters to exist. In general, unsupervised classifiers do not perform as well as supervised classifiers when using well-matched training and testing data sets; however, obtaining good training data may not always be possible. Further, when the environment is not stationary, which is usually the case in operational environments, unsupervised classifiers are necessary. Supervised QDA and k-NN classifiers, as well as unsupervised GMM and k-means clustering-based classifiers, are used in this section.

As with features, all classifiers can further be partitioned into parametric or nonparametric. Parametric classifiers assume some probabilistic model, while nonparametric classifiers do not. Parametric classifiers tend to perform better than nonparametric classifiers, but often, an accurate probabilistic model is unattainable or the design of an optimal classifier, given a particular statistical model, is mathematically intractable. QDA and GMM are parametric classifiers that are founded on the assumption that the distribution of the feature vectors from all the classes is multivariate Gaussian distributed. The k-NN rule and k-means clustering are nonparametric classifiers.

As shown in the previous section, voices/unvoiced/background segments do not occur independently of one another. To incorporate the contextual information between speech frames, an HMM was used. Since the features were approximately multivariate Gaussian distributed for each of the classes, the HMM was implemented in conjunction with QDA and GMM. However, a correlation distance measure was used for the k-means clustering classifier, which incorporates frame dependencies.

19.2.5.1 Supervised Approaches

The k-NN rule is probably the most well-known nonparametric classification rule. It is nonparametric in the sense that although training data is still required, no assumptions are made regarding the probability distribution of the feature vector used in the classification process. For example, if one used a two-element feature vector to decide between two classes, the k-nearest rule would compare this test vector to the k-closest (in terms of Euclidean distance) neighbors obtained from the training data and assign it to the class that generated the majority of the k-closest training vectors. To further improve the performance of the k-NN rule, weighting the k-neighbors by their distance from the test vectors was performed. It can be shown that the total probability of error using the k-NN approach will be no greater than twice that of the optimal parametric classifier.

To gage the performance of the classification techniques, several confusion matrices are generated using 12 female and 12 male utterances. These utterances are segmented into 5187 128-sample contiguous frames. The 16-element feature vector is then extracted from each one of these frames, and the first three principal components are used as the input to the classifiers. Using additional principal components as input to the classifiers produced only modest performance gains.

Table 19.3 shows the results for the k-NN algorithm using all 24 files with 10 dB white background noise added to each file. If the classifier correctly classified all of the speech frames, each diagonal element of the matrix would be 100% and all other elements would be zero. One can observe that for voiced speech frames and background noise frames, reliable classifications are obtained; however, the k-NN algorithm grouped 36% of the unvoiced speech frames into the background noise cluster and 11% into the voiced speech cluster.

Unlike the k-NN rule, discriminant analysis is a parametric approach to classification. The goal of discriminant analysis is to partition the space spanned by the features into classes, via a parametric discriminant function, such that some objective function is maximized. The objective function is usually chosen in order to obtain a prescribed percentage of misses for a given false alarm rate. In this section, a zero-one cost is used, that is, the cost of obtaining a miss is equal to that of obtaining a false alarm. Hence, the total probability of making a detection error is minimized. The following example should clarify the discriminant analysis concept.

When the feature vectors are multivariate Gaussian distributed with the same covariance matrix for both classes, the resulting discriminant function is a hyperplane. This situation is referred as LDA. LDA uses a linear transformation to project the set of raw testing data vectors onto a vector space of lower dimension, such that the metric of class discrimination is maximized. The metric is the ratio of the between-class scatter (variance) S_b to the within-class scatter (variance) S_w:

$$\text{trace}\{S_w^{-1} S_b\} \tag{19.34}$$

The result of this optimization problem for the two-class problem is the following linear transformation (matrix equation):

$$\hat{y} = (\mu_1 - \mu_2)^T S_w^{-1} x \tag{19.35}$$

where
μ are the two mean vectors of the two classes
x is a test data feature vector

TABLE 19.3

Confusion Matrix of k-NN Classifier for 10 dB White Background Noise

k-NN	Labeled as Background	Labeled as Unvoiced	Labeled as Voiced
Classified as background	88%	11%	2%
Classified as unvoiced	9%	76%	11%
Classified as voiced	3%	13%	87%

The mean vectors and within-class-scatter matrix are estimated using the sample mean and sample variance of the training feature vectors.

When the feature vectors are multivariate Gaussian distributed with different covariance matrices for each class, the resulting discriminant function is a quadratic hypersurface, that is, a parabola, hyperbola, or ellipse for a 2D feature space. In this situation, the optimal classifier having the minimum total detection error is also relatively easy to implement and is referred to as QDA.

In general, the decision curve that minimizes the total probability of error would have to satisfy the following equality between the likelihood ratio and the ratio of the class probabilities for all pairs of classes:

$$\frac{f(w_i \mid X)}{f(w_j \mid X)} = \frac{P(w_j)}{P(w_i)} \tag{19.36}$$

The functions are the *a posteriori* pdfs, the Ps are the prior class probabilities, the X is the feature vector, and the $w_{i/j}$ represent each possible combination of two classes. The decision curve is a function defined over the feature space spanned by the feature vector X. When the likelihood ratio was greater than the ratio of the priors, the data would more likely be from the ith class, and when the likelihood ratio was less than the ratio of the priors, the data would more likely be from the jth class. This result is intuitively appealing.

With linear and QDA, because of the exponential form of the involved Gaussian densities, it is preferable to work with the monotonically increasing logarithmic function of the two density functions, that is, the log-likelihood ratio. This function is the discriminant function, which for Gaussian densities yields

$$g_i(x) = \ln(f(x \mid w_i)P(w_i)) = -\frac{1}{2}(x - \mu_i)^T \sum_i^{-1} (x - \mu_i) + \ln P(w_i) + C_i \tag{19.37}$$

where
C is a constant equal to $-(l/2)\ln 2$
μ_i is the mean vector of the feature vectors for the ith class

$$f(x \mid w_i) = \frac{1}{2\pi^{l/2} |\Sigma_i|^{1/2}} \exp\left(-\frac{1}{2}(x - \mu_i)^T \sum_i^{-1}(x - \mu_i) \right) \tag{19.38}$$

$$\Sigma_i = E[(x - \mu_i)(x - \mu_i)^T] \tag{19.39}$$

It should be noted that the population mean vectors μ_i and the covariance matrices Σ_i are not known and must be estimated from the training data. Using the ML estimates of these parameters has been shown to yield consistent estimates of the corresponding decision curves. The ML estimators for the mean vector and the covariance matrix for the multivariate Gaussian distribution are, respectively,

$$\bar{x} = \sum_{i=1}^{n} \frac{x_i}{n} \quad \text{and} \quad S = \sum_{i=1}^{n} \frac{(x_i - \bar{x})(x_i - \bar{x})^T}{n-1} \tag{19.40}$$

where n is the number of observations in the training data. Observe that the variance of these estimates decreases with increases in sample size. The prior probabilities were estimated in the

TABLE 19.4

Confusion Matrix of QDA Classifier for 10 dB White Background Noise

QDA	Labeled as Background	Labeled as Unvoiced	Labeled as Voiced
Classified as background	92%	9%	3%
Classified as unvoiced	6%	79%	8%
Classified as voiced	2%	12%	89%

previous section. Table 19.4 shows the results for the QDA algorithm using all 24 files with 10 dB white background noise added to each file. Half of the male and female files are used to estimate the necessary mean vectors and covariance matrices. The remaining half of the files are used to test the performance of the QDA classifier.

19.2.5.2 Unsupervised Approaches

The k-means clustering algorithm generates k disjoint clusters, which presumably represent the k pattern classes. It is well suited to generating globular clusters, that is, clusters that form convex regions in the feature space. The problem of clustering data can be set up as an integer programming problem, but since solving integer programs with a large number of variables is time-consuming, clustering algorithms are often computed using a heuristic method that generally produces good, but not necessarily optimal, solutions. The k-means algorithm is one such method. There are other clustering algorithms available, but for small k, the k-means algorithm is very computationally efficient.

The k-means algorithm

- The data set is initially partitioned into k disjoint sets and the data points are randomly assigned to these sets. This results in clusters that have roughly the same number of data points.
- For each data point, calculate the distance from the point to each cluster. If the data point is closest to its own cluster, leave it where it is; however, if the data point is not closest to its own cluster, move it into the closest cluster.
- Repeat the previous step until a complete pass through all the data results in no data point moving from one cluster to another. At this point, the clusters are stable and the clustering process ends.

Since there are only three classes in voice/unvoiced/background segmentation, k is set to three. One should note that the choice of an initial partition can greatly affect the final clusters that result, in terms of intercluster and intracluster distances and cohesion. In this section, the medians of the training data are used to create the initial partition.

The k-means algorithm requires the definition of two distance measures: one that defines distance between points and another that defines distances between points and clusters. The between-point distance is measured using Pearson's correlation, which measures the similarity in shape between two profiles. The formula for the Pearson correlation distance, for two feature vectors x and y, is the dot product of the z-scores of the vectors x and y. The z-score of x is constructed by subtracting from x its mean and dividing by its standard deviation. An average linkage distance was used to define the distance between points and clusters, which is simply the average of the distances between all the points in the cluster.

To test the algorithm, the k-means algorithm was used to determine three clusters for each utterance in the data set. Once the clusters were determined, the frames in the cluster that had the smallest mean for the first principal component were classified as background noise, the frames in the cluster that had the largest mean for the first principal component were classified as voiced speech, and the remaining frames were classified as unvoiced speech. Although the clusters shift for different speakers and environments, the order of the clusters in the first principal component does not change (see Figure 19.20). Table 19.5 shows the confusion matrix of the k-means classifier with 10 dB SNR white noise added to each file.

One can observe that the QDA classifier outperforms the k-means classifier. It is interesting to note that even QDA does not perform as well at classifying unvoiced speech as the other classes. From the scatter plots in Section 19.2.4, the confusion in classifying unvoiced speech appears to result from the fact that the unvoiced speech frames tend to cluster between the background noise and voiced speech clusters. It should be noted that the QDA classifier is only optimal if the features are multivariate Gaussian distributed.

The ideal situation of having matching noiseless training and testing data is rarely encountered in practice. To observe what one would expect in an operational environment, where one of the data sets is corrupted by a different kind of noise, Table 19.6 shows the confusion matrix for the same QDA classifier where white noise at 10 dB SNR was added to the training data and testing was done with 20 dB SNR. Clearly, the supervised QDA classifier performs poorly at classifying all three of the classes. The reason for this poor performance stems from the fact that the noise shifts the locations of the clusters. Hence, one would not want to use a supervised classifier when the training and testing environments are mismatched.

A parametric version of the unsupervised k-means algorithm can be found in the GMM classifier. Formally, a random vector x that is described by a mixture of M Gaussians has a pdf of the form

$$f(x) = \sum_{i=1}^{M} \lambda_i N(\mu_i, \Sigma_i) \tag{19.41}$$

where the $N(\mu_i, \Sigma_i)$ are multivariate Gaussian distributions having mean vector μ_i and covariance matrix Σ_i. The λ_i sum to one and indicate the relative weight of each Gaussian component in the mixture. The mean vector μ_i has the same number of components as the feature vector. It can be shown that any multivariate distribution can be approximated with arbitrary precision using a mixture of enough Gaussians. In this section, it is assumed that $M = 3$, that is, three components are

TABLE 19.5

Confusion Matrix of k-Means Classifier for 10 dB White Background Noise

k-Means	Labeled as Background	Labeled as Unvoiced	Labeled as Voiced
Classified as background	82%	13%	4%
Classified as unvoiced	13%	68%	12%
Classified as voiced	5%	19%	84%

TABLE 19.6

Confusion Matrix of QDA Classifier for 10 dB White Background Noise Added to the Training Data and 20 dB White Background Noise Added to the Testing Data

QDA	Labeled as Background	Labeled as Unvoiced	Labeled as Voiced
Classified as background	53%	22%	23%
Classified as unvoiced	26%	61%	28%
Classified as voiced	21%	17%	49%

TABLE 19.7

Confusion Matrix of GMM Classifier for 10 dB White Background Noise

GMM	Labeled as Background	Labeled as Unvoiced	Labeled as Voiced
Classified as background	84%	13%	2%
Classified as unvoiced	13%	72%	10%
Classified as voiced	3%	15%	88%

used to represent the three classes. A clustering procedure that uses a GMM should yield better results, since in theory, the variables are approximately distributed as a mixture of three multivariate Gaussians.

To obtain the parameters λ_i, μ_i, and Σ_i, the expectation maximization (EM) algorithm is typically used, which is an iterative implementation of ML estimation. An iterative implementation is necessary, since incomplete information about the underlying probability distributions exists. Table 19.7 shows the confusion matrix of the k-means classifier with 10 dB SNR white noise. One can see that the GMM classifier performs slightly better than the k-means classifier.

Figure 19.23 further demonstrates the effectiveness of unsupervised classification. The utterance is the sentence *We'll serve rhubarb pie after Rachelle's talk* spoken by a female in dialect region 1 (see Appendix 19.A for more information on the TIMIT database). The top panel of the Figure 19.23 corresponds to the labeled data used in Figure 19.21. The bottom panel displays the same data, but with the addition of white noise at 20 dB SNR, and the colors now correspond to what the frames were classified as using the k-means algorithm earlier. Most of the errors occur when unvoiced speech is classified as background noise, but rarely is background noise classified as unvoiced speech.

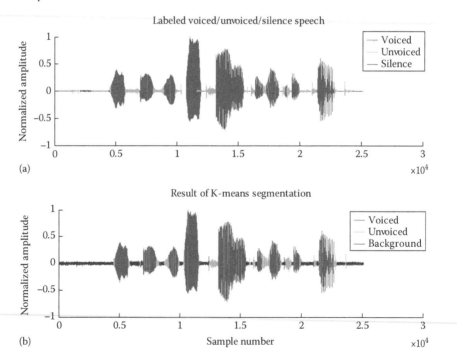

FIGURE 19.23 The utterance *We'll serve rhubarb pie after Rachelle's talk* spoken by a female: (a) Corresponds to the labeled data used in Figure 19.21 and (b) has the same data with noise added and the colors correspond to how the frames were classified.

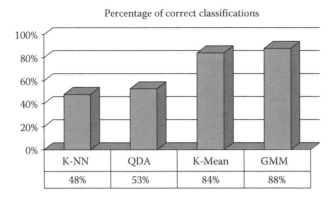

Percentage of correct classifications

	K-NN	QDA	K-Mean	GMM
	48%	53%	84%	88%

FIGURE 19.24 Total percentage of correct classifications for all four algorithms; for the supervised algorithms, 10 dB white noise is added to the training data and 20 dB white noise is added to the testing data.

19.2.6 CONCLUDING REMARKS

To compare the overall performance of the classifiers earlier, the total percentage of correct classification can be calculated using

$$P_c = 100\% \times \sum_{i=1}^{3} P(\text{Classifying Class}_i \mid \text{Class}_i) P(\text{Class}_i) \qquad (19.42)$$

where the relative frequencies of occurrence are used to estimate the conditional and prior probabilities. Alternatively, one can compute P_c by weighting the percentages along the main diagonal of the confusion matrix by the corresponding prior probabilities and adding the results. The prior probabilities for background noise, unvoiced speech, and voiced speech are estimated to be 0.23, 0.30, and 0.47 respectively. Figure 19.24 shows the results of these calculations using mismatched training and testing data. To find the total percentage of error P_e for each classifier, one can simply use the relation

$$P_e = 1 - P_c \qquad (19.43)$$

It should be borne in mind that for the k-NN and QDA classifiers, half of the data is used to determine the parameters and the remaining half is used to evaluate the classifier's performance. Other ways of using the data, such as the *leave one out method*, would probably yield different and possibly better estimates of the actual performance of the k-NN and QDA classifiers.

19.A APPENDIX TIMIT DATABASE

The TIMIT database was designed to provide data for the acquisition of acoustic–phonetic knowledge and for the development and evaluation of automatic speech recognition systems. The National Institute of Standards and Technology (NIST) prepared the data.

The 630-speaker database is broken down into eight dialect regions (Table 19.A.1). A speaker's dialect region is the geographical area of the United States where they lived during their childhood years. The geographical areas correspond with recognized dialect regions in the United States (Language Files, Ohio State University Linguistics Department, 1982), with the exception of the Western region dr7 in which dialect boundaries are not known with any confidence and dialect region dr8 where the speakers moved around a lot during their childhood.

TABLE 19.A.1
TIMIT Database Dialect Regions

Dialect Regions

dr1: New England
dr2: Northern
dr3: North Midland
dr4: South Midland
dr5: Southern
dr6: New York City
dr7: Western
dr8: Army Brat (moved around)

TABLE 19.A.2
TIMIT Database Speaker List

Female Speakers	Male Speakers	Additional Speakers in the Training Set
faks0	mabw0	mmdm2
fcmr0	mbjk0	mpdf0
fdac0	mccs0	mpgl0
fdrd1	mcem0	mrcz0
felc0	mdab0	mreb0
fjas0	mdbb0	mrgg0
fjem0	mdld0	mrjo0
fjre0	mgwt0	msjs1
fjwb0	mhpg0	mstk0
fpas0	mjar0	mtas1
fram1	mjsw0	mtmr0
fslb1	mmdb1	mwbt0
		mwew0
		mwvw0

Note: Original 16 kHz 16 bit speech downsampled to 8 kHz.

There were two dialect sentences (SA sentences) that were read by each speaker. These were meant to expose the dialect variants of the speakers. The phonetically compact sentences (SX sentences) were designed to provide a good coverage of pairs of phones, with extra occurrences of phonetic contexts thought to be either difficult or of particular interest. Each speaker read five of these sentences, and seven different speakers spoke each sentence. Table 19.A.2 on the following page lists the speakers that were randomly chosen for this research.

ACKNOWLEDGMENTS

This effort was sponsored by the Air Force Research Laboratory, Air Force Material Command, and the United States Air Force under agreement numbers: F30602-00-1-0517, F30602-02-2-0501, F30601-03-1-0216, and FA8750-04-1-0146. The US Government is authorized to reproduce and distribute reprints for government purposes notwithstanding any copyright annotation thereon.

BIBLIOGRAPHY

Altincay H. and Demirekler M., An information theoretic framework for weight estimation in the combination of probabilistic classifiers for speaker identification, *Speech Commun.* 30, 255–272, 2000.

Astola J. and Kuosmanen P., *Fundamentals of Nonlinear Digital Filtering*, Boca Raton, FL: CRC Press, 1997.

Barkat M., *Signal Detection and Estimation*, Boston, MA: Artech House, 1991.

Chandra N., Yantorno R. E., Benincasa D. S., and Wenndt S. J., Usable speech detection using the modified spectral autocorrelation peak-to-valley ratio using the LPC residual, *Fourth IASTED International Conference on Signal and Image Processing*, Honolulu, HI, August 2002, pp. 146–149.

Childers D. G., *Speech Processing and Synthesis Toolboxes*, New York: John Wiley, 2000.

Flury B. D., *A First Course in Multivariate Statistics*, New York: Springer, 1997.

Freund R. J. and Wilson W. J., *Regression Analysis: Statistical Modelling of a Response Variable*, San Diego, CA: Academic Press, 1998.

Godsill S. J. and Rayner P. J. W., *Digital Audio Restoration: A Statistical Model Based Approach*, New York: Springer, 1998.

Hall D. L., *Mathematical Techniques in Multisensor Data Fusion*, Boston, MA: Artech House, 1992.

Hamming R. W., *The Art of Probability: For Scientists and Engineers*, Reading, MA: Addison-Wesley, 1991.

Huang X. D., Ariki Y., and Mervyn J. A., *Hidden Markov Models for Speech Recognition*, Edinburgh, Scotland: Edinburgh University Press, 1990.

Jackson B. W. and Thoro D., *Applied Combinatorics with Problem Solving*, Reading, MA: Addison-Wesley, 1990.

Kay S. M., *Fundamentals of Statistical Signal Processing*, Englewood Cliffs, NJ: Prentice Hall, 1998.

Kittler J., Hatef M., Duin R. P. W., and Matas J., On combining classifiers, *IEEE Trans. Pattern Anal. Mach. Intell.* 22(3), 226–239, March 1998.

Kizhanatham A. R., Yantorno R. E., and Wenndt S. J., Co-channel speech detection approaches using cyclostationarity or wavelet transform, *Fourth IASTED International Conference on Signal and Image Processing*, Honolulu, HI, August 2002, pp. 126–129.

Krishnamachari K. R., Yantorno R. E., Benincasa D. S., and Wenndt S. J., Spectral autocorrelation ratio as a usability measure of speech segments under co-channel conditions, *IEEE International Symposium on Intelligent Signal Processing and Communication Systems*, Honolulu, HI, November 5–8, 2000.

Kubin G., Nonlinear processing of speech, in *Speech Coding and Synthesis*, Amsterdam, the Netherlands: Elsevier, 1995.

Lim J. S., ed., *Speech Enhancement*, Englewood Cliffs, NJ: Prentice Hall, 1983.

Lovekin J., Krishnamachari K. R., Yantorno R. E., Benincasa D. S., and Wenndt S. J., Adjacent pitch period comparison (APPC) as a usability measure of speech segments under co-channel conditions, *IEEE International Symposium on Intelligent Signal Processing and Communication Systems*, November 2001a.

Lovekin J., Yantorno R. E., Benincasa D. S., Wenndt S. J., and Huggins M., Developing usable speech criteria for speaker identification, *IEEE International Conference on Acoustics, Speech and Signal Processing 2001*, Salt Lake city, UT, May 7–11, 2001b, pp. 421–424.

Luo F. L. and Unbehauen R., *Applied Neural Networks for Signal Processing*, New York: Cambridge University Press, 1997.

McLachlan G. J. and Basford K. E., *Mixture Models: Inference and Applications to Clustering*, New York: Marcel Dekker, 1988.

Nikias C. L. and Shao M., *Signal Processing with Alpha-Stable Distributions and Applications*, New York: Wiley, 1995.

O'Shaughnessy D., *Speech Communications: Human and Machine*, New York: Institute of Electrical and Electronics Engineers, 2000.

Petruccelli J. D., Nandram B., and Chen M., *Applied Statistics for Engineers and Scientists*, Upper Saddle River, NJ: Prentice Hall, 1999.

Quatieri T. F., *Discrete-Time Speech Signal Processing: Principles and Practice*, Upper Saddle River, NJ: Prentice Hall, 2002.

Rabiner L. R. and Schafer R. W., *Digital Processing of Speech Signals*, Englewood Cliffs, NJ: Prentice Hall, 1978.

Rahman M. and Mulolani I., *Applied Vector Analysis*, Boca Raton, FL: CRC Press, 2001.

Ricart R., Speaker identification technology, RL-TR-95-275: Final technical report, Sponsored by AFRL/IF, Rome, NY, 1996.

Roberts S. and Everson R., eds., *Independent Component Analysis: Principles and Practice*, New York: Cambridge University Press, 2001.

Seber G. A. F. and Wild C. J., *Nonlinear Regression*, New York: Wiley, 1989.

Sheskin D., *Statistical Tests and Experimental Design: A Guidebook*, New York: Gardner Press, 1984.

Smolenski B. Y., Yantorno R. E., Benincasa D. S., and Wenndt S. I., Co-channel speaker segment separation, *IEEE International Conference on Acoustics, Speech and Signal Processing*, Orlando, FL, May 2002a, pp. 125–128.

Smolenski B. Y., Yantorno R. E., and Wenndt S. J., Fusion of co-channel speech measures using independent components and nonlinear estimation, *IEEE International Symposium on Intelligent Signal Processing and Communication Systems*, November 2002b.

Smolenski B. Y., Yantorno R. E., and Wenndt S. J., Fusion of usable speech measures quadratic discriminant analysis, *IEEE International Symposium on Intelligent Signal Processing and Communication Systems*, Awaji Island, Japan, December 7–10, 2003.

Stark H. and Woods J. W., *Probability, Random processes, and Estimation Theory for Engineers*, Englewood Cliffs, NJ: Prentice Hall, 1994.

Takens F., Detecting strange attractors in turbulence, in Rand D.A. and Young L.-S., eds., *Dynamical Systems and Turbulence*, Lecture Notes in Mathematics, Vol. 898, Berlin, Germany: Springer, 1981.

Talpaert Y., *Differential Geometry: With Applications to Mechanics and Physics*, New York: Marcel Dekker, 2001.

Theodoridis S. and Koutroumbas K., *Pattern Recognition*, San Diego, CA: Academic Press, 1999.

Thomas G. B., Jr. and Finney R. L., *Calculus and Analytic Geometry*, Reading, MA: Addison Wesley, 1979.

Varshney P. K. and Burrus C. S., eds., *Distributed Detection and Data Fusion*, New York: Springer, 1997.

Yantorno R. E., Co-channel speech and speaker identification study, Final report for Summer Research Faculty, Sponsored by AFRL/IF Laboratory, Rome, NY, 1998.

Yantorno R. E., Co-channel speech study, Final report for Summer Research Faculty, Sponsored by AFRL/IF Laboratory, Rome, NY, 1999.

Yantorno R. E., A study of the spectral autocorrelation peak valley ratio (SAPVR) as a method for identification of usable speech and detection of co-channel speech, Final report for Summer Research Faculty, Sponsored by AFRL/IF Laboratory, Rome, NY, 2000.

Yantorno R. E., Fusion—The next step in usable speech detection, Final report for Summer Research Faculty Program, Research Laboratory AFRL/IF, Speech Processing Lab, Rome Labs, New York, 2001.

20 Integrated Geographic Information System for Real-Time Observations and Remote Sensing Data Acquisition

Nikolaos P. Preve

CONTENTS

20.1 INTRODUCTION

With the increasing usage of various types of management information systems (MISs) in both industry and government for numerous applications, there has been a tremendous increase in demand for remote sensing information systems in order to enable the monitoring, retrieval, and elaboration of environmental data in real time. Information systems and network infrastructures that are linked together in order to provide remote sensing data are characterized as geographical information systems (GISs). The main advantages from the usage of a GIS are based on the creation of a cost-effective infrastructure and on a rapid data collection in real time from geographically disperse areas that are supported by digital media such as wireless sensor networks (WSNs) are.

The WSNs consist of groups of small, inexpensive, low-power, and self-contained sensor nodes with sensing, data processing, and wireless communication capabilities. They are also able to provide direct monitoring from of the environmental conditions and observations in real time from geographically disperse locations. On the other hand, the promising technology of grid computing is based on the concept of unlimited shared computer resource in order to provide to its users a vast amount of computational and storage resources. However, the coordination of the distributed and heterogeneous resources supports grid users to collaborate together, via the formed dynamic virtual organizations (VOs), in order to solve large-scale problems.

The integration of the aforementioned network technologies develops a mixed infrastructure that successfully incorporates their advantages and their produced result is called sensor grid network. The advantage of this integrated network is to overcome any existing computational and storage limitations of a WSN. Also, it extends the capabilities of a grid infrastructure to the real world supporting the incorporation of remote sensing capabilities. These mixed characteristics can be achieved only by the integration and utilization of both network technologies. The main goal of the proposed sensor grid infrastructure is to be able to collect real-time data from inhospitable and hazardous locations related to environmental protection, climate monitoring and forecasting, and other earth sciences. Nevertheless, we have to mention that in our implementation, the core of this infrastructure relies on the grid computing network but its eyes are the WSN.

The uninterrupted monitoring of a remote location has resulted in the generation of tremendous amount of sensed data, which have to be elaborated in a short-time period. Thus, the creation and usage of a sensor grid network is compulsory in order to overcome the biggest issue that geo-physicists have to confront in order to elaborate and forecast successfully environmental changes. Besides that, the sensor grid technology is a new and emerging field of research with many important unresolved issues. The most important is the interoperability that occurs from the network heterogeneity structures and the second is the data management.

In this chapter, we present a novel integrated network architecture that is flexible and scalable because it can integrate heterogeneous wireless networks, such as WSN with the traditional wired grid infrastructure, achieving to deal with the previously mentioned unresolved issues. The main goal of this chapter is to demonstrate the integration features and to handle the dynamic sensor data into a sensor grid implementation. Another goal is to ensure the uninterrupted processing, monitoring, and storing of the results into a sensor grid implementation that provides us with the ability to measure the environmental pollution in real time and to forecast the climate changes. Also, this network is user-friendly and available to any user independently of his geographical location. However, the proposed sensor grid implementation demonstrates a novel architecture that increases and extends the usability of a grid network through the Pan-European e-Infrastructure. This infrastructure can be the basis for uninterrupted monitoring of the environmental protection across the earth.

The rest of the chapter is organized as follows. Section 20.2 presents the related work on the area of sensor networks and grid infrastructure. Section 20.3 analyzes the technological requirements of a geo-infrastructure. In Section 20.4, a framework that provides resources coordination is developed. Section 20.5 presents the application of the proposed implementation and methodologies developing the interoperable environmental protection system. Section 20.6 demonstrates the test bed of the developed environmental monitoring application. Finally, Section 20.7 concludes the chapter.

20.2 RELATED WORK

In recent years, efforts have been made by the geophysicists to monitor and predict environmental incidents on earth. This scientific field is still challenging for the scientific community. The usage of environmental applications and computer technologies, such as sensor networks and grid infrastructures, is increasing rapidly. The main characteristics of an environmental monitoring and forecasting system are the capacity of processing, the huge data volume, and the autonomous operation [1,2].

The term of a distributed sensor network (DSN) is defined as a collection of a large number of heterogeneous intelligent sensors that are distributed geographically over an environment and connected through a communication network [3]. Also, a different approach to the definition of a sensor network is given in [4] because this kind of network can monitor phenomena in a geographic space in which the geospatial content of the collected, aggregated, analyzed, and monitored information can be implemented by a network called geosensor network (GSN). In order to clarify the aforementioned definitions, we should mention that a DSN without georeferenced sensor nodes is not a GSN. Thus, we define a GSN as a specialized sensor network compromising georeferenced geosensor nodes. If the geosensor nodes are actively or passively moving through the space, they form a mobile GSN.

However, a sensor network can be utilized for environmental monitoring applications [1]. For example, there is microclimate monitoring [5], habitat monitoring [6], Glacsweb project [7], and PODS project [8]. The microclimate monitoring applications check the climate data such as radiant light, relative humidity, barometric pressure, and temperature throughout the volume of giant trees [5]. Glacsweb project monitors the behavior of ice caps and glaciers for understanding the climate of the earth [7]. The PODS project monitors the rare and endangered species of plants in a volcano neighboring with high-resolution cameras, temperature, and solar radiation sensors [8]. Sensor network is also utilized in the flood monitoring to provide warnings and in the monitoring of coastal erosion around small islands [9]. The Automated Local Evaluation in Real Time (ALERT) was developed for providing important real-time rainfall and water level information to evaluate the possibility of potential flooding [10].

Regarding the side of grid computing in earth science, several projects have been deployed across Europe. The Deployment of Remote Instrumentation Infrastructures (DORII) project [11] aims to deploy an e-Infrastructure for new scientific communities. On one hand, ICT technology is still not present at the appropriate level today. On the other hand, it is required to improve the communities' daily work. The DORII project focuses on groups of scientific users with experimental equipment and instrumentation that are currently not or only partially integrated in the European e-Infrastructures.

The GEOGrid [12] project integrates virtually a wide variety of data sets, such as satellite imagery, geological data, and ground sensed data. The integration is enabled by the grid technology, and data can be accessed and processed on user demand through standardized web services interfaces on basic OGSA services. GEOGrid is based on four layers: hardware, virtual storage, application and data services, and user interface. But, the GEOGrid project does not integrate sensors into the grid. The Cowbridge project [13] tries to predict and manage flooding in a river valley. The system incorporates an adaptive, resilient sensor network that feeds real-time sensor data to a computationally intensive flood-prediction algorithm running on a general-purpose computational cluster in a customized grid middleware environment. Also, this system enables real-time prediction of flooding, including detailed predictions of what areas flooding will most likely affect. It also provides timely alerts to local stakeholders when it perceives an imminent flooding. An interesting technical aspect of this system is that it supports selectively and dynamically assigning computation tasks to either the sensor network itself or the remote cluster. This approach evolves wired sensor networks only raising new challenges such as power management, sensors discovery, communications safety, and routing protocols.

Sensors Anywhere (SANY) project [14] deals with sensor networks research for environmental applications and tries to improve the interoperability of in situ sensors and sensor networks. This project aims to allow quick and cost-efficient reuse of data and services from currently incompatible sources in future environmental risk management applications. To achieve this goal, they (1) specified a standard open architecture for fixed and moving sensors and SNs, (2) developed and validated reusable data fusion and decision support service building blocks and a reference implementation of the architecture, and (3) tried to contribute to future standards applicable to Global Monitoring for Environment and Security (GMES). SANY does not focus on the resources utilization of the grid computing such as our proposed infrastructure.

In our network architecture, we propose a new approach in order to integrate WSNs with a grid infrastructure, developing a fully interconnect implementation that successfully involves more than one different applications of these new technologies. The usage of this implementation aims to achieve better performance than the mentioned networks monitoring and predicting environmental changes while it will support engineering researchers to have more accurate results about geophysics phenomena in real time.

20.3   BUILDING A GEOGRAPHIC CYBERENVIRONMENT

Scientific communities nowadays are increasingly collaborative and multidisciplinary, and it is not unusual for teams to span institutions, states, countries, and continents. Web-based technologies provide basic mechanisms that allow various research groups to work together. But, if they link their data, computers, sensors, and other resources into a single virtual laboratory, they will have a better outcome. So we propose a novel sensor grid infrastructure that is consisted by sensors and a grid network in order to solve this problem. The proposed integrated infrastructure can provide high-performance computing, shared databases, and various software tools necessary for geophysics science. This functionality provides an environmental character to the analyzed infrastructure. Researchers in different geographic areas may be able to access and analyze the online data with the available software through the web browser with the help of high-performance computers.

20.3.1   Sensor Web

A sensor web infrastructure consists of wireless, intracommunicating, spatially distributed sensor pods that can be easily deployed to monitor and explore new environments. According to the Open Geospatial Consortium (OGC) [15], a sensor web refers to web-accessible sensor networks and archived sensor data that can be discovered and accessed using standard protocols and application program interfaces [16]. In a sensor web, each pod consists of two primary modules. The first module comprises the transducers that physically interact with the environment and convert environmental parameters into electrical signals. The second module represents the infrastructure of the sensor web itself. This module includes telecommunication capabilities, power sources and energy-harvesting devices, and computation devices to run the protocol schemes. The results of this process are provided for local data analysis.

The wireless communication between pods is omnidirectional. Unlike star-network configurations where data collected from all pods are passed directly to a central point, information within the sensor web is passed to an uplink point, denoted as a prime pod, by hopping it from pod to pod. In other words, data from various pods are shared as well as communicated throughout the entire web. The overall protocol is quite simple. Information is obtained at each pod via two routes: (1) direct measurements taken by local sensors at that pod and (2) information gathered by other pods and communicated throughout the web.

The key concept is that there is no artificial differentiation between the two types of information. The protocol then is to simply rebroadcast the data or information to any pod within communication range. Any information received at a prime pod is not rebroadcasted to any secondary pods and disappears from the web at this point although it is accessible to an outside user or another prime pod. Each measurement begins with the pods taking in sensor data. After a measurement is taken, each individual pod in the system broadcasts the information it has taken or received from others in an omnidirectional manner to all pods in communication range. Each pod then processes and analyzes the information it has received and the cycle repeats. In this way, information is hopped pod to pod and spread throughout the entire sensor web. The entire system becomes a coordinated whole by possessing this internal, continuous data stream, drawing knowledge from it.

The sensor web enablement (SWE) standard has been defined by the OGC [15], which is composed of a set of specifications, including Sensor Markup Language (SensorML), Observation and Measurement, Sensor Collection Service, Sensor Planning Service, and Web Notification Service. OGC has also proposed a reusable, scalable, extensible, and interoperable service-oriented sensor web architecture, which conforms to the SWE standard, integrates sensor web with grid computing, and provides middleware support for sensor webs.

20.3.2 GRID COMPUTING

The usage of a grid infrastructure aims to achieve efficient support to a WSN in side of processing power while it provides vast storage capabilities. Thus, the necessity to incorporate a grid network in a GSN is compulsory because it can store the huge observational data inputs that are produced by the sensors. A grid network is a collection of web services that provide capabilities such as securely accessing supercomputers, accessing information services that describe available resources, accessing databases, and services for accessing scientific applications.

The three major sharable grid components in our proposed integrated environment are the software, data, and computational resources that collaborate ensuring the network interoperability. Due to the environmental characteristics of our network, the monitored data consist of the GSN contents such as air pollution, temperature, and humidity. This occupies the major chunk of the grid. The grid can host all the widely used engineering software for various European environmental experiments, environmental data analysis processing, and simulations. Also the researchers can develop various software modules that can be plugged into this facility and make it available to the other researchers for use. The third of the aforementioned components is the computational resources that can be used to solve challenging problems. The submitted jobs through web-based interfaces are in queue and automatically run on these compute nodes depending on their availability. The output files generated during the run are stored in a temporary space of the data server and will be deleted after transferring those to the users.

The identification, integration, and interconnection of various grid infrastructures across Europe can achieve firstly a short-scale interoperable integrated infrastructure building the basis to an extendable global integrated infrastructure that will be based on two different network infrastructures, the grid network and sensor networks. This global computer environment will be accessible by any researcher through the integrated grid infrastructure interconnecting him with any remote places through various and sensor networks. However, the user will be able to retrieve data related to environmental observations in real time across the network.

20.3.3 CYBERIMPLEMENTATION

In order to distinguish the cyberimplementation, cyberenvironment, and cyberinfrastructure, we should define the appropriate terms. Regarding the term cyberimplementation, we define the interoperability between a cyberinfrastructure and a cyberenvironment ensuring the uninterrupted operation of this network.

The term cyberinfrastructure defines the combination of distributed computers, large-scale data storage, high-speed networks, high-throughput instruments and sensor networks, and associated software that, as a ubiquitous, persistent infrastructure, has and will continue to have a direct impact on scientific and engineering productivity [17]. This kind of infrastructure supports real-time information and asynchronous collaboration between different communities. The definition of the term cyberenvironment is obvious to the representation of the software infrastructure and needed interfaces. This realizes the vision of cyberinfrastructure as a systemic catalyst for transformative change of the shared resources in research practice and end-to-end productivity [17]. Also, cyberenvironments provide an interface to local and shared instruments and sensor networks, data stores, computational resources and capabilities, and analysis and visualization services within a secure framework. Their combination enables the management of complex projects, development and automation of processes in groups and community-scale collaboration, and coordination with geographically dispersed users. Cyberenvironments emphasize the integration of shared resources, hardware, and knowledge, into end-to-end scientific processes and the continuing development and dissemination of new resources and new knowledge.

However, the research community cannot yet fully bridge the gap as far as the cyberimplementations are concerned. The cyberenvironment is the key to overcoming any limitations of a cyberimplementation, because it is an interface acting as a gateway server providing coordination, interoperability, and access to any resources of a cyberinfrastructure. From now on, we will refer to the previously mentioned terms in this chapter emerging the unique features of the proposed implementation.

20.4   NECESSITY OF A FRAMEWORK

Numerous of scientific resources are accessible via portals, gateways, and grids supporting researchers to solve environmental problems, but these infrastructures are still far from the goal of an ubiquity environment. Although, cyberimplementations are still at an early stage of development and they already have emerged their significant impact on the researchers' work. In order for us to be led to an interoperable cyberimplementation supporting the community to solve various scientific and environmental problems, we have to design suitable collaborative suites. The lack of design in internetworking and software stages will not ensure the uninterrupted resources coordination. As a result, the adaption and adoption of these integrated tools will frustrate the users into larger communities. Recognizing and emerging the limitations of these development parameters is a beginning to conduct us out of the limited model of the cyberimplementation usage.

Figure 20.1 illustrates a framework that deploys the integration requirements. It also provides a reliable solution overcoming any current limitations of a cyberimplementation while it focuses on the following:

- The support and utilization of the distributed network resources ensuring an effective collaboration between various shared computational systems and applications.
- A user-friendly interface that ensures the security by preventing unauthorized users to have access to it.
- Supporting the diversity of users' applications in a large-scale complex experiment.
- The need of effective integration design techniques between different high-performance computing centers of laboratories, institutions, etc., around the world is compulsory.

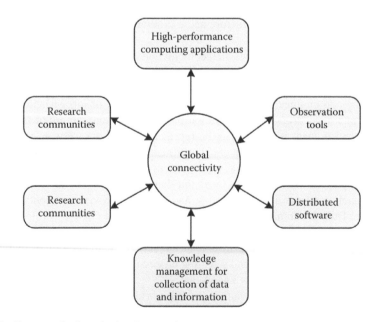

FIGURE 20.1   Framework of a cyberimplementation.

The proposed framework can fully cover the users' requirements while it supports the interaction of research communities. Additionally, it provides design and development methodologies supporting very large-scale deployment and reducing the implementation costs. This framework can be a basis of the next-generation computer systems, in case we ensure the bidirectional and the continuous flow of knowledge and information between various global research communities.

20.5 ENVIRONMENTAL CYBERIMPLEMENTATION AND REMOTE SENSING SCENARIO

The environmental scientific research is characterized by a number of attributes that makes a cyberimplementation important for this field of scientific efforts. Many environmental research activities are observationally oriented, rely on the integration and analysis of many kinds of data, and are highly collaborative and interdisciplinary. Much of the relevant data needs to be geospatially indexed and referenced, and there is a host of currently non-interoperable data formats and data manipulation approaches.

The development of principles is compulsory because it provides sustainability and adaptability using current and emerging techniques such as web and grid services, integrating/translating middleware, global unique identifiers and metadata, workflow, meta-workflow and provenance, and semantic descriptions of resources and data. These types of technologies lower the architectural coupling of cyberinfrastructure and cyberenvironment components while maintaining end-to-end capabilities. As a result, we are led to integrate distributed and heterogeneous resources into the initial infrastructure.

Figure 20.2 illustrates this integration process with the environmental sensor networks around a wired grid infrastructure. The integration between these two infrastructures is achieved by the usage of a gateway server that acts as an interface hiding their heterogeneity. Moreover, the involvement of heterogeneous shared resources and complex applications enables new ways of organizing, applying, and performing scientific research. Analyzing the utilization of the deployed cyberimplementation, an authenticated user can log into the system through an application client, which is based on a user's browser.

After his authentication from a grid node, he is able to retrieve in real time the sensed data across the cybernetwork. The usage of an application client is necessary in order to achieve and validate the conversion of sensor data format using the data format conversion service and the

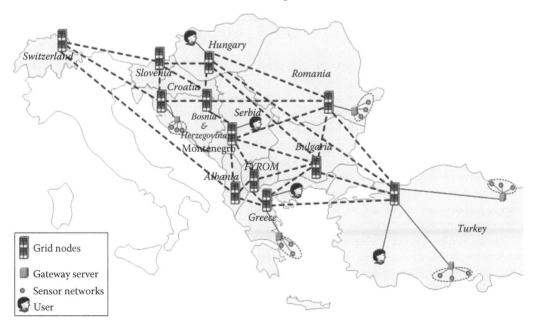

FIGURE 20.2 Environmental protection in the cyberimplementation.

conversion of sensor data units using the data unit conversion service. The grid network apart from the authentication process provides access to vast processing and storage resources. Moreover, the role of the gateway server is to act as a proxy hiding any differences between the different networks of this cyberimplementation. It also succeeds to ensure uninterruptedly interoperability between the networks providing 24/7 accessibility to cyberimplementation users.

The large amount of the produced sensed data can be stored in the grid providing the capability to a user to duplicate the sensed data in a different location, that is, in his account, or to elaborate the data without being concerned about the computational resources, while he will be able to utilize the grid simulating any scientific environmental issues. When a user logs into the system, he has to provide his credentials in order to access the sensed data and the cyberinfrastructure. Thus, a user can log into the system from Hungary, and through this cyberimplementation, he is able to observe, collect, and process the sensed data from the distributed GSN of Greece or any other country, which is connected in the cyberimplementation.

Consequently, the usage of a grid network in such implementations is the key to a successful development and integration of an extendable distributed infrastructure that is based on different ethernet and wireless networks, while it is accessible from any user across different countries. The grid infrastructure is the base of this large-scale implementation because it provides vast network extendibility through the integration of grid networks with sensor networks form different countries. As a result, we cope with any network limitations that lead us to global cyberimplementation and give the opportunity to any user to observe and collect data in real time from any geographically remote location.

20.5.1 Air Quality Monitoring System

A sensor grid network is extremely valuable in the geographical area of environmental protection because it continuously monitors and collects environmental measurements due to the high density of sensors. Sensor data monitoring system receives the measured data from a GSN providing useful information for users by understanding the condition of the remote place. Consequently, a GSN is a useful tool for the earth scientists because it enables environmental protection and monitoring by measuring changes in climate and ocean systems and atmospheric pollution and supporting earthquake prediction and assessment. Our GSN implementation is related to atmospheric pollution while it uses two systems for controlling and monitoring air quality into the observation area. The main system is constituted by (1) GSN control system and (2) air pollution monitoring system.

As it is illustrated in Figure 20.3, the grid users can access the GSN system through the web by inquiring a gateway server of the cyberenvironment that translates the standardized protocol to the proprietary protocol for both sides. Several early integration methodologies have also been proposed [18–21], but our implementation is based on the service-oriented approach (SOA) with standard open architecture technologies such as web services [22]. This approach provides a common information and communication format to facilitate the integration. The Web Services Description Language (WSDL) and Simple Object Access Protocol (SOAP) are used for describing the services and formatting messages used by the underlying communication protocol [22]. The control system supports the operators that control sensor network such as sampling interval change and network status check. The operators are useful for keeping the good status of data transmission in GSN.

Our air quality monitor infrastructure aims at making various types of web-resident sensors, instruments, image devices, and repositories of sensor data, discoverable, accessible, and controllable via the World Wide Web. The air quality monitoring system supports sensor data abstraction and pollution prevention models for understanding the pollution level of an area. The abstracted data is used for defining the pollution and the potential pollution area with the air pollution prevention model. The models are used for providing alarm message and safety guideline for people in a pollution area. The measurements of each sensor can be published in a registry using standardized

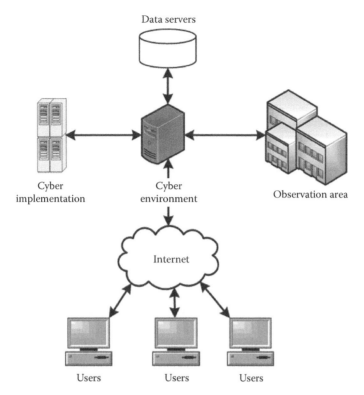

FIGURE 20.3 Accessing an observation area.

methods such as the SensorML [23] allowing the sensor data as well as metadata describing the sensor properties to be accessed and retrieved using standardized protocols.

The effective acquisition is required for trade-off between battery lifetimes and sampling rates [24]. Thus, the definition of the sampling interval is very important because of battery limitation. If the interval is short, the system can recognize the conditions of the remote place promptly. However, the batteries of sensors could have gone out in a short time. If the interval is long, it can keep the electronic power for a long time. However, the system cannot promptly react for the detected events. We change the sampling interval depending on the situation that is derived from the abstraction data model for the sensors, so as to control the sampling interval for keeping a sleep mode as long as it can. The "power-saving" mode must require less power than a mode for active vigilance.

Nevertheless, the interval cannot escape beyond the user-defined interval boundary for the environmental monitoring. When the sensors in the network receive the order for changing the interval, all of the sensors will be in the sleep mode until the ordered time. Only the timer is alive in the sensors. When it is time for wake up, all of the sensors wake up and send their measured values all at once to the GSN control system. After data transmission is completed, the sensors sleep again and wait the next awake time.

The interval time can vary depending on the pollution level and on the administrator's specifications. Although, when the system recognizes that it is an indication of air pollution after checking the observed condition, it will change the interval time making it shorter, because the probability for air pollution is high. If the system confirms a dangerous pollution level when it measures the air quality, then it will provide us an alarm. After this process, it minimizes the interval time measuring continuously the pollution level until it is reduced again. When the pollution level is reduced or stabilized according to our predefined environmental values, the system changes the interval time because the probability of air pollution is also reduced. If the system does not confirm any further high pollution level, then it changes the interval time to a continuously increasing value in order to save the batteries of the geosensors.

20.5.2 Air Pollution Prevention Models

We developed an air pollution prevention model in order to manage effectively the abstracted sensed data from the polluted areas. This model checks the dangerous rate for the polluted areas with each area type and schedule. The usefulness of this model is obvious when we observe different types of air pollution areas, the current dangerous area, and the future dangerous area, because we can set different rules for each observation area.

Data abstraction has been considered to be an effective method for accommodating relationships between data values. In data abstraction, a specific data value is generalized into an abstract value. Figure 20.4 illustrates an overview of the constructed data abstraction hierarchy focusing on relationships between abstract values and specific values. Also, the definition rules for each area separately are illustrated in Figure 20.4. Two abstraction hierarchies exist in our prevention model: (1) a value abstraction hierarchy and (2) a domain abstraction hierarchy. In the value abstraction hierarchy, a specific value in a lower abstraction level is generalized into an abstracted value in a higher abstraction level and the abstracted value can be generalized further into a more abstract value. Conversely, the specific value is considered as a specialized value of the abstract value. For instance, the dangerous level is related to an area, while an area belongs to the domain of dangerous areas.

The domain abstraction hierarchy consists of domains that include all individual values in the value abstraction hierarchy and there exist relationships between the domains and values, that is, area is the class of longitude. The abstraction relationships between values and domains are represented in Table 20.1. The cardinal relationship between two adjacent domains is assumed to be one to one and a class is called n-level class of the subdomain according to the abstraction level difference n. In Table 20.1, D_i denotes a domain at the abstraction level i and v_i^{ji} is a specific value of the domain D_i. The relationships (1) and (2) represent 1-level value and domain abstraction relationships. On the basis of the 1-level abstraction relationships, (3) and (4) represent general n-level abstraction relationships.

The sensed data are transmitted from sensors to the air pollution monitoring system through the control system. Figure 20.5 illustrates our data model that abstracts sensed data and then

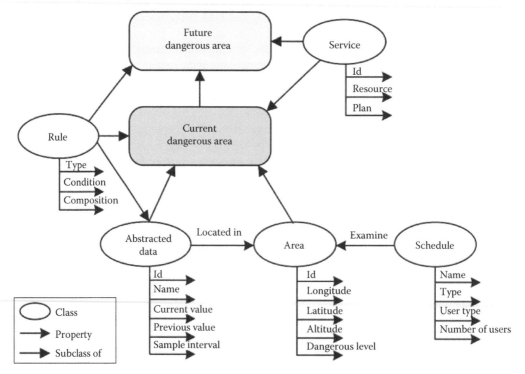

FIGURE 20.4 Air pollution prevention model.

TABLE 20.1

Formal Representation of Value and Domain Abstraction Relationships

Relationship	Formal Notation
1. 1-level domain abstraction	$D_i \Rightarrow D_{i+1}$ where D_i is a subdomain and D_{i+1} is its superdomain
2. 1-level value abstraction	$v_i^{ji} \in {}^* v_{i+1}^{j_{i+1}}$ where v_i^{ji} is a specific value and $v_{i+1}^{j_{i+1}}$ is its abstract value
3. n-level domain abstraction	$D_i \overset{n}{\Rightarrow} D_{i+n}$ s.t. $v_i^{jj} \overset{n}{\in} {}^* v_{i+n}^{j_{i+n}}, v_i^{jj} \in D_i, v_{i+1}^{j_{i+1}} \in D_{i+1}, \dots, v_{i+n}^{j_{i+n}} \in D_{i+n} \forall j_i$
4. n-level value abstraction	$v_i^{ji} \in {}^* v_{i+1}^{j_{i+1}}$ iff $\exists v_{i+1}^{j_{i+1}}, \dots, \exists v_{i+n-1}^{j_{i+n-1}}$ s.t. $v_i^{ji} \in {}^* v_{i+1}^{j_{i+1}} \in {}^* v_{i+2}^{j_{i+2}} \in {}^* \dots v_{i+n-1}^{j_{i+n-1}} \in {}^* v_{i+n}^{j_{i+n}}$

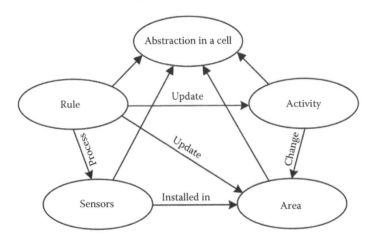

FIGURE 20.5 Data abstraction model.

it uploads and stores them at intervals to the grid infrastructure. The sensed data are accessible form SQL databases. During this process, the GSN does not discontinue to monitor the specified areas. In order to support the system interoperability and queries of multiple databases, we have to encode the abstracted geosensor data by adopting the geography markup language (GML) specification. GML is an extensible markup language (XML) grammar for expressing geographical features [25]. GML also serves as a modeling language for geographic systems as well as an open interchange format for geographic transactions on the web facilitating data exchange across the grid infrastructure.

The wealth of abstracted data allows detailed examination of the area being monitored. With the support of the designed scientific tools, we are able to perform a distributed analysis of the data. The process of abstraction, upload, storage, analysis, security, and identification of sensors is based on the grid security infrastructure (GSI) [26].

20.6 EXPERIMENTAL RESULTS

In this section, we describe the proposed system while we demonstrate and illustrate the usage of the proposed methodology. The selected area is covered by 20 sensors that are scattering. They sense various types of measurements such as temperature, humidity, micrograms per cubic meter of air, dust, carbon dioxide, ultraviolet, wind direction, wind speed, air pressure, altitude, and illumination. We launch the created application that is based on the aforementioned parameters, as it is illustrated in Figure 20.6, in order to test the functionability and interoperability of the proposed environmental cybersystem, which is based on the followed methodology. Firstly, the application checks the network availability and retrieves the sensed data in real time. Then the data is stored in the databases of the grid network for further process.

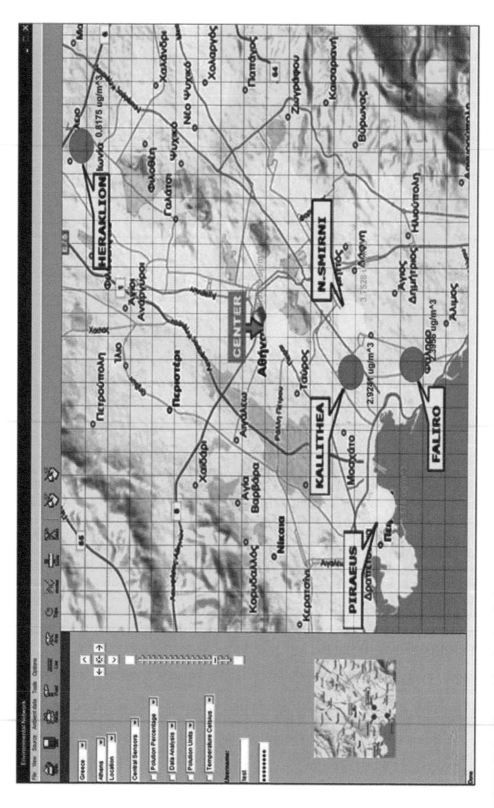

FIGURE 20.6 Observed data for air pollution.

The increased pollution values according to the prevention model are indicated with yellow. When the observed data is higher than usual or in a dangerous level, our application automatically checks the area type, previous sensed data, and compares observed data from the nearest sensors.

On the one hand, we have the current dangerous area that is defined by combining the current dangerous types and levels in the local areas with some rules for pollution. On the other hand, we consider as a future dangerous area an area that is near to high pollution level preventing further predicted pollution damage. In order to define a future dangerous area, we consider the detected data, the gradient, and the dangerous level from this area. The data are processed by the user-defined rule with other factors such as the priority of space, the constant danger probability, and the reaching probability to a critical point. However, the definition of a predicted area occurs from the domain knowledge.

In our environmental cyberinfrastructure monitoring system, it is essential to support the frequent updates for reacting promptly in order to notify the system users. The measured data of heterogeneous geosensors are sending according to the defined rules in the grid database. As we mentioned, it is hard to constantly keep the air pollution description, because the frequent data transmission makes the batteries of the geosensors to go out rapidly. After installing the sensors on the observation area, the environmental monitoring system recognizes the locations, types, and accuracies of the installed sensors by importing the SensorML [23], which describes the properties of geosensors. It also connects the sensor network control system that operates the sampling interval change, network status check, and the communication control.

In case of a sensor malfunction that produces a mistaken high measurement, our proposed air pollution prevention model checks the area type around the sensor trying to define any other parameter related to this malfunction. After it defines the current pollution area, it also checks the potential pollution area in the near future with the related factors such as the pollution level gradient, the area type, wind direction, and speed. When it finds a factor to make a dangerous condition in the near future, it shows an alarm message about that until the factor is gone. The alarm message includes the pollution level and type and safety guideline.

20.7   CONCLUSION

WSNs and grid computing are promising technologies that are being adopted in the industry. By integrating WSNs and grid computing, sensor grids greatly enhance the potential impact of these technologies for new and powerful applications, such as environmental monitoring. Thus, we believe that sensor grids will attract growing attention from the research community and the industry from various domains.

In this chapter, we have successfully achieved to integrate various WSNs with a traditional fabric grid infrastructure in order to establish a distributional control system, which enables the system user to be allowed to monitor, elaborate, store, and share with other users the retrieved data. The previously mentioned functions are available for the system users on real time across the implemented network.

Also, we have coined the term cyberimplementation and we present a novel architecture that provides integration and coordination between heterogeneous networks such as WSNs and traditional fabric grid infrastructure. We also demonstrate via the proposed framework and the architectural pattern a small-scale scenario that supports the vision of a global-scale cyberimplementation for environmental protection around the earth in real time. Moreover, we developed according to the OGC standards an environmental protection system that can be integrated into any cyberimplementation and is constituted by two systems, the network control system and the monitoring system. In order to achieve accurate measurements and better data management, we deployed two models: on the one hand, the pollution prevention model that is defined by specific rules in the observation areas and, on the other hand, the data model that ensures interoperability and exchange of the abstracted data across the cyberimplementation. Besides, we achieved to manage successfully the current issues in this scientific field, demonstrating the balanced network capabilities such as robustness, adaptability, and scalability.

REFERENCES

1. K. Martinez, J.K. Hart, and R. Ong, Environmental sensor networks, *IEEE Comput.*, 37, 50–56, 2004.
2. A.R. Ilka, C. Gilberto, A. Renato, and M.V.M. Antonio, Data aware clustering for geosensor networks data collection, in *Proceedings Anais XIII Simposio Brasileiro de Sensoriamento Remoto*, 2007, pp. 6059–6066.
3. S.S. Iyengar, A. Tandom, Q. Wu, E. Cho, N.S.V. Rao, and V.K. Vaishnavi, *Deployment of Sensors: An Overview*, CRC Press, Boca Raton, FL, 2004, pp. 483–504.
4. S. Nittel, and A. Stefanidis, *GeoSensor Networks and Virtual Georeality*, CRC Press, Boca Raton, FL, 2004, pp. 1–9.
5. D. Culler, D. Estrin, and M. Srivastava, Overview of sensor networks, *IEEE Comput.*, 37, 41–49, 2004.
6. A. Mainwaring, J. Polastre, R. Szewczyk, D. Culler, and J. Anderson, Wireless sensor networks for habitat monitoring, in *Proceedings of the First ACM International Workshop on Wireless Sensor Networks and Applications*, Atlanta, GA, 2002, pp. 88–97.
7. J.K. Hart and J. Rose, Approaches to the study of glacier bed deformation, *Quatern. Int.*, 86, 45–58, 2001.
8. E. Biagioni and K. Bridges, The application of remote sensor technology to assist the recovery of rare and endangered species, *Int. J. High Performance Comput. Appl.*, 16, 315–324, 2002.
9. Envisense-Secoas, Self-organizing collegiate sensor networks, http://envisense.org/secoas.htm (accessed Feb 2013).
10. R. Hartman, 2011, ALERT, http://www.alertsystems.org (accessed Feb 2013).
11. Deployment of Remote Instrumentation Infrastructure: The DORII project, http://www.dorii.eu (accessed Feb 2013).
12. N. Yamamoto, R. Nakamura, H. Yamamoto, S. Tsuchida, I. Kojima, Y. Tanaka, and S. Sekiguchi, GEO Grid: Grid infrastructure for integration of huge satellite imagery and geoscience data sets, in *Proceedings of the Sixth IEEE International Conference on Computer and Information Technology (CIT'06)*, Seoul, Korea, 2006, p. 75.
13. G. Coulson, D. Kuo, and J. Brooke, Sensor networks + grid computing = A new challenge for the grid? *IEEE Distrib. Syst.* Online, 7, 2–2, 2006.
14. D. Havlik, G. Schimak, R. Denzer, and B. Stevenot, Introduction to SANY (Sensors ANYwhere) integrated project, 2006, www.sany-ip.eu/filemanager/active?fid = 19 (accessed Jan 2013).
15. B. Domenico and S. Nativi, 2013, Open Geospatial Consortium (OGC), http://www.opengis.net/doc/is/netcdf-data-model-extension/1.0 (accessed April 2013).
16. M. Botts, G. Percival, C. Reed, and J. Davidson, OGC sensor web enablement: Overview and high level architecture, Open Geospatial Consortium Inc., White Paper, OGC 07-165, 2007.
17. J.D. Myers and R.E. McGrath, Cyberenvironments: Adaptive middleware for scientific cyberinfrastructure, in *Proceedings of the Sixth International Workshop on Adaptive and Reflective Middleware*, Newport Beach, CA, 2007, pp. 1–3.
18. V. Hingne, A. Joshi, E. Houstis, and J. Michopoulos, On the grid and sensor networks, in *Proceedings of the IEEE Fourth International Workshop on Grid Computing*, Phoenix, AZ, 2003, p. 166.
19. M. Gaynor, S.L. Moulton, M. Welsh, E. LaCombe, A. Rowan, and J. Wynne, Integrating wireless sensor networks with the grid, *IEEE Internet Comput.*, 8, 82–87, 2004.
20. J. Humble, C. Greenhalgh, A. Hamsphire, H.L. Muller, and S.R. Egglestone, *A Generic Architecture for Sensor Data Integration with the Grid*, S. Perez and V. Robles (eds.), Lecture Notes in Computer Science Series, vol. 3458, Springer, New York, 2005, pp. 99–107.
21. C.K. Tham, and R. Buyya, Sensor grid: Integrating sensor networks and grid computing, Invited paper in CSI Communications, Special Issue on Grid Computing, Computer Society of India, 2005.
22. R. Chinnici, J.J. Moreau, A. Ryman, and S. Weerawarana, Web Services Description Language (WSDL) Version 2.0 Part 1: Core WSDL, World Wide Web Consortium (W3C), 2007, http://www.w3.org/TR/2007/REC-wsdl20-20070626.
23. M. Botts, and A. Robin, OpenGIS sensor Model Language (sensorML), Open Geospatial Consortium Inc., White Paper, OGC 07-000, 2007.
24. I.F. Akyildiz, S. Weilian, Y. Sankarasubramaniam, and E. Cayirci, A survey of sensor networks, *IEEE Commun. Mag.*, 40, 102–114, 2002.
25. C. Portele, OpenGIS Geography Markup Language (GML) encoding standard, Open Geospatial Consortium Inc., White Paper, OGC 07-036, 2007.
26. V. Welch, F. Siebenlist, I. Foster, J. Bresnahan, K. Czajkowski, J. Gawor, C. Kesselman, S. Meder, L. Pearlman, and S. Tuecke, Security for grid services, in *Proceedings of the 12th IEEE International Symposium on High Performance Distributed Computing*, Seattle, WA, 2003, pp. 48–57.

21 Versatile Recognition Using Haar-Like Features for Human Sensing

Jun Nishimura and Tadahiro Kuroda

CONTENTS

21.1 INTRODUCTION

This chapter motivates our work toward a study on versatile recognition using Haar-like features for human sensing. Firstly, we discuss several potential target applications of our work and explain the significance of context recognition. Then, we discuss on the characteristic constraints of a system comprising wireless sensor nodes. Furthermore, we mention the significance of smart sensing and introduce our approach. Based on these discussions and a literature survey, our work is outlined.

21.1.1 SENSOR NETWORKS

The advances in microelectromechanical systems (MEMS) technology, wireless communications, and digital electronics have enabled the development of low-cost, low-power, multifunctional sensing device that are small in size and communicate untethered in short distances [1]. These tiny sensing devices, generally referred to as *motes* [2], are autonomous sensor unit that consist of sensing, data processing, and communicating components. The collaborative interconnection of the sensing device acts as network node and forms wireless sensor networks (WSNs).

More recently, the availability of low-cost hardware such as CMOS cameras and microphones has fostered the sensing devices to be able to ubiquitously retrieve multimedia content such as video and audio streams, still images, and scalar sensor data from the environment [3,4]. When one of the sensing devices is worn by the user, sensor network is closely related to the research field of wearable computing.

The wide range of applications includes environmental monitoring, condition-based maintenance, habitat monitoring, multimedia surveillance, inventory tracking, health care, and home automation. In realization of various applications, multiple types of sensors are integrated into resource-constrained platform. Therefore, the sensor nodes should no longer be just data collectors, and perform *smart sensing*, in which signal processing and decision making are done locally.

21.1.2 POTENTIAL APPLICATION

Target applications utilize the capability of context recognition using sensor nodes. Context information includes location, users' activity, surrounding environment, and social situation [5]. Context recognition can be described as the function of a system to detect and recognize automatically what the user is doing and what is happening around the subject. Here, context refers to environmental information around a sensing device [6]. When the subject is wearing the sensing device, the context includes the user activity and the surrounding environment.

The sensornet is an emerging technology platform where sensors and networking technology merge together to enable a novel human sensing and understanding. Often, the sensornet is expressed as a *macroscope* for its capability of visualizing the large-scale environment and phenomenon by connecting the microscopic information captured by a number of small sensor nodes.

Business microscope utilizes various recognition results using the signals captured by multiple sensors embedded in the wearable sensor node such as microphone, accelerometer, and image sensors to sense the workers' communication status and visualizes their interactions during the work hours in a bird's eye perspective (Figure 21.1) [7–10]. In the age of knowledge, enhancing the productivity of knowledge workers, individuals with the specialized knowledge, is considered as a major issue. As opposed to the manual works, a quality of the knowledge works highly depends on the combined power of a group of knowledge workers [11]. The sensor node is developed in

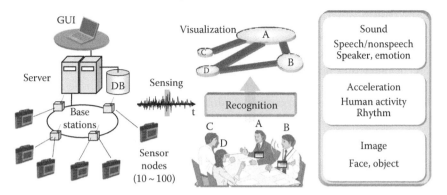

FIGURE 21.1 Business microscope.

the shape of the worker identification tag to extract the information on workers' communication. The wearable sensor nodes are used to detect human behavior in organizational settings for next-generation human resource management.

For the quantitative analysis of the communication among workers, object and face detection, speech/nonspeech classification, speaker detection, and human activity recognition are used to measure the face-to-face encounter, the vocal activity, and the working status. In the work of [12], dominance, or the level of influence a person has in a conversation, is recognized by using face detection and tracking visual cues.

In this application, sensors run in low duty ratio [7]. The microphone in the sensor node, for example, has the limited energy budget of 10 mAh. In such a condition, the microphone can continuously sample the data only for 1 h. In order to limit energy consumption of the microphone to 10 mAh and run for 24 h for all day observation, 0.4 mA is the limit. Since average energy consumption of continuously running microphone is 10 mA, the limit of duty ratio becomes 0.04. The duty ratio is set to 1%. Because the sensor node is battery powered and recharged in the end of day, the lifetime is set to approximately 1 day.

Another concern is privacy. The use of a conventional microphone would require the recording and transmission of the actual sounds to be processed at a central node. In the work of [7], the length of each data taken by microphone is kept very short (0.1 s) and intermittently sensed in a long period (10 s). In this way, the recorded signals cannot tell its contents. This type of sensing procedure was taken to ensure the privacy in the transmission of raw sound signals. Wyatt et al. also proposed a method of only using features such as log energy, spectral entropy, and auto-correlation peak that intelligible speech cannot be reconstructed [21,22]. Having smart sensors programmed to perform processing and recognition locally can contribute to the privacy issue management.

Summarizing the potential target applications shown previously, the characteristic properties of the target applications include the following:

- Long-term sensing, 1 day ~
- Multiple sensor inputs, sound/image/accelerations
- Low duty ratio, 1%–10%
- Smart sensing, execute context recognition locally
- Low calculation cost
- Local processing for privacy

The devices running context recognition application are becoming smaller and smaller. We are to have a tiny wearable sensor node that performs detection and recognition tasks, with very low power consumption of sub-mW and very few RAM.

21.1.3 Conventional Approach in Context Recognition

Many types of sensors are embedded in the sensor node to capture multimodal signals such as image, sound, and acceleration signals. The context recognition based on different sensor signals include face/object recognition using image sensor data, speech/nonspeech classification using sound signals, and human activity recognition using acceleration signals. In conventional studies, these recognition problems have been studied based on completely different perspective. Each type of sensor signals has been seen and understood differently in the sense that each sensor is different from the others.

Figure 21.2 shows a general structure of context recognition. To carry out certain recognition task, some features, which can be used to distinguish the target, must be extracted from the sensor signals. The feature extraction process is executed at the front-end part of the recognition, so-called recognition front end. Since each type of sensor signals has been studied in totally different concept, the recognition front end has been designed differently. The extracted features are then sent to recognition process. In general, they are compared against pretrained classifiers stored in the database to make certain decision. Simply putting the conventional recognition front ends together can result in the large redundancy. In this section, we will look into the different methods of context recognition based on each type of sensor signals. Here, we focus on sound, acceleration, and image signals.

21.1.3.1 Sound Signals

Sound signals can be a rich source of contextual information. The contextual information includes a presence of vocal activity (speech/nonspeech classification) [23–25], gender [24,26], emotion [28], human activity [30,31], and location or surrounding environment [32–33]. In our previous works [7], we used sound signals to visualize the aggregate vocal activity of the workers in the office environment.

Zhao et al. [23] used sound signals for audio surveillance network application. They used speech/nonspeech classification to distinguish normal utterance, scream, footsteps, and glass breaking sound. Considering the limited hardware capability and energy supply, they evaluated the computational cost of the fast Fourier transform (FFT)-based acoustic features and compared with mel frequency cepstrum coefficient (MFCC)-based acoustic features. They employed support vector machine (SVM)-based classifier that requires high computational cost. Kwon et al. [24,25] also studied on the real-time acoustic scene analysis for sensor networks. They extracted pitch and RASTA-PLP features for gender and emotion classification and transmitted to the base station PC for real-time calculation of Gaussian mixture model (GMM) and SVM-based classifier. Kinnunen et al. [26,28] also used MFCC-based acoustic features with vector quantization algorithm. In the work of Ravindran [35], new acoustic feature named noise-robust auditory features (NRAFs) is proposed for accurate audio classification with a cost comparable to that of implementing MFCC. Istrate et al. and Lukowicz et al. [30,31] used sound signals for human activity recognition. Even though their work presented very accurate sound recognition performance with over 90% accuracy, they required PC to do the sound processing in real time for high computational complexity.

For sound, MFCC or FFT is used as the most robust and versatile sound feature extraction method to detect speech and recognize speaker, gender, emotion, and human activity. The conventional recognition

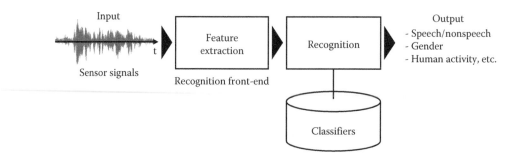

FIGURE 21.2 General structure of context recognition.

front ends for sound signals are motivated to model (or imitate) human auditory and articulatory property. Due to their conceptual background, the complex computations are required for extracting MFCC, and it is considered to be not suitable for resource-constrained platform of the sensornet.

These conventional studies also did not consider about the temporal locality of the target signals. For example, in speech detection scenario, there are only 30% of speech signals within usual conversation recordings. Therefore, calculating every input frame of signals with equal computational complexity can result in high calculation cost.

21.1.3.2 Acceleration Signals

Another facet of the user's context information is his/her activity. In wearable computing scenarios, human activities such as walking, standing, and sitting can be inferred from data provided by body-worn triaxial acceleration sensor [37]. A triaxial acceleration sensor is a sensor that returns a real-valued estimate of acceleration along the x, y, and z axes from which velocity and displacement can also be estimated.

Human activity recognition can be literally explained as the task to distinguish each target activity by using acceleration signals. The acceleration signals are obtained through the accelerometer worn by the users. Ravi et al. [38] used a single triaxial accelerometer. They employed basic statistical features such as mean, standard deviation, energy, and correlation. Huynh et al. [39] used FFT-based features for activity recognition. Bao et al. [40] introduced the use of spectral entropy. In their work, many types of classifiers are compared in terms of recognition accuracy, and it is concluded that C4.5 decision tree classifier achieved the highest accuracy. Their work also suggests that a mobile computer and small wireless accelerometers placed on an individual's thigh and dominant wrist can be used to detect common human activities using FFT-based feature computation and a decision tree algorithm.

21.1.3.3 Image

Context information includes any information explaining the surrounding situation of the device. When image sensor is used, additional information can be extracted. Image can be used to detect information such as human face and object, pedestrian, facial feature point, and facial attributes (gender, ethnicity, and expression). Viola et al. [41] proposed a state-of-the-art face/object detection using Haar-like feature and cascaded classifier algorithm.

In image recognition field, Haar-like feature (Figure 21.3) is known as a state-of-the-art recognition front end for face/object detection. The feature value is defined by the difference of the sum of positive area and the sum of negative area in grayscale image. It does not require any multiplication operation.

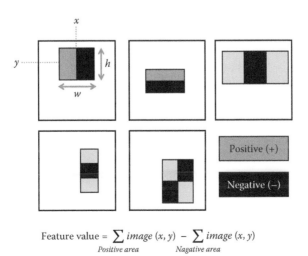

$$\text{Feature value} = \sum image\,(x, y) - \sum image\,(x, y)$$
$$\quad\quad\quad\quad\quad\text{\textit{Positive area}} \quad\quad\quad \text{\textit{Nagative area}}$$

FIGURE 21.3 Haar-like features for image.

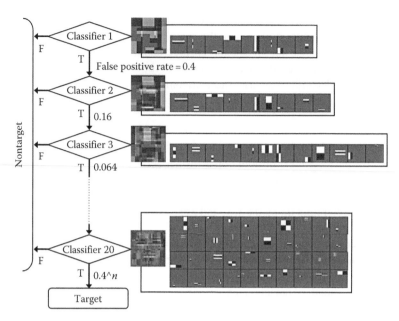

FIGURE 21.4 Cascaded classifiers for face detection.

In addition, Viola et al. proposed the cascade structure of classifier. As shown in Figure 21.4, multiple of the classifiers are connected in series to form a simple classifier with a small number of Haar-like features for coarse filtering, which are then connected to more complex classifier with a larger number of Haar-like features for fine filtering. A false output at any point of the cascade results in the immediate conclusion of the evaluation of the instance. Because target object is typically localized (or concentrated) spatially, the cascade structure is employed to reduce the total calculation cost of detection process dramatically.

Image can provide context-rich information. Siala et al. [42] applied Viola's method to pedestrian detection. Shakhnarovich et al. [43] employed Haar-like feature for gender and ethnicity classification. The detected faces are passed to a demographic classifier that uses the same architecture as the face detector. Zhao et al. extended Paul Viola's Haar-like feature to detect facial features such eye, nose, and mouth [44]. Furthermore, Omron applied it to estimate the smile intensity in real time [45]. Furthermore, Jung et al. [46] extended Haar-like feature for efficient facial expression recognition [46].

The idea behind the use of Haar-like filters is that simple difference filters possess the degrees of freedom, which can be trained to be adapted to specific recognition problem. The degrees of freedom are provided by simple filter parameters such as basic shapes, size, and position relative to the detection window. In training phase, optimal set of features are selected by choosing the optimal value for each trainable parameter based on the training error of the classifier. In face detection, the best set of Haar-like filters that distinguish the face and nonface training samples is selected.

The disadvantages of this approach are as follows:

- Collection of training samples and labeling procedure can be costly.
- Time-consuming training process.

The advantages of this approach are as follows:

- Computationally low-cost recognition front end can be obtained using very simple filters.
- No need for the sophisticated knowledge on the image processing.

Summarizing the conventional studies on context recognition, the following can be said:

- Different feature extraction method is applied on each type of signal: FFT- or MFCC-based features for sound, basic statistical features for acceleration signals, and simple rectangular feature such as Haar-like feature for image.
- Different recognition based on different signals (sound, accelerations, image) can share same types of classifiers (decision boundary based, C4.5 decision tree, boosting, SVM; distribution based, vector quantization, GMM, etc.).
- Temporal locality of the target signals is not considered.
- Cascade structure is used to utilize a spatial locality of target object for reduction of overall calculation cost.

Consequently, the naive integration of these conceptually and structurally different methods may result in higher calculation cost, larger chip area, and more complex system architecture. In sensor network applications, the computation platform is highly limited with the resource.

21.1.4 OVERVIEW OF OUR APPROACH: VERSATILE RECOGNITION

By looking into the conventional studies on context recognition, we considered that a different recognition based on different signals can share same architecture if we could unify the recognition front end. This chapter brings together new algorithms and insights to construct a versatile recognition framework that can be used to recognize patterns from sound, acceleration signals, and image at low calculation cost.

The versatile recognition front end using Haar-like feature for images, sound, and acceleration signals is proposed, as shown in Figure 21.5. Haar-like feature is a simple difference filter of low calculation cost that requires only addition and subtraction to extract a feature value. In our preliminary studies [47–50], we have shown that 1-D Haar-like features can also be applied to recognition problems in other signals, such as speech/nonspeech classification in sound and human activity recognition in acceleration signals. A 1-D Haar-like feature applied to signals in temporal dimension can be considered as very rough band-pass filter. This idea of versatile recognition brings the idea of unifying the different recognition front ends by Haar-like filtering framework. In this way, the redundancy resulted by using completely different feature extraction methods on different sensor signals can be avoided.

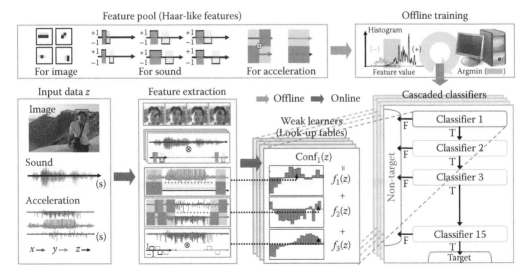

FIGURE 21.5 Versatile recognition framework using Haar-like features.

By using Haar-like features for a feature pool, the discriminative features are selected in an offline training process. A weak classifier is made of a lookup table (LUT) of the Haar feature. By forming a linear combination of weak classifier, a strong classifier (stage classifier) of high recognition accuracy is designed. The classifier can be employed for various recognition problems by only changing the linear combination and hence is versatile.

For the classifier, the cascade structure is employed to utilize the temporal locality of the target signals for calculation cost reduction. In the cascaded classifier, dozens of the classifiers are connected in series to form a simple classifier with a small number of Haar-like features for coarse filtering, which are then connected. A false output at any point of the cascade results in the immediate conclusion of the evaluation of the instance. However, a true output with enough confidence does not result in a conclusion until the full classifiers are used in the conventional cascade. Positive estimation (PE) technique is introduced to allow a sensor node to become content-aware, computes finely only when the inputs are target-like and difficult to classify, and stops computing when inputs obtain enough confidence. When the sensor node continuously senses the signals for long periods of time, the content-aware property allows the sensor node to save as much power as possible based on the difficulty of the classification. Since the target to be recognized is typically localized in the space and time, the cascaded classifier instantly excludes the nontargets from the further processing and thus reduces a total calculation cost significantly. In addition to the influence on the calculation cost, PE allows the training process of cascaded classifier to focus on relatively difficult samples yielding higher classification accuracy.

In addition to the PE, two techniques are proposed to construct a compact cascaded classifier. Redundant feature selection (RFS) is proposed to allow stage classifiers in the cascade to share the features to reduce the calculation cost. As another technique, we propose a dynamic lookup table (DLUT), the LUT-based weak classifiers with the smallest possible data size. Since all the data cannot be stored in internal cache memory due to its size limitation, most data have to be stored in external DRAM and have to be loaded via memory bus repeatedly for the detection. This transfer radically increases overall power consumption of the system. By reducing classifier data, the bandwidth of memory bus is also reduced, resulting in power reduction of the overall system.

21.2 DESIGNING HAAR-LIKE FEATURES FOR SOUND

21.2.1 Characteristic Feature of Sound Signals

In order to design the feature extraction method for the pattern recognition based on sound signals, we first look at a simplified model of the human speech production system and explain the characteristic property of the speech signals.

In the simplified model of the human speech production, the lungs push air through the glottis to create either a periodic pulse, forcing it to open and close, or a pulse just enough to hold it open in a turbulent flow. This periodic movement acts as an excitation signal. Then, the signal flows through the vocal tract, which acts as a time-varying linear filter shaping the resulting quasiperiodic or flat spectrum [51].

The spectrum of the excitation signal consists of harmonics of the fundamental frequency of glottal vibration. This property results in a narrowband spectrogram of voiced speech. The response of vocal is defined as the ratio of sound pressure in front of the speaker to the sound pressure at the source of excitation. The pole locations define the resonance frequencies of the vocal tract that correspond to the formant structure observed in spectrograms [52].

In pattern recognition studies on temporal signals such as sound signals, a shape of spectrum is mainly used as key features. Speech signals provide convex and concave formant shapes in frequency bands between 400 Hz and 4 kHz. The concave and convex formant patterns were observed in speech, but not noise samples. In the work of Hoyt et al., the only nonspeech signal evaluated that produced formant shapes that met the same criteria was a police siren [53].

For nonspeech signals, most segments do not have well-defined prosodic structures as speech segments [54]. However, Chu et al. showed that each class of nonspeech signals can be well clustered in frequency and scale feature space. Each class of nonspeech sound signals is decomposed into sinusoid waveform with certain frequency and the scale, which indicates the temporal width of the localized sinusoid waveform.

21.2.2 DESIGNING 1-D HAAR-LIKE FEATURES

21.2.2.1 Conventional Haar-Like Features

Haar-like features are the simple filters that calculate the difference between the sums of the pixels in the different regions. They are popularized by the work of Viola et al. [41], and many of the works have extended the idea to other applications [42–46]. Haar-like features possess two important properties.

The main advantage of using these base functions is that the inner product of a data vector with each of them can be performed by several integer additions, instead of N floating point multiplications, where N is the dimension of the base vectors. This can be achieved by computing the integral image of the original image that is defined as

$$f_i(i,j) = \sum_{m=1}^{i} \sum_{n=1}^{j} f(m,n) \tag{21.1}$$

The dot product of an image with a one-box base function is the summation of a rectangular area of the image, which can be computed efficiently as

$$\sum_{i=\text{top}}^{\text{bottom}} \sum_{j=\text{left}}^{\text{right}} f(i,j) = f_i(\text{bottom},\text{right}) - f_i(\text{bottom},\text{left}-1)$$
$$-f_i(\text{top}-1,\text{right}) + f_i(\text{top}-1,\text{left}-1) \tag{21.2}$$

where
$f(i,j)$ is the image function
$f_i(i,j)$ is the integral image

Using the integral image, any rectangular sum can be computed in four array references. The difference between two rectangular sums can be computed in eight references. When the two-rectangle features defined involve adjacent rectangular sums, they can be computed in six array references, eight in the case of the three-rectangle features, and nine for four-rectangle features.

Haar-like features provide versatility. Haar-like features possess trainable parameters that can be used to adapt to various recognition problems. For example, 2-D Haar-like features for face detection are generated by selecting type of basic patterns, scaling width and height, and changing the location of pattern relative to detection window [41]. Higashijima et al. generalized the conventional Haar-like feature and proposed long Haar-like filter, having the form of M-dimensional vector $(g_1,...,g_M)^T$ where $g_i = 1$ or -1, for face recognition purpose [57]. The characteristic property of Haar-like bases is that they are not orthogonal to each other. The dot product is not zero when Haar-like box functions have overlapping areas. Because the features do not possess complete orthogonality, a very large and varied overcomplete set of rectangle features are generated to provide a rich image representation that supports effective learning [41]. In this way, the features can act to encode knowledge that is difficult to learn using a finite quantity of training data.

Cui et al. proposed a 3-D Haar-like feature for pedestrian detection to extract motion pattern feature from video [58]. In order to capture the long-term motion patterns among multiple frames and record the person's appearance features at the same time, temporal dimension is added to the conventional 2-D Haar-like features. In this way, *cubic filters* are formed in a space–time volume. This suggests that Haar-like features can be extended to temporal signals such as sound and acceleration signals for various pattern recognition problems.

21.2.2.2 Basic Concept

In pursuit of extracting characteristic features of sound signals, many methods specifically designed for sound signals are proposed. These methods include linear predictive cepstrum coefficient (LPCC) and MFCC [55]. They are applied to speech/nonspeech classification, gender recognition, speaker recognition, emotion recognition, and environmental sound recognition. To extract feature from temporal signals for various recognition, information on a shape of spectrum must be captured.

The conventional methods basically relied on FFT and/or precisely designed filter banks to obtain the feature, requiring large memory and energy consumption. They focused on minimizing an intraclass signal reconstruction error by decomposing signals into orthogonal basis.

Each filter bank can be viewed as a localized sinusoid wave in the temporal dimension. Fundamentally, the correlation between input signals and the localized sinusoid wave of a certain frequency can provide spectral information that is useful for pattern recognition. In this work, we propose 1-D Haar-like feature patterns to simplify the localized sinusoid wave to form a rough sine-like wave, as shown in Figure 21.6. The 1-D Haar-like feature is a simple difference filter that requires only addition and subtraction. It can be calculated at extremely low computational cost. Although the trained classifier is a combination of various Haar-like patterns, each pattern can be calculated as virtually the same cost.

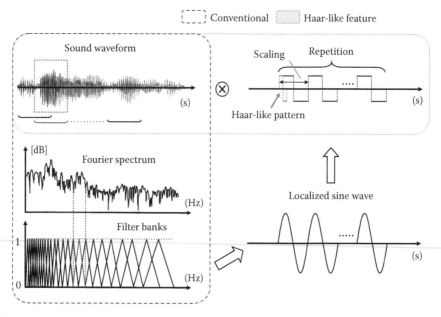

FIGURE 21.6 Comparison of Haar-like feature and the conventional feature extraction framework.

Since the primary focus in the pattern classification is to minimize an interclass discrimination error, we can search for more simple yet the most discriminative and not necessarily orthogonal filters from the overcomplete feature pool as suggested in [41]. Scaling and repeating different basic Haar-like patterns provides the feature pool for training a discriminative classifier.

21.2.2.3 1-D Haar-Like Feature Design

Haar-like feature patterns are used to simplify the localized sine wave to form a rough sine-like wave to obtain a rough observation of spectrum. The filter coefficients are restricted to -1, 0, 1 for a fast evaluation. A 1-D Haar-like feature can be seen as a rough band-pass filter. Haar-like features are not orthogonal to each other and only weakly discriminative when each feature is alone. However, this problem can be considered trivial. The feature selection process using Real AdaBoost searches for relatively noncorrelative features out of overcomplete feature pool to form a strongly discriminative combination. The feature pool is generated by scaling and repeating different basic Haar-like patterns. The features are parameterized by a type of basic pattern, scaling rate, and number of repetition.

Each basic pattern is designed as to minimize a square root error of frequency response against the localized sine wave of a certain target frequency f_0. The peak frequency of the basic pattern is

$$f_0 = \frac{F_S}{W_{\text{type}}} \tag{21.3}$$

where
F_S is a sampling frequency
W_{type} is a width of basic pattern

The frequency response of Haar-like feature $h(n)$ is given by

$$|H(\omega)| = \sqrt{\left(\sum_{n=0}^{W_{\text{Haar-like}}-1} h(n)\cos n\omega \right)^2 + \left(\sum_{n=0}^{W_{\text{Haar-like}}-1} h(n)\sin n\omega \right)^2} \tag{21.4}$$

where $W_{\text{Haar-like}}$ is a Haar-like feature width. $W_{\text{Haar-like}}$ is calculated by

$$W_{\text{Haar-like}} = W_{\text{type}} \cdot \alpha \cdot N_{\text{repeat}} \tag{21.5}$$

where
α is an scaling rate
N_{repeat} is a number of repetition
$W_{\text{Haar-like}}$ and α are related to the peak frequency, whereas N_{repeat} refers to the bandwidth

To generate the basic patterns, α is set to 1. The frequency response of the sinusoid wave localized in the width of $W_{\text{Haar-like}}$ is given by

$$|F(\omega)| = \left| \frac{\sin(\omega-\omega_0)W_{\text{Haar-like}}/2}{(\omega-\omega_0)} + \frac{\sin(\omega+\omega_0)W_{\text{Haar-like}}/2}{(\omega+\omega_0)} \right| \tag{21.6}$$

The square root error of frequency response can be calculated as

$$\varepsilon = \sum_{\omega=0}^{\pi} \left(|F(\omega)| - |H(\omega)| \right)^2 \tag{21.7}$$

TABLE 21.1

Basic Haar–Like Feature Patterns

Basic Pattern	W_{type}	$h(n)$	f_{peak}[Hz]
	2	$\{1,-1\}$	$4000/\alpha$
	3	$\{1,0,-1\}$	$2666/\alpha$
	4	$\{1,1,-1,-1\}$	$2000/\alpha$
	5	$\{1,1,0,-1,-1\}$	$1600/\alpha$
	6	$\{1,1,0,-1,-1,0\}$	$1333/\alpha$
	7	$\{1,1,1,0,-1,-1,0\}$	$1142/\alpha$
	8	$\{1,1,1,0,-1,-1,-1,0\}$	$1000/\alpha$
	9	$\{1,1,1,1,0,-1,-1,-1,0\}$	$888/\alpha$
	10	$\{1,1,1,1,0,-1,-1,-1,-1,0\}$	$727/\alpha$
	11	$\{1,1,1,1,0,0,-1,-1,-1,-1,0\}$	$666/\alpha$
	12	$\{1,1,1,1,1,0,0,-1,-1,-1,-1,-1\}$	$615/\alpha$
	13	$\{1,1,1,1,1,0,0,-1,-1,-1,-1,-1,-1\}$	$571/\alpha$

In order to generate the basic Haar-like feature patterns, we searched for the pattern that minimizes (21.7) for each W_{type}. During the pattern selection procedures, we did not take into account the phase information.

The basic patterns of 1-D Haar-like feature can be given by −1, 0, 1 patterns that minimize ε for each width of the basic pattern. The generated patterns are depicted in Table 21.1. There is a freedom in the phase of the basic patterns, since the frequency-domain information is meaningful in the recognition. In our experiments, W_{type} is set to 2–14, α is set to 1–40, and N_{repeat} is set to 1–20, providing 9600 overcomplete nonorthogonal features for training.

As shown in Figure 21.7, the output of various 1-D Haar-like features roughly estimates frequency information. The speech signals of utterance "/a//i//u//e//o/" are shown on top, and their spectrogram is shown in the middle. A 1-D Haar-like feature output is shown with respect to peak frequency of its impulse response.

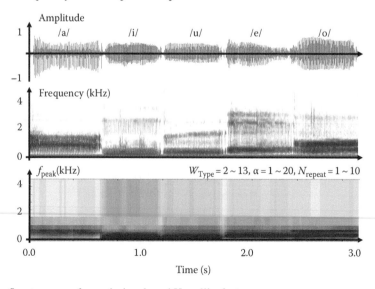

FIGURE 21.7   Spectrogram of speech signals and Haar-like features.

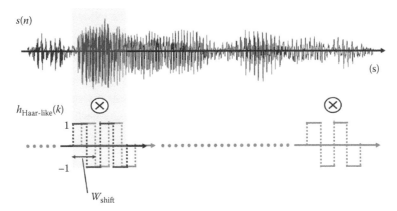

FIGURE 21.8 Basic calculation procedure of 1-D Haar-like feature value.

21.2.3 CALCULATING HAAR-LIKE FEATURE VALUE

A 1-D Haar-like feature is calculated by

$$x_{\text{Haar-like}} = \sum_{n=0}^{N} \left| \sum_{k=0}^{W_{\text{Haar-like}}} h_{\text{Haar-like}}(k) s(n \cdot W_{\text{shift}} - k) \right| \tag{21.8}$$

as shown in Figure 21.8, where N is a number of the filtering computations within a frame width W_{frame} of input signals $s(t)$. N is given by

$$N = \frac{W_{\text{frame}} - W_{\text{Haar-like}}}{W_{\text{shift}}} \tag{21.9}$$

Because the filter coefficients are restricted to −1, 0, 1, the calculation is limited to add and subtract integer operations. In typical digital signal filtering, W_{shift} is set to 1.

21.2.4 CALCULATION COST REDUCTION OF HAAR-LIKE FEATURE

21.2.4.1 Filtering Shift Width

W_{shift} is considered as a parameter for calculation cost reduction. It is given by

$$W_{\text{shift}} = \alpha_{\text{shift}} \cdot W_{\text{Haar-like}} \tag{21.10}$$

When $\alpha_{\text{shift}} = 0$, W_{shift} is set to 1. In conventional filtering procedure of FIR digital filter, W_{shift} is set 1. By controlling α_{shift}, the observation resolution is decreased while reducing the calculation cost dramatically.

21.2.4.2 Integral Signal

A 1-D Haar-like feature *is* can be calculated efficiently by using intermediate signal representation called integral signals. Integral signals is(n) can be given by

$$\text{is}(n) = \sum_{k \leq n} s(k) \tag{21.11}$$

The actual calculation of the integral signal is(n) is based on the following recurrence:

$$is(n) = is(n-1) + s(n) \qquad (21.12)$$

is(-1) = 0 and the integral signal can be calculated efficiently over the original sound signals. By using is(n), the sum of the input signals within the shaded region for each convolution process can be simplified to a single subtraction as

$$\sum_{n=n_1}^{n_2} s(n) = is(n_2) - is(n_1) \qquad (21.13)$$

Using the integral signal, each filter calculation can be reduced to three array references for W_{type} = 1. In this way, memory access and add calculations during the feature calculation decrease dramatically:

$$\left[is(n + W_{\text{Haar-like}}) - is\left(n + \frac{W_{\text{Haar-like}}}{2} \right) \right] - \left[is\left(n + \frac{W_{\text{Haar-like}}}{2} \right) - is(n) \right]$$

$$= is(n + W_{\text{Haar-like}}) - 2 \cdot is(n + W_{\text{Haar-like}}) + is(n) \qquad (21.14)$$

However, when the filter width is less than 4, the use of integral signal is not effective on the calculation cost reduction. When the integral signal is not used, the amount of add calculation is equaled to $W_{\text{Haar-like}} - 1$, while 3 add calculations are always needed when the integral signal is used for W_{type} = 2 as an example.

21.2.4.3 Δ-Integral Signal Recycling

Another intermediate signal representation is called Δ-integral signals. Since Haar-like features are overcomplete or highly redundant and nonorthogonal features, some of the selected Haar-like features may share some of the feature parameters and have similar patterns. To make use of such a situation to calculation cost reduction, Δ-integral signals are proposed to share the intermediate result of the previous feature calculation. Δ-Integral signals can be given by

$$\Delta(n_1, n_2) = is(n_2) - is(n_1) = \sum_{n=n_1}^{n_2} s(n) \qquad (21.15)$$

21.2.5 SOUND DATA SETS FOR EVALUATION

To evaluate the versatile recognition, various data sets are prepared. For speech/nonspeech, gender classification, speaker classification, and environmental sound recognition, the following data sets are prepared.

Speech data set includes 11 (10 males, 1 female) workers' utterances of Japanese phoneme-balanced sentences in office environment using wearable sensor node [7,59]. Each worker contributed about 5 min. Nonspeech data set includes typical office/home environmental and activity sounds such as ventilation, running, walking, touching/pulling up and down the blinds, shutting/opening desk drawers, dragging the chair, phone ring, printer working, electric shaver, vacuum cleaner, server room, typing, and NOISEX-92 noise database (car, hf channel, white, pink, factory noise) that are down-sampled to 8 kHz with 8 bit [60]. There are 24 types of nonspeech sounds including both stationary and nonstationary sounds, and each sound contributed about 5 min. The data sets are recorded in 8 kHz sampling frequency with 8 bit resolution. Train data include 60% of the whole data set and the remaining is used for the test. In this work, we assume the sound signals to be recognized as dominant and relatively stationary sound signals.

For emotional speech classification, there is an emotional speech database from Linguistic Data Consortium [61]. There are four classes of emotion: hot anger, happy, sadness, and neutral. The utterances are spoken by seven actors (three males, four females). We evaluated in a speaker-independent manner to avoid the effect of the individuality. The utterances of six actors are used for the training and one actor for the test.

The dominant sound signals refer to the loudest sound received by the system. Hence, the recognition does not have to deal with separating the relevant signal from background noise. Stationary sound refers to the sound that is relatively constant over time. The spectrum of the sound is identical in all time frames regardless of their position and length. Sound recognition problem is reduced to pattern matching of the spectral information.

21.3 DESIGNING HAAR-LIKE FEATURES FOR ACCELERATION SIGNALS

21.3.1 INTRODUCTION

Acceleration signals obtained from wearable sensors contain valuable information originated from human activity. Human activity recognition can be used to transform the raw sensor data into higher-level descriptions of people's behaviors to form a structured data set describing people's daily patterns of activities. The potential applications of such data sets include mapping patterns of information flow within an organization, predicting the spread of disease within a community, monitoring the health and activity levels of elderly patients as well as healthy adults, allowing *smart environments* to respond proactively to the needs and intentions of the users, and life recording for an external memory [63,64].

Considering the WSN applications such as our work of *business microscope* [6], in which sensor node is driven by strictly limited power and run on the low duty ratio in the order of 0.1, only recognition results processed in the sensor node can be sent to a server once in a few seconds. This is because transmitting raw signals consumes a lot of energy. Hence, low calculation cost yet highly reliable human activity recognition system is needed.

Combining with other signal modality is also considered important. The audio captures the sounds produced during the various activities, whereas the accelerometer data are sensitive to the movement of the body [65]. In this chapter, we propose a novel feature extraction method that allows us to handle acceleration signals using Haar-like feature extraction framework.

21.3.2 CONVENTIONAL STUDIES ON HUMAN ACTIVITY RECOGNITION

21.3.2.1 Technological Trends

Making wearable devices aware of the activity of the user fits into the framework of context awareness. Ubiquitous computing is centered around the idea of providing services to the user in a seamless manner. Bringing such pervasive services to the user based on activity is an active research area. Attempts have been made at recognizing user activity from accelerometer data [37–40]. In the exhaustive work of Bao [40], subjects wore five biaxial accelerometers on different body parts as they performed a variety of activities like walking, sitting, standing still, watching TV, running, bicycling, eating, and reading. Data generated by the accelerometers were used to train a set of classifiers, which included decision trees (C4.5), naive Bayes classifier, and nearest-neighbor algorithm found in the Weka Machine Learning Toolkit [66]. Decision tree classifiers showed the best performance, recognizing activities with an overall accuracy of 84%. Maurer et al. also employed decision tree classifier for its good balance between accuracy and computational complexity [67].

However, the previous works on human activity recognition using accelerometer did not focus on calculation cost. In these works, same features such as mean, standard deviation, energy, correlation, and frequency-domain entropy are adopted. Some of them require FFT. In [39], they concluded FFT coefficients are effective for the classification. Except for mean, large amount of multiplication is needed to calculate these feature values. Furthermore, since these methods are not specifically designed to extract features from acceleration signals that can contribute to human activity recognition, there are many redundancies in features themselves. This can cause inefficient classifier to be generated by training with such features.

21.3.2.2 Basic Features

This section describes four major feature extraction methods for human activity recognition and discusses their calculation cost. The feature extraction of *mean* is the lowest calculation cost among previous methods, whose actual computation equation is given by

$$\text{Mean} = \frac{1}{W_{\text{Frame}}} \sum_{t=0}^{W_{\text{Frame}}} s(t) \tag{21.16}$$

where

$s(t)$ denotes the raw signal input value

W_{Frame} denotes the filtering frame width, which is 512 (2.56 s) in our experiment, thus, 512 times addition is needed for the calculation per frame

In this application, division is not required because W_{Frame} is a fixed value. Although the calculation cost is low, the modeling ability of activity is also low. This feature extracts *rough moving direction*, but that's it.

The feature *standard deviation* has relatively low calculation cost of the four. The actual calculation of the feature *standard deviation* is given by

$$\sigma_x = \sqrt{\frac{1}{N} \sum_{n=0}^{N} \left(x(n) - \mu_x \right)^2} \quad \mu_x, \text{mean value for } x\text{-axis signals} \tag{21.17}$$

In this case, the actual computation is equivalent to its definition because removing square root causes different distributions of the feature, thus generating different classifiers. The calculation cost roughly corresponds to W_{Frame} times addition and W_{Frame} times multiplication because the division and the square root operation are not dominant. This feature extracts *rough energy*.

The feature *correlation* is calculated between each pair of axes as the ratio of the covariance and the product of the standard deviation, thus relatively high calculation cost. It is given by

$$\sigma_{xy} = \frac{1}{N} \sum_{n=0}^{N} \left(\frac{x(n) - \mu_x}{\sigma_x} \right) \cdot \left(\frac{y(n) - \mu_y}{\sigma_y} \right) \quad \mu_y, \text{mean value for } x\text{-axis signals} \tag{21.18}$$

This feature is useful for classifying activities that involves more than 1-D acceleration signal such as ascending and descending. The feature *energy* is extremely high-calculation-cost method. It is the sum of the squared discrete FFT component magnitudes of the signal and given by

$$\text{Energy} = \frac{1}{W_{\text{Frame}}} \sum_{i=1}^{W_{\text{Fraame}}} X_i^2 \qquad (21.19)$$

where X_i denotes each FFT component. Even FFT, which is the order of $O(W_{\text{Frame}} \log W_{\text{Frame}})$, is adopted, large amount of floating point multiplication such as $\exp(x)$ is required.

21.3.3 SENSOR SPECIFICATIONS AND HUMAN ACTIVITY DATA SET

The 200 Hz 3-D acceleration data were collected from four people for over 120 min using off-the-shelf wireless accelerometer marketed by ATR-Promotions [68]. The amplitudes range from [–3G, +3G]. This specification will be sufficient to recognize various movement patterns. The accelerometer is attached to name tag and placed on the chest position. According to the empirical studies on sensor position, chest is one of the effective wearing positions for activity recognition and best for classifying activities such as squatting and standing [70]. The data are transmitted to PC via Bluetooth and include five actions:

- Walking
- Running
- Standing
- Ascending
- Descending

No postprocessing was carried out and classes were labeled manually.

21.3.4 EVALUATION CRITERION

To generate classifier of decision tree with pruning (C4.5), the Weka Machine Learning Toolkit was used. A 10-fold cross validation was carried out for the evaluation. Human activity recognition accuracy is defined as the average of each target action's classification accuracy as in

$$\text{Accuracy} = \frac{1}{\#\text{of actions}} \sum^{\#\text{of actions}} \left(\frac{\#\text{of correct inputs}}{\#\text{of inputs/actions}} \right) \times 100 \ \% \qquad (21.20)$$

This definition is used throughout the experiments in this work.

21.3.5 INTEGRATING STATISTICAL FEATURES

The statistical features such as standard deviation and correlation are playing an important role in the human activity recognition. In this section, two new techniques are proposed to integrate these statistical features into Haar-like filtering framework to gain higher human activity recognition performance in terms of the accuracy and calculation cost.

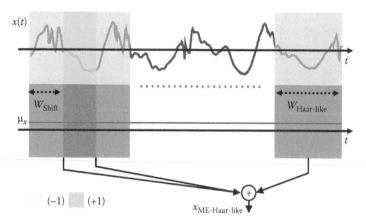

FIGURE 21.9   Mean-embedded Haar-like feature for 3-D acceleration signals.

21.3.5.1   Mean-Embedding Haar-Like Features

In the conventional human activity recognition studies, standard deviation is used to capture the fact that the range of possible acceleration values differ among different activities such as walking and running. Standard deviation is calculated as the Equation 21.17. It is calculated as the average of the difference between the signal amplitude and dc component of the signals as in

$$\sigma_x = \sqrt{\frac{1}{N}\sum_{n=0}^{N}\left(x(n)-\mu_x\right)^2} \tag{21.21}$$

The most fundamental part of Haar-like filtering is to take the difference between the signals located in certain areas. To integrate the concept of the standard deviation feature into Haar-like filtering framework, the mean values of the signals are *embedded* into the one side of Haar-like filter during the calculation as shown in Figure 21.9. The feature value is calculated by

$$x_{\text{ME-Haar-like}} = \sum_{n=0}^{N}\left|\sum_{k=0}^{W_{\text{Haar-like}}}\left\{x\left(n\cdot W_{\text{shift}}-k\right)-\mu_x\right\}\right| \tag{21.22}$$

where μ_x is mean value of the input signals $x(t)$. Mean-embedded haar-like feature (MEH) calculates the overall difference between the mean value and the input signals in the filter width. Therefore, the periodicity is considered in the method. In this way, *standard-deviation*-like feature can be extracted from the signals in Haar-like filtering framework. For MEH, $W_{\text{Haar-like}}$ is set to $1 \sim W_{\text{frame}}$. W_{frame} is 400, when the input length is set to 2.0 s for 200 Hz sampling frequency.

21.3.5.2   Biaxial Haar-Like Features

Correlation can be especially useful when recognizing the human activities that involve a translation in just one dimension. For example, walking and running can be distinguished from stair climbing using correlation.

To extract information on the correlation in the Haar-like feature extraction framework, biaxial Haar-like feature (BH) is proposed. The fundamental idea of Haar-like feature is to take the

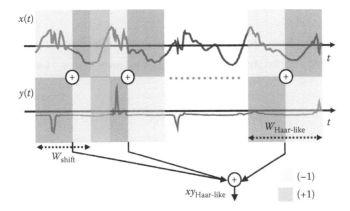

FIGURE 21.10 Biaxial Haar-like feature for 3-D acceleration signals.

difference between the sums of the signals in certain regions. As shown in Figure 21.10, the difference between the Haar-like filter outputs in two axes is calculated. The feature value is calculated by

$$xy_{\text{Haar-like}} = \sum_{n=0}^{N} \left| \sum_{k=0}^{W_{\text{Haar-like}}} h_{\text{Haar-like}}(k) x(n \cdot W_{\text{shift}} - k) \right.$$

$$\left. - \sum_{k=0}^{W_{\text{Haar-like}}} h_{\text{Haar-like}}(k) y(n \cdot W_{\text{shift}} - k) \right| \quad (21.23)$$

where $x(t)$ and $y(t)$ correspond to x and y axes of a 3-D accelerometer. This can then be simplified into the basic 1-D Haar-like feature on the interaxial difference signals as

$$xy_{\text{Haar-like}} = \sum_{n=0}^{N} \left| \sum_{k=0}^{W_{\text{Haar-like}}} h_{\text{Haar-like}}(k) \Delta_{xy}(n \cdot W_{\text{shift}} - k) \right| \quad (21.24)$$

$$\Delta_{xy}(n \cdot W_{\text{shift}} - k) = x(n \cdot W_{\text{shift}} - k) - y(n \cdot W_{\text{shift}} - k) \quad (21.25)$$

Integral signals of the interaxial difference signals can be prepared beforehand to reduce the calculation cost. For 3-D acceleration signals, biaxial filtering is carried out for each of two axes, respectively (e.g., xy, yz, and zx are for xy). Using this filter, the recognition problems on 3-D signals can be solved much more efficiently.

21.3.5.3 Initial Evaluation

Several results based on the conventional state-of-the-art classifier (C4.5 decision tree) merit discussion on the discriminative property of the proposed Haar-like features. The calculation cost for feature extraction on 2.56 s analysis frame is evaluated along with the human activity recognition accuracy. The calculation cost is estimated by the total amount of add and multiplication operations, where 16 bit short multiplication is approximated by 16*add operations according to Booth's multiplication algorithm [62].

Table 21.2 shows the performance comparison of human activity recognition among the different feature extraction configurations. M refers to mean, MS refers to M plus standard deviation, and $MSEC$ refers to MS plus energy and correlation. To generate the decision tree classifiers with pruning (C4.5), the Weka Machine Learning Toolkit was used. Due to the memory usage limitation of the toolkit, the feature pool is limited to only the first basic feature pattern $\{-1, 1\}$.

The proposed features yielded the highest accuracy with much lower calculation cost than the conventional methods. Among the previous methods, MS achieves both high accuracy of 92.6% and low

TABLE 21.2
Trained Classifier Specifications and Classification Performance

	Accuracy [%]	# of Leaves	Calc. Cost [kopf]
Mean	76.0	191	1.556
MS	92.6	83	16.614
MSEC	93.1	83	100.571
Basic Haar-like (H)	92.7	81	3.924
H+BH	93.9	65	3.314
H+BH+ME	94.2	63	3.597

calculation cost of 16.61 k operations per frame (opf). Using basic Haar-like feature (H), high accuracy of 92.7% is achieved with 3.08 kopf, which is lowest among the methods with over 90% accuracy. By adding Haar-like biaxial feature (HHB), the accuracy of 93.9% is yielded with 3.31 kopf. By adding mean-embedded Haar-like feature (HHB+Me), the highest accuracy of 94.2% is achieved with 3.59 kopf.

The number of leaves in the decision tree is one of the important factors that decide the calculation cost of the classifier. Haar-like feature methods achieve relatively compact decision trees. Especially, Haar-like + Haar-like Biaxial + Mean-Embedding (H+HB+ME) configuration achieves the highest accuracy of 94.2% with the smallest number of leaves and lower calculation cost relative to the conventional methods. Due to higher discriminative property of the proposed features, the number of leaves in the C4.5 decision tree classifier is 63, which is the lowest when the proposed methods are used. Therefore, memory usage can also be reduced as well as the total calculation cost. This is due to the fact that Haar-like features themselves possess the flexibility to fit into the problems. In the conventional methods, the feature extraction methods are fixed and do not have a capability of adapting themselves to specific recognition problem.

By integrating the correlation into Haar-like feature by using biaxial Haar-like feature, the confusion between ascending and descending, which involves a translation in multiple axes, is suppressed. Moreover, the confusion among ascending, descending, and walking is also mitigated by integrating the standard deviation by using MEH. These results reflect our aim for proposing the novel features for 3-D acceleration signals.

21.4 DESIGNING COMPACT CLASSIFIER

21.4.1 INTRODUCTION

In this chapter, the classifier design for versatile recognition is investigated. As we have looked at the conventional pattern recognition studies, the following can be said on the design of classifier:

- Different recognition based on different signals (sound, accelerations, image) can share same types of classifiers (decision boundary based, C4.5 decision tree, boosting, SVM; distribution based, vector quantization, GMM, etc.).
- Temporal locality of the target patterns is not considered.
- Cascade structure is used to utilize a spatial locality of target object for reduction of overall calculation cost in image recognition field.

Based on these findings, cascade classifier is used as the basic structure of the classifier in our study. Cascade classifier is a degenerated decision tree where at each stage, a classifier is trained to detect almost all objects of interest (e.g., typically >99%, frontal faces) while rejecting a certain fraction (typically 50%–60%) of the nonobject patterns [65]. The multiple of the strong classifiers are connected in series. Each strong classifier is comprised of several weak classifiers. Each weak classifier literally classifies the target weakly using single discriminative feature. At the front end of the cascade, simple classifiers that consist of a small number of Haar-like features are formed for coarse filtering as shown in Figure 21.4. Those simple classifiers are then connected to more complex classifiers with a larger number of Haar-like features for fine filtering. A false output at any point of the cascade can result in the immediate conclusion of the evaluation of the instance. Simpler classifiers are used to reject the majority of subwindows before more complex classifiers are called upon to achieve low false-positive rates.

The target signal pattern is usually localized in space and time. In face detection scenario, there are at most a few dozen faces among the 50,000 subwindows in an image as stated in [66]. In speech and nonspeech classification scenario, there were only about 30% of speech signals contained in the typical conversations. Even when speaker continuously reads 10 phoneme-balanced sentences [59], there were about 24 s of nonspeech signals in 45 s of recording time. Moreover, people usually do not talk all day. The same can be stated for human activity recognition problem. Since the target pattern can be considered as a rare event, the classifier should be designed to be able to emphasize the computational resource only on target-like patterns.

As a preliminary study, Ravindran et al. [67] applied the cascade classifier to speech detection. He first compared the cascade classifier with single-stage classifier and showed that classification performance can be improved by using the cascade structure. Using 262 features (256 NRAF [32] + 6 simple features such as volume, fluctuation, width, and skewness of amplitude histogram) as the feature pool, features were selected in *ad hoc* manner where the number of features in each stage was predefined to a certain value. In the study, it is stated that a cascade of classifiers approach gives high accuracy while reducing the computation time and power by allocating resources proportional to the classification difficulty of the example being considered. As for the choice of classifier, Ravindran et al. [67] and Matsuda et al. [68] compared the boundary-based classifier using AdaBoost against distribution-based classifier such as GMM and obtained comparable results in simple sound classification.

In our work, cascade structure is employed with Haar-like features to construct versatile recognition classifier. By changing the LUTs of the weak classifiers and types of features to use, common classifier design can be adapted to perform different recognitions using different types of sensor signals.

In order to construct the classifier, discriminative features must be selected from the pool of over-complete features [72]. For this purpose, Real AdaBoost [76] is employed as the feature selection method as shown in Figure 4-2. This algorithm is an improved version of Discrete AdaBoost [75] proposed in the work of Yoav Freund and Robert E. Schapire.

AdaBoost is an adaptive boosting algorithm in which the rule for combining the weak classifiers adapts on the problem. Real AdaBoost algorithm deals with a confidence-rated weak classifier that is a map from a sample space to a real-valued space instead of prediction as adopted in [77].

Each stage in the cascade is a linear combination of the confidence-rated weak classifiers. Because each confidence value in each bin of the feature space is stored as a LUT, it is often called as LUT-based weak classifier. To represent the distributions of positive and negative data, the domain space of the feature value is evenly partitioned into disjoint bins as in Figure 4-3. Each bin has a real-valued confidence, which is calculated according to the ratio of the training data input to the bin. The weight of each sample is updated using the LUT of the selected feature. The positive samples with high confidence result in lower weight, whereas those with low confidence result in higher weight. In this way, training samples, which are difficult to distinguish using the previously selected features, are emphasized. For each round of boosting, sample distribution is calculated using the updated weights, and new feature is selected using the updated distribution. After selecting T, features for the stage classifier j in the cascade can give the confidence value $\mathrm{Conf}_j(x)$ by

$$\mathrm{Conf}_j(x) = \sum_{t=1}^{T} h_t(x) - \theta_j \qquad (21.26)$$

where θ_j is a threshold. If $\mathrm{Conf}_j(x)$ is positive, then the input data are passed onto the next classifier. If the value is negative, classification is terminated immediately and detected as negative.

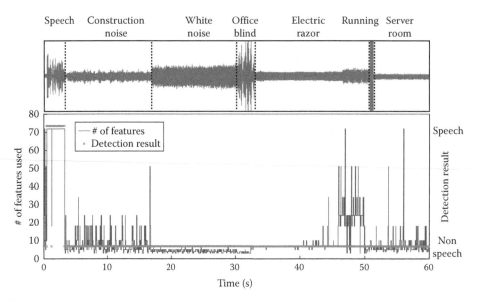

FIGURE 21.11 Continuous classification of speech and nonspeech using cascaded classifier.

Figure 21.11 shows the result of continuous classification of speech and nonspeech using cascaded classifier. During the speech inputs, more features are used to distinguish as compared with nonspeech inputs. Considering the long-term sensing, cascaded classification is suitable for sensornet applications.

21.4.2 Positive Estimation

In sensor network applications, the sensor nodes continuously sense the signals for long periods of time to provide a pervasive service without any human interventions. However, the target pattern to be recognized is typically localized in the space and time. For example, among the millions of image patches generated from an input image, only very few contain faces. For a speech input, even when speaker continuously reads 10 phoneme-balanced sentences [59], there were about 24 s of nonspeech signals in 45 s of recording time. Moreover, people usually do not talk all day. The same can be stated for human activity recognition problem. In cascaded classifier, classifier is designed to reject the majority of negative instances using simple classifiers before more complex classifiers are called to achieve low false-positive rates. By focusing attention on the promising regions of image, the algorithm reduced the calculation cost dramatically.

However, the conventional cascaded classifier requires computations by using full classifiers to finalize the positive detection as shown in Figure 4-5(a) [71–73], because each stage is trained to detect 99% of the positive instances and reject 50% of the negative instances. The subsequent stages neglect how well the detection is made by the previous stages. However, a true output with enough confidence does not result in a conclusion until the full classifiers are used in the conventional cascade.

In this aspect, the accumulated confidence value is proposed to focus the training on the difficult samples. The accumulated confidence value at the classifier k in the cascade is defined as

$$ACV_k(x) = \sum_{j=1}^{k} \text{Conf}_j(x) \tag{21.27}$$

By setting threshold θ_k^{PE} according to the maximum $ACV_k(x)$ of the negative samples as shown in Figure 4-6, the training samples for the subsequent classifiers are focused only in the range of $[\theta_j, \theta_j^{\mathrm{PE}}]$, where the detection is most difficult.

During the detection, when the input data have $ACV_k(x) > \theta_k^{\mathrm{PE}}$, then the detection is terminated to finalize the evaluation. We call this method as PE as depicted in Figure 4-5(b). In this way, the sensor node can compute thoroughly only on the difficult samples and terminates the classification process immediately on the relatively easy samples. The detailed training procedure of the algorithm is explained in Figure 21.12.

Figure 21.13 shows the calculation cost using integral signals and Δ-integral signals in Haar-like feature extraction. The use of the integral signal reduced 54% of the calculation cost. The use of Δ-integral signals reduced additional 7.3%. However, the effect depends on the combination of selected features. Most of the positive samples are detected positive as the final decision in the early stages of the classification by using PE. The use of PE reduces another 74.8% of the calculation cost.

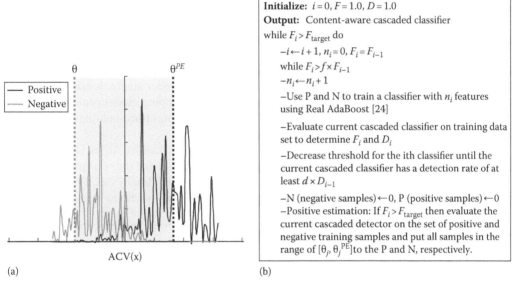

Initialize: $i = 0, F = 1.0, D = 1.0$
Output: Content-aware cascaded classifier
while $F_i > F_{\mathrm{target}}$ do
 $-i \leftarrow i+1, n_i = 0, F_i = F_{i-1}$
 while $F_i > f \times F_{i-1}$
 $-n_i \leftarrow n_i + 1$
 −Use P and N to train a classifier with n_i features using Real AdaBoost [24]
 −Evaluate current cascaded classifier on training data set to determine F_i and D_i
 −Decrease threshold for the ith classifier until the current cascaded classifier has a detection rate of at least $d \times D_{i-1}$
 −N (negative samples) ← 0, P (positive samples) ← 0
 −Positive estimation: If $F_i > F_{\mathrm{target}}$ then evaluate the current cascaded detector on the set of positive and negative training samples and put all samples in the range of $[\theta_j, \theta_j^{\mathrm{PE}}]$ to the P and N, respectively.

(a)   (b)

FIGURE 21.12 Cascaded classifiers using PE. (a) Accumulated confidence value. (b) Training algorithm using positive estimation.

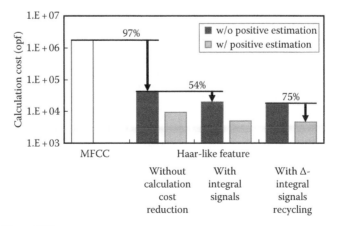

FIGURE 21.13 Effect of PE.

21.4.3 REDUNDANT FEATURE SELECTION

In conventional method, the features already selected by the previous classifiers from the feature pool are omitted from the training of the subsequent classifiers. In the training process, new negative training samples are provided by bootstrapping the data sets, and weights of training samples are initialized for each layer. Therefore, each classifier in the cascade can be considered as a different classifier with relatively loose correlation. Hence, discriminative features that are already selected in the previous classifiers of cascade should not be omitted from the successive training process. In other words, feature space can be shared among all the strong classifiers in the cascade.

RFS incorporates features that are already selected in the previous classifiers. RFS allows each classifier in the cascade to share some features and reduce the calculation cost-effectively while maintaining the recognition performance.

21.4.4 DYNAMIC LOOKUP TABLE

Each stage of the cascaded classifier is made up of a linear combination of weak classifiers. Figure 21.14 shows various representations of the weak classifier. At the first proposal of the cascaded classifier, Viola et al. employed a simple single thresholding (or decision stump) as a weak classifier. The decision stump determines the optimal threshold classification function, so that the minimum numbers of examples are misclassified. It is too simple to fit complex distributions and intraclass differences of samples are ignored. In later empirical studies, Overett et al. showed that more than 90% of features are nondiscriminative under single threshold discrimination. This is usually due to multimodal distributions or overlapping modal peaks [83]. As the related works, Kim et al. [84] and Rasolzadeh [85] proposed multithresholding to model the feature response.

Huang et al. [86] proposed the use of a real-valued LUT weak classifier. To build a LUT-based weak classifier, the feature space is evenly partitioned into subranges, and the real value confidence is calculated on each bin. Generally, histogram is considered as poor density estimator. Histogram suffers from the binning problem, that is, a tendency to underfit and overfit the true distribution at

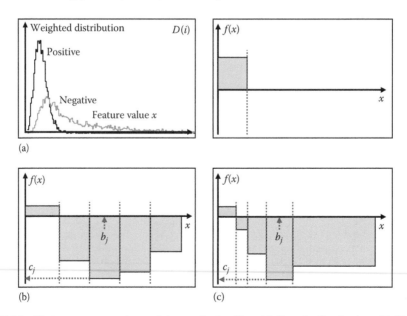

FIGURE 21.14 Various representations of the weak classifier. (a) Sample distribution. (b) Fixed lookup table. (c) Dynamic lookup table.

regions of high and low density, respectively. Overfitting is caused by the lack of training data falling in certain histogram bins, while underfitting occurs where there are too few bins to model the changes in positive and negative distributions [87,88]. Since the number of bins is fixed to a certain number with an equal partition width, we call this method as fixed lookup table (FLUT). However, when training the cascaded classifier, the complexity of the training data can change by which features are selected at each stage.

Le et al. [89] proposed that *Ent-Boost* algorithm that employs the relative entropy between probability distributions of positive and negative samples is used to select the best weak classifier. However, the selection of weak classifier is not globally optimal. Xiao et al. [90] proposed *Bayesian stump* that merges adjacent bins with similar value. This method is also suboptimal in binning procedure. To mitigate the problem of binning procedure, DLUT is proposed.

Suppose that $X_0, X_1,\ldots, X_N$ is a partition of the Haar-like feature value domain Z. Given a data set $\{(x_1, y_1),\ldots,(x_m, y_m)\}$, where $x_i \in X$ and class label $y_i \in \{-1, 1\}$, the weighted fraction of examples that fall in bin j with label l can be calculated as

$$W_l^j = \sum_{i:x_i \in X_j \wedge y_i = l} D(i) = \Pr_{i \sim D}\left[x_i \in X_j \wedge y_i = l\right], \qquad (21.28)$$

where $D(i)$ is the sample distribution. Then, each prediction in the bin j can be given by

$$c_j = \frac{1}{2}\log\left(\frac{W_{+1}^j + \varepsilon}{W_{-1}^j + \varepsilon}\right)$$

where ε denotes a smoothing value and its typical choice is approximately $1/m$ where m is the number of training examples [91]. Given the input data and its feature value x, the weak classifier output $f(x)$ maps feature value x to $\{c_1, \ldots, c_N\}$. During the feature selection process, the feature that yields the weak classifier with the smallest train loss is selected. The train loss Z for each weak classifier is given by

$$Z = \sum_j \sum_{i:x_i \in X_j} D(i)\exp(-y_i c_j)$$

$$= \sum_j \left(W_{+1}^j \exp(-c_j) + W_{-1}^j \exp(c_j)\right) \qquad (21.29)$$

By plugging Equation 21.28 into 21.29, Z can be viewed as a smoothed Bhattacharyya distance [81], which measures a discriminative property of the feature value by approximating the amount of overlap between positive and negative sample distributions.

One important property is that the train loss Z depends only on which partition X_j a given instance falls into. Therefore, the train loss is expressed as a sum of the train loss for each partition X_j. The domain partitioning based on dynamic programming requires its cost function to be expressed as a sum of contributions, each depending only on a single partition [92,93]. Hence, the train loss function can be directly incorporated as the cost function of dynamic programming based domain partitioning to construct DLUT.

Supposing the bit width of the feature space is $b_{f(x)}$, we have $M = 2^{b_{f(x)}}$ possible number of bins in the histogram of the feature space. Let $th_0 < th_1 < \cdots < th_K$ be a sequence of thresholds partitioning

Haar-like feature space, where $th_0 = 0$ and $th_K = M$. If $H(th_{j-1}, th_j)$ is the measure of the quality of the particular interval X_j, then the overall measure of the partitions is given by

$$H\left(th_0, \ldots, th_K\right) = \sum_{j=1}^{K} H\left(th_{j-1}, th_j\right) + n\beta \tag{21.30}$$

where β is the penalty for each interval in the partition, enabling the domain partitioning to choose as few numbers of partitions as possible.

Let $H^*(th_b)$ be the score of the optimal partition of the signals $(0, th_b)$. The optimal partition of $(0, th_b)$ must either be the unpartitioned interval or be composed of the optimal partition of $(0, th_a)$ with the additional interval (th_a, th_b), where $th_a < th_b$. The dynamic programming procedure can be expressed by

$$H^*\left(th_b\right) = \min_{th_a}^{th_b - 1} H^*\left(th_a\right) + H\left(th_a, th_b\right) + \beta \tag{21.31}$$

where $H^*(0) = 0$. The quality measure of each partition is given by train loss

$$H\left(th_{j-1}, th_j\right) = \sum_{i:x_i \in (th_{j-1}, th_j)} D(i)\exp(-y_i c_j') \tag{21.32}$$

where each prediction in the partition (th_{j-1}, th_j) is given by

$$c_j' = \frac{1}{2}\log\left(\frac{\sum_{i:x_i \in (th_{j-1}, th_j)^\wedge y_i = +1} D(i) + \varepsilon}{\sum_{i:x_i \in (th_{j-1}, th_j)^\wedge y_i = -1} D(i) + \varepsilon}\right) \tag{21.33}$$

In this way, DLUT can be constructed using dynamic programming to select the optimal number of bins by globally optimizing the train loss function. The optimal solution of this problem can be found by using dynamic programming with the computational complexity of $O(KM^2)$ [94,95]. This complexity does not affect the dedicated hardware, because the training phase is supposed to be executed on the PC. However, given the large M, overall training time is largely increased since the dynamic programming-based domain partitioning must be carried out for every feature selection. Therefore, FLUT of $L \ll M$ bins is generated as an intermediate representation to speed up the optimization process when training the cascaded classifier. The computational complexity is reduced to $O(KL^2)$. In the experiments, L is set to 100.

The effect of using DLUT on the cascaded classifier is evaluated based on the data size of the trained classifier. The data size of one FLUT can be given by

$$s_{\text{FLUT}} = Kb_{\text{output}} + 2b_{\text{threshold}} \tag{21.34}$$

where

    K is the number of bins
    b_{output} represents the bit width of output constants
    $b_{\text{threshold}}$ denotes the bit width of thresholds

FLUT is bounded by the minimum and maximum feature values. The data size of one DLUT can be represented by

$$s_{\text{DLUT}} = Kb_{\text{output}} + \left(K - 1\right)b_{\text{threshold}} \tag{21.35}$$

where $b_{\text{threshold}}$ denotes the bit width of the thresholds partitioning the feature value domain. b_{output} and $b_{\text{threshold}}$ are set to 1 to evaluate the normalized data size for a comparison, because the bit width can be empirically optimized.

21.4.5 PERFORMANCE EVALUATION

Figures 21.15 and 21.16 compare the maximum accuracy and the calculation cost of the proposed methods with various conventional methods in speech/nonspeech classification and human activity recognition.

In human activity recognition, our method achieved the highest accuracy of 96.1% with 2.65 kopf using RFS, which is the lowest calculation cost among the methods with recognition accuracy over 90.0%. As shown in Figure 21.15, the calculation cost is reduced by 84% (= (16.61 − 2.64)/16.61 * 100%) in terms of calculation cost relative to the conventional state-of-the-art method based on *MS* with C4.5 decision tree classifier. By using basic 1-D Haar-like feature (H), accuracy of 95.8% is achieved with calculation cost of 3.62 kopf. By adding BH to the feature pool, the calculation cost dropped to 3.21 kopf with the slight increase in the accuracy. By also adding MEH to the feature pool, the calculation cost is further dropped to 3.11 kopf with slightly increased accuracy of 96.1%. By integrating the basic statistical features into the Haar-like feature extraction framework, the calculation cost is reduced without degradation in recognition accuracy.

In speech/nonspeech classification accuracy, the proposed methods achieved the calculation cost of 12.24 kopf with 96.9% accuracy. The result is compared with the state-of-the-art sound feature extraction method called MFCC [55] and other conventional methods such as linear frequency cepstrum coefficient (LFCC) [96], linear frequency power coefficient (LFPC) [97], and LPCC [55]. For MFCC, each signal is multiplied with a 20 ms Hamming window, with an overlap of 10 ms. Two-hundred fifty-six points of FFT is computed on each window, and the magnitude spectrum is filtered with a bank of triangular filters spaced linearly on the mel scale. When mel scale is not used, LFCC can be extracted. The log-compressed filter outputs are converted into cepstral coefficients by discrete cosine transform (DCT). When DCT is not applied on LFCC, LFPC is extracted instead. In addition, we also compared with the Walsh–Hadamard Transform, which is the square wave equivalent of the FFT. For the comparison, we used fast Walsh–Hadamard transform (FWT) to extract the feature value [98]. For the classifier, LBG vector quantization-based classifier [36] with a model size of 4 to 512 is used [28], and the maximum accuracy is plotted.

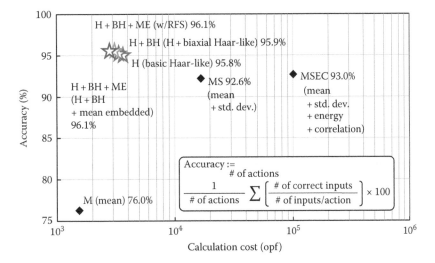

FIGURE 21.15 The comparison of human activity recognition performance among different features.

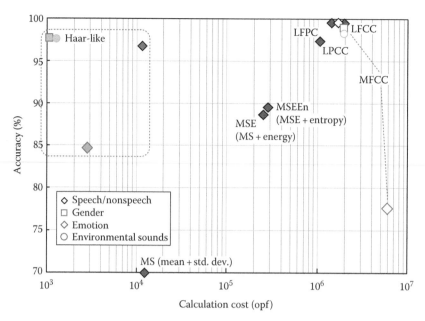

FIGURE 21.16 The comparison of speech/nonspeech classification among different features.

MFCC resulted in the highest speech/nonspeech classification accuracy of 99.8% with the calculation cost of 1880 kopf. In addition, FFT, DCT, and filter bank analysis require large amount of memory for sine and cosine LUT and filter bank coefficients. When the input frame length is 0.1 s, nine windows are computed. LFCC yielded 99.9% accuracy with 1940 kopf. LFPC yielded 99.9% accuracy with 1665 kopf. LPCC resulted in the accuracy of 97.5% with 1176 kopf. FWT yielded 88.0% accuracy with 140 kopf.

The methods based on the conventional human activity recognition studies are also evaluated. *MS* achieves only 68.8%, *MSE* achieves 88.2%, and *MSEEn (MSE + Entropy)* achieves 89.2% speech/nonspeech classification accuracy. By using the proposed Haar-like feature with RFS, the accuracy of 96.9% with 12.2 kopf is achieved. This is only 2.9% error relative to MFCC, while the calculation cost is 99% (= (1880 − 12.2)/1880 * 100%) lower.

For gender recognition, the accuracy of 97.5% (relative accuracy, 98%) is achieved with 99% efficient calculation cost. The input frame length is set to 0.1 s. For emotion recognition, emotional-versus-neutral speech classifier on each class of emotion (sadness, happiness, and anger) is trained. As a result, 84.6% is for accuracy (relative accuracy, 109%), while 99% is for efficiency. This accuracy improvement can be considered as a result of flexibility in the choice of features. For environmental sound recognition, target-versus-nontarget classifier is trained on each of 21 sounds with the frame length of 0.1 s. The accuracy of 97.3% is achieved (relative accuracy, 98%). For most of recognition problems, only small degradation in accuracy is observed while achieving high efficiency in the computational cost.

Table 21.3 shows the comparison of the sizes of the cascaded classifiers between FLUT- and DLUT-based weak classifier. DLUT achieves 7.7 bins on average, which is smaller than 20.0 bins FLUT at maximum speech/nonspeech classification accuracy of 97.1%. The size of the cascaded classifier is reduced by 54.3% (= (1166 − 533)/1166 * 100%) using DLUT as compared with FLUT. The penalty value β for each interval in the partition in Equation 21.31 is experimentally found to be 0.02. As a result, DLUT enabled the cascaded classifier to be more compact than FLUT-based conventional classifier.

TABLE 21.3
Classifier Size Comparison (Speech/Nonspeech Classificaiton)

	Size	Bin$_{AVE}$	# of features	Accuray (%)	Calculation cost [kopf]
FLUT	1166	20.0	53	96.9	12.2
DLUT	533	7.7	37	97.1	11.7

21.5 CONCLUSIONS AND DISCUSSIONS

A versatile recognition algorithm processes image, sound, and 3-D acceleration signals with the common framework at low calculation cost. There are two main contributions.

Firstly, 1-D Haar-like feature is proposed to roughly estimate frequency information of the temporal signals. In addition, biaxial and mean-embedded Haar-like features are proposed to extract standard deviation and interaxial correlation from 3-D acceleration signals in Haar-like feature framework.

Secondly, three techniques are proposed to build the compact cascaded classifier. The cascaded classifier with PE is introduced to allow a sensor node to be able to compute finely only when the inputs are target-like and difficult to recognize and stop computing when inputs obtain enough confidence. RFS incorporates the features that are already selected in the previous stage classifiers to reduce the calculation cost. In addition, DLUT is proposed to construct the LUT-based weak classifiers with the smallest possible number of bins by globally optimizing the train loss function.

Our algorithm was tested on sound recognition and human activity recognition and outperformed conventional feature extraction methods. The versatile recognition algorithm was also applied on face detection and achieved the detection rate of 81% on the false-positive rate/image of 0.05. This is comparable result with OpenCV, a notable computer vision library [99].

Our versatile recognition algorithm has been employed to build a versatile recognition processor, which is the first solution to perform multiple recognition tasks while dissipating sub-mWs per frames per second. The processor is fabricated in a 90 nm CMOS technology. It runs at 54 MHz clock frequency with a 0.9 V supply voltage. For speech/nonspeech classification using Haar-like features and the cascaded classifier, the power consumption per frame is 0.28 µJ/frame. For human activity recognition, the power consumption per frame rate is 0.15 µW/fps. Because the sensor nodes need versatility for various types of the input signals with limited battery power, the versatile recognition algorithm can be considered as a suitable solution.

REFERENCES

1. I. Akyildiz, W. Su, Y. Sankarasubramaniam, and E. Cayirci, Wireless sensor networks: A survey, *Computer Networks*, 38(4), 393–442, March 2002.
2. Th. Arampatzis, J. Lygeros, and S. Manesis, A survey of applications of wireless sensors and wireless sensor networks, *IEEE International Symposium on Intelligent Control*, Limassol, Cyprus, June 27–29, 2005, pp. 719–724.
3. I.F. Akyildiz, T. Melodia, and K.R. Chowdhury, A survey on wireless multimedia sensor networks, *Computer Networks*, 51, 921–960, 2007.
4. I.F. Akyildiz, T. Melodia, and K.R. Chowdhury, Wireless multimedia sensor networks: A survey, *IEEE Wireless Communications*, 14(6), 32–39, December 2007.
5. B. Schilit, N. Adams, and R. Want, Context-Aware computing applications, in *Proceedings of IEEE Workshop on Mobile Computing Systems and Application*, Santa Cruz, California, December 8–9, 1994, pp. 85–90.

6. A.J. Eronen, V.T. Peltonen, J.T. Tuomi, A.P. Klapuri, S. Fagerlund, T. Sorsa, G. Lorho, and J. Huopaniemi, Audio-based context recognition, *IEEE Transactions on Audio, Speech, and Language Processing*, 14(1), 321–329, January 2006.

7. J. Nishimura, N. Sato, and T. Kuroda, Speech "Siglet" detection for business microscope, in *Proceedings of IEEE Pervasive Computing and Communication*, Hong Kong, March 17–21, 2008, pp. 147–152.

8. J. Nishimura, N. Sato, and T. Kuroda, Speaker siglet detection for business microscope, in *Proceedings of AMLA/IEEE International Conference on Machine Learning and Applications*, San Diego, CA, December 11–13, 2008, pp. 376–381.

9. J. Nishimura and T. Kuroda, Speaker recognition using speaker-independent universal acoustic model and synchronous sensing for business microscope, in *Proceedings of International Conference on Wireless Pervasive Computing*, Melbourne, Australia, December 11–13, 2009, pp. 1–5.

10. K. Ara, N. Kanehira, D.O. Olguin, B.N. Waber, T. Kim, A. Mohan, P. Gloor et al., Sensible organizations: Changing our businesses and work styles through sensor data, *Journal of Information Processing*, 16, 1–12, April 2008.

11. K. Yano and H. Kuriyama, Human x sensor: How sensor information will change human, organization and society, *Hitachi Hyouron*, 89(07), 62–67, 2007.

12. S. Escalera, R.M. Martinez, J. Vitria, P. Radeva, and M.T. Anguera, Dominance detection in face-to-face conversations, in *Proceedings of IEEE Computer Vision and Pattern Recognition*, June 20–25, 2009, pp. 97–102.

13. A.R. Doherty and A.F. Smeaton, Combining face detection and novelty to identify important events in a visual, in *Proceedings of IEEE International Conference on Computer and Information Technology Workshops*, Sydney, Australia, July 8–11, 2008, pp. 348–353.

14. E.H. Spriggs, F. De La Torre, and M. Hebert, Temporal segmentation and activity classification from first-person sensing, in *Proceedings of IEEE Computer Vision and Pattern Recognition Workshops*, June 20–25, 2009, pp. 17–24.

15. L. Han, Z. Li, H. Zhang, and D. Chen, Wearable observation supporting system for face identification based on wearable camera, in *Proceedings of IEEE International Conference on Computer Science and Information Technology*, Chengdu, China, July 9–11, 2010, Chengdu, China, pp. 91–95.

16. M. Chan, E. Campo, and D. Esteve, Monitoring elderly people using a multisensory system, in *Proceedings of Second International Conference on Smart Homes and Health Telematic*, 2004, pp. 162–169.

17. D.V. Anderson and S. Ravindran, Distributed acquisition and processing systems for speech and audio, in *Proceedings of Forty-Fourth Annual Allerton Conference*, Monticello, September 27–29, 2006, pp. 1150–1154.

18. R.V. Kulkarni, A. Förster, and G.K. Venayagamoorthy, Computational intelligence in wireless sensor networks: A survey, *IEEE Communications Survey and Tutorials*, 13(1), 68–96, 2011.

19. G.J. Pottie and W.J. Kaiser, Wireless integrated network sensors, *Communications of the ACM*, 43(5), 51–58, May 2000.

20. R. Kleihorst, B. Schueler, and A. Danilin, Architecture and applications of wireless smart cameras, in *Proceedings of International Conference on Acoustics, Speech and Signal Processing*, Honolulu, vol. IV, April 15–20, 2007, pp. 1373–1376.

21. D. Wyatt, J. Bilmes, H. Kautz, and T. Choudbury, A privacy-sensitive approach to modeling multi-person conversations, in *Proceedings of International Joint Conferences on Artificial Intelligence*, January 6–12, 2007, Hyderabad, India, pp. 1769–1775.

22. D. Wyatt, J. Bilmes, T. Choudbury, and J.A. Kitts, Towards the automated social analysis of situated speech data, in *Proceedings of ACM International Conference on Ubiquitous Computing*, Seoul, South Korea, September 21–24, 2008, pp. 168–171.

23. D. Zhao, H. Ma, and L. Liu, Event classification for living environment surveillance using audio sensor networks, in *Proceedings of IEEE International Conference on Multimedia and Expo*, July 19–23, 2010, Singapore, pp. 528–533.

24. H. Kwon, H. Krishnamoorthi, V. Berisha, and A. Spanias, A sensor network for real-time acoustic scene analysis, in *Proceedings of IEEE International Symposium on Circuits and Systems*, May 24–27, Taipei, Taiwan, 2009, pp. 169–172.

25. H. Kwon, V. Berisha, and A. Spanias, Real-time sensing and acoustic scene characterization for security applications, in *Proceedings of IEEE International Symposium on Wireless Pervasive Computing*, Santorini, Greece, May 7–9, 2008, pp. 755–758.

26. T. Bocklet, A. Maier, J.G. Bauer, F. Burkhardt, and E. Nöth, Age and gender recognition for telephone applications based on GMM supervectors and support vector machines, in *Proceedings of IEEE International Conference on Acoustics, Speech, and Signal Processing*, Las Vegas, March 31–April 4, 2008, pp. 1605–1608.

27. T. Kinnunen, E. Chernenko, M. Tuononen, P. Fränti, and H. Li, Voice activity detection using MFCC features and support vector machine, in *Proceedings of Speech and Computer*, vol. 2, Moscow, Russia, October 15–18, 2007, pp. 556–561.
28. N. Sato and Y. Obuchi, Emotion recognition using Mel-frequency cepstrum coefficients, *Journal of Natural Language Processing*, 14(4), 83–96, 2007.
29. T. Kinnunen, E. Karpov, and P. Fränti, Real-time speaker identification and verification, *IEEE Transactions on Audio, Speech, Language Process*, 14(1), 277–288, January 2006.
30. P. Lukowicz, J.A. Ward, H. Junker, M. Stäger, G. Tröster, A. Atrash, and T. Starner, Recognizing workshop activity using body worn microphones and accelerometers, in *Proceedings of International Conference on Pervasive Computing*, Linz, Austria, April 21–23, 2004, pp. 18–22.
31. D. Istrate, E. Castelli, M. Vacher, L. Besacier, and J.-F. Serignat, Information extraction from sound for medical telemonitoring, *IEEE Transactions on Information Technology in Biomedicine*, 10(2), 264–274, April 2006.
32. S. Chu, S. Narayanan, and C.C. Jay Kuo, Environmental sound recognition with time–frequency audio features, *IEEE Transactions on Audio, Speech, and Language Processing*, 17(6), 1142–1158, August 2009.
33. C. Parker, An empirical study of feature extraction methods for audio classification, in *Proceedings of International Conference on Pattern Recognition*, Istanbul, Turkey, August 23–26, 2010, pp. 4593–4596.
34. D.A. Reynolds et al., Speaker verification using adapted Gaussian mixture models, *Digital Signal Processing*, 10, 19–41, 2000.
35. S. Ravindran, D. Anderson, and M. Slaney, Low-power audio classification for ubiquitous sensor networks, in *Proceedings of IEEE International Conference on Acoustics, Speech, and Signal Processing*, vol. 4, Montreal, Quebec, Canada, May 17–21, 2004, pp. 337–340.
36. Y. Linde, A. Buzo, and R.M. Gray, An algorithm for vector quantizer design, *IEEE Transactions on Communications*, 20, 84–95, 1980.
37. P. Nurmi, P. Floréen, M. Przybilski, and G. Lindén, A framework for distributed activity recognition in ubiquitous systems, in *Proceedings of International Conference on Artificial Intelligence*, Las Vegas, NV, June 27–30, 2005, pp. 650–655.
38. N. Ravi, N. Dandekar, P. Mysore, and M.L. Littman, Activity recognition from accelerometer data, in *Proceedings of AAAI Conference on Artificial Intelligence*, Pittsburg, PA, July 9–13, 2005, pp. 1541–1546.
39. T. Huynh and B. Schiele, Analyzing features for activity recognition, in *Proceedings of Joint Conference on Smart Objects Ambient Intelligence*, Grenoble, France, October 12–14, 2005, pp. 159–163.
40. L. Bao and S. Intille, Activity recognition from user-annotated acceleration data, in *PERVASIVE*, Vienna, Austria, April 21–23, 2004, pp. 1–17.
41. P. Viola and M.J. Jones, Robust real-time face detection, *International Journal of Computer Vision*, 57(2), 137–154, 2004.
42. M. Siala, N. Khlifa, F. Bremond, and K. Hamrouni, People detection in complex scene using a cascade of boosted classifiers based on Haar-like-features, *IEEE Intelligent Vehicle Symposium*, Shaanxi, China, June 3–5, 2009, pp. 83–87.
43. G. Shakhnarovich, P. Viola, and B. Moghaddam, A unified learning framework for real time face detection and classification, in *Proceedings of IEEE International Conference on Automatic Face and Gesture Recognition*, Washington, D.C., May 21, 2002, pp. 14–21.
44. X. Zhao, X. Chai, and Z. Niu, Context constrained facial landmark localization based on discontinuous Haar-like feature, in *Proceedings of IEEE International Conference on Automatic Face and Gesture Recognition*, California, March 21–25, 2011, pp. 673–678.
45. Y. Konishi, K. Kinoshita, S. Lao, and M. Kawade, Real-time estimation of smile intensities, *IEEE Asian Conference on Computer Vision* (demo), 2007.
46. S.-Uk. Jung, D.H. Kim, K.H. An, and M.J. Chung, Efficient rectangle feature extraction for real-time facial expression recognition based on AdaBoost, in *Proceedings of IEEE/RSJ International Conference on Intelligent Robots and Systems*, Edmonton, Alberta, Canada, August 2–6, 2005, pp. 1941–1946.
47. J. Nishimura and T. Kuroda, Low cost speech detection using Haar-like filtering for sensornet, in *Proceedings of IEEE International Conference on Signal Processing*, Beijing, China, vol. 3, October 26–29, 2008, pp. 2608–2611.
48. J. Nishimura and T. Kuroda, Haar-like filtering based speech detection using integral signal for sensornet, in *Proceedings of International Conference on Sensing Technology*, Tainan, Taiwan, November 30–December 3, 2008, pp. 52–56.

49. J. Nishimura and T. Kuroda, Haar-like filtering with center-clipped emphasis for speech detection in sensornet, in *Proceedings of IEEE DSP/SPE Workshop*, Marco Island, FL, January 4–7, 2009, pp. 1–4.
50. Y. Hanai, J. Nishimura, and T. Kuroda, Haar-like filtering for human activity recognition using 3-D accelerometer, in *Proceedings of IEEE DSP/SPE Workshop*, Marco Island, FL, January 4–7, 2009, pp. 675–678.
51. J.W. Pitton, K. Wang, and B.-H. Juang, Time-frequency analysis and auditory modeling for automatic recognition of speech, *Proceedings of the IEEE*, 84(9), 1199–1125, September 1996.
52. J.W. Picone, Signal modeling techniques in speech recognition, *Proceedings of the IEEE*, 81(9), 1215–1247, September 1993.
53. J.D. Hoyt and H. Wechsler, Detection of human speech in structured noise, in *Proceedings of IEEE International Conference on Acoustics, Speech, and Signal Processing*, Adelaide, South Australia, Australia, vol. 2, April 19–22, 1994, pp. 237–240.
54. Y. Tian, Z. Wang, and D. Lu, Nonspeech segment rejection based on prosodic information for robust speech recognition, in *Proceedings of IEEE Signal Processing Letters*, 9(11), 364–337, 2002.
55. S.B. Davis and P. Mermelstein, Comparison of parametric representations of monosyllabic word recognition in continuously spoken sentences, *IEEE Transactions on Speech Audio Processing*, 28, 357–366, 1980.
56. Y. Higashijima, S. Takano, and K. Niijima, Face recognition using long Haar-like filter, in *Proceedings of Image and Vision Computing*, Dunedin, New Zealand, November 28–29, 2005, pp. 43–48.
57. X. Cui, Y. Liu, S. Shan, X. Chen, and W. Gao, 3D Haar-like features for pedestrian detection, in *Proceedings of IEEE International Conference on Multimedia & Expo*, July 2–5, 2007, Beijing, China, pp. 1263–1266.
58. T. Kobayashi, S. Itahashi, S. Hayamizu, and T. Takezawa, ASJ continuous speech corpus for research, *Journal of the Acoustical Society of Japan*, 48, 888–893, 1992.
59. NOISEX-92 [Online]. http://spib.rice.edu/spib/select_noise.html.
60. M. Liberman, K. Davis, K. Grossman, N. Martey, and J. Bell, Emotional prosody speech and transcripts [Online]. http://www.ldc.upenn/edu/Catalog/CatalogEntry.jsp?catalogId = LDC2002S28.
61. A.D. Booth, A signed binary multiplication technique, *Quarterly Journal of Mechanics and Applied Mathematics*, 4(2), 236–240, 1951.
62. T. Choudhury, M. Philipose, D. Wyatt, and J. Lester, Towards activity databases: Using sensors and statistical models to summarize people's lives, *IEEE Data Engineering Bulletin*, 29, 49–58, 2006.
63. N. Kern, B. Schiele, H. Junker, P. Lukowicz, and A. Schmidt, Context annotation for a live life recording, in *Proceedings of Pervasive, Workshop on Memory and Sharing of Experiences*, Vienna, Austria, April 20, 2004.
64. J. Lester, T. Choudhury, and G. Borriello, A practical approach to recognizing physical activities, *Pervasive Computing*, 3968, 1–16, 2006.
65. I.H. Witten and E. Frank, *Data Mining: Practical Machine Learning Tools and Techniques*, 2nd edn., Morgan Kaufmann, Burlington, MA, 2005.
66. U. Maurer, A. Smailagic, D.P. Siewiorek, and M. Deisher, Activity recognition and monitoring using multiple sensors on different body positions, in *Proceedings of International Workshop on Wearable and Implantable Body Sensor Networks*, April 3–5, 2006, pp. 113–116.
67. ATR-Promotions [Online]. http://www.atr-p.com/sensor01.html.
68. M. Beigl, A. Krohn, T. Zimmer, and C. Decker, Typical sensors needed in ubiquitous and pervasive computing, in *Proceedings of International Workshop on Networked Sensing Systems*, Tokyo, Japan, June 22–23, 2004, pp. 153–158.
69. D. Olguín Olguín and A. Pentland, Human activity recognition: Accuracy across common locations for wearable sensors, in *Proceedings of International Symposium on Wearable Computers*, Montreux, Switzerland, October 11–14, 2006, pp. 11–13.
70. R. Lienhart, A. Kuranov, and V. Pisarevsky, Empirical analysis of detection cascades of boosted classifiers for rapid object detection, MRL Technical Report, Intel Labs, May 2003.
71. P. Viola and M. Jones, Fast and robust classification using asymmetric AdaBoost and a detector cascade, *Advances in Neural Information Processing Systems*, 14, 1311–1318, 2002.
72. S. Ravindran and D.V. Anderson, Cascade classifiers for audio classification, in *Proceedings of IEEE Digital Signal Processing Workshop*, Taos Ski Valley, 1–4 Aug, 2004, pp. 366–370.
73. H. Matsuda, T. Takiguchi, and Y. Ariki, Voice activity detection with real AdaBoost, in *Proceedings of Acoustical Society of Japan Fall Meeting*, 2006, pp. 117–118.
74. Y. Freund and R.E. Schapire, A decision-theoretic generalization of on-line learning and an application to boosting, *Journal of Computer and System Sciences*, 55, 119–139, 1997.

75. C. Huang, B. Wu, H. Ai, and S. Lao, Omni-directional face detection based on real AdaBoost, in *Proceedings of IEEE International Conference on Image Processing*, October 24–27, 2004, pp. 593–596.

76. B. Wu, H. Ai, C. Huang, and S. Lao, Fast rotation invariant multi-view face detection based on real AdaBoost, in *Proceedings of IEEE International Conference on Automatic Face and Gesture Recognition*, May 17–19, 2004, pp. 79–84.

77. P. Viola and M. Jones, Rapid object detection using a boosted cascade of simple features, in *Proceedings of IEEE Conference on Computer Vision and Pattern Recognition*, December 8–14, 2001, pp. 1–9.

78. S. Yan, S. Shan, X. Chen, and W. Gao, Fea-Accu cascade for face detection, in *Proceedings of IEEE International Conference on Image Processing*, November 7–10, 2009, pp. 1217–1220.

79. J. Wu, S. Brubaker, M.D. Mullin, and J.M. Rehg, Fast asymmetric learning for cascade face detection, *IEEE Transactions on Pattern Analysis and Machine Intelligence*, 30(3), 369–382, March 2008.

80. T. Kailath, The divergence and Bhattacharyya distance measures in signal selection, *IEEE Transactions on Communication Technology*, COM-15(1), 52–60, 1967.

81. G. Xuan, X. Zhu, P. Chai, Y.Q. Shi, and D. Fu, Feature selection based on the Bhattacharyya distance, in *Proceedings of International Conference on Pattern Recognition*, August 20–24, 2006, pp. 1–4.

82. G. Overett and L. Petersson, Improved response modelling on weak classifiers for boosting, in *Proceedings of IEEE International Conference on Robotics and Automation*, April 10–14, 2007, pp. 3799–3804.

83. J.H. Kim, B.G. Kwon, J.Y. Kim, and D.J. Kang, Method to improve the performance of the AdaBoost algorithm by combining weak classifiers, in *Proceedings of International Workshop on Content-Based Multimedia Indexing*, June 18–20, 2008, London, U.K., pp. 357–364.

84. B. Rasolzadeh, L. Petersson, and N. Petersson, Response binning: Improved weak classifiers for boosting, in *Proceedings of IEEE Intelligent Vehicles Symposium*, Tokyo, Japan, June 13–15, 2006, pp. 344–349.

85. C. Huang, H.Z. Ai, Y. Li, and S. Lao, High-performance rotation invariant multi-view face detection, *IEEE Transactions on Pattern Analysis and Machine Intelligence*, 29(4), 671–686, 2007.

86. G. Overett and L. Petersson, On the importance of accurate weak classifier learning for boosted weak classifiers, in *Proceedings of IEEE Intelligent Vehicles Symposium*, June 4–6, 2008, pp. 816–821.

87. G. Overett and L. Petersson, Boosting with multiple classifier families, in *Proceedings of IEEE Intelligent Vehicles Symposium*, October 11–13, 2007, pp. 1039–1044.

88. D. Le and S. Satoh, Ent-boost: Boosting using entropy measure for robust object detection, in *Proceedings of International Conference on Pattern Recognition*, August 20–24, 2006, pp. 602–605.

89. R. Xiao, H. Zhu, H. Sun, and X. Tang, Dynamic cascades for face detection, in *Proceedings of IEEE International Conference on Computer Vision*, October 14–21, 2007, pp. 1–8.

90. R.E. Schapire and Y. Singer, Improved boosting algorithm using confidence-rated predictions, *Machine Learning*, 37, 1999, Kluwer Academic Publishers, pp. 297–336.

91. J. Himberg, K. Korpiaho, H. Mannila, and J. Tikanmaki, Time series segmentation for context recognition in mobile devices, in *Proceedings of IEEE International Conference on Data Mining*, November 29–December 2, 2001, pp. 203–210.

92. E. Nichols and C. Raphael, Globally optimal audio partitioning, in *Proceedings of International Society for Music Information Retrieval Conference*, Victoria, British Columbia, Canada, October 8–12, 2006, pp. 202–205.

93. B. Jackson et al., An algorithm for optimal partitioning of data on an interval, *IEEE Signal Processing Letters*, 12(2), 105–108, 2005.

94. R. Bellman, On the approximation of curves by line segments using dynamic programming, *Communications of the ACM*, 4(6), 284, 1961.

95. S.G. Koolagudi, S. Nandy, and K.S. Rao, Spectral features for emotion classification, in *Proceedings of IEEE International Advance Computing Conference*, Patiala, India, March 6–7, 2009, pp. 1292–1296.

96. T.L. New, S.W. Foo, and L.C. De Silva, Detection of stress and emotion in speech using traditional and FFT based log energy features, in *Proceedings of International Conference on Information, Communications and Signal Processing*, December 15–18, 2003, pp. 1619–1623.

97. W. Ouyang and W. Cham, Fast algorithm for Walsh Hadamard transform on sliding windows, *IEEE Transactions on Pattern Analysis and Machine Intelligence*, 32(1), 165–171, 2010.

98. Open CV library [Online]. http://www.opencv.org/.

22 Informatics of Remote RF Sensing

John Kosinski

CONTENTS

22.1 INTRODUCTION

Most efforts in improving sensors are focused on the technical aspects of sensors involving device physics, engineering design, materials science, fabrication technology, etc. The tasks of sensor data processing and sensor data fusion are handled as a separable problem set involving signals, signal processing, information, and information processing, and in many cases, these are handled as a separable problem set from the ultimate assessment and logical processing of the output data. We consider here the fullness of the remote sensing problem and the relationships and interactions between the various levels of the underlying information problem, with examples from the field of remote RF sensing. We consider the *informatics* of sensors and sensor systems, taking the framework of communication elucidated by Shannon and Weaver as a launching point and identify additional elements specific to the sensor problem that must be joined into the framework. In particular, we find that concepts in mutual information, observation, observables, and logical inference all play a role in the overall framework of remote RF sensing.

22.2 BACKGROUND

Many may find it surprising that in their classic book *The Mathematical Theory of Communication*, Claude E. Shannon and Warren Weaver lay out a broad philosophy of *communication* as "all of the procedures by which one mind may affect another [1]. This, of course, involves not only written and oral speech, but also music, the pictorial arts, the theatre, the ballet, and in fact all human behavior." Surprising indeed, considering the widespread use of Shannon's mathematics in calculating data rates and channel capacities in communication systems. But the philosophical framework is indispensable if advances are to be made in communication and *information* systems. In fact, the same is true of *sensor* systems, whose very purpose is to in some fashion affect something other than the sensor itself, typically tracing back to whoever or whatever put the sensor in place. And so, we consider the *informatics* of sensors and sensor systems, taking the framework of Shannon and Weaver as a launching point and identifying additional elements specific to the sensor problem that must be joined into the framework. In particular, we find that concepts in mutual information, observation, observables, and logical inference all play a role in the overall framework of remote RF sensing.

22.2.1 Overall Framework of Shannon and Weaver

The framework of Shannon and Weaver decomposes the subject of communication into problems at three levels. These are, in ascending order, the technical problem at level A, the semantic problem at level B, and the effectiveness problem at level C.

The *technical problem* concerns the transfer of a signal or symbols from a sender to a receiver with a prescribed level of fidelity. The technical problem falls squarely into the province of engineering with clearly defined performance criteria and objective metrics. A message is generated at the sender and transmitted by some means to a receiver. The technical problem centers on the degree of exactness to which the received message is identical to the message that was transmitted by the sender. The communication system is considered to include five parts as shown in Figure 22.1. These are as follows: an information source that generates the information-bearing messages, a transmitter that acts to transduce the messages onto a communication channel, the communication channel itself, a receiver that acts to transduce the received signal from the form used for transmission into the form of a received message intelligible by the destination, and a destination for whom the message is intended. The channel includes not only the transmission mechanism but also any noise processes that alter the signal on its way to reception at the receiver. The key technical result is Shannon's equation for channel capacity given by [2]:

$$C = B \log_2 \left(1 + SNR\right) \tag{22.1}$$

The channel capacity C (measured in bits per second or binary digits per second) is the maximum rate at which symbols can be transmitted reliably between the source and destination over a channel subject to additive, white, Gaussian noise. The channel capacity depends upon the bandwidth B (measured in Hz) of the channel and the signal-to-noise ratio (SNR) at the output of the channel. While the source may hold an unbounded amount of information content, that content cannot be transferred to the destination at a rate $R > C$ without the introduction of random errors. Significantly, this principle applies just as well to the remote sensing problem: information about the remote object cannot be transferred at a higher rate than the channel capacity of the link being used to exfiltrate the data from the remote object, and analysis of that link is therefore central to the remote RF sensing problem.

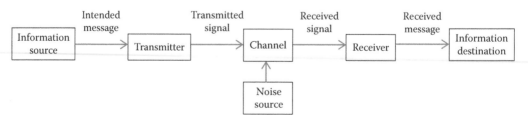

FIGURE 22.1 Basic block diagram of the communication system as considered by Shannon and Weaver.

The *semantic problem* concerns the interpretation by the receiver of whatever it has received. This involves a number of interesting challenges in defining and delimiting common meaning for symbols, concepts, and words and the language in which awareness, logic, and reason are expressed. The semantic problem is less squarely an engineering problem, at least as far as current practice is concerned, and is complicated by the natural evolution of language as leading to changes in meaning over time.

The *effectiveness problem* concerns the degree to which communication has affected the awareness, state, behavior, etc. of the destination in a fashion deemed as positive relative to the purposes of the sender and/or receiver. This is an engineering problem to the extent that it involves clear statements of objective expectations from the sender and receiver and to the extent that defining requirements is foundational to engineering. It is less squarely an engineering problem when the expectations involve subjective concepts such as situational awareness and when the outcomes depend upon inherently subjective human-in-the-loop decision making.

22.2.2 MUTUAL INFORMATION

The technical problem at level A is readily described in mathematical terms and subject to quantitative analysis. The semantic problem at level B and the effectiveness problem at level C are less easily placed on a quantitative basis. This is particularly true in RF remote sensing. While the RF probe signal may be both known and tightly controlled, the returned response signal is essentially unbounded with respect to its potential variability. While it is certainly useful to understand the fidelity with which the response signal is transferred, the larger and more significant questions center on the informatics of the response signal, that is, how much information is encoded, how is it encoded, and how can it be accessed both in the clear and in the presence of noise? These questions fall within the domain of *mutual information*.

Mutual information describes the reduction in uncertainty about a source of information as a result of receiving a message from that source. The central tenet—that information gain is measured as the reduction in uncertainty about the source—is in common with Shannon and Weaver. However, the essence of mutual information involves a number of mathematical properties directly applicable to the remote sensing problem. Mutual information enables quantitative assessment of the information gain from a message of arbitrary and unknown content as is often the case in remote RF sensing and optimization of the return signal in the case of a known or expected returned signal.

In all cases, the measure of information is taken as the entropy of the subject under consideration. Mathematically, the entropy H of a set of N events is calculated as

$$H = \sum_{i=1}^{N} p_i \log_2 \left(\frac{1}{p_i} \right) = -\sum_{i=1}^{N} p_i \log_2 \left(p_i \right) \tag{22.2}$$

where the ith event has probability p_i, and the units of entropy are bits. The entropy is greatest for a set of equally likely events and is reduced when some of the events are more likely than others. The entropy is the fundamental basis for quantifying the information of a source, the information content of a set of messages, or the mutual information between source and receiver through the use of various direct, joint, and conditional entropies [1]. These principles are directly applicable to the problem of remote sensing in describing the potential information encoded in an object, encoded in a specific aspect of that object, encoded in the returned RF waveform, and the mutual information between the remote object and the sensor system.

22.2.3 OBSERVABLES, OBSERVATIONS, AND LOGICAL INFERENCE

Radio detection and ranging, that is, radar, is perhaps the premier use of remote RF sensing. The name radar conveys two aspects of the effectiveness problem, namely, the (remote) detection of an object when one is present and the determination of its range from the radar system. On the surface, it seems uncomplicated to consider that an object exists, that the object is illuminated by an RF probe signal transmitted by the radar transmitter, that the object reflects a portion of the probe signal back toward the radar, that the reflected returned signal is received and measured by the radar receiver, that the object is detected by the presence of the reflected returned signal, and that the object is ranged by measurement of the round-trip delay time. And, in fact, this analysis is effective the vast majority of the time. A deeper look, however, finds that the situation is actually much more complex: the radar receiver does not measure the presence or absence of an object, nor does it directly measure the range to an object! The *observable* that the radar receiver actually measures is the complex RF voltage or waveform at a particular point or points within an RF electronic system. Neither the object nor its range is measured directly. Declaration arises as a *logical inference* drawn from *observations* of the RF voltage, and the declaration of range is necessarily dependent on that baseline inference. The remote RF sensing system can know or measure its state and properties (antenna gain and direction, receiver noise figure, etc.), the probe signal that it transmitted, and the received complex RF voltage, all as a function of time. The information gained in the remote RF sensing process rests entirely on the inferential value of the measured RF voltage as conditioned by knowledge of the state of the system and the probe signal that was transmitted, and any available contextual information of the interaction. As examples of critical context, consider the direction in which the radar is pointed (toward presumably empty space as compared to at the horizon where there could easily be a large building) and the likelihood that someone is broadcasting a jamming signal. The critical role of logical inference is often overlooked in remote RF sensing, especially where the system is intended for use in a benign and commonly occurring environment such as a civilian air traffic corridor.

There are, of course, many uses for remote RF sensing beyond simply detection and ranging. These commonly include determining whether an object is in motion and the nature of any such motion as inferred from variations in the center frequency of the received signal, the size, shape, and characteristics of a remote object as inferred from temporal and spectral characteristics of the received signal, and the development of images as inferred from the phase history of the received signal jointly with the motion of the sensor system. We emphasize here that the system operates on the received signal, which may or may not be the returned signal. The received signal could always be some form of unintended interference and, in the case of military systems, is likely to be a jamming signal. The system may be reporting exactly what it is designed to report for a given RF voltage measurement, and the report could be exactly wrong if the RF voltage was generated by something other than the returned signal. In either case, the remote RF sensor fails as a system, not because the system is failing to work properly but because the logical inference fails.

The central role of logical inference in processing and interpreting sensor data places important requirements on potential observables that may or may not be met in practice. The most basic is that of being logically invertible. In the context of remote RF sensing, specific characteristics of the received RF signal are taken to be observables. These include, of course, the amplitude of the received signal relative to the thermal noise as used in declaring signal detection and the time delay between the transmission of the probe signal and the detection of an assumed returned signal as used in ranging. Signal amplitude relative to the thermal noise is a reasonable choice for an observable: if an object is present, then some level of reflection will occur and a returned signal may be present with amplitude larger than that of the thermal noise; conversely, if an object is not present, no reflection will occur and the measured RF voltage will correspond to that expected for thermal noise. The challenge is that the logical propositions are not invertible: "if object, then reflection and return signal" does not yield "if signal, then *necessarily* object," and "if RF voltage is greater than noise alone, then a signal is present" does not yield "if RF voltage is at the level of noise alone, then *necessarily* no signal is present." In fact, this challenge is captured implicitly in the probability of detection versus false alarm trade-offs made when setting a detection threshold: the system correctly measures the RF voltage and some predictable fraction of the time infers the wrong result, whether a false positive or a missed detection. Since the failure is in the inference, potential solutions and opportunities for advancement lie there also.

The forward problem of identifying the set of potential observables that would be associated with a remote object of interest is well understood. The theoretical response to an RF probe signal can be calculated from the material properties, dimensions, and state of an assumed object, and the potential informatics of the response signal can be assessed using a variety of maximum mutual information (MMI) techniques. We expect that different segments and components of the returned signal will have different SNR and therefore different capacities for information transfer. In this case, the identification of preferred observables corresponds to the identification and selection of waveform *features* with higher SNR, and the process of observation involves both the actual measurement and its processing to isolate the higher SNR and therefore potentially more informative features.

As noted, the real challenge lies in understanding the inverse problem and its power for logical inference. As a general rule, remote RF probing of an uncontrolled spatial region leads to an open-set problem for the potential response. The only way to bound the problem for a single sensor with a single probe signal is to make explicit assumptions about the presumed nature of what might be in that volume and to accept that some level of inferential failures will occur. The alternative is to use additional sensors or additional probe signals such as to reduce the likelihood of a failure in the inference by means of multiple independent observations.

22.3 THE REMOTE RF SENSING PARADIGM

The remote sensing paradigm expands upon the communication paradigm in several ways as it involves first stimulating and then gaining information from a noncooperative potential information source. The most significant are

1. The sensing system generates a controlled RF probe signal. If desired, the characteristics of the probe signal can be tailored toward a specific *query* of a particular remote object.
2. The probe signal *interacts with* the remote object, and a return signal is generated using any or all of several interaction mechanisms. Information about the remote object and its state is transferred to the probe signal during the interaction.

These features are illustrated in Figure 22.2, along with additional features related to the two-way nature of the remote RF sensing link.

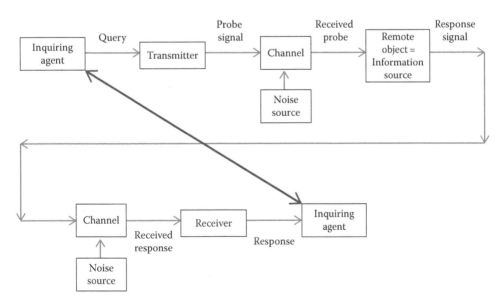

FIGURE 22.2 Expanded block diagram of the remote RF sensing problem.

In the RF remote sensing case, an inquiring agent forms a query that both has relevance to its interests and can be effected using an RF sensor. The query is instantiated as a probe signal by a transmitter and travels over some noisy channel to the remote object. The remote object acts as an information source that interacts with the incoming noisy probe signal. The interaction transfers information to the probe signal by any of several mechanisms, and an information-bearing response is emitted. The response signal travels over some noisy channel, and the noisy received response is captured by the receiver, converted into a format suitable to answering the query, and delivered to the inquiring agent.

Note that noise is, strictly speaking, introduced by both the forward channel of the probe signal and the return channel of the response signal. In practice, the probe signal is typically broadcast at a much higher power level than that of the forward channel noise power, and as such, the additive noise of the forward channel can be neglected with no noticeable impact on the overall analysis of the remote RF probe system. A similar comment can be made with respect to the calibration of the probe signal: we presume that the as-transmitted probe signal is well calibrated and that we do not have to include an error signal in the as-transmitted signal description.

22.3.1   The Effectiveness Problem and the Inquiring Agent

As we consider remote RF sensing, it is helpful to make explicit some implicit and therefore easily overlooked points. The existence of the sensor system implies that it is there for a purpose:

1. Whoever or whatever put it there did so to obtain answers to a set of questions that the sensor is capable of answering. This implies both that the sensor performs as expected (the technical problem) and that its output is formatted appropriately for answering the questions (the semantic problem).
2. The potential contribution of the answers is of sufficient importance as to justify the time, expense, and other costs of acquiring, placing, and operating the sensor (the effectiveness problem).

As a first example, consider a military organization tasked with engaging and defeating enemy forces while minimizing collateral damage. The military organization has the higher-level purposes of accomplishing its mission and of survival. The effectiveness problem of the sensor is measured at this level—does the sensor contribute to accomplishing these purposes and with what level of contribution? As a second example, consider a civilian organization tasked with managing natural resources. The civilian organization has the higher-level purposes of fair allocation of resources among potentially contradictory and competing priorities. Fair allocation requires accurate and timely knowledge of the resources and their usage—does the sensor meet this need in a way that others cannot?

In both cases, the effectiveness problem needs to be interpreted and parsed into an actionable query by an *inquiring agent*. Thus, the higher-level purposes are analyzed and decomposed into a set of specific queries that can be implemented by the sensor or sensors available. In the military case, this would involve working through the logical sequence of "in order to survive, we need to detect any threats; in order to detect threats, we need to know if something (anything) is out there and if whatever is out there is a threat or not; if it is a threat, then what type and how dangerous is it, and exactly what is it doing now?" The inquiring agent needs to parse this into an executable query for an RF system such as "transmit a probe waveform capable of reliably detecting cylindrical metal objects with the size and shape of typical gun barrels." In the civilian case, this might be "transmit a probe waveform whose response is expected to be proportional to the percentage of live foliage."

Thus, the inquiring agent is tasked with converting the higher-level purposes into specific queries (technical problems) that are matched to the capabilities of the sensor. To this end, it is important to note that other considerations are also in play. The speed with which a particular sensor can obtain an answer to an initial query and the rate at which an answer can be updated may determine whether a sensor is a candidate at all for use in collision avoidance or feedback and control applications.

It is important to note that the informatics of the effectiveness problem is tied directly to the logical processes used in assessing, deciding, and controlling actions. In order to properly specify what is expected of the sensor in the technical problem, one must understand fully how the logical processes are executed and how the characteristics of various inputs affect the correct operation of the logic. Formally, one would benefit from clear specifications on characteristics such as accuracy (are numerical data correctly calibrated?), precision (how finely are numerical data resolved?), reliability (the chance that the data are erroneous), temporal relevance (sampled at the required time and obtained quickly enough for the time constraints of the process), etc. However, the complexity of the logical processes can quickly become prohibitive of detailed analysis. Specifications typically can be found for a simple system with a single decision threshold (1 bit, "yes or no," based on exceeding a single fixed threshold), but are harder to find for more complex systems dealing with multiple logical comparison states (3.2 bit, based on comparisons to "gigantic, huge, very large, large, normal, small, very small, tiny, miniscule"). Notice that the more complex logic involves a richer semantics.

22.3.2 The Semantic Problem as the Interface between Effectiveness and Technical Problems

The effectiveness problem of purpose and the technical problem of objective measurement can be brought into harmony by the semantic problem. The semantic challenge is to extract the language used in the reasoning process and then to map the objective measurements of the technical problem into the semantic space. This is, in fact, a place where informatics and information theory can be used to great effect, and this will be illustrated by some simple examples.

Consider, for example, the military problem of detecting potential threats. Take this one step further, and consider that tanks, artillery, and missiles constitute significant threats that can strike from a fairly long range. The purpose of detecting threats gains specificity as detecting tanks and self-propelled artillery, wheeled–towed artillery, and missiles. We then consider whether these have distinguishing characteristics: they are all made of metal; they are all about 10 m long (well more than 1 m and well less than 100 m); they are all longer than they are wide or tall; they all have a prominent cylindrical component; the dimensions of the cylindrical components are well known. Tanks and self-propelled artillery typically have treads, wheeled–towed artillery has wheels, missiles have steering fins, etc. The semantic problem then involves understanding the definition of objects such as "tank" and events such as "movement" to identify both general characteristics that may be common to many objects but are useful in enabling detection and specific characteristics that are useful in discrimination. These characteristics are then mapped into the effectiveness problem of logical classification.

Specifically in the example here, a tank is a metallic object; it is typically longer than it is wide or tall; it rides on tracks not wheels; it has a long, cylindrical gun barrel; and the length and diameter of the gun barrel are known for many types of tank. Now consider the various depths to which the technical problem can detect and distinguish a tank. The remote RF sensor can detect a large reflection as typically associated with a metallic object; the overall size of the object can be about 10 m; reflections from the metal treads and road wheels can be observed; the cylindrical gun barrel can be observed; the dimensions of the gun barrel can be measured. Logically then, the semantics encompass detecting a metallic object that might or might not be a tank as well as anything else; detecting a metallic object whose size is that of a tank; whose size and shape is that of a tank; that has the size, shape, and treads like a tank; matches the size, shape, and treads of a tank and has what could be a gun barrel; and matches the size, shape, and treads and has a cylindrical tube that matches the dimensions of some type of gun barrel. To the effectiveness problem, these are equivalent to stating that something is out there but we can't tell what; something is out there that can't be ruled out as a threat; something is out there that looks a little bit like a threat; something is out there that looks a lot like a threat; something is out there that looks exactly like a known threat.

All of this, of course, seems like an obvious exercise in common sense, so why spend so much time in developing this example? Because it exposes the semantic problem as mapping objective technical measurements into a classification problem and that the semantic problem is they key to designing an effective classifier for the remote RF sensor. The semantics immediately expose issues in logical inference. For example, detection systems are often designed to select between one of two options as noisy measurements are compared against a decision threshold. The maximum-likelihood hypothesis testing result is mathematically provable and optimum for the case that is assumed: a binary decision. The flaw is in the assumption that the decision is binary: the case is actually ternary, with an easily defined region wherein a decision cannot and should not be declared if errors in inference are to be avoided. The designated binary approach forces a decision where one is not warranted by the nature of the inference being drawn. The correct semantics are immediately clear as case A and case B and too close to call given the noise in the measurement. The corresponding classifier maps into the technical problem quite simply using two thresholds instead of one, where each threshold is displaced from the simpler single threshold by three times the variance of the noise, that is, a no-decision region that is six standard deviations of the noise wide. Implementing the ternary classifier in turn forces the implementation of more complicated, but ultimately more *robust*, decision logic for the effectiveness problem.

Many examples can be given with regard to the richness of the semantic descriptions that inform awareness and decision logics. Consider the case of comparing a returned signal against a set of templates and selecting which, if any, of the templates matches the returned signal. Using a semantic approach, we might consider the spectrum of possibilities as follows: looks exactly like one of the templates, is very close to one of the templates, looks a lot like one of the templates, looks only a little like one of the templates, hardly looks at all like one of the templates, and doesn't look at all like any of the templates. And, of course, many of these statements could apply to multiple templates at the same time. The decision logic would be informed properly by semantics such as it could be either of these two but looks a little more like one than the other. Compare this to the output of a mathematically rigorous but logically inappropriate classifier that simply declares a match based on the closest Mahalanobis distance: the classifier will declare a match for the closest distance template, but this may just as easily be a very poor match to any of the templates, and the inferentially correct result of looks only vaguely like one of the possibilities is not even an option.

In the context of informatics, development of the semantics is necessary to draw out and articulate the full range of possible events that can be distinguished, and this allows for calculation of the corresponding entropy associated with the query. But more significantly, there is a natural correspondence between MMI techniques and the design of optimum classifiers. With nuanced differences between the various alternatives, we can consider that all of the MMI techniques in some fashion maximize the separability of a data set into underlying classes. The nature of MMI then is directly appropriate for the task of mapping an objective measurement into an appropriate semantic class. This is, of course, more easily said than done and details in supervised versus unsupervised learning, linear and nonlinear features, Gaussian and non-Gaussian data statistics, etc. all must be mastered. The work is well worth the effort, however, as MMI techniques can be used to extract features that are both optimum with respect to classification and maximize the SNR per feature, thus making the technical problem easier.

22.3.3 The Technical Problem

The technical problem in remote RF sensing centers on stimulating and receiving an information-bearing signal under a given set of circumstances. A given set of circumstances involves both the configuration of the sensor and the nature and configuration of any remote object. The technical problem involves understanding when the system will work (i.e., when an information-bearing signal will be detected and can be processed), how well it will work (i.e., the SNR and SNR-dependent performance parameters), and how to optimize the system for a particular task. These goals are achievable by means of ordinary RF and communication systems engineering given the recognition that the interaction of the probe signal with the remote object is an information-bearing modulation of the returned signal. The practicalities of generating an RF probe signal, propagation over the forward and reverse links, scattering from the remote object, reception by and retransmission from the remote object, and reception of the returned RF signal all fall within ordinary RF systems engineering. Similarly, specifying a probe signal to drive an information-bearing modulation from a presumed object, analyzing the modulations expected in response to given probe signal, and the reception of any returned signals all fall within ordinary communications theory. The informatics challenge is in mapping the content of the information-bearing signal to the semantic and effectiveness problems.

The technical problem for radar is fairly well understood and is commonly described using the "radar equation" [3]. The radar equation describes the SNR of the received signal as a function of the radar system parameters (transmit power, antenna gain, noise figure), the radar signal parameters (bandwidth, modulation), the propagation channel (noise temperature), and the properties of the remote object (radar cross section). Using this as a starting point, it is relatively easy to develop a more general analysis of the technical problem for remote RF sensing including higher-order interactions [4]. The key insight is in recognizing the radar cross section as being just a special case of a more general *interaction* term, where that interaction may be linear or nonlinear depending upon the specifics of the remote object. Thus, we obtain a general form for the SNR of the received signal returned from an nth-order interaction with a remote object (the remote RF sensing equation) as [4]

$$SNR = \frac{n!\big((1/n)P_T G_T\big)^n G_R \lambda^2 \sigma_n}{kTBF\left(4\pi\right)^{n+2} R^{2n+2}} \cdot \frac{k_x}{R_e^{(n+1)u}} \tag{22.3}$$

where in the leading term

P_T is the transmit power of the RF probe signal

G_T and G_R are respectively antenna gains in the direction of the remote object at frequencies f and nf

R is the range from the sensing system to the remote object

λ is the wavelength of the RF probe signal

σ_n is the nth-order interaction term

k is Boltzmann's constant

T is the absolute temperature

B is the sensor bandwidth

F is the noise factor

and in the postmultiplied term

R_e is the apparent excess range

u is the unknown additional loss exponent found from measurements

k_x has magnitude 1 and units of distance $(n+1)u$

The leading term calculates the SNR under idealized conditions of free-space propagation. The postmultiplied term is a correction term that describes the excess propagation loss observed under "real-world" conditions of propagation in varying terrain, among buildings, trees, etc. Equation 22.3 provides significant insight into the potential for nonlinear remote RF sensing. Whereas radar ($n = 1$) exhibits at best $1/R^4$ propagation loss, nonlinearity adds at least an additional $1/R^2$ propagation loss for every integer increase in the order of the interaction n. Thus, while nonlinear sensing may be information rich with respect to the modulation of the returned waveform, it is distinctly range limited and rate challenged as compared to conventional radar sensing. In either case, the bandwidth and SNR combine via Shannon's channel capacity as given in Equation 22.1 to determine the maximum rate at which information can be delivered from the remote object.

The remote RF sensing paradigm can be broken down into five basic elements:

1. Specifying an informatically relevant RF probe signal
2. Generation of the RF probe signal and its propagation to the remote object
3. Interaction of the probe signal with the remote object
4. Generation of the returned RF signal, its propagation from the remote object, and its final reception
5. Processing of the returned signal to extract relevant information

Analyses of the first and last elements are clearly communication and information theory tasks, while analyses of the second and fourth elements are clearly RF engineering tasks. Analysis of the third element is a hybrid task as electromagnetics and RF engineering techniques are used to examine the information-bearing modulation impressed on the returned signal by the interaction.

The interaction of the RF probe signal and the remote object can involve any of several mechanisms, depending upon the nature of the remote object. These include scattering (reflection), absorption and retransmission at the same frequency, and nonlinear conversion to generate returns at other frequencies. Scattering is primarily a linear phenomenon determined by the materials that compose the remote object. Under ideal conditions, the overall modulation from scattering can be resolved into a distribution of significant scattering centers, and this can be interpreted in terms of the likely shape and materials of the remote object whether natural or man made. Absorption and retransmission is also a linear phenomenon, but characteristic of man-made objects such as passive RF tags. Such systems receive the over-the-air RF signal and convert it into a current within a circuit that is then reflected back by the circuit to the antenna and retransmitted. Nonlinear conversion can occur naturally as "the rusty bolt effect" from the junction of dissimilar materials but more often is seen when an RF probe signal is received by an electronic system and the induced current interacts with a transistor, diode, or other nonlinear circuit element. The nonlinear conversion leads to the generation of harmonics and other mixer products, some portion of which appears at the antenna and is transmitted as part of the returned signal. In fact, a number of papers have been published recently analyzing candidate RF probe signals for the purpose of detecting and identifying remotely placed electronic systems [5–9]. The trade-off to be had with a specially designed probe signal is one of efficiency versus specificity. A probe can be optimized for a specific type of remote object such that the SNR of the returned signal is maximized, but this is gained at the expense of reduced SNR against any other type of object.

So overall, the technical problem is an objective exercise in the reception and measurement of a complex RF voltage either as a single snapshot or as a segment over some time interval. The technical problem joins to the semantics and effectiveness problems as the measured RF voltage is somehow converted into meaningful information that addresses the underlying query.

22.3.4 Application of Information Theory toward Analyzing Various Aspects of the Problem

Information theory can be applied on many levels in the analysis of the remote RF sensing problem. Where the effectiveness problem can be clearly articulated and translated into a semantic space, one can examine the entropy of the effectiveness problem and its decision logic and the entropy of the semantic space (i.e., how many semantics terms are used to span the available space and what the individual probability of a given semantic term is). These entropies provide insight into the higher-level information requirements placed on the remote RF sensing system.

Similarly, the entropy of the query can be examined in terms of the number of possible answers that can occur, and this provides insight into the information requirements for the specific query.

As regards the specific query, the function of the remote RF sensing system falls into one of two general categories: remote measurement of a physical quantity against a metric scale and remote detection and description in a semantic space. It is important to note that accuracy and precision are two distinctly different things in remote measurement against a metric scale. The entropy associated with precision depends upon the number of resolvable states over the dynamic range of the scale that is covered by the measurement system. The entropy associated with the accuracy depends upon a different set of factors associated with calibration offsets and measurement noise. In practice, for example, the entropy associated with the precision of something like a 16-bit digitizer capable of resolving 65,536 states far exceeds the entropy associated with the accuracy of the measurement that may only be within ±10% or just over 3 bits as limited by calibration and noise. The information requirement then depends upon whether the query is to make an absolute measurement (accuracy) or to resolve the difference between two measurements (precision). The semantic space has a different nature and this will be reflected in its entropy. The partition of semantic space is almost certainly not uniform, and the boundaries are fuzzy. Nevertheless, it should be possible to obtain a reasonable estimate of the semantic entropy for a thorough analysis.

As regards the remote object, the potential entropy is essentially infinite in theory with perfect measurements of infinite precision, since the object can induce a continuum of responses over the complex numbers. The object represents an arrangement of matter at a location in space and time and under particular conditions of pose, motion, and energy states. In practice, the remote RF sensor will be able to interact with some macroscopic observable that arises from the ensemble behavior of the object or some part of it (e.g., the electrical wiring in a circuit). It may be possible to solve the forward problem of how the object will respond to a particular probe signal and how the response will vary depending upon specific parameters of the remote object. The source entropy of the object for the case considered is then found from the potential parameter variations, while the entropy actually transferred to the information-bearing waveform is found from the potential waveform variations.

The waveform variations are particularly amenable to analysis using MMI methods, and these are especially helpful in cases where there is no *a priori* model for the expected waveform variations. In such cases, the waveforms can be decomposed using any number of techniques to extract linear or nonlinear features and with the decomposition matched to the statistics of the returned waveforms [10]. The most general case involves nonlinear features extracted from non-Gaussian data and is fairly processing intensive. More often than not, the statistics of the data are "adequately Gaussian," and simpler techniques such as principal components analysis/singular value decomposition (PCA/SVD) provide good performance in extracting features that maximize the mutual information about the object. The MMI features have the desirable property of maximizing the SNR per feature, and this in turn maximizes both the range and rate at which they can be exploited. And where linear features can be extracted by PCA/SVD, these features can be processed using matched filter techniques. One other aspect of note regarding MMI techniques is that they aggregate energy across multiple frequencies. Coherent harmonic content from a common source is aggregated into a single, higher-energy feature rather than appearing as separate features.

What remains heuristic at this point is the mapping from an MMI feature set into a semantic space used in the effectiveness problem. For example, consider a set of unknown but similar objects such that no analytical model exists as to what the returned signals might look like and how one can be distinguished from another. The objects can be probed and the returned signals aggregated into a data set that is then decomposed using MMI techniques. A new response from an unknown target can be compared against the MMI feature vectors. The heuristics come in designing a classifier that best declares the result of the comparison. Often as not, classifiers implement a decision based upon a "closest-match" measurement such as Mahalanobis distance. However, the combination of MMI feature vectors and semantic space representation enable a much richer classification scheme. The MMI feature vectors will capture the different levels of information content of the various components, in descending order of information content. That is to say, that the primary feature vector component is the most informative, the secondary component less so, on down to the smallest and least informative component. The semantic space allows a richer comparison as exactly alike (every component within one standard deviation of the template vector), strongly alike (the largest components within one standard deviation and none more than three standard deviations away from the template vector), very much alike (every component within three standard deviations of the template vector), etc. The heuristics arise in defining the set of objective measurements that map to a particular semantic category. The semantic mapping has a particular strength in its simple definition of "no match found" such as no component is within three standard deviations of the template vector.

22.4 DISCUSSION

An informatics interpretation exists for every aspect of remote RF sensing. For example, the simplest of remote RF sensing scenarios involves the measurement of a returned signal against some scale such as a clock for determining the range from the radar to the assumed remote object. It is well known that more precise measurements can be obtained through coherent integration over multiple radar pulses. The increased SNR of the coherently integrated signal has a higher channel capacity than a single pulse and thus is capable of transferring more information, in this case greater precision about the range to the remote object.

In all cases, the remote RF sensor provides an objective measurement of a received RF voltage, and the ultimate use of that information involves an inference as to the cause and significance of that RF voltage relative to some higher-level purpose of whomever or whatever employed the sensor. We believe that the inferential nature of remote RF sensing and the higher-level purposes are often overlooked to the detriment of the user. We believe that at the higher level, the concept of *regret* in taking a decision is always operative and that minimizing the probability of regret is an unstated but implicit imperative that must be pulled forward into the sensor design. What causes regret will vary with the situation, but avoiding regret is always operative at some level. Returning to a military example, consider a system such that a weapon is fired whenever a target is positively identified. Using a conventional maximum-likelihood threshold for a presumed binary decision (identified/not identified) results in an erroneous decision some percentage of the time. One can choose false alarms over missed detections or vice versa, but errors in inference are guaranteed to occur and these necessarily will cause some level of regret. Consider instead the ternary decision of (identified/not identified/too close to call given the noise). In a free-fire zone where attack by hostiles is expected, the regret falls in not taking action and getting killed. In contrast, in a permissive fly zone where civilians are also known to be operating, the regret is in taking action with less than near absolute certainty in the identification. The more complex inference allows for managing the probability of regret by choosing to include or exclude the "too close to call" region in the decision to fire.

One final point should be made in regard to practical approaches to handling the open-set problem. The open-set problem arises because it is always possible that the received signal is not actually the returned signal as expected and was generated under unexpected conditions or by something not covered by the system design. The open-set problem makes it impossible, in theory, to ever know anything with absolute certainty based on measurements from a remote RF sensor. The practical reality, however, is somewhat different, and a high degree of certainty can be achieved by managing the conditions under which the sensor is operated. The logical inferences involved are thus conditional inferences whose strength depends on the certainty that the assumed conditions are met, and the reliability of the RF sensor reports becomes dependent on the availability of auxiliary sensors and inputs that validate whether the assumed conditions are met.

22.5   TOWARD THE FUTURE

Two significant open questions remain with respect to the informatics of remote RF sensing and to some extent informatics in general. The first is, of course, formalization of the heuristic mapping from MMI techniques into the semantic space. The second, which is somewhat different and certainly more general, is to formalize the concept of *value* and specify a value metric that captures the importance of a particular piece of information as regards the effectiveness problem. For instance, it is certainly useful for a law enforcement officer to know that shots are being fired in a general area, but it is much more important to know that he or she is the immediate target of the shots being fired. "Target not you" and "target you" both convey the same amount of information, but the one is clearly more significant and requires immediate defensive action that the other does not. It would seem that the concept of value is ultimately derived from the effectiveness problem, as each piece of information is prorated by its significance in the total decision process and the significance of the decision itself. Perhaps the coefficients used in prorating are a good starting point toward metrics for value. However, in the long run, it seems more likely that value will be defined first in terms of eliminating the probability of regret and then in maximizing return over cost.

REFERENCES

1. C. E. Shannon and W. Weaver. *The Mathematical Theory of Communication*. Urbana, IL: University of Illinois Press, 1949.
2. C. E. Shannon Communication in the presence of noise. *Proceedings of the Institute of Radio Engineers*, 37(1), 10–21, 1949.
3. M. I. Skolnik, ed. *Radar Handbook*, 3rd edn. New York: McGraw-Hill, 2008.
4. J. A. Kosinski, W. D. Palmer, and M. B. Steer. Unified understanding of RF remote probing. *IEEE Sensors Journal*, 11(12), 3055–3063, 2011.
5. K. M. Garaibeh, K. G. Gard, and M. B. Steer. Estimation of co-channel nonlinear distortion and SNDR in wireless systems. *IET Microwave Antennas and Propagation*, 1(5), 1078–1085, 2007.
6. F. P. Hart and M. B. Steer. Modeling the nonlinear response of multitones with uncorrelated phase. *IEEE Transactions on Microwave Theory and Techniques*, 57(10), 2147–2156, 2007.
7. A. F. Martone. Forensic characterization of RF circuits. PhD thesis. Purdue University, West Lafayette, IN, 2007.
8. G. J. Mazzaro, M. B. Steer, K. G. Gard, and A. L. Walker. Response of RF networks to transient waveforms: Interference in frequency-hopped communications. *IEEE Transactions on Microwave Theory and Techniques*, 56(12), 2808–2814, 2008.
9. G. J. Mazzaro, M. B. Steer, and K. G. Gard. Filter characterisation using one-port pulsed radio-frequency measurements. *IET Microwave Antennas and Propagation*, 3(2), 303–309, 2009.
10. S. Petridis and S. J. Perantonis. On the relation between discriminant analysis and mutual information for supervised linear feature extraction. *Pattern Recognition*, 34(5), 857–874, 2004.

23 Reliability of Integrated Overtemperature Sensors in Electromagnetic Polluted Environment

Orazio Aiello and Franco Fiori

CONTENTS

23.1 INTRODUCTION

High-performance high-reliable integrated circuits (ICs) show an increasing importance not only in the traditional segments as information and signal processing but also in the fields of power electronics and sensor systems. Integrated smart-power circuits enable innovative solutions moving toward a system integration. The combination of analog, digital, and power building blocks on a single chip allows the design and production of less-volume-and-weight electronics systems for different applications in the several fields. Smart-power ICs are usually employed in the automotive industry, where higher safety standards, tighter environmental normative, and the demand for increasing comfort on board lead to a constantly increasing amount of microelectronic components built into modern cars [1]. Therefore, high performance has to be guaranteed with the highest level of reliability and critical parameters have to be detected, and proper countermeasures have to be taken immediately. Thus, for safety reasons, specific integrated sensors are designed to report status signals to the central unit and to protect the overall system-on-chip. It means that monitoring and sensing building blocks have to work properly in every operating condition in order to prevent malfunctioning or damaging of the overall integrated system-on-chip.

Among the monitoring and control building blocks, the overtemperature sensors are specifically required in every electronic system-on-chip that operates in harsh environment where power devices generate heat depending on the power dissipation [2]. Although the design of IC temperature sensors have been extensively considered in academic and industrial fields research especially in the last decades and high-valuable results have been reached [3–27], the reliability of integrated overtemperature sensors to electromagnetic interference (EMI) has still not been investigated.

Electromagnetic (EM) fields generated by radio-frequency transmitters and the disturbances delivered by the electronic modules nearby are increased in the last decades due to the widespread diffusion of radio and television broadcasting and wireless systems for military and civil communications. This EM pollution is collected by printed circuit board (PCB) traces, cables, and wiring harnesses that connect the electronic units with the surrounding environment behaving like unwanted antennas. As a result, every part of the electronic system as sensors and actuators can be corrupted by EMI. Usually, the propagation of such disturbances to the ICs is avoided including EMI filters at the connector level of electronic modules and/or into the sensor packages. In some cases, such EMI filters can impair the system operation or the performance can be reduced, while in some other, filters are not effective because the EMI directly couples with the wiring interconnects (flat cables, PCB traces) at the module level. Such unwanted coupling can be attenuated using EM shields and additional filters at the PCB level that show the drawback of increasing size and cost. In order to perform ICs included in electronic modules immune to EMI as well as filters and EM shields not necessary, in the last decades, the effects of radio-frequency interference (RFI) on ICs have been investigated [28,29]. In the ICs of modern high-volume applications, the interference collected from the off-chip interconnects can propagate also through parasitic paths such as the silicon substrate and the metal-to-metal capacitances. Researches have been mostly focused on the susceptibility of basic analog and digital blocks (e.g., operational amplifiers, simple logic gates) to the RFI superimposed onto the input signals and/or to the power supply. Analog circuits rectify the RFI and their output signals can be corrupted by the demodulated RFI (base-band interference) that cannot be distinguished from the nominal signals [30–35]. Furthermore, it has been shown that the RFI affecting digital circuits can induce timing failures [35,36]. In addition, the RFI can affect the ancillary circuits of the power transistor that protect it from possible failures such as those due to overcurrent, overvoltage, or overtemperature [37–39]. In this context, this chapter analyzes the effect of RFI on the overtemperature detection referring to a thermal shutdown circuit.

The chapter is organized as follows. Section 23.2 presents the nominal operation of a common thermal shutdown circuit. Section 23.3 shows the way the RFI superimposed onto a power transistor signals couples with the thermal shutdown circuit of a power device, and the effect of such disturbances on the operation of this circuit is analyzed. Finally, Section 23.4 shows a solution to be used to increase the immunity to RFI of this circuit and the results of the experimental tests carried out on a test chip are presented. Concluding remarks are drawn in Section 23.5.

23.2 THERMAL SHUTDOWN CIRCUIT: NOMINAL OPERATING CONDITION

The power dissipation generated by power device increases the die temperature until the die temperature overcomes the maximum IC operating temperature, which usually drops in the range 150°C–180°C. Whenever this phenomenon occurs, the breakdown of the circuit is experienced. In order to avoid permanent damaging of the system-on-chip due to an overtemperature condition, a thermal shutdown circuit is usually integrated on chip as close as possible to the heat source. Such thermal protections sense the power transistor temperature and compare it with a given threshold. When the die temperature overcomes this threshold, the circuit provides to the electronic system a fault signal and the power transistor is switched off.

In order to build up an integrated thermal sensor, the relationship of the bipolar transistor base–emitter voltage with temperature is exploited and its output voltage (current) is compared with that provided by a constant voltage (current) reference. Then, the output logic signal is used to switch off the primary heat source avoiding the IC to be destroyed.

Figure 23.1 shows the commercially available thermal shutdown circuit considered in this chapter. It comprises a base–emitter referenced current source composed by a bipolar junction transistor (BJT) T1, a resistor (R_0), and an n-type Metal-Oxide-Semiconductor (MOS) transistor (M1). The output current I_{OUT} is replicated by the current mirror M3–M4 and compared with the reference

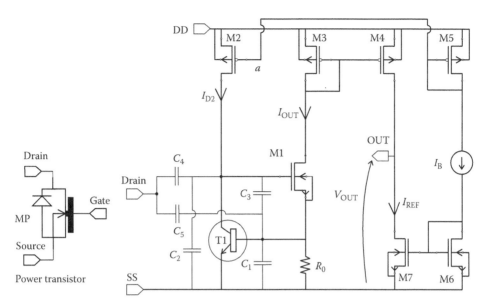

FIGURE 23.1 Thermal shutdown circuit with the power MOS transistor drain to BJT capacitive parasitic coupling evidenced.

current I_{REF}. The power transistor MP is electrically isolated from the thermal shutdown circuit and the bipolar transistor T1 (the sensing element) is buried into the power device. As long as the temperature of T1 is such that $I_{OUT} > I_{REF}$, the transistor M4 is in the triode region and the output voltage V_{OUT} is high (close to the power supply voltage). On opposite, for $I_{OUT} < I_{REF}$, M7 is in the triode region and the output voltage is low (close to the ground voltage).

The output current I_{OUT} can be defined as

$$I_{OUT} = \frac{V_{BE}(T)}{R_0}. \tag{23.1}$$

where

$V_{BE}(T)$ is the base–emitter voltage of the bipolar transistor (T1)
R_0 is the resistor of the current source

For instance, Figure 23.2 shows that the current I_{OUT} decreases linearly over temperature and crosses the reference current I_{REF} (dashed line) at about 140°C. Therefore, the output voltage V_{OUT} is high for temperature below this threshold and it is low otherwise. However, the output current is also affected by the R_0 thermal drift and the technology spread (fabrication tolerances), but these errors can be compensated using a dc source (I_B) whose output depends on the value of an integrated resistor similar to R_0 (matched resistors must be used). This makes the shutdown turn-on point (see Figure 23.2) stable with temperature. Moreover, it is worth mentioning that the bipolar transistor that senses the heat generated by the power ICs is usually placed within the power transistors, where the heat dissipation is less effective. Even though this layout choice reduces the thermal shutdown activation delay, it implies a parasitic coupling of the bipolar transistor interconnects with the power transistor metal grids. Therefore, disturbances affecting the power transistor signals can propagate to the thermal shutdown circuit and its nominal operation can be impaired as shown in Section 23.3. Such a parasitic coupling is represented in Figure 23.1 by a set of capacitances (from C_1 to C_5) that connect the bipolar transistor (T1) terminals with those of the power transistor.

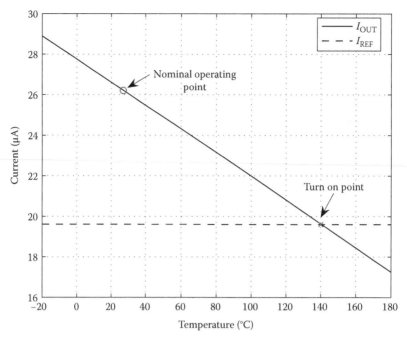

FIGURE 23.2   Thermal shutdown output current versus temperature.

23.3   THERMAL SHUTDOWN CIRCUIT IN THE PRESENCE OF RADIO-FREQUENCY INTERFERENCE

The susceptibility to RFI of the thermal shutdown circuit introduced in Section 23.2 (Figure 23.1) can be analyzed assuming the power transistor drain-source voltage corrupted by the RFI. The capacitive coupling of the on-chip drain interconnect with the base and the emitter metal traces of the thermal sensing transistor (T1) propagates the drain disturbances to the base–emitter voltage of T1. As a result, the output current (I_{OUT}) is modulated accordingly and a false overtemperature fault is delivered whenever the amplitude of the output current drops below the turn-on point (see Figure 23.2). In order to avoid false signaling, the output voltage (V_{OUT}) is usually filtered. Based on this consideration, the probability of having the power transistor switched off by the thermal shutdown circuit seems to be low and erroneous fault signaling seems to take place for very high level of RFI. However, the aforementioned conclusions are wrong because the RFI superimposed onto the gate–source voltage of M1 and in turn to the base–emitter voltage of T1 induces a variation of the average currents flowing through T1 and M1 because of distortion phenomena.

The dc shift of the output current I_{OUT} due to the RFI superimposed onto the power transistor terminals can be evaluated referring to the equivalent circuit shown in Figure 23.3. Only the RFI distortion in the transistors M1 and T1 is taken into account and it is represented by the dc sources ΔI_D and ΔI_C, respectively [40,41]. Assuming that the operating region of each transistor in the circuit is not changed by the RFI presence, the analysis of the equivalent circuit in Figure 23.3 highlights that the average output current $\bar{I}_{OUT}$ is mostly affected by ΔI_C rather than ΔI_D. This result can be explained observing that the average collector current is equal to I_{D2}, which is constant; hence, the RFI-induced offset of the collector current ΔI_C is compensated by the feedback loop that reduces the base–emitter voltage, then the average output current (I_{OUT}) is reduced as well. On vice versa, the RFI rectification in M1 (ΔI_D) cannot change the dc output current because the M1 average current must be equal to that flowing through R_0 (in this reasoning, the dc base current of T1 is neglected). As a result, in the presence of RFI, the average gate–source voltage of M1 is changed by the loop to keep the dc output current equal to that flowing through R_0.

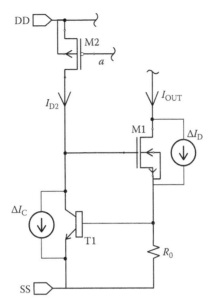

FIGURE 23.3 Equivalent circuit of the thermal protection circuit to calculate the output offset current induced by RFI distortion in M1 and T1.

Based on this, the average output current can be evaluated referring only to the RFI distortion of T1. To this purpose, let's assume this transistor biased in the active region so that its collector current can be approximated as

$$I_C = I_S \left[\exp\left(\frac{v_{BE}}{\eta V_T} \right) - 1 \right], \qquad (23.2)$$

where
I_S and η are the model parameters
$V_T = k_B T/q$ is the temperature equivalent voltage
k_B is the Boltzmann constant
T is the junction temperature
q is the electron charge

Moreover, let's assume the base–emitter voltage affected by continuous wave RFI so that

$$v_{BE} = V_{BE} + v_{RF} \cos(2\pi f t), \qquad (23.3)$$

where
V_{BE} is the average base–emitter voltage
v_{RF} is the RFI peak value
f is the interference frequency

Using this simple model, the average base–emitter voltage in the presence of RFI can be expressed as

$$V_{BE} = \eta V_T \ln\left[\frac{I_{C0}}{I_S I_0 \left(\dfrac{v_{RF}}{\eta V_T} \right)} \right], \qquad (23.4)$$

where I_{C0} is the BJT dc bias current, which is provided by the MOS current source M2 while I_0 is the modified Bessel function of the first kind [40,41]. Under the assumption that the average value of the M2 drain current (I_{D2}) is not affected by the RFI and being I_0 a monotonic function of the RFI magnitude (v_{RF}), the average voltage across R_0 decreases while increasing v_{RF}, and the average output current decreases as well. Depending on the RFI amplitude and frequency, the average output current of this circuit can be shifted down below the reference current (I_{REF}) so that the output voltage can switch from high to low as it happens in the presence of real overtemperature. Such a false signaling cannot be filtered and propagates into the logic block shutting down the power transistors and/or the overall integrated system.

This effect has been highlighted by several computer analyses carried out on a thermal shutdown circuit like that shown in Figure 23.1, whose bipolar sensing transistor was placed into the center of a power MOS, and it was connected to the rest of the circuit by means of three metal strips of length $L_t = 450$ μm. The design was modeled on a commercially available smart-power IC.

The parasitic capacitances used in the computer analyses were extracted referring to the metal cross section shown in Figure 23.6a and using the quasi-static simulator ANSOFT Q3D [42]. These capacitances are listed in the second column of Table 23.1. Performing the small-signal analysis of the circuit shown in Figure 23.1, the effect of a drain terminal of the power transistor driven by an ideal RF voltage source (v_{drain}) on the base–emitter voltage v_{RF} can be found. To this purpose, Figure 23.4 (continuous line) shows the magnitude versus frequency of the transfer functions v_{RF}/v_{drain} that resulted from the analysis of a circuit designed using a 0.35 μm 60 V CMOS technology. Furthermore, the continuous line shown in Figure 23.4 can be used to calculate the dc shift of the output current induced by the RFI. To this purpose, the average output current can be expressed as

$$I_{OUT} = \frac{V_T}{R_0} \ln\left[\frac{I_{C0}}{I_S I_0\left(\frac{|H(f)|\,v_{drain}}{\eta V_T}\right)}\right],\tag{23.5}$$

where $H(f)$ is the transfer function that relates the base–emitter voltage v_{RF} and the drain RF voltage v_{drain} [40,41].

The dc output current (I_{OUT}) versus the RFI amplitude has been evaluated referring to this model. The results of this analysis carried out for RFI magnitude in the range 0–3 V and frequencies of 30 MHz, 100 MHz, 300 MHz, 600 MHz, and 1 GHz are shown in Figure 23.5 by the continuous lines. A more accurate prediction of the RFI-induced upset can be obtained through time-domain computer simulations that however show the drawback of being time consuming. Several time-domain analyses were carried out referring to a circuit that does not comprise the power transistor model nor the IC package model. These further investigations are aimed to prove the correctness of the assumptions at the basis of the approximate model. The results of these analyses are shown in

TABLE 23.1

Set of Parasitic Capacitances Extracted by ANSOFT Q3D Referring to the Cross Section Shown in Figure 23.6a and b

Capacitance	Value: Figure 23.6a	Value: Figure 23.6b
C_1	100 fF	108 fF
C_2	100 fF	108 fF
C_3	7 fF	3 fF
C_4	42 fF	400 aF
C_5	42 fF	400 aF

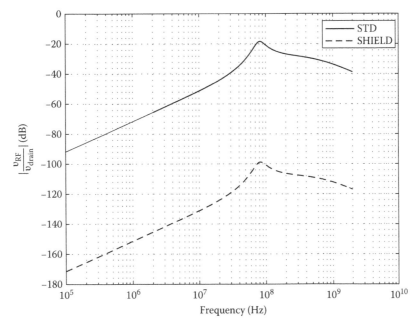

FIGURE 23.4   Magnitude of the bipolar transistor base–emitter voltage over drain voltage v_{RF}/v_{drain} obtained from small-signal computer analysis. The circuit is driven at the drain terminal to ground by an ideal voltage source.

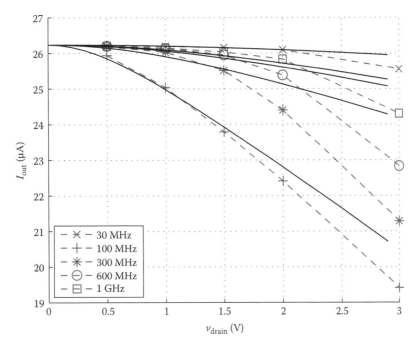

FIGURE 23.5   Average output current ($\bar{I}_{OUT}$) versus drain input voltage magnitude (v_{drain}). The continuous lines have been obtained using (23.5), whereas marks have resulted from time-domain computer simulations. These analyses have been carried out at 30 MHz, 300 MHz, 600 MHz, and 1 GHz.

Figure 23.5 by dashed lines and different symbols. The prediction of the approximate model is in good agreement with time-domain simulations for low RFI magnitude, that is, below a few volts, whereas, over this level, further nonlinear effects not included in the previously considered model of the bipolar transistor take place and the RFI distortion of the other transistors of the circuit is not negligible anymore; hence, the prediction error increases. However, these analyses show that the output current I_{OUT} is reduced by increasing the RFI magnitude. This means that disturbance can lead the output voltage (V_{OUT}) to switch as it happens in case of overtemperature.

In order to increase the immunity to RFI of such a kind of circuits, a reduction of the parasitic coupling of the power transistor interconnects with the thermal shutdown metal strips that connect the sensing transistor to the rest of the circuit is needed. To this purpose, the cross section shown in Figure 23.6a has been reworked in order to minimize the capacitive coupling of the base and collector metal strips with the power transistor interconnects while increasing those with the emitter metal strip. The cross section that resulted from this analysis is that shown in Figure 23.6b, where the emitter interconnect is routed by means of the M1 and the M3 layers shorted through vias, the base and the collector are in the M2 layer, while the drain metal interconnects are routed in the M4 layer. In practice, the base and the collector metal strips are surrounded by the emitter and are decoupled from the power transistor gate and drain interconnects. The parasitic capacitances extracted from this cross section are listed in the third column of Table 23.1. A comparison between the capacitance coupling value of the metal strip routing with the cross section in Figure 23.6a and b points out a reduction of the parasitic coupling with the new solution. The computer analyses carried out on the thermal shutdown circuit that comprises the interconnects, where cross section of which is shown in Figure 23.6b, have put on evidence a strong reduction of the RFI affecting the bipolar transistor base–emitter voltage and the circuit output current, as it is shown by the dashed lines in Figure 23.4. However, these results do not assure by their own that the thermal protection circuit cannot be affected by the RFI because, in actual devices, high-frequency disturbances injected into the power transistor terminals are spread throughout the chip via the silicon substrate. In order to reduce the direct coupling of the circuit transistors with the substrate, the body of the nMOS transistors has been connected to the ground net (SS) and that of the pMOS transistors has been connected to the power supply net (DD). Furthermore, the circuit layout has been optimized to reduce as much as possible the parasitic coupling of the metal interconnects with the substrate. Finally, it is worth mentioning that the immunity of the considered thermal shutdown circuit could be also increased filtering the

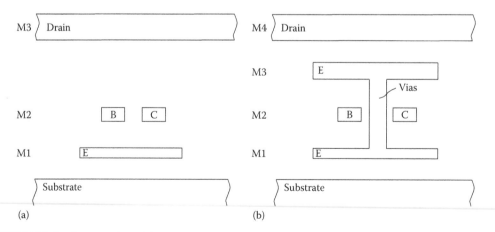

(a)　　　　　　　　　　　　　　　　(b)

FIGURE 23.6 Cross section of the metal strips connecting the bipolar transistor of the thermal shutdown circuit to the remaining part of the circuit. (a) The metal strips C, B, and E are electrically connected to the BJT (T1) collector, base, and emitter, respectively. (b) The emitter interconnect is used to decouple the base and the collector interconnect from the power transistor drain (M4).

base–emitter and the collector–emitter voltages of T1, but this requires additional components to be integrated on chip resulting in increased silicon area and fabrication costs.

In Section 23.4, the results of the measurements carried out on a test chip to investigate the susceptibility to the RFI of the previously considered thermal shutdown circuit are presented and discussed.

23.4 EXPERIMENTAL RESULTS

The effectiveness of the proposed integrated metal shield has been proved, fabricating a test chip in order to verify the presented analysis. The test chip comprises the aforementioned thermal shutdown protection circuit and nMOS power transistor, a gate driver, and some other analog cells designed referring to a 0.35 μm 60 V CMOS technology. In particular, the test chip comprises two thermal shutdown circuits like that shown in Figure 23.1 that differ each other only for the metal strips that connects the transistor T1 to the rest of the circuit. In the following, the circuit that comprises the metal strips, where cross section of which is shown in Figure 23.6a, is named *STD cell*, whereas that comprising the metal bus, where cross section of which is shown in Figure 23.6b, is named *SHIELD cell*. Moreover, these circuits comprise a current reference source, which is represented in Figure 23.1 by the ideal current source I_B. This circuit compensates the R_0 thermal drift and the fabrication spread.

A microphotograph of the fabricated test chip is shown in Figure 23.7, where the nMOS power transistor concerned by the RFI, the STD cell, and the SHIELD cell is highlighted. The test chip has been encapsulated in an LQFP-80 ceramic package and has been mounted on a small test board to be used in the test bench shown in Figure 23.8, which has been set up according with the international standard IEC 62132-3 [43]. In this test bench, the output signal of the RF source Agilent E8257D is provided to the RF power amplifier AR-10W1000A hence to the drain–source port of the power transistor by means of a 1 mm pitch microprobe. This power transistor is driven

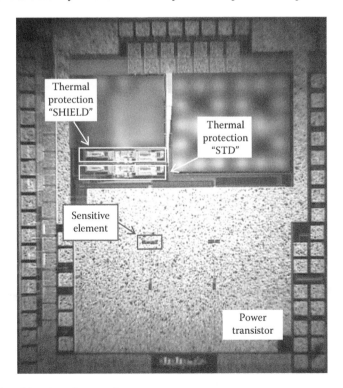

FIGURE 23.7 Test chip microphotograph.

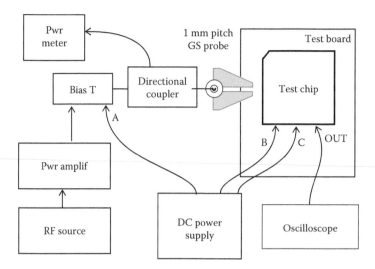

FIGURE 23.8 Experimental test setup.

at the gate–source port to be switched off and the dc drain–source voltage is set to 20 V (line A in Figure 23.8) through a bias tee and a directional coupler. Using this test setup, the level of the RF incident power superimposed onto the power transistor drain nominal voltage can be increased up to 32 dBm for frequencies in the range 1 MHz–1 GHz.

Furthermore, the power supply is provided to the thermal shutdown circuits by means of an external dc power supply (line B in Figure 23.8) through the on-chip analog power supply lines. Also, the power section is biased separately (line C). The output signals (OUT) of the two thermal

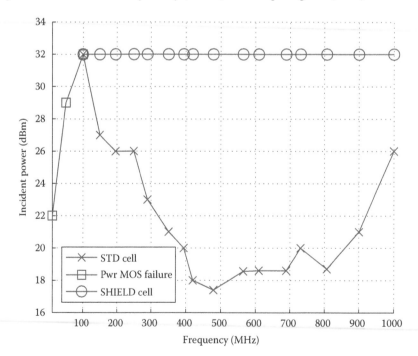

FIGURE 23.9 Experimental results. Crosses refer to thermal shutdown circuit whose bipolar transistor is connected to the rest of the circuit by the STD interconnects. Circles refer to the thermal shutdown circuit that uses the SHIELD interconnect, while squares indicate the magnitude of the RF incident power that switched on the nMOS power transistor without false signaling at the thermal shutdown logic output.

shutdown circuits under investigation (STD, SHIELD) have been monitored using the oscilloscope Agilent MSO 6104A.

Measurements have been carried out as follows: the magnitude of the continuous wave delivered by the RF amplifier, which is superimposed onto the dc drain–source voltage of the power transistor is increased step by step while monitoring the thermal shutdown logic output signals (V_{OUT}). When the output logic level changes steadily to the low state, the RFI-induced failure is detected. This procedure was repeated for a set of frequencies in the range 1 MHz–1 GHz obtaining the results shown in Figure 23.9. In particular, crosses refer to the STD cell, circles refer to the SHIELD cell, while squares indicate the maximum RF injected power that switches on the nMOS power transistor (this effect is analyzed in [44]), but no failure in the thermal shutdown circuits under test were observed.

The aforementioned measurement results highlight that the temperature sensing is affected by the RFI that propagates through the metal parasitic capacitances rather than that of other parasitic paths. Furthermore, it has been shown the effectiveness of the proposed metal shield that makes the thermal protection circuit fully compliant with the IEC 62132-4 requirements.

23.5　CONCLUSIONS

In this work, the reliability of integrated overtemperature sensors employed in smart-power system-on-chip operating in harsh environment has been pointed out. In particular, the susceptibility of a thermal shutdown circuit to the RFI injected into the power device has been investigated. Since the temperature-sensing element is buried into the power device, the RFI can reach the components of the protection circuit through parasitic paths such as the silicon substrate, the metal-to-metal parasitic capacitances, and the package parasitic elements. The susceptibility of this cell to the RFI has been investigated under the assumption that the power device and the protection circuit are mainly coupled through the metal parasitic capacitances. Through approximate analyses and computer simulations, it has been shown that the rectification of the RFI can lead to false overtemperature fault signaling. The operation of the thermal shutdown circuits can be impaired causing false overtemperature signaling that cannot be distinguished from the true signaling. To avoid this, a specific cross section for the metal strips that connect the temperature-sensing element to the rest of the protection circuit has been proposed and its effectiveness has been proved through measurements carried out on a test chip. Moreover, since experimental tests have shown that using the shielded metal strips the thermal protection circuit meets the IEC-62132-4 requirements, the other RFI parasitic paths are negligible in the frequency range 1 MHz–1 GHz.

REFERENCES

1. B. Murari, F. Bertotti, G. Vignola, *Smart Power ICs: Technologies and Applications*. New York: Springer, 2002.
2. M.A.P. Pertijs, J. Huijsing, *Precision Temperature Sensors in CMOS Technology*. New York: Springer, 2006.
3. K.Souri, C. Youngcheol, K.A.A. Makinwa, A CMOS smart temperature sensor with a voltage-calibrated inaccuracy of ±0.15°C (3 σ) from 55°C to 125°C, *IEEE Journal of Solid State Circuits*, 48(1), 292–301, 2013.
4. S. Hwang, J. Koo, K. Kim, H. Lee, C. Kim, A 0.008 mm² 500 µW 469 kS/s frequency-to-digital converter based CMOS temperature sensor with process variation compensation, *IEEE Transactions on Circuits and Systems I*, PP(99), 1–8, 2013.
5. A.L. Aita, M.A.P. Pertijs, K.A.A. Makinwa, J.H. Huijsing, G.C.M. Meijer, Low-power CMOS smart temperature sensor with a batch-calibrated inaccuracy of from to 130, *IEEE Sensors Journal*, 13(5), 1840–1848, 2013.
6. G. Chowdhury, A. Hassibi, An on-chip temperature sensor with a self-discharging diode in 32-nm SOI CMOS, *IEEE Transactions on Circuits and Systems II*, 59(9), 568–572, 2012.
7. C. Ching-Che, Y. Cheng-Ruei, An autocalibrated all-digital temperature sensor for on-chip thermal monitoring, *IEEE Transactions on Circuits and Systems II*, 58(2), 105–109, 2011.

8. R.P. Fisk, S.M. Rezaul Hasan, A calibration-free low-cost process-compensated temperature sensor in 130 nm CMOS, *IEEE Sensors Journal*, 11(12), 3316–3329, 2011.

9. K.C. Souri, K.A.A. Makinwa, A 0.12 mm 7.4 W micropower temperature sensor with an inaccuracy of 0.2°C (3 σ) from 30°C to 125°C, *IEEE Journal of Solid-State Circuits*, 46(7), 1693–1700, 2011.

10. C. Poki, C. Chun-Chi, P. Yu-Han, W. Kai-Ming, W. Yu-Shin, A time-domain SAR smart temperature sensor with curvature compensation and a 3 σ inaccuracy of –0.4°C 0.6°C over a 0°C to 90°C range, *IEEE Journal of Solid-State Circuits*, 45(3), 600–609, 2010.

11. C. Jimin, L. Jeonghwan, L. Inhee, C. Youngcheol, Y. Youngsin, H. Gunhee, A single-chip CMOS smoke and temperature sensor for an intelligent fire detector, *IEEE Sensors Journal*, 9(8), 914–921, 2009.

12. A.C. Paglinawan, Y.-H. Wang, S.-C. Cheng, C.-C. Chuang, W.-Y. Chung, CMOS temperature sensor with constant power consumption multi-level comparator for implantable bio-medical devices, *Electronics Letters*, 45(25), 1291–1292, 2009.

13. C. Poki, C. Tuo-Kuang, W. Yu-Shin, C.-C. Chen, A time-domain sub-micro watt temperature sensor with digital set-point programming, *IEEE Sensors Journal*, 9(12), 1639–1646, 2009.

14. H. Lakdawala, Y.W. Li, A. Raychowdhury, G. Taylor, K. Soumyanath, A 1.05 V 1.6 mW, 0.45 C 3 resolution based temperature sensor with parasitic resistance compensation in 32 nm digital CMOS process, *IEEE Journal of Solid-State Circuits*, 44(12), 3621–3630, 2009.

15. M.K. Law, A. Bermak, A 405-nW CMOS temperature sensor based on linear MOS operation, *IEEE Transactions on Circuits and Systems II*, 56(12), 891–895, 2009.

16. Z. Bin, F. Quan-Yuan, A novel thermal-shutdown protection circuit, *Proceedings of the Third International Conference on Anti-Counterfeiting, Security, and Identification in Communication*, Hong Kong, China, 2009, pp. 535–538.

17. P. Ituero, J.L. Ayala, M. Lopez-Vallejo, A nanowatt smart temperature sensor for dynamic thermal management, *IEEE Sensors Journal*, 8(12), 2036–2043, 2008.

18. M. Sasaki, M. Ikeda, K. Asada, A temperature sensor with an inaccuracy of C using 90-nm 1-V CMOS for online thermal monitoring of VLSI circuits, *IEEE Transactions on Semiconductor Manufacturing*, 21(2), 201–208, 2008.

19. K.A.A. Makinwa M.F. Snoeij, A CMOS temperature-to-frequency converter with an inaccuracy of ± 0.5°C (3 σ) from –40 to 105°C, *IEEE Journal of Solid State Circuits*, 41(12), 2992–2997, 2006.

20. M.A.P. Pertijs, A. Niederkorn, M. Xu, B. McKillop, A. Bakker, J.H. Huijsing, A CMOS smart temperature sensor with a 3s inaccuracy of ± 0.5°C from –50°C to 120°C, *IEEE Journal of Solid-State Circuits*, 40(2), 455–461, 2005.

21. C. Poki, C.-C. Chen, T. Chin-Chung, L. Wen-Fu, A time-to-digital-converter-based CMOS smart temperature sensor, *IEEE Journal of Solid-State Circuits*, 40(8), 1642–1648, 2005.

22. M.A.P. Pertijs, K.A.A. Makinwa, J.H. Huijsing, A CMOS smart temperature sensor with a 3 σ inaccuracy of ± 0.1°C from 55°C to 125°C, *IEEE Journal of Solid State Circuits*, 40(12), 2805–2815, 2005.

23. B. Krabbenborg, Protection of audio amplifiers based on temperature measurements in power transistors, *International Solid State Circuits Conference (ISSCC)*, San Francisco, CA, 2004, pp. 374–375.

24. M.H. Nagel, M.J. Fonderie, G.C.M. Meijer, J.H. Huijsing, Integrated 1V thermal shutdown circuit, *Electronics Letters*, 28(10), 369–370, 1992.

25. K. Sakamoto, I. Yoshida, S. Otaka, H. Tsunoda, Power MOSFET with hold-type thermal shutdown function, *Proceedings of 1992 International Symposium on Power Semiconductor Devices and ICs*, Tokyo, Japan, 1992, pp. 238–239.

26. R. Amador, A. Polanco, H. Hernsindez, E. Gonziilez, A. Nagy, Technological compensation circuit for accurate temperature sensor, *Sensors and Actuators*, A 69, 172–177, 1998.

27. C. Gerard, M. Meijer, An IC temperature transducer with an intrinsic reference, *IEEE Journal of Solid State Circuits*, SC-15(3), 370–373, 1980.

28. J.M. Redoute, M. Steyaert, *EMC of Analog Integrated Circuits*. New York: Springer, January 2010.

29. M. Ramdani, E. Sicard et al., The electromagnetic compatibility of integrated circuits—Past, present and future, *IEEE Transactions on Electromagnetic Compatibility*, 51(1), 78–100, 2009.

30. S. Graffi, G. Masetti et al., Criteria to reduce failures induced from conveyed electromagnetic interferences on CMOS operational amplifier, *IEEE Transactions on Electromagnetic Compatibility*, 51(1), 78–100, February 2009.

31. E. Orietti, N. Montemezzo, S. Buso, G. Meneghesso, A. Neviani, G. Spiazzi, Reducing the EMI susceptibility of a Kuijk bandgap, *IEEE Transactions on Electromagnetic Compatibility*, 50(4), 876–886, 2008.

32. F. Fiori, Design of an operational amplifier input stage immune to EMI, *IEEE Transactions on Electromagnetic Compatibility*, 49(4), 834–839, 2007.

33. J.G. Tront, J.J. Whalen, C.E. Larson, J.M. Roe, Computer-aided analysis of RFI effects in operational amplifiers, *IEEE Transactions on Electromagnetic Compatibility*, 21(4), 297–306, 1979.

34. S. Graffi, G. Masetti, D. Golzio, New macromodels and measurements for the analysis of EMI effects on 741 op-amp. circuit, *IEEE Transactions on Electromagnetic Compatibility*, 33, 2534, 1991.

35. J.J. Laurin, S.G. Zaky, K.G. Balmain, EMI-induced failures in crystal oscillators, *IEEE Transactions on Electromagnetic Compatibility*, 33(4), 334–342, 1991.

36. J.-J. Laurin, S.G. Zaky, K.G. Balmain, On the prediction of digital circuit susceptibility to radiated EMI, *IEEE Transactions on Electromagnetic Compatibility*, 37(4), 528–535, 1995.

37. O. Aiello, F. Fiori, A new mirroring circuit for power MOS current sensing highly immune to EMI, *Sensor*, 2, 1856–1871, 2013.

38. O. Aiello, F. Fiori, A new MagFET-based integrated current sensor highly immune to EMI, *Microelectronics Reliability*, 53(4), 573–581, 2013.

39. O. Aiello, F. Fiori, On the susceptibility of embedded thermal shutdown circuit to radio frequency interference, *IEEE Transactions on Electromagnetic Compatibility*, 54(2), 405–412, 2012.

40. P. Wambacq, W. Sansen, *Distortion Analysis of Analog Integrated Circuits*. Norwell, MA: Kluwer, 1998.

41. F. Fiori, V. Pozzolo, Modified Gummel-Poon model for susceptibility prediction, *IEEE Transactions on Electromagnetic Compatibility*, 42(2), 206–213, 2000.

42. Ansoft Q3D Extractor [Online]. Available: http://www.ansoft.com/products/si/q3d _ extractor/, 2009.

43. *Integrated Circuits, Measurement of Electromagnetic Immunity—Part 4: Direct RF Power Injection Method*, IEC 62132-4, 2002.

44. C. Bona, F. Fiori, A new filtering technique that makes power transistors immune to EMI, *IEEE Transactions on Power Electronics*, 26(10), 2946–2955, 2011.

24 Coupled Chemical Reactions in Dynamic Nanometric Confinement

II. Preparation Conditions for Ag$_2$O Membranes within Etched Tracks

Dietmar Fink, G. Muñoz Hernandez, H. García Arellano, W.R. Fahrner, K. Hoppe, and J. Vacik

CONTENTS

24.1 INTRODUCTION

Irradiation of polymer foils by swift heavy ion beams is known to leave trails of radiation damage in the material (the so-called latent tracks) that usually are more sensitive to dissolution by adequate aggressive chemicals than the neighboring unirradiated regions. In this way, one can create parallel, straight, and long nanopores (the so-called etched tracks) that exhibit large aspect (i.e., length/width) ratios of up to ~1000 [1]. Such nanostructures can be filled with various solids or liquids to transform them into functional materials [2] that are useful for electronics [2–8], medicine [9], or biosensing [10–17].

The materials can either be attached to etched track walls as nanotubules, or they can fill the tracks partly or completely to form massive nanosized rods [18]. In this work, another form of structures embedded within etched tracks is examined, which are membranes (or plugs) that can be situated at any position within the tracks.* The strategy adopted for this purpose is to combine two chemical reactions, etchant–polymer and etchant–reactant solution, with each other within the dynamic confinement of etched ion tracks [19]. Here, the criterion for the selection of the reactant solution is that it forms a solid precipitate with the etchant. By a proper selection of the three decisive parameters, etchant concentration (C), etching temperature (T), and applied test voltage (V), the formation of tiny membranes (or plugs) of the precipitated material within these nanopores is possible, so that an individual etched ion track can be separated into two independent individual sections.

It is well known that the etching of a swift heavy ion track from one side only leads to the formation of an asymmetric nanopore that is conical in a crude first approximation and funnel-like in a much better description [20]. When following the etchant processes by measuring a test current through the tracks upon application of a constant dc or ac voltage, one starts recording remarkable currents only from the moment of etchant breakthrough on, as from this moment on the electrolyte's ions are enabled to pass freely across the foil from one side to the other.

In the case of etching of a latent track from both sides, the two etched cones will meet in the center, thus forming double cone (or, more precisely, double funnel [20])–type nanopores. When, shortly before merging of both etched tracks (*breakthrough*), the etchant in one pore is replaced by a liquid that reacts with the etchant under formation of a solid precipitation (e.g., of Ag_2O [19], LiF, CaO, $BaCO_3$ [to be published]), then it will be possible to create a membrane of that reaction product at the place of intersection of the two etched cones. Prolonged materials deposition will transform these membranes into plugs. Whereas in [19], the fundaments for the corresponding coupled chemical reactions in the dynamic confinement of track etching were laid to create such membranes, in this paper their preparation conditions are elaborated to some more extent.

24.2   EXPERIMENT: FORMATION OF ETCHED TRACKS WITH EMBEDDED Ag_2O MEMBRANES

24.2.1   Pre-Etching Step

Polyethylene terephthalate (PET) foils that are 12 μm thick had been irradiated by Kr ions at 250 MeV energy at the Joint Nuclear Research Institute (JNRI), Dubna, up to fluencies of 4×10^6 and 5×10^7 cm^{-2}. Around a dozen 1 cm^2 large pieces of this foil were cut out and then inserted into the center of a measuring chamber (made after the example of Ref. [21]) with two adjacent compartments and then etched as previously reported in [19] with 9 M KOH from both sides at ambient (~25°C) temperature. In conventional track etching experiments the etching continues up to the full etchant breakthrough in the foil centers. In this work, the etching was interrupted shortly before this expected etchant breakthrough would take place, by removing the KOH etchant and washing the foil thoroughly. This pre-etching stage is necessary to let the two etched track cone tips arrive rapidly, sufficiently, and closely to each other near the foil center in due time, where the membrane formation should preferentially take place.

---

* For the sake of clearness, let us define here membranes as the thin regions (with thicknesses below some 100 nm or so) of solid precipitation within etched tracks in polymer foils (with thicknesses in the order of typically some 10 μm) that separate the etched tracks into two compartments. More extended solid precipitation regions are denoted by us as plugs.

24.2.2 MEMBRANE FORMATION STEP

Thereafter, 1 M AgNO$_3$ solution was introduced on one side (let us call it here the *left* side) and 1 M KOH introduced on the other side (the *right* side) of the membrane. Thus, the etching continued at a slower speed from the right side only, until etchant breakthrough occurred. As reported in [19], simultaneously AgOH is formed at the point of etched track intersection, which readily transforms to Ag$_2$O. As these silver compounds are insoluble in water, they will precipitate within the etched track at its narrowest place and thus form a plug or a membrane. The fact that Ag$_2$O is rather impermeable for both ions and electrons (except at very high applied electric field strengths or frequencies) gives us a tool for the detection of the Ag$_2$O membrane formation.

24.2.3 ELECTRONIC CHARACTERIZATIONS

Both the control of the track etching and membrane formation processes and the subsequent membrane characterization were performed electrically, by applying a voltage across the measuring chamber (including both polymer foil and electrolytes) and determining the passing currents. This was achieved by means of a Velleman PCSGU250 pulse generator/oscilloscope combination.

- During the pre-etching stage, a sinusoidal ac voltage U of 5 V$_{peak-peak}$ at a frequency of ~0.5 Hz was applied through Ag electrodes to the system. Both the voltage and the corresponding electrical current were continuously measured in the *Transient Recording* mode of that equipment (settings: dc, 0.3 V/div for measuring the applied voltage and 10 mV/div for measuring the current via a 1 MΩ probe resistance; time resolution: 0.1 s/div) as a function of time. This measurement served only for control that the etching proceeded reliably; in the standard case, no current signal emerged. Deviations from this behavior indicated erroneous production situations (such as the leakage of etchant due to incomplete sealing of the reaction chamber with eventual subsequent shortcuts), after which the experiment had to be repeated.
- During the membrane formation stage, the ac voltage setting was reduced to U = 1 V$_{p-p}$, to minimize the influence of the corresponding electric field on the membrane formation, and the measurement took place in the ac mode. As described in the previous text, at the moment of etchant breakthrough from one track side to the other, a strongly spiky current emerges that vanishes rather abruptly when stable membranes have been formed. These *quiet phases* served as fingerprints for the Ag$_2$O membrane formation. Therefore, the corrosion was stopped whenever such a quiet phase could be identified unambiguously.

24.2.4 DETAILS OF THE MEMBRANE FORMATION

The Ag$_2$O membranes were found to be inert against alkaline (NaOH, KOH) attack and sufficiently stable upon prolonged sample use, if replacing the etchant by water or buffer solution. If, however, the etchant is kept, track etching from the right side will proceed and the right half cone will become larger until the etchant can pass the Ag$_2$O membrane and establish a new connection with the left half cone. This is recorded via the emergence of current spikes [19,22]. The reaction of the etchant with the Ag$^+$ ions on the left side leads to new Ag$_2$O formation, hence to a lateral membrane growth that initiates again current blocking. This means one will record a series of alternating current spikes, followed by *quiet phases* and renewed spiky currents.

This simple picture is, however, unfortunately disturbed by the superposition of another current component that arises from the chemical potential differences (<1 V) between the $AgNO_3$- and KOH-filled compartments as driving force. Depending on the actual state of both the track etching and the membrane formation, the overall track resistivity may vary considerably with time. Though the membranes largely prevent the passage of ionic currents, they do not hinder the passage of dielectric currents through the Ag_2O dielectric, as each of the parallel membranes—in the initial stage still very thin—acts as a powerful capacitor. Therefore, the overall current through the foil during the membrane formation is not exactly zero even during the *quiet phases*, but characterized rather by large-scale irregular current fluctuations; see, for example, Figure 24.1.

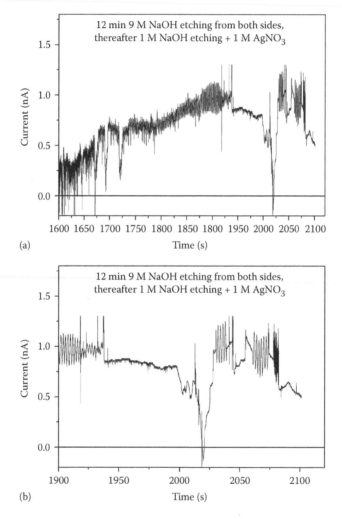

FIGURE 24.1 Protocol of Ag membrane formation (a: overview and b: detail), after pre-etching of the latent tracks with 9 M KOH for 12 min (not shown here), washing of both sides, and replacing one side with 1 M $AgNO_3$ solution and the other side with 1 M KOH solution (denoted here as time t = 0). The gradual thinning of the residual polymer region between the two etched cones and the emerging membrane formation upon track breakthrough led to the occurrence of many electrical breakthrough events (spikes) as well as to the emergence of ac leakage currents that follow the applied 0.5 Hz voltage (up to ~1930 s). These signals are superposed by currents stemming from the chemical potential difference between both solutions. At t ~ 1940 s, the first longer *quiet phase* without any oscillating current is seen, indicating that the membrane formation has been accomplished. This is followed by three new etchant breakthroughs at t ~ 2045, 2075, and 2084 s, when it was decided to stop the Ag_2O membrane formation process for its further characterization.

The overlapping of a strongly fluctuating current with another current component character-ized by alternating spike-rich and rather *quiet* time intervals (with no or very small current only) is clearly seen. For performing the experiments described in the following text, the observed alternat-ing sequences between current spikes and quiet phases were interrupted either during one of the first quiet phases (thus obtaining rather small and narrow membranes) or at later stages of the mem-brane formation (with larger and thicker membranes), by removing the etchant and washing the foil thoroughly with water. From the conditions for electrical breakthrough and ionic electrodiffusion through the membranes, one can estimate that the membranes/plugs should have widths of typically some 10–100 nm [19].

In a forthcoming paper of this series, the results of biosensors will be described that were pro-duced from these structures, by inserting enzymes into the two etched track sides separated by the membranes.

24.3 RESULTS AND DISCUSSION

24.3.1 INFLUENCE OF PRE-ETCHING TIME ON THE EMERGENCE OF Ag₂O MEMBRANES WITHIN ETCHED TRACKS

The first pre-etching step serves only to arrive with both cone tips sufficiently, rapidly, and closely near the foil's center. As hitherto the etching time necessary for this step was only badly defined, we performed here a systematic study on the influence of the pre-etching times on the Ag₂O membrane formation. The reason for this is that if $AgNO_3$ is inserted into one track side *after* the two etched cones have already merged, a completely different mechanism sets in that is explained by the specific behavior of nanofluids [19]. In this case, the Ag^+ ions will migrate as counterions through the tracks from the left side to the opposite side along the negatively charged etched track walls (the charge stemming from COO^- groups in the polymer surface in alkaline ambient) within the electrical double layer, whereas the OH^- ions rather diffuse through the track centers from the right to the left side (if the etched tracks have sufficiently large diameters). Thus, the two anionic and cationic currents are geometrically well separated from each other within the nanopores, so that they can meet and react with each other only in their contact region (which is given by the limit of the Debye layer). The Ag₂O molecules or molecular clusters created there will be easily swept out from the track by osmosis and/or electrophoresis and accumulate in the electrolytes on both foil sides. This signifies that the precondition for the membrane formation is to care for good anchorage of the Ag₂O molecules on the track walls. This is given best in the very moment when two etched track cones meet and merge with each other, but it is not given any longer thereafter within flowing nanoliquids within open tracks.

Therefore, we found it reasonable to study the transition from one mechanism (the membrane formation) to the other one (nanofluidic motion without membrane formation) in more detail before continuing with the membrane characterization. As here, the pre-etching time with concentrated KOH has a decisive influence on whether Ag₂O membranes are formed or not, we performed a series of experiments with increasing pre-etching times with concentrated etchant, before replacing the latter by Ag salt solution on one side and by weak etchant on the other.

There is yet another factor playing a role in such experiments. This is the buildup of chemical potentials across the foils, as we work with different liquids on both foil sides. The potential dif-ferences can be recorded via current measurements through the foils as long as the foil is insu-lating or weakly conducting. However, more conducting tracks (as emerging during the etching process) will act as shortcuts, which makes these potential differences collapse. This means when recording the current through polymer foils during track etching (or during membrane formation within the etched tracks), one will record a current background that always depends on the actual foil conductivity. As the latter is highly variable especially in the case of membrane formation, it is expected that this constant current background will also vary strongly, in dependence on the track's actual internal geometry.

Four measurements were undertaken, with (1) 12 min, (2) 15 min, (3) 18 min, and (4) 22 min pre-etching times each with 9 M KOH at ambient temperature (~22°C), before the latter was replaced by 1 M silver nitrate solution on one side and 1 M KOH on the other side to arrive at etchant breakthrough. The results are depicted in Figures 24.1 through 24.4.

Table 24.1 summarizes the effects obtained qualitatively from Figures 24.1 through 24.4. It appears that some transition regime exists between the formation of stably anchored intratrack Ag_2O membranes and the osmotic/electrophoretic sweeping out of Ag_2O nanoparticles from the etched tracks. In the first case, *quiet phases* emerge that stem from complete and prolonged blocking of transmitted alternating currents and that can therefore be understood as macroscopic fingerprints for stable intratrack membrane formation. During continued track etching from one side, the membranes grow by sideways addition of more Ag_2O during every intermediate track breakthrough. The basic requirement for stable membrane formation is a firm anchorage of the membranes within the etched tracks. This can apparently be only achieved by forming them already in the initial phase of etched track breakthrough.

When the membrane-forming reaction (considered here: $2Ag^+ + 2OH^- \rightarrow 2AgOH \rightarrow Ag_2O + H_2O$) is initiated only *after* the two etched track half-cones have already merged, the two ion types show the typical nanofluidic behavior within the tracks, with positive ions migrating closely to the negatively charged track walls and the negative ions diffusing through the track centers outside the outer Helmholtz plane [23]. This leads to chemical reactions between anions and cations only in their interface planes at the limit of the Debye layer, so that the emerging reaction products cannot anchor at the track walls but are swept out of the tracks. Thus, one might consider such arrangements as possible chemical reactors for nanoparticles.

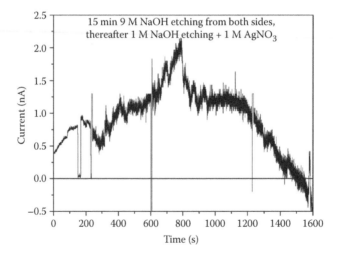

FIGURE 24.2 Protocol of a similar measurement as shown in Figure 24.1, however, after pre-etching of the latent tracks with 9 M KOH for 15 min, washing of both sides, and replacing one side with 1 M AgNO_3 solution and the other side with 1 M KOH solution (at time t = 0). As in Figure 24.1, the measurement shows a gradually increasing alternating current, superposed by a slowly changing constant current due to the chemical potential difference between AgNO_3 and KOH. No *quiet phases* are seen, indicating that no stable Ag_2O membrane formation took place. However, the strongly spiky behavior of that current might indicate that at least short-living intermediate membrane fragments might have been formed. The abrupt current changes at 152, 232, 603, and 1230 s are artifacts stemming from transient switching for control to other measuring modes.

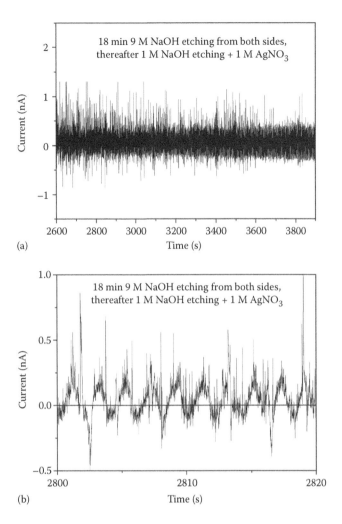

(a)

(b)

FIGURE 24.3 Protocol of a similar measurement as shown in Figure 24.1, however, after pre-etching of the latent tracks with 9 M KOH for 18 min, thereafter washing of both sides, and replacing one side with 1 M AgNO$_3$ solution and the other side with 1 M KOH solution (at time t = 0). (a) For hours, the measurement showed a highly spiky alternating current (detail in (b)) following the applied ac voltage. After some ~2000 s, no more dc current from chemical potential differences was superposed. No *quiet phases* are seen, indicating that no Ag$_2$O membrane formation took place.

From Figures 24.2 and 24.3, it appears that there exists some kind of transition regime between the purely membrane-forming strategies and the purely nanoparticle-forming track reactors that is characterized by the frequent occurrence of current spikes and strongly varying chemical potential differences. It cannot be excluded that under such conditions, very short-living membranes or membrane fragments emerge that cannot yet be recorded unambiguously by the present current/voltage approach. The strong spikes emerging under these conditions might be a hint for this.

We find it amazing that the pre-etching time must be relatively short to allow for membrane formation. This may be a consequence of the peculiar double funnel–type shape of symmetrically etched tracks [20]. Apparently, track pre-etching must stop before the formation of the long and narrow funnel tip region is completed, so that the membranes still find good anchorage before smooth nanofluidic anion/cation intermigration emerges.

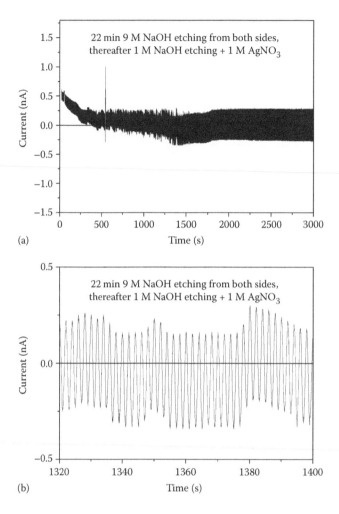

(a)

(b)

FIGURE 24.4   Protocol of a similar measurement as shown in Figure 24.1, however, after pre-etching of the latent tracks with 9 M KOH for 22 min, thereafter washing of both sides, and replacing one side with 1 M AgNO$_3$ solution and the other side with 1 M KOH solution (at time t = 0). (a) Overview, (b) detail. The measurement showed an ac current with gradually increasing amplitude, superposed by a slowly decreasing dc due to the decreasing chemical potential difference between AgNO$_3$ and KOH with increasing pore opening. No *quiet phases* are seen, indicating that no Ag$_2$O membrane formation took place. Note that the current changes are smaller than in Figure 24.3, indicating that even short-living intermediate membrane formation can be excluded here.

24.3.2   Destruction of Ag$_2$O Membranes within Etched Tracks in PET Foils

The samples described here were transformed to glucose biosensors with two different compartments in each track, by immobilizing the enzyme glucose oxidase (GOx) on the etched track walls as described earlier [14]. The results of these studies will be presented in a forthcoming paper. After the enzymes started deteriorating, it was decided to remove the Ag$_2$O membranes, by attacking the foil with concentrated HNO$_3$ for 1 min at room temperature. The comparison of a sample foil before and after membrane removal and subsequent thorough washing with double deionized water indicated that the overall current through the foil had increased by about 10 times after the membrane removal, and also more signals were transmitted at both very low and very

TABLE 24.1

Qualitative Comparison of the Strategies for Membrane Formation, Nanoparticle Reactors, and Intermediate Stages

Pre-Etching Time (min), Figure	Transmitted Alternating Currents	Current Spikes	Currents due to Chemical Potential Differences	Formation of Stable Ag₂O Membranes
12, Figure 24.1	Often interrupted by *quiet phases*	Often preceding the *quiet phases*	Strong	Yes
15, Figure 24.2	No *quiet phases*	Very frequent	Strong	No
18, Figure 24.3	No *quiet phases*	Very frequent	Small or absent	No
22, Figure 24.4	No *quiet phases*	Largely absent	Small or absent	No

high frequencies (not shown here). In accordance with our expectations, the new structures closely resembled to ion tracks etched from both sides up to regular etchant breakthrough.

24.4 SUMMARY

A peculiarity of experiments on coupled chemical reactions in dynamic nanometric confinement is the transient formation of stable, unsoluble, and impermeable membranes of the precipitating material (here: Ag_2O) within the etched tracks that separate the latter into two compartments. Especially at sufficiently low etching speed, membranes of good quality may emerge. The growth of the membrane thickness is expected to proceed first via current spike emission and later via Ag^+ ion electrodiffusion through the membranes, with subsequent Ag_2O precipitation whenever the ions arrive at the opposite foil side. Membrane growth in lateral size is expected to proceed via new etchant breakthrough processes at the membrane rims and subsequent lateral addition of new material there.

Basic requirement for the formation of stable membranes along the tracks is a good anchorage of the emerging Ag_2O nanoparticles on the track walls. This is only given by the two-step etching process suggested in a previous work [19], with sufficiently short pre-etching times. If they are too long so that a transparent connection between both foil sides emerges, nanofluidic peculiarities will prevent permanent membrane formation. This was examined here, where a maximum etching time was established that still allows stable membrane formation.

Etchant breakthrough and membrane growth are nearly always accompanied by the onset of strong current spikes. Their partly amazing heights might indicate that they do not always stem from individual tracks only, but that they are possibly the result of synchronized charge emission through a (statistically varying) number of neighbored tracks. The existence of stable membranes of sufficient thickness is indicated by a dramatic current decrease, denoted by us as *stable phase*.

ACKNOWLEDGMENT

D. Fink is grateful to the Universidad Autónoma Metropolitana-Cuajimalpa, Mexico City (UAM-C), for the guest professorship in the frame of the Cathedra *Roberto Quintero Ramírez* and to the Grant Agency of the Czech Republic (P108-12G-108) and the Nuclear Physics Institute, Řež, for providing support. We are especially obliged to Profs. S. Revah, R. Quintero R., and M. Sales Cruz from UAM-C for their continuous help, encouragement, and discussions and for providing us with adequate working facilities. We are further obliged to Dr. P. Apel from JNRI, Dubna, Russia, for providing us with the ion-irradiated polymer foils.

REFERENCES

1. R.L. Fleischer, P.B. Price, and R.M. Walker, *Nuclear Tracks in Solids: Principles and Applications*. University of California, Berkeley, Berkeley, CA, 1975.

2. D. Fink, P. Yu Apel, and R.H. Iyer, Ion track applications, in: Fink, D. ed., *Transport Processes in Ion Irradiated Polymers; Springer Series in Materials Science*, Vol. 65, pp. 269, 300, Chapter II.5. Springer Verlag: Berlin, Germany, 2004, and references therein.

3. A. Biswas, D.K. Avasthi, B.K. Singh, S. Lotha, J.P. Singh, D. Fink, B.K. Yadav, B. Bhattacharya, and S.K. Bose, Resonant tunnelling in single quantum well heterostructure junction of electrodeposited metal semiconductor nanostructures using nuclear track filters. *Nucl. Instrum. Methods Phys. Res. B* 151 (1999) 84–88.

4. L. Piraux, J.M. George, J.F. Despres, C. Leroy, E. Ferain, R. Legras, K. Ounadjela, and A. Fert, Giant magnetoresistance in magnetic multilayered nanowires. *Appl. Phys. Lett.* 65(19) (1994) 2484–2486.

5. M. Lindeberg, L. Gravier, J.P. Ansermet, and K. Hjort, Processing magnetic field sensors based on magnetoresistive ion track defined nanowire cluster links, in: *Proceedings of the Workshop on European Network on Ion Track Technology*, Caen, France, February 24–26, 2002.

6. K. Hjort, The European network on ion track technology, in: Presented at the *Fifth International Symposium on "Swift Heavy Ions in Matter"*, Giordano Naxos, Italy, May 22–25, 2002.

7. D. Fink, A. Petrov, K. Hoppe, and W.R. Fahrner, Characterization of "TEMPOS": A new tunable electronic material with pores in oxide on silicon, in: *Symposium R—Radiation Effects and Ion Beam Processing of Materials*, Boston, MA, December 1–5, 2003, 2003 MRS Fall Meeting, Vol. 792.

8. D. Fink, A. Petrov, H. Hoppe, A.G. Ulyashin, R.M. Papaleo, A. Berdinsky, and W.R. Fahrner, Etched ion tracks in silicon oxide and silicon oxynitride as charge injection channels for novel electronic structures. *Nucl. Instrum. Methods Phys. Res. B* 218 (2004) 355–361.

9. M. Tamada, M. Yoshida, M. Asano, H. Omichi, R. Katakai, R. Spohr, and J. Vetter, Thermo-response of ion track pores in copolymer films of methacryloyl-L-alaninemethylester and diethyleneglycol-bis-allylcarbonate (CR-39). *Polymer* 33(15) (1992) 3169–3172.

10. C.G.J. Koopal, M.C. Feiters, R.J.M. Nolte, B. de Ruiter, and R.B.M. Schasfoort, Glucose sensor utilizing polypyrrole incorporated in track-etch membranes as the mediator. *Biosens. Bioelectron* 7 (1992) 461–471; S. Kuwabata and C.R. Martin, Mechanism of the amperometric response of a proposed glucose sensor based on a polypyrrole-tubule-impregnated membrane. *Anal. Chem.* 66 (1994) 2757–2762.

11. Z. Siwy, L. Trofin, P. Kohl, L.A. Baker, C.R. Martin, and C. Trautmann, Protein biosensors based on biofunctionalized conical gold nanotubes. *J. Am. Chem. Soc.* 127 (2005) 5000–5001; Z.S. Siwy, C.C. Harrell, E. Heins, C.R. Martin, B. Schiedt, C. Trautmann, L. Trofin, and A. Polman, Nanopores as ion-current rectifiers and protein sensors, in: Presented at the *Sixth International Conference on Swift Heavy Ions in Matter*, Aschaffenburg, Germany, May 28–31, 2005 (unpublished).

12. L. Alfonta, O. Bukelman, A. Chandra, W.R. Fahrner, D. Fink, D. Fuks, V.Golovanov et al., Strategies towards advanced ion track-based biosensors. *Radiat. Eff. Defect. Solids* 164 (2013) 431–437.

13. C.R. Martin and Z.S. Siwy, Learning nature's way: Biosensing with synthetic nanopores. *Science* 317 (2007) 331–332.

14. D. Fink, I. Klinkovich, O. Bukelman, R.S. Marks, A. Kiv, D. Fuks, W.R. Fahrner, and L.Alfonta, Glucose determination using a re-usable ion track membrane sensor. *Biosens. Bioelectron* 24 (2009) 2702–2706.

15. D. Fink, G. Muñoz H., and L. Alfonta, Highly sensitive ion track-based urea sensing with ion-irradiated polymer foils. *Nucl. Instrum. Methods Phys. Res. B* 273 (2012) 164–170.

16. D. Fink, G. Muñoz H., J. Vacik, and L. Alfonta, Pulsed biosensing. *IEEE Sens. J.* 11 (2011) 1084–1087.

17. Y. Mandabi, S.A. Carnally, D. Fink, and L. Alfonta, Label free DNA detection using the narrow side of conical etched nano-pores. *Biosens. Bioelectron.* 42 (2013) 362–366.

18. D. Fink, Ion track manipulations, in: Fink, D. ed., *Transport Processes in Ion Irradiated Polymers; Springer Series in Materials Science*, Vol. 65, p. 227, Chapter II.6. Springer Verlag: Berlin, Germany, 2004, and references therein.

19. G. Muñoz H., S.A. Cruz, R. Quintero, D. Fink, L. Alfonta, Y. Mandabi, A. Kiv, and J. Vacik, Coupled chemical reactions in dynamic nanometric confinement: Ag₂O formation during ion track etching. *Radiat. Eff. Defect. Solids* 168 (2013) 675–695.

20. P.Yu. Apel, I.V. Blonskaya, O.L. Orelovitch, B.A. Sartowska, and R. Spohr, Asymmetric ion track nanopores for sensor technology. Reconstruction of pore profile from conductometric measurements. *Nanotechnology* 23 (2012) 225503.

21. M. Daub, I. Enculescu, R. Neumann, and R. Spohr, Ni nanowires electrodeposited in single ion track templates. *J. Optoelectron Adv. Mater.* 7 (2005) 865–870.

22. D. Fink, S. Cruz, G. Muñoz H., and A. Kiv, Current spikes in polymeric latent and funnel-type ion tracks. *Radiat. Eff. Defect. Solids* 5 (2011) 373–388.

23. H.-J. Butt, K. Graf, and M. Kappl, *Physics and Chemistry of Interfaces*. Wiley-VCH: Weinheim, Germany, 2006.

Index